SCIENCE 5

Teacher's Edition

Fourth Edition

bju press
Greenville, South Carolina

Note
The fact that materials produced by other publishers may be referred to in this volume does not constitute an endorsement of the content or theological position of materials produced by such publishers. Any references and ancillary materials are listed as an aid to the student or the teacher and in an attempt to maintain the accepted academic standards of the publishing industry.

SCIENCE 5 Teacher's Edition
Fourth Edition

Authors
Peggy S. Alier
Debra White

Coordinating Authors
Susan Burkholder
Joyce Garland
Donald Jacobs
Janet E. Snow

Consultant
Brad Batdorf

Bible Integration
Wesley Barley
Bryan Smith

Project Editor
Kristin McClanahan

Designer
Duane Nichols

Concept Designer
Drew Fields

Page Layout
Rosemaria Rosier

Project Coordinator
Gina Stewart

Cover Design
Craig Oesterling

Permissions
Sylvia Gass
Lilia Kielmeyer

Illustrators
Preston Gravely
Dave Schuppert

Teacher's Edition photograph credits:

Front Matter: **v**, **xi**, **xiv** BJU Photo Services; **viii** (left), **ix** (left) ©2005 JUPITERIMAGES/photos.com; **viii** (top right) BJU Press; **viii** (bottom right) iStockphoto/Thinkstock; **ix** (right) © iStockphoto.com/timeless; **xiii** © Kojihirano | Dreamstime.com

10 © Kojihirano | Dreamstime.com; **18** BJU Photo Services; **81** BJU Photo Services; **160** BJU Press; **161** Image courtesy United States Geological Survey; Image source: Earth Science World Image Bank; **189** © Map Resources; **210**, **293**, **294** iStockphoto/Thinkstock; **211** Biosphoto / SuperStock; **263** John Bavosi/Science Source; **345** © iStockphoto.com/Hocus Focus Studio

Photograph credits for *SCIENCE 5 Student Text* appear on pages A26–A27.

Greenville, South Carolina 29609

Printed in the United States of America

ISBN 978-1-60682-602-7 (Teacher's Edition with CD-ROM)

15 14 13 12 11 10 9 8

Contents

Appendix

Teacher's Toolkit CD

Goals for *Science 5*

Develop a knowledge of God

- Inculcate an understanding of God as the Creator and Sustainer of the universe (Colossians 1:16–17)
- Identify the precision of God's creation (Ecclesiastes 3:1–8)
- Inspire curiosity, wonder, and appreciation of God's creation (Psalm 19:1)

Encourage Christian growth

- Promote good stewardship of God's creation (Genesis 2:15)
- Promote Christ-honoring character through cooperative group activities (2 Peter 1:5–8)
- Challenge students to use science to show the love of Christ to others (Matthew 22:39–40)

Promote scientific literacy

- Establish foundational science facts and skills for further science instruction
- Balance presentation of facts with active participation
- Teach processes involved in a scientific method
- Promote disciplined and orderly approaches to problem solving (1 Thessalonians 5:21)
- Show the integration of science into the student's everyday life
- Affect a positive attitude toward science through active participation and relevant discussions

Christian Worldview in *Science 5*

Science is one of the most hotly contested areas of human culture. Many scientists today do not believe in God and try to explain everything from a naturalistic point of view. The Bible presents us with a view of the world we live in and the tools we should use to study it. This biblical view is called a Christian worldview. There are only two worldviews—one based on God's Word and one based on man's own word and philosophies.

What is a Christian worldview in science?

A Christian worldview is based on the authority of the Bible. God made humans in His own image. We are to declare His glory by being like Him. Tragically, the disobedience of the first man and woman brought death into the world. But God did not abandon His image bearers. Even while God was pronouncing a curse upon His perfect creation, He promised to redeem the world to Himself by sending His own Son into the world to accomplish salvation (Genesis 3:15). The story of the human race is the story of God's redemptive acts to restore His kingdom and rescue His people. God reveals Himself in the Bible and through nature. As we study science and learn how God's world works, we also learn more about the God who created it all.

Your *worldview* refers, simply, to your way of viewing the world. It is the set of assumptions and beliefs used to interpret everything that happens. Everybody has a way of looking at the world. Your way is evident when you answer questions like these: Where did we come from? Are humans more important than animals? What happens to us when we die? What is truth? The study of science raises many of these questions and so becomes a powerful way of learning and applying a Christian worldview. A Christian worldview as it appears in the field of science has several key ideas.

God is the Creator of all things

The fact that God is Creator stands in direct contradiction to the evolutionary theory about the earth. Genesis 1 describes the creation of the stars, the sun, the moon, the earth, and all the plants and animals that live on the earth. Each plant and animal is categorized according to its "kind." There is no room in the biblical story for animals evolving into higher life forms. Neither is there room for billions of years. The "days" in Genesis 1 are normal 24-hour days. This is emphasized by the repetition of "evening" and "morning." God declared that everything he created was "very good" (Genesis 1:31). There was no sin, and people lived in perfect harmony with the created world and with God Himself. Yet because of Adam's sin, all that God made has fallen under a curse (Genesis 3). Today the study of science clearly reveals the effects of sin. Lots of scientific study deals with suffering and death. A Christian has the foundation to glorify God in science as he recognizes the source of good and bad that is seen around him.

People are important to God

A naturalistic perspective on science views humans as a part of the evolutionary process and ultimately no more important than the animals on the earth. A Christian worldview identifies mankind as resulting from a separate act of creation, made in the "image of God" (Genesis 1:26). This means that people are very important. They are not just another species of animal but are image bearers of God. Because people are God's special creation, He instituted the death penalty for anyone who murdered another human (Genesis 9:6). Jesus also said that the second most important commandment was to love your neighbor as yourself (Matthew 22:39). It was God's own love for His image bearers that led Christ to pay such an awful price to redeem mankind from sin (2 Corinthians 5:21). Because of all this, Christians have a powerful motivation to study science. When a Christian helps poor children get clean drinking water or develops a new medication for a harmful disease, he glorifies God by reflecting the value that God places on people.

God placed people on the earth to rule in His place

From the beginning, God gave people a way to bring glory to Him. He gave people authority over the created world. This means that we have authority over living things, such as fish, birds, and cattle (Genesis 1:26). God's plan includes that people are to have families, spread out, and live in every part of the earth. People are to bring the vast resources and potential of the earth under their careful stewardship (Genesis 1:28). Just as Adam was told to dress and keep the garden (Genesis 2:15), so all of mankind is to be a steward over the earth. Much of our work in science today fulfills this mission given to us by God. Science is used every day to build and design things that help people to steward, or manage, the potential that the earth has. When Christians participate in this work, they are obeying the Lord and bringing glory to Him.

Instructional Materials

Student Text

The *SCIENCE 5 Student Text* provides grade-appropriate information through text, diagrams, graphs, charts, and annotated photographs and illustrations. The Student Text also includes special interest boxes and Quick Check questions as well as instructions for experiments, activities, and projects. The last page of each chapter provides lists of vocabulary and key ideas. A glossary and index are included at the end of the book.

Teacher's Edition

The *SCIENCE 5 Teacher's Edition* contains 180 lessons. These lessons are grouped into six units, with each unit having two interrelated chapters. Each chapter has an Answers in Genesis lesson that relates to the topic discussed in that chapter. Most lessons include additional background information, Bible application, cross-curricular links, and scientific activities.

The Teacher's Edition also includes useful information about science process skills, the management of activities, and grading. A materials list and the alignment of *SCIENCE 5* with national standards are located in the back of the Teacher's Edition.

Activity Manual

The *SCIENCE 5 Activity Manual* provides a variety of pages to aid understanding. Study Guide pages provide a systematic review of concepts. Pages for recording information for Activity and Exploration lessons reinforce scientific methods. Preview, Reinforcement, Expansion, and Answers in Genesis pages are also included.

Answers are located in the Activity Manual Answer Key on the Teacher's Toolkit CD. The CD is part of the Teacher's Edition.

Tests

The *SCIENCE 5 Tests* include twelve chapter tests. Tests include objective items, the use and labeling of charts and diagrams, and essay questions. An answer key for the tests is available separately.

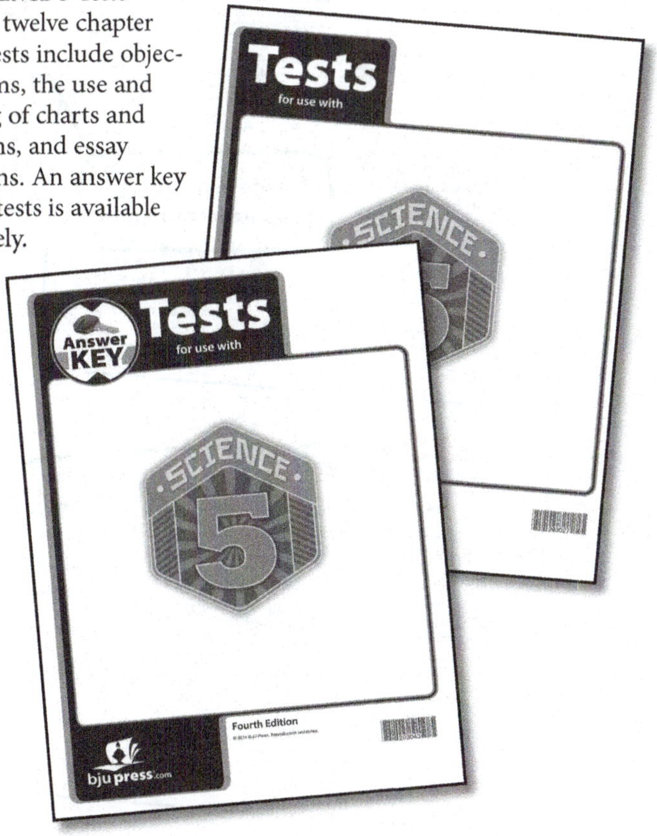

Teacher's Toolkit CD

The *SCIENCE 5 Teacher's Toolkit* is located inside the back cover of the Teacher's Edition. It includes the Activity Manual Answer Key, a game bank, rubrics for assessing Activity and Exploration lessons, quizzes, instructional aids, visuals, and information about science fairs and two-day lessons.

Text Lesson Features

Use the **materials list** to determine the materials needed to teach the lesson.

Begin the lesson with an **introductory discussion** or **activity**.

Guide student reading with questions to give a purpose for reading found in **Preparation for Reading**.

Use the **Teacher's Toolkit CD list** to determine the Instructional Aids, Quizzes, Rubrics, or Visuals needed to teach the lesson.

Evaluate the understanding of science concepts through **discussion** found in Teach for Understanding.

Include discussion of **graphic information**.

Apply **biblical truth and principles** with discussions.

Enrich the student's knowledge by incorporating **cross-curricular links, activities,** and **demonstrations**.

Locate the necessary **materials** needed for enrichment with the activity description.

81/82

Student Text pages 146–49
Activity Manual pages 98–100

Objectives
- Describe basic characteristics of the coniferous forest
- Describe basic characteristics of the deciduous forest
- Differentiate between conifers and deciduous trees
- Name two ways that animals in the deciduous forest survive the changing seasons

Materials
- small branch of evergreen needles (leaves)
- leaves from a deciduous tree
- globe or world map

Teacher's Toolkit CD
- Visuals 6.1–6.2: *Land Biomes of the World; Biome Characteristics*
- Quiz 6-A

Vocabulary
- conifer
- deciduous

Two days have been allotted for this lesson. You may want to teach Lesson 81 the first day and use ideas from Lesson 82 on the second day. You may also incorporate the ideas from Lesson 82 into the text materials, making the lesson two days.

Introduction
- Display the two types of leaves.
  What are some differences that you notice between these leaves? Answers may include differences in size, shape, texture, and color.
  Which of these leaves belongs to a tree that you think could live in a very cold climate? the evergreen needles Why? Possible answers: The other leaves fall off trees in cold weather. Evergreen trees are often in pictures with snow.
- Explain that these leaves belong to trees that are often found in different forest biomes.

Preparation for Reading
- Display the words *coniferous, taiga* (TIE guh), and *deciduous*. Pronounce the words together.
- Direct the students to read silently pages 146–49 to find the answers to these questions.
  What kind of trees are most of the trees in the coniferous forest?
  What does *deciduous* mean?

Teach for Understanding
- Refer to the globe and the *Land Biomes of the World* visual as you discuss the forest biomes. Refer the students to the *Land Biomes of the World* map on Student Text pages 142–43. Continue use of the *Biome Characteristics* organizer on Activity Manual page 98.

168 · Lessons 81–82

Coniferous Forest

The coniferous forest is the largest land biome. This biome, also called the *taiga*, is located just south of the tundra. The biome is so large that it stretches across most of Canada and into some of the northern United States. It also covers large parts of Europe and Asia.

The coniferous forest has two main seasons: summer and winter. Spring and autumn do occur. They are so short, ho...

During part of the winter the sun is up for only four to six hours!

In contrast, the sun may shine for twenty hours a day during the summer. The growing season lasts longer in the coniferous forest than in the tundra. The summer is still short and cool, however. Most of the year's precipitation occurs during the summer.

Circumpolar map of the tundra and

Science Misconceptions
Identifying a Coniferous Forest

Conifers grow in places other than the coniferous forest. A group of conifers growing together does not necessarily mean that the biome is a coniferous forest. The coniferous forest biome is also characterized by its latitudinal location and specific climate conditions.

Background
Coniferous Forest Names

The northern coniferous forest has many different names. It is often called the *taiga* or the *boreal forest*. The term *taiga* is a Russian term for a swampy, marshy forest. The term *boreal forest* is most often used in North America. *Boreal* means "north-ern." This biome is also nicknamed the "spruce-moose" biome. Spruce is a common tree found in the forest, and the moose is one of the ...

FANTASTIC FACTS

The taiga has many ponds, bogs, and lakes. One type of bog, a *muskeg*, can be found in some areas of the taiga where water does not drain through the soil. This swamplike ground may appear firm because of the moss, grass, or trees growing on it. But, underneath, the land is soggy and wet. In some areas in Manitoba, Canada, stepping on certain parts of the ground causes the trees to move.

Snowshoe hare

Most of the trees in the coniferous forest are **conifers**, or cone-bearing trees. Some of the most common conifers are spruce, fir, and pine trees. Many conifers are evergreens. These trees do not lose their leaves during the winter. This enables evergreens to make food even during the long winter season.

Winter temperatures in the coniferous forest are usually so cold th... the snow does not melt between storms. So the trees must be able... support a lot of heavy snow. Oth...wise, the weight of the snow wou... break the tree branches.

Many conifers are shaped like cones. This shape helps some of the snow to slide off the tree without causing damage. Other trees are able to bend with the weight of the snow. Some trees can bend so much that they become almost flat to the ground. But when the snow melts, the trees stand straight again.

God provided some animals with the ability to adapt to extreme environments. For example, the snowshoe hare's fur changes to brown ... white in winter. The hare also ... feet that act like snowshoes, helping it stay on top of the snow. Stiff hairs on the bottom of its feet form fur pads that keep the hare's feet warm and insulated.

147

heat over a g...
Does the coniferous fores... tion as rain or snow? rain How do you kn... its precipitation falls during the summer months.
What type of tree is most common in the coniferous forest? conifers
What are conifers? trees that have cones
What term describes trees that do not lose their leaves during the winter? evergreen
How does the cold climate affect snowfall in the coniferous forest? Winter temperatures are so cold that snow does not melt between storms. This causes the snow to pile up.
How are most conifers shaped? like cones
How does the shape of a conifer help it during the winter? The snow can slide off without breaking the branches.
How did God design some of the other trees to handle the weight of snow? They bend with the weight of the snow and then stand straight again when the snow is gone.
What special characteristics does the snowshoe hare have that help it survive during the winter? Possible answers: Its white fur provides camouflage, and the air-filled spaces in its fur provide insulation for warmth. Its large back feet act like snowshoes, and the stiff hairs on the bottom of its feet help keep its feet warm.
How do the changing colors of the snowshoe hare's fur help camouflage it? The fur helps the hare to blend in with its surroundings in both summer and winter for protection from enemies.
- Discuss *Fantastic Facts*.
  What is a muskeg? a type of bog that is found in the taiga
  Why does the surface of a muskeg appear firm? Moss, grass, or trees can grow on it.
  What happens in some muskegs when a person walks on them? The trees move.

Demonstration
How Rays of Sunlight Hit the Earth

Materials: globe, flashlight

Direct a student to hold the globe so that it is tilted at a slight angle. If the globe you are using is already tilted, you could just place the globe on a desk or table. Darken the room and stand about 1 m away from the globe. Holding the flashlight parallel to the floor, shine the flashlight toward the equator.
If the flashlight represents the sun, which part of the earth receives the most sun? the equator
Is the light as bright at the poles? no
Rotate the earth on its axis while continuing to shine the light toward the equator. Locate the areas that have tundra and coniferous forest biomes.
As the earth turns, look at the light near the North Pole. Does the light reach every area during each rotation? no
How do you think the slant of the light and its brightness affect the growing seasons for the tundra and the coniferous forest? Possible answers: The growing season would be shorter. Less light from the sun means that the climate is cooler.

Lessons 81–82 · 169

Chapter 6: Biomes

81/82

Where are deciduous forests found? along the eastern coast of North America and in other parts of the earth such as Europe and Asia

Would deciduous forests most likely be found in a polar region, a temperate region, or a tropical region? a temperate region

In each hemisphere, where is the temperate zone in relation to the polar region and the tropical region? It is between them.

How did the deciduous forest get its name? Most of the trees there are deciduous.

What are deciduous trees? trees that lose their leaves in the winter

Are deciduous forests classified as a wet biome or a dry biome? a wet biome Why? A deciduous forest receives more rain than any biome except a rain forest.

How many seasons do deciduous forests have? four

Which biome has a longer growing season, the coniferous forest or the deciduous forest? the deciduous forest

Why does the deciduous forest have a longer growing season than the coniferous forest? The deciduous forest receives more direct sunlight, causing it to have a warmer climate, more rainfall, and a longer growing season than the coniferous forest.

How long does the growing season in the deciduous forest last? about half of the year

Other than trees, what kinds of plants are commonly found in a deciduous forest? shrubs, berry bushes, wildflowers

On the globe or world map, locate the countries and continents where deciduous forests are found. Deciduous forests in the Northern Hemisphere are found in the eastern United States, Great Britain, France, central Europe, and in parts of Russia, China, Korea, and Japan. The Southern Hemisphere has deciduous forests in South America, Australia, and Africa. The deciduous forests of different countries have different species of trees.

Deciduous Forest

Deciduous forests line the eastern coast of North America. They are also found in other parts of the earth, such as in Europe and in Asia. This biome gets its name from the trees that make up the forest. Most of these trees are **deciduous**, meaning that they lose their leaves in the winter. Deciduous trees usually have large, wide leaves.

The weather in the deciduous forest is not constant but changes with the seasons. Spring is usually warm and rainy. Summer is hot and humid. Autumn is cool and crisp. Winters are often cold and snowy.

The deciduous forest is a wet biome. It receives precipitation throughout the whole year. In fact, a deciduous forest gets more rain than any other type of biome except a rain forest.

The warm growing season lasts about half of the year. This longer time allows the biome to get warmer than those biomes with shorter summers. This longer growing season provides time for plants and trees to grow. Many shrubs, berry bushes, flowers, and other plants cannot survive in colder biomes. But these plants flourish in the warmer climate of the deciduous forest.

The plants and trees in a deciduous forest must change with the seasons. Deciduous trees lose their leaves in cold weather. If deciduous trees kept their broad leaves throughout the winter, the water in the leaves would freeze. This could damage the tree.

Deciduous trees in Quebec, Canada

148

Clarify and enhance the lesson with information from **Background** and **Science Misconceptions**.

Background

Autumn Leaves

Plants need sunlight to produce the green-colored chlorophyll. The seasonal changes, especially decreased daylight, signal deciduous trees to begin growing a layer of cells at the base of each leaf. This layer limits nutrients and the leaf's production of chlorophyll. As the green decreases, other colors within the leaf are seen. The layer of cells at the base of the leaves gradually dries, and the leaves separate and fall from the trees.

Hibernation

Hibernation is discussed in Chapter 7, *Interactions in an Ecosystem*.

Promote **higher-level thinking skills** with questions marked with a gear icon. Answers to these questions are not taken directly from the Student Text.

Review each text lesson with a **Quick Check**.

Science 5

170 · Lessons 81–82

...ther animals remain in the forest ...round. Some, such as the deer, ... thick winter coats of fur. They ...hese winter coats when the ...rature gets warmer. Other ...s, such as the woodchuck, ...te.

...no longer ...lorophyll, so th[illegible] green color begins to fade. The other c[illegible]lors that had been hidden by the green [illegible]egin to show. Leaves fall to the ground, d[illegible]cay, and add nutrients back into the soi[illegible]. In the spring the trees use their stored food energy to form new leaves.

Some birds, butterflies, and large mammals move to warmer areas for the winter. The animals then return to the forest in the spring when it is warm again.

A deer grows a thick winter coat to stay warm in the deciduous forest during cold weather.

QUICK CHECK

1. What is the largest land biome?
2. How are the trees in deciduous forests different from the trees in coniferous forests?

149

81/82

What would happen if broad-leaved trees kept their leaves through the winter? The water in the leaves would freeze. This could damage the tree.

What conditions in autumn cause deciduous trees to block nutrients and water from their leaves? the shorter days and cooler temperatures

What causes the leaves of deciduous trees to change colors? The leaves can no longer make chlorophyll. This causes the colors hidden by the green chlorophyll to appear.

How do deciduous trees improve soil in the forest? Fallen leaves decay and add nutrients back into the soil, helping new plants to grow.

Name some animals that may move to another biome in the winter. Possible answers: some birds, butterflies, large mammals

How do some animals that remain in the forest year-round survive the changing seasons? Possible answers: grow thick fur, hibernate

ANSWERS

1. the coniferous forest, or taiga
2. Most of the trees in a deciduous forest lose their leaves every winter. Most of the trees in a coniferous forest are evergreens.

82

- The *How Rays of Sunlight Hit the Earth* demonstration, the *Reveal the Colors* [illegible] *Leaves Masked by Chlorophyll* activity, and the *Make* [illegible] *Rubbings* activity found at the bottom of the pages [illegible] Lessons 81–82 may be incorporated into the lesson.
- Suggestions for additional activi[illegible]s for this two-day lesson are on the Teacher's Toolkit [illegible]D.

Reinforce, review, and enrich student learning with pages from the **Activity Manual**.

Activity

Reveal the C[illegible]

Materials: pap[illegible] or acetone nail [illegible] ...g alcohol

Cut the coffee filter into 2.5 × 9 cm strips. Rub a green leaf against the middle of one filter strip until the leaf makes a green smudge. Pour a small amount of rubbing alcohol or nail polish remover into a jar. Hook a paper clip through one end of the filter, and hang it from a toothpick balanced across the top of the jar so that the lower end of the filter strip touches the liquid. Repeat with the other leaves. Set the jars aside and check on them every 15 minutes. As the liquid is absorbed, other colors in the green smudges will be revealed. These different bands of colors are pigments other than chlorophyll that also function in photosynthesis.

Make Leaf Rubbings

Materials: deciduous leaves, newspaper, white paper, pencil [illegible]rayon

Collect leaves from several different decidu[illegible]rees. Place one leaf on newspaper with the underside of the lea[illegible]. Cover the leaf with a thin sheet of white paper. Rub a [illegible] crayon lightly over the area of the paper that is over the leaf. Label the leaf rubbing with the type of tree the leaf came from. You may choose to cut out the leaf rubbings and use them to design note cards, stationery, or bookmarks.

Activity Manual

Study Skill and Reinforcement [illegible]

Review, pages 99–1[illegible]

These pages review L[illegible]

Extend lesson content with **two-day lesson** ideas.

Assessment

Quiz 6-A

The quiz may be given any time after completion of this lesson.

Assess the student's work and knowledge by using a **quiz, test** or **rubric**.

Chapter 6: Biomes

Lessons 81–82 · 171

Activity Lesson Features

Identify a **problem**. Most activities state a problem that needs a solution.

Make teaching easier by using the management ideas and other information included in **the shaded boxes or under the Background** information.

Record specific **materials** and **measurements** used for each activity on the Activity Manual pages.

Use the **materials list** in the Student Text to identify the basic resources needed for each activity.

Emphasize a specific science **process skill** in each activity.

The student formulates a **hypothesis** that states what he is trying to prove in the activity.

The student records observations and measurements in the **Procedure** section. Data is sometimes recorded using graphs and charts.

56

Student Text pages 100–101
Activity Manual pages 71–72

Objectives

- Predict which type of insulation will best keep hot water warm
- Test different types of insulation to determine which is the most effective
- Measure and use numbers in an activity

Materials

Materials needed for this lesson are listed on the Student Text page.

Teacher's Toolkit CD

- Rubrics: *Keeping Warm*, for each student

Cut 12 cm squares of cotton batting, craft foam, bubble wrap, and aluminum foil for each group. Some 3 oz cups have a wider bottom than others do. Cups with a wider bottom will be less likely to tip over when wrapped with insulation.

Introduction

What are some ways that people keep picnic foods or drinks cold? Possible answers: put them in ... thermos; ... a box ...

Wha... for a... it un... it in...

Prep...

- Direct ... 100–101 ... Activity Manual pages 71–72 before beginning the activity.

Teach for Understanding

Process Skill

The process skill emphasized in this activity is measuring and using numbers, which is using standard measuring devices and techniques to quantify information.

What kinds of measurements are being used in this experiment? milliliters, degrees

Why is it important to measure the amount of water that we put in each cup? If the amount of water is not the same in each cup, the temperature measurements will not show which insulation worked best.

Why is it important to measure the temperature? Without measuring, we are simply guessing which insulation works best.

ACTIVITY

Keeping Warm

One of the most important reasons to study heat is to protect people. Often that means using insulation. Insulation keeps something from getting hot. Many cooking pots have insulated handles. The handles stay cool and keep you from burning your hand.

Insulation can also keep hot things warm, though. For example, an insulated thermos can keep hot foods warm or cold foods cool. Many animals, such as polar bears, have air-filled spaces in their fur. The air helps the animals keep warm during the winter. Likewise, insulation in the ceilings or walls of buildings keeps heat inside during cold weather.

In this activity you will test several kinds of insulation to determine which is the most effective.

Process Skills
- Hypothesizing
- Predicting
- Inferring
- Collecting and recording data
- Communicating

Problem

Which kind of insulation will keep hot water warm the best?

Procedure

1. Wrap cotton batting around one of the cups. Be sure to cover the bottom and the sides of the cup. Use a rubber band to secure the cotton batting against the cup.
2. Prepare three more cups: one wrapped with craft foam, one with bubble wrap, and one with a double thickness of aluminum foil. Use rubber bands to secure each material. Do not wrap anything around the fifth cup.
3. Predict which cup will best keep the hot water warm. Complete the hypothesis in your Activity Manual.
4. Pour 50 mL of hot water into each cup. Place a thermometer in each cup.
5. Cover the top of each cup with plastic wrap. Do not cover the thermometer with plastic wrap. You should be able to read the temperature without removing the plastic wrap.

Materials
- 5 plastic cups, 3 oz
- cotton batting
- rubber bands
- craft foam
- bubble wrap
- aluminum foil
- 250 mL hot water
- metric measuring cups
- 5 thermometers
- plastic wrap
- Activity Manual

Keeping Warm

Name ___________

Problem

Which kind of insulation will keep hot water warm the best?

Materials

5 plastic cups, 3 oz
aluminum foil
cotton batting
250 mL hot water
rubber bands
metric measuring cups
craft foam
5 thermometers
bubble wrap
plastic wrap

Hypothesis

___________ will keep the hot water warm the best.

Procedure

1. Record the water temperature for each cup.

	No Insulation	Foam	Bubble Wrap	Aluminum Foil	Cotton Batting
Temperature at beginning					
Temperature after 5 min					
Temperature after 10 min					

2. Calculate and record the temperature difference for each cup.

	No Insulation	Foam	Bubble Wrap	Aluminum Foil	Cotton Batting
Temperature difference					

6. Measure and record the starting temperature for each cup.
7. Leave the cups undisturbed for five minutes. Then measure and record the water temperature in each cup. Measure and record the temperatures again after another five minutes.
8. Calculate the difference between the starting and ending temperatures for each cup.

Conclusions

- Did your results support your hypothesis?
- Which cup had the greatest change in temperature? Why?
- Which type of insulated cup would you choose to hold hot chocolate? Why?

Follow-up

- Test other materials as insulation.
- Use ice cubes instead of hot water. Determine which insulation is the best for keeping ice cubes from melting.

101

Chapter 4: Energy and Heat

56

- Emphasize the importance of measuring accurately. If we fail to measure accurately, how useful is the information that we gain from the experiment? The information is unreliable because it has no specific data that can be quantified.

Procedure

- If needed, help the students prepare the cups. It is easier to cover a cup with a square if the cup is first placed upside down on a table. The cup wrapped with craft foam will be less likely to tip over if it is wrapped similar to the way you would wrap a gift. If necessary the students can prop the cups against other objects for support.

- The students should prepare all the cups before adding the hot water. This should keep the starting temperatures about the same.
- Remind the students not to cover the thermometer with the plastic wrap. They should be able to read the temperature on the thermometers without disturbing the plastic wrap.
- Direct the students to check the temperatures after 5 minutes and then after 10 minutes.

Conclusions

Which cup had the great... temperature?

Was your hypothesis co...

Which type of insulated... hot chocolate? Why?

Activity Manual

Activity, pages 71–...

Assessment

Rubrics

Select the prepared r... your chosen criteria...

Use the **Conclusions** questions to analyze the results of the activity and discuss what the student has learned.

Extend the activity by using suggestions in the **Follow-up** section.

The student records in his Activity Manual his responses to the questions to support his **conclusions**.

Conclusions

1. Did your results support your hypothesis?
2. Did the water in each container have the same starting temperature? If not, what could be a reason for the difference?
   Possible answer: The water in the cups poured first may have cooled more before the starting temperature was checked.
3. Which cup had the greatest change in temperature?
4. Which cup had the least change in temperature?
5. Which insulation seemed to work the best for keeping hot things warm?
6. Did any types of insulation have the same change in temperature? If so, what does that indicate about those types of insulation?
   They are probably equally good at insulating.
7. Which type of insulated cup would you choose to hold your hot chocolate? Why?
8. How can you use your knowledge about insulation to help you serve others better?
   Answers will vary.

72 Chapter 4 | Lesson 56, pages 100–101
Activity

Science 5

Teaching a Text Lesson

Overview

The introduction of each lesson provides a short activity or relevant questions to focus the student on the main point. Most lessons cover four pages of the Student Text. The Preparation for Reading questions should be presented before the student begins to read the text. The student should then read the material silently on his own. Assign the student to read the text as part of the lesson or prior to the lesson at home or in class. Oral reading of the text should be reserved for short passages only as they relate to the discussion.

The lesson plan provides questions to help you identify the student's level of comprehension of the material he has read. This discussion provides an auditory reinforcement of key ideas. The higher-level thinking questions (marked with the gear icon) help the student apply the information that he has read by relating it to previous knowledge and everyday situations.

Cross-curricular information, background information, activities, and demonstrations located at the bottom of the lesson pages provide additional material for you to use with the lesson based on the class dynamics and available time. Cross-curricular information in the lessons includes art, Bible, geography, history, language, math, technology, and writing. An activity may reinforce a concept taught in the lesson or extend a concept. Enrichment information, demonstrations, and activities may also be used to apply the science concepts to other subjects.

Reading for Information

The teaching instructions focus on reading for information, which is an essential skill for reading success. It is also a skill emphasized in the standards. The following elements can assist the student in developing and refining his informational reading skills.

- The Preparation for Reading questions in the Teacher's Edition give you opportunity to direct the student's focus as he reads the material.

Preparation for Reading

- Display the words *paleontology* and *preparators*. Pronounce the words together.
- Direct the students to read silently pages 40–43 to find the answers to these questions.
  What does a preparator do?
  What is an index fossil?

- The Quick Check at the end of each lesson in the Student Text helps the student evaluate how well he has comprehended what he read.

QUICK CHECK

1. What is all matter made of?
2. What is an element?
3. What is a chemical change?

- The strategies for instruction incorporated in the *Science 5* program allow you to guide the student in interpreting graphic information. Graphic information, such as charts, graphs, maps, photos, illustrations, and diagrams, are forms of communication that enhance the student's reading and understanding.
- The PQ3R method is presented in Lesson 12. This is one method that may be used throughout the year to aid the student in gathering information from the text.

Science 5 also incorporates several strategies drawn from standards as they apply to science. These standards are meant to help the student meet the specific challenges of reading, writing, speaking, listening, and using language in the science fields.

Vocabulary and Materials

The terms that the student needs to know are found in the vocabulary section at the beginning of each lesson. These terms are bold in the Student Text. They may also be found on the Chapter Summary page at the end of each chapter in the Student Text. The terms and their definitions will be reviewed on the study guides in each chapter of the Activity Manual.

The materials needed to teach *Science 5* can be found in the following sections of the Teacher's Edition:

- The Instructional Materials section on page vii identifies the essential items available for purchase to teach *Science 5*.
- The Materials section at the beginning of each lesson lists the materials needed to teach that particular lesson.
- The Materials List in the Appendix and also on the Teacher's Toolkit CD specifies by chapter all the materials needed to teach the entire program.

School supplies typically found in a student's desk (glue stick, scissors, pencil, paper, Bible, etc.) are not usually listed in the materials list of each lesson. Pages on the Teacher's Toolkit CD are noted under the Teacher's Toolkit CD section at the beginning of each lesson.

Discussion

In the Teach for Understanding section, you will find questions and comments to help you evaluate the understanding of concepts. The questions listed are not exhaustive but are a springboard to help you begin the discussion. A gear icon ( ) is used to help you identify the higher-level thinking questions. The answer to a higher-level thinking question is not taken directly from the pages being discussed. Supply any prompts or background needed to guide the student to the answer.

Two-Day Lessons

Two days have been allotted to cover some content. You may want to teach the text pages the first day and use activities, demonstrations, or other ideas the second day. You may also want to incorporate the activities and demonstrations into the text material, extending the discussion to two days. Activities and demonstrations are located at the bottom of many of the lesson pages. There is also a Two-Day Lesson section on the Teacher's Toolkit CD, which provides a list of ideas for each two-day lesson as well as a bank of teaching suggestions.

36

- The *Compression* demonstration and the *Using Physical Properties, Demonstrate Different Strengths of Attraction*, and *Writing a Fictional Story* activities found at the bottom of the pages in Lessons 35–36 may also be incorporated into the lesson.
- Suggestions for additional activities for this two-day lesson are on the Teacher's Toolkit CD.

Bible Integration

Science 5 incorporates a Christian worldview by showing God's nature and mankind's responsibility as revealed through creation. Biblical truth is woven throughout the Student Text. In addition, both the *Creation Corner* and the *Science & The Bible* interest boxes help the student develop a Christian worldview.

Bible integration is included as part of the text discussion in the Teacher's Edition. Some lessons provide additional Bible links for further discussion. Bible Action Truths [BATs] provide Christians with specific goals for their actions and attitudes. A list of BATs, as well as Bible Promises, is on pages A2–A4 of the appendix. Both sections provide additional Scripture resources. Take advantage of teachable moments to apply biblical truth whenever the occasion arises.

Christian worldview is also presented with an Answers in Genesis lesson in each chapter. These lessons teach the students to look at a chapter-related topic using the Bible as a set of "glasses" to understand the world. There are two accompanying pages in the Activity Manual for each Answers in Genesis lesson.

One of the greatest desires of Christian teachers is to lead children to the Savior. The teacher may find the Explaining the Gospel outline helpful. It is located on page A5 of the appendix.

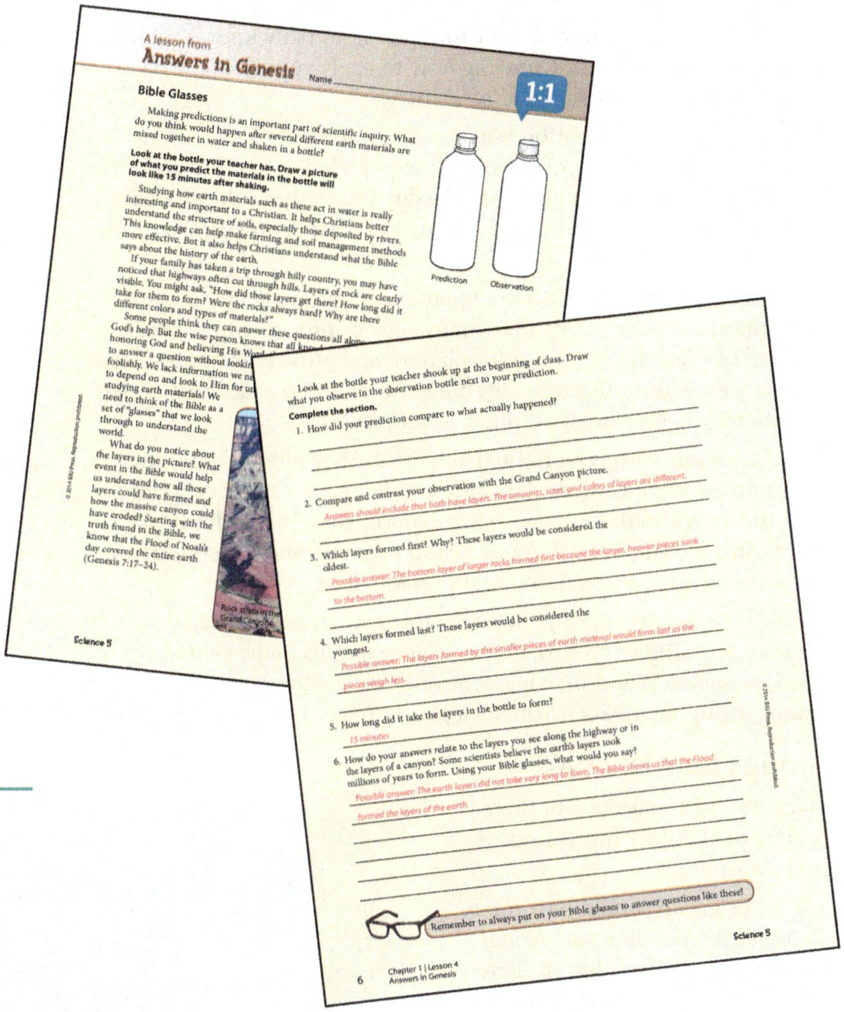
A lesson from
Answers in Genesis Name
1:1
Bible Glasses
Making predictions is an important part of scientific inquiry. What do you think would happen after several different earth materials are mixed together in water and shaken in a bottle?
Look at the bottle your teacher has. Draw a picture of what you predict the materials in the bottle will look like 15 minutes after shaking.
Prediction Observation
Studying how earth materials such as these act in water is really interesting and important to a Christian. It helps Christians better understand the structure of soils, especially those deposited by rivers. This knowledge can help make farming and soil management methods more effective. But it also helps Christians understand what the Bible says about the history of the earth.
If your family has taken a trip through hilly country, you may have noticed that highways often cut through hills. Layers of rock are clearly visible. You might ask, "How did those layers get there? How long did it take for them to form? Were the rocks always hard? Why are there different colors and types of materials?"
Some people think they can answer these questions all alon… God's help. But the wise person knows that all kno… honoring God and believing His Wor… to answer a question without lookin… foolishly. We lack information we ne… to depend on and look to Him for un… studying earth materials! We need to think of the Bible as a set of "glasses" that we look through to understand the world.
What do you notice about the layers in the picture? What event in the Bible would help us understand how all these layers could have formed and how the massive canyon could have eroded? Starting with the truth found in the Bible, we know that the Flood of Noah's day covered the entire earth (Genesis 7:17–24).
Rock strata in the Grand Canyon
Science 5
© 2014 BJU Press. Reproduction prohibited.

Look at the bottle your teacher shook up at the beginning of class. Draw what you observe in the observation bottle next to your prediction.
Complete the section.
1. How did your prediction compare to what actually happened?
2. Compare and contrast your observation with the Grand Canyon picture.
Answers should include that both have layers. The amounts, sizes, and colors of layers are different.
3. Which layers formed first? Why? These layers would be considered the oldest.
Possible answer: The bottom layer of larger rocks formed first because the larger, heavier pieces sank to the bottom.
4. Which layers formed last? These layers would be considered the youngest.
Possible answer: The layers formed by the smaller pieces of earth material would form last as the pieces weigh less.
5. How long did it take the layers in the bottle to form?
15 minutes
6. How do your answers relate to the layers you see along the highway or in the layers of a canyon? Some scientists believe the earth's layers took millions of years to form. Using your Bible glasses, what would you say?
Possible answer: The earth layers did not take very long to form. The Bible shows us that the Flood formed the layers of the earth.
Remember to always put on your Bible glasses to answer questions like these!
Science 5
6 Chapter 1 | Lesson 4 Answers in Genesis
© 2014 BJU Press. Reproduction prohibited.

Review and Test Lessons

A review lesson is planned for each chapter. Use of the suggested game or an alternate game from the Game Bank on the Teacher's Toolkit CD enables the teacher to reinforce concepts in a fun way.

Included in the Activity Manual is a Thinking It Through page for each review lesson. This page requires written responses to application questions from the chapter.

The review lesson also gives the teacher an opportunity to check to see that each student has accurately completed the study guides for the chapter. The material for each chapter test is taken from the Activity Manual review pages for that chapter. A student who knows the material (not just the answers) covered on the review pages will be adequately prepared to take the test.

The test day has been given its own lesson number. There is no material assigned to be taught in this lesson. (**Note**: The tests must be purchased separately.)

Science Notebook

It is highly recommended that the student keep a loose-leaf notebook for science. The Activity Manual pages are three-hole punched to accommodate this suggestion. A notebook will allow the student to keep study guides, activity records, and other useful information organized. A section of the notebook should serve as a science journal for recording notes, observations, and thoughts about concepts and activities.

Teaching Activity and Exploration Lessons

It is important to emphasize the value of the Activity and Exploration lessons in *Science 5*. In order for the student to truly know science, he must be able to use it. Knowing how to apply science is the basic premise behind the term *scientific literacy*.

The Activity and Exploration lessons allow the student to demonstrate his understanding of science concepts. However, these activities also allow the student to apply knowledge from other subject areas, such as measurement skills from math, and writing and communication skills from English.

Many activities in *Science 5* are open-ended experiments. These require the student to incorporate many science process skills. Because the goal of these activities is to teach a mental process as opposed to specific procedures, the activities may not follow an exact format. An activity may have more than one correct result.

Explorations are project-oriented activities. They often require the student to work outside of class and usually require more creativity than the activities that involve basic experiments. While doing explorations, the student will develop and use skills such as conducting research, writing, and making oral presentations.

Most of the Activity and Exploration lessons can be completed by students working in groups. Using science groups helps students learn cooperation and management skills. Brainstorming ideas among group members is usually a good problem-solving technique.

Groups for Activities

Place students in groups of three or four to allow for maximum participation.

A science group should be a mixture of students from all achievement levels. You may find that students who are not good "book learners" are much better at doing hands-on activities.

At the beginning of the year, you may choose to assign tasks to each group member to ensure active participation by all. Later the students within each group can decide the tasks of each member.

Groups for Explorations

Using groups for explorations may require extra input from you. Because the Exploration lessons are less structured, students may have a difficult time assigning themselves appropriate tasks. However, since these lessons often require a high level of creativity, students may be less intimidated in a group setting.

Management Tips

Being prepared is the key to a successful activity. Materials lists are given in the Student Text and in the Activity Manual. The Teacher's Edition also includes a comprehensive materials list on pages A6–A14. A printable copy of this list is included on the Teacher's Toolkit CD. You will need to determine specific quantities based on the number of science groups you have in your class. You may need an additional quantity of some materials in order for the student to follow up the activity with new variables.

Have newspaper or plastic on hand to cover work surfaces and buckets or containers in which to dispose of waste material. Paper towels and damp rags are helpful also.

Science Process Skills

Science is not just a collection of facts. It is also a demonstration of processes that show understanding of how science works. Each Activity lesson in *Science 5* allows the student to practice multiple science process skills. Additionally, one particular process skill is highlighted for discussion in each Activity lesson.

Although process skills are emphasized in the Activity lessons, they are also used in class discussions, demonstrations, and Exploration lessons. The following is a list of process skills that are used throughout *Science 5*.

Hypothesizing	formulating a statement that can be tested by experimentation
Predicting	forecasting an expected result based on prior experience or knowledge
Measuring and Using Numbers	using standard devices and techniques to quantify information
Experimenting	setting up and following a procedure to test a hypothesis
Making and Using Models	creating a physical or mental representation to explain or clarify ideas, objects, or events
Observing	using the senses to gather information regarding objects and events
Inferring	suggesting explanations based on previous knowledge or observation
Classifying	grouping or ordering objects based on similarities or differences
Identifying and Controlling Variables	recognizing the changing and unchanging factors in an investigation and adjusting one factor to obtain data
Collecting, Recording, and Interpreting Data	gathering information about objects and events in an organized and systematic manner
Communicating	using written, oral, or graphic means to transmit information to others
Defining Operationally	explaining an object or event properly but in terms of the student's observation and experience

Assessment and Grading

Science 5 provides a variety of tools that teachers may use for assessment and grading. Frequent assessment enables you to adjust your instructional plan to better meet the needs of each student. The chart identifies suggested percentages you may use for calculating the student's total grade.

Tests and Quizzes	Written Work	Activities & Explorations	Participation
50%	25%	25%	
33%	33%	33%	
50%	20%	20%	10%
40%	25%	25%	10%

Tests and Quizzes

The *Science 5 Tests* can serve as the objective part of the evaluation of a student's progress. The most effective tests are an outgrowth of the teaching process. Accordingly, these tests should not replace your individual assessment of a student's understanding and application. The tests can be adapted to meet the teaching emphasis and direction as well as the student's maturation level. You may find it necessary to eliminate some items, provide additional test items, or do both. Students will find the mastery-level information for the *Science 5 Tests* on the Study Guide and Thinking It Through pages in the Activity Manual and the Chapter Summary pages in the Student Text.

Optional quizzes are provided for each chapter on the Teacher's Toolkit CD. The quizzes correlate with the Study Guide pages in the Activity Manual. You may choose one of a variety of ways to grade the quizzes. You could use point values or percentages for individual scores. The scores could then be compiled and averaged, treating them as one test grade, or quiz scores could be assigned a designated percentage of the total grade. Quiz scores could also be included within the category of Written Work.

Written Work

Written work may include daily assignments, Quick Check answers, Activity Manual pages, essay and application questions, and journal entries. Scores may be given based on the completion of some assignments and the accuracy of content on others.

Activities and Explorations

A rubric is a useful tool for assessing work that is not objective, such as the Activity and Exploration lessons. Specific rubrics for each Activity and Exploration, as well as instructions for their use, are located on the Teacher's Toolkit CD. In a group activity it is often beneficial to give not only the group as a whole a grade but also to give each student an individual grade. You may choose to give the rubric to the student or to the group for self-assessment as the activity or project is completed. Your scores and comments on a rubric allow the student to see why he received a particular grade and show him areas in which he can improve.

Participation

You may choose to evaluate the student's participation during lesson discussions. This subjective assessment should be tracked. You could use checklists, rubrics, point values, or letter grades.

Decomposers at Work Name Jon Clater

Category	Possible points	Points earned
Instructions Student followed instructions	4	4
Materials Student handled materials properly	5	5
Measurements Student took accurate and appropriate measurements	5	5
Recorded Information Student recorded necessary information	5	3
Process Skills: Identifying and Controlling Variables Student identified the changing and unchanging factors and controlled them to obtain data	5	4
Content Student showed understanding of content	4	4
Completed Task Student completed task in a timely manner	3	3
Independent/Group Work Student worked independently when required or made a worthwhile contribution to a group	3	2
Total	34	30

Grade 88%

Comments

Jon, you showed good organizational skills. Remember to list all of the materials you use in the activity.

You participated a lot, but you need to allow others equal opportunity. Try to work on sharing materials.

Rubrics • For use with Lessons 115–16
Science 5

Scheduling

Science 5 provides material for a year-long program. Test and review days are included, and some activities have been allotted two lesson days for completion. The instruction of all textual information, activities, and explorations will adequately fill a full-year science program.

A program that includes less than a full year of science instruction must adjust the use of the materials to best fit both the individual and the school's needs. The following are additional ways to adjust the program to fit an alternate schedule.

- The material covered in Chapter 11, *Respiratory System*, and Chapter 12, *Circulatory System*, may not appear on the achievement tests. You may choose to include these chapters as part of your health program.
- You may choose to combine lessons. For example, you may choose to combine two text lessons into one lesson or three text lessons into two lessons.
- Each chapter has at least one Exploration and two Activity lessons. If needed, you may introduce the exploration and assign it to be completed for homework. Many of the exploration projects may be completed outside of class.
- Since many of the explorations require the use of research and study skills, they can be incorporated into another subject area, such as language arts.
- Explorations and enrichment activities from many lessons can be adjusted for use in learning centers.

Lesson Plan Overviews

Lesson	TE pages	ST pages	AM pages	Objectives and Christian Worldview	Process Skills
1		1–2	1	• Define *worldview* • Recognize that everyone has a worldview • Identify characteristics of a Christian worldview	
Chapter 1: Minerals and Rocks					
2	2–5	3–5	2	• Recognize the interrelationship of science concepts • Distinguish facts and assumptions in the evolution/Creation debate • Evaluate evolutionary assumptions from a Christian worldview *Mankind's imitation of God's creation* *The Bible as the final authority* *God as the only Creator*	
3	6–9	6–9	3–4	• Identify and locate the layers of the earth • Describe features of the core, mantle, and crust • Explain how weathering and erosion affect sediment • Define *humus* *The Flood's effect on the earth* *Fall of mankind* *Mankind's use of God's resources*	
4	10–11		5–6	**Answers in Genesis** • Explain why it is necessary to look at the world with a biblical perspective • Justify from a biblical viewpoint that the layers of the earth did not take millions of years to form	
5–6	12–17	10–15	7	• Define *mineral* • Identify crystal structure, luster, hardness, color, and cleavage as characteristics of minerals • Explain how the Mohs scale is used to determine hardness *God's design for the earth's resources* *God's design for the human body*	
7	18–19		8	**Activity: Measuring Mass and Volume** • Measure mass to the nearest gram • Measure volume to the nearest milliliter	
8–9	20–21	16–17	9–10	**Activity: Salty Crystals** • Follow directions • Observe the formation of Epsom-salt crystals • Collect and record observation data	Measuring Experimenting Observing Identifying and controlling variables Collecting, recording, and interpreting data
10	22–26	18–22	11–13	• Differentiate between characteristics of precious and semiprecious stones • List some common uses of minerals • Recognize that some minerals are metals • Identify where minerals are found *God's creation for mankind's enjoyment* *God's salvation through Christ*	
11	27	23	14	**Exploration: Munching Minerals** • Research a mineral found in foods or beverages • Display foods or beverages that contain the mineral • Prepare an oral presentation *God's design for the human body*	
12	28–29		15–18	**Study Skill: PQ3R** • Use the PQ3R method to read informational text	
13	30–33	24–27	19	• Define *rock* • Identify three types of rock and explain how each is formed • List examples of igneous rock, sedimentary rock, and metamorphic rock *Consequences of sin* *The Bible as the final authority*	
14	34–35	28–29	20–21	**Activity: Rock Hounding** • Label rocks in a collection • Classify rocks according to chosen criteria	Observing Classifying Communicating Defining operationally
15	36	30	22	**Chapter Review** • Recall concepts and terms from Chapter 1 • Apply knowledge to everyday situations	
16	36			**Chapter 1 Test** • Demonstrate knowledge of concepts taught in Chapter 1	

Lesson	TE pages	ST pages	AM pages	Objectives and Christian Worldview	Process Skills
Chapter 2: Fossils and Dinosaurs					
17	37	31	23	• Evaluate evolutionary assumptions from a Christian worldview *Faith in the Word of God* *God's orderly design*	
18	38–40	32–34		• Define *fossil* • Compare and describe some types of fossils that form in sediment: petrified fossil, mold, cast, carbon film, trace fossil • Identify other materials in which fossils are sometimes preserved *The Flood's effect on the earth*	
19	41	35	24–25	• Compare beliefs of evolutionists and Creationists *The Flood's effect on the earth* *Faith in the Word of God*	
20	42–43	36–37	26	**Activity: Fact or Theory?** • Identify phrases or statements that indicate a Creationist or evolutionist viewpoint • Make inferences as to the viewpoint from which literature is written *Discerning what is true* *The Bible as the final authority*	Inferring Collecting and interpreting data Communicating Defining operationally
21–22	44–45	38–39	27–28	**Activity: Molds and Casts** • Make models of fossils • Relate models to fossils	
23–24	46–49	40–43	29–30	• Define *paleontology* • Describe how fossils are excavated and reconstructed • Explain why rock layers do not indicate the age of a buried fossil • Describe how paleontologists use carbon dating to guess the age of fossils *The Flood's effect on the earth* *Faith in the Word of God*	
25–26	50–51	44–45	31–32	**Exploration: Fossil Dig** • Model the procedures a paleontologist uses while excavating • Complete a site map	
27	52–55	46–49	33	• Recognize that what is known about dinosaurs is based on the observations of fossils • Name some of the types of information that are known from fossils • Recognize some of the types of information that can be inferred from fossils *Mankind's God-given curiosity* *Faith in the Word of God* *God's perfect creation*	
28	56–59	50–53	34	• Realize that man and dinosaurs lived at the same time • Recognize that some dinosaurs survived the Flood • Identify biblical animals that may have been dinosaurs • Name some causes of extinction • Identify reasons why dinosaurs may have become extinct *Faith in the Word of God* *God's orderly design*	
29	60–61		35–36	**Answers in Genesis** • Justify from a biblical viewpoint that dinosaurs existed and that dinosaurs and people lived together • Examine scientific evidence to show that dinosaurs are thousands of years old and not millions	
30	62	54	37–38	**Chapter Review** • Recall concepts and terms from Chapter 2 • Apply knowledge to everyday situations	
31	62			**Chapter 2 Test** • Demonstrate knowledge of concepts taught in Chapter 2	

Lesson	TE pages	ST pages	AM pages	Objectives and Christian Worldview	Process Skills
Chapter 3: Matter					
32	64–67	55–57	39	• Recognize that God created different kinds of matter to melt at different temperatures • Recognize that learning about matter and how it works is important to glorify God and serve others • Give an example of how God's design of the properties of matter benefits people *Christian behavior as showing God's love to others* *Christians as a reflection of God*	
33	68–71	58–61	40	• Define *matter* • Explain how to find the volume of a solid and of a liquid • Differentiate between mass and weight • Recognize that volume, mass, and weight are ways by which matter can be measured • Explain how density is related to mass and volume *Mankind's use of wisdom to serve others* *God's provision for mankind* *God's perfect design*	
34	72–73		41	**Activity: Measuring Length, Volume, and Temperature** • Measure length to the nearest millimeter • Measure volume using cubic centimeters • Measure temperature to the nearest degree	
35–36	74–79	62–67	42–44	• Identify and describe the three states of matter • List examples of solids, liquids, and gases • Define *physical change* • Recognize that a change of state is a physical change • Differentiate among melting, freezing, vaporization, and condensation *God's orderly design*	
37	80–81		45–46	**Activity: A Science Experiment** • Use a scientific method *Discerning what is true*	
38	82–83	68–69	47–48	• Identify atoms as small particles of matter • Differentiate between elements and compounds • Contrast chemical changes and physical changes	
39	84–85	70–71	49–50	**Activity: Separating a Mixture** • Plan a procedure for separating the parts of a mixture • Apply the physical properties of the items that make up a mixture • Experiment to test predictions • Infer how to physically remove a dissolved item from water	Predicting Experimenting Observing Inferring Communicating
40	86–89	72–75	51	• Define *mixture* • Explain the difference between a mixture and a compound • Give some examples of mixtures • Identify some ways that substances in a mixture can be separated using physical properties	
41	90–93	76–79	52–54	• Identify a solution as a type of mixture • Identify the parts of a solution • Define *concentration* • Explain ways to increase the rate of dissolving *Mankind's demonstration of God's love*	
42	94–95		55–56	**Answers in Genesis** • Recognize that God created the matter in the universe from nothing • Provide examples from Scripture of how the universe was created • Identify the object of faith for materialists (matter) and Christians (God and the Bible)	
43	96–97	80–81	57–58	**Activity: A Disappearing Act** • Predict how surface area will affect the rate of dissolving • Relate results to other situations	Hypothesizing Experimenting Observing Inferring Defining operationally
44–45	98–99	82–83	59	**Exploration: Float a Boat** • Design a clay boat that will float • Demonstrate buoyancy *God overruling His natural laws*	
46	100	84	60	**Chapter Review** • Recall concepts and terms from Chapter 3 • Apply knowledge to everyday situations	
47	100			**Chapter 3 Test** • Demonstrate knowledge of concepts taught in Chapter 3	

Lesson	TE pages	ST pages	AM pages	Objectives and Christian Worldview	Process Skills
Chapter 4: Energy and Heat					
48	101	85	61	• Explain the importance of energy and heat in designing useful technology *God's provision for His creation* *Mankind's use of wisdom to serve others*	
49–50	102–5	86–89	62	• Define *energy* • Differentiate between potential energy and kinetic energy • Recognize that energy is often classified as either potential or kinetic • Recognize that the amount of thermal energy depends on the temperature and mass of a substance • Differentiate between thermal energy and temperature *People as stewards of God's creation*	
51	106–7	90–91	63–64	**Activity: Rock Heaters** • Predict how the mass of a substance affects the amount of thermal energy it can transfer • Experiment to test a hypothesis	Hypothesizing Measuring and using numbers Collecting and recording data Defining operationally
52	108–10	92–94	65–66	• Recognize that increasing or decreasing thermal energy can cause matter to change to a different state • Explain what happens during thermal expansion • Define *calorie* • Recognize that substances differ in their ability to store thermal energy	
53	111	95	67–68	**Exploration: Energy for Your Body** • Recognize that a food Calorie is also called a kilocalorie • Calculate the resting metabolic rate • Track Calorie consumption for three days	
54–55	112–15	96–99	69–70	• Define *heat* • Recognize that heat always flows from a warmer substance to a cooler substance • Identify and describe three ways that heat occurs • Differentiate between conductors and insulators	
56	116–17	100–101	71–72	**Activity: Keeping Warm** • Predict which type of insulation will best keep hot water warm • Test different types of insulation to determine which is the most effective • Measure and use numbers in an activity	Hypothesizing Predicting Inferring Collecting and recording data Communicating
57	118–20	102–4	73–75	• Identify some common fuels • Distinguish between renewable and nonrenewable resources • Name some ways fuel is used • Give examples of unwanted heat *God's design for the human body*	
58	121–24	105–8	76	• Explain why controlling heat is necessary • Explain how scientists controlled heat for the reentry of space capsules • Name two types of insulation used on space shuttles • Name some ways that thermal energy is part of our everyday lives *Mankind's imitation of creation* *Mankind's responsibility to glorify God*	
59	125–26		77–78	**Answers in Genesis** • Show how Christian scientists can do operational science in order to exercise biblical dominion • Give examples of discoveries that show that operational science does not need to refer to evolutionary principles to be successful • Explain why biomimicry is an example of exercising dominion to love our neighbor and to glorify God	
60	127	109	79	**Exploration: Moon Station** • Design a piece of equipment for a moon station • Research equipment developed for the space program	
61	128	110	80	**Chapter Review** • Recall concepts and terms from Chapter 4 • Apply knowledge to everyday situations	
62	128			**Chapter 4 Test** • Demonstrate knowledge of concepts taught in Chapter 4	

Lesson	TE pages	ST pages	AM pages	Objectives and Christian Worldview	Process Skills
Chapter 5: Weather					
63	130–31	111–13	81	• Recognize the interrelationship of science concepts • Recognize, from a Christian worldview, reasons for studying climate • Understand the role of meteorology in preserving human life • Apply the biblical teaching on the value of human life to everyday situations	
64–65	134–37	114–17	82–83	• Describe the atmosphere • Define *air pressure* • Recognize that gravity pulls the atmosphere toward the earth • Name an instrument that measures air pressure • Identify and describe the two lower layers of the atmosphere *Mankind's God-given ability to observe* *God's design for the human body* *God's orderly design*	
66–67	138–43	118–23	84–86	• Compare and contrast high-pressure air masses and low-pressure air masses • Define *front* and describe three types • Explain how temperature affects wind • Differentiate between global winds and local winds • Name examples of global winds and local winds	
68	144–45	124–25	87	**Activity: Temperature Changes** • Predict whether water and soil will warm or cool at the same rate • Identify and control variables • Measure and record temperatures • Relate temperature changes to the ability of each substance to hold and give off heat	Measuring Observing Inferring Recording data
69–70	146–51	125–31	88	• Define *precipitation* • Differentiate among rain, sleet, snow, and hail • Define *humidity* • Identify and describe three basic shapes of clouds *God's provision for His creation* *Christian behavior as showing God's love to others*	
71	152–54	132–34		• Describe characteristics of thunderstorms, tornadoes, and hurricanes • Differentiate between a weather watch and a weather warning *Mankind's God-given dominion* *Christian behavior as showing God's love to others*	
72	155	135		**Exploration: Dangerous Extremes** • Research the safety precautions for a type of severe weather • Make and present a poster or pamphlet	
73	156–57	136–37	89–90	• Describe the job of a meteorologist • Read and interpret basic symbols on a weather map *Mankind's use of wisdom to serve others* *Christian behavior as showing God's love to others*	
74–75	158–59	138–39	91–92	**Activity: Weather Observatory** • Make working weather instruments • Correctly use the instruments to gather information about the weather • Record data • Use data to make weather predictions	Measuring and using numbers Making and using models Observing Collecting, recording, and interpreting data
76	160–61		93–94	**Answers in Genesis** • Explain how clouds form • Defend a biblical view of evidence for one ice age against a secular view of evidence for multiple ice ages	
77	162	140	95	**Chapter Review** • Recall concepts and terms from Chapter 5 • Apply knowledge to everyday situations	
78	162			**Chapter 5 Test** • Demonstrate knowledge of concepts taught in Chapter 5	

Lesson	TE pages	ST pages	AM pages	Objectives and Christian Worldview	Process Skills
Chapter 6: Biomes					
79	163	141	97	• Appreciate the effect of human intervention on a wetland biome • Apply the Bible's teaching of stewardship of creation to biomes • Generate possible solutions to the concerns about destroying or using biomes *Mankind as steward of God's creation* *Mankind's use of wisdom to serve others*	
80	164–67	142–45	98	• Differentiate between a biome and the biosphere • Identify climate as a major influence on land biomes • Describe basic characteristics of the tundra • Name some ways that animals and plants survive on the tundra *God's provision for His creation*	
81–82	168–71	146–49	99–100	• Describe basic characteristics of the coniferous forest • Describe basic characteristics of the deciduous forest • Differentiate between conifers and deciduous trees • Name two ways that animals in the deciduous forest survive the changing seasons *God's provision for His creation*	
83–84	172–75	150–53	101–2	• Describe basic characteristics of grasslands • Compare and contrast prairies and savannas • Name ways some savanna grasses and trees survive the dry season • Describe characteristics that all deserts have in common • Name some ways that desert animals and plants survive the extreme temperatures and dryness *God's provision for His creation*	
85	176–77	154–55	103–4	**Activity: Help Prevent Water Loss!** • Identify some characteristics of water-efficient plants • Predict how waxy surfaces on plants affect water loss • Relate the effectiveness of a petroleum-jelly coating on a sponge to the waxy surfaces on some leaves and stems *God's provision for His creation*	Predicting Measuring Making and using models Inferring Recording data
86–87	178–180	156–58	101, 105–7	• Describe basic characteristics of a tropical rain forest • Identify the layers of the rain forest • Name ways that roots benefit the rain forest trees • Recognize that biomes are only a general way to classify sections of the biosphere • Explain how a mountain can have several biomes	
88–89	181	159		**Exploration: Build a Biome** • Research a biome • Create a model of that biome	
90–91	182–87	160–65	108–10	• Name the two categories of aquatic biomes • Explain why coral reefs are called "the rain forests of the sea" • Identify the force that keeps river water moving • Describe kinds of wetlands • Recognize that people have the God-given responsibility to be good stewards of the earth *God's provision for His creation* *Mankind as steward of God's creation*	
92	188–89		111–12	**Answers in Genesis** • Compare the description of the Garden of Eden to a map of modern-day Iraq • Explain why the climate and biomes changed after the Flood	
93	190–91	166–67	113	**Activity: From Dirty to Clean** • Demonstrate how wetlands purify water • Infer how the activity models the purifying process of a real wetland *God's provision for His creation*	Making and using models Observing Inferring
94	192	168	114	**Chapter Review** • Recall concepts and terms from Chapter 6 • Apply knowledge to everyday situations	
95	192			**Chapter 6 Test** • Demonstrate knowledge of concepts taught in Chapter 6	

Lesson	TE pages	ST pages	AM pages	Objectives and Christian Worldview	Process Skills
Chapter 7: Interactions in an Ecosystem					
96	194–97	169–71	115	• Recognize the interrelationship of science concepts • Explain the relationship between the study of ecosystems and Genesis 1:28 • Apply the Bible's teaching of stewardship to creatures in a habitat	
97–98	198–201	172–75	116–18	• Identify the two parts of an ecosystem • Explain the relationships between individuals, communities, and populations • Identify the functions of producers, consumers, and decomposers • Explain why scavengers and decomposers are important to an ecosystem *Mankind as steward of God's creation*	
99	202–3	176–77	119–20	**Activity: Habitat Investigation** • Investigate a habitat • Distinguish between living things and nonliving things • Identify producers and consumers • Record interactions	Observing Classifying Collecting and recording data Defining operationally
100	204–7	178–81	121–22	• Identify the predators and prey in a food chain • Differentiate between a food chain and a food web • Describe the transfer of energy from one organism to another • Explain how competition affects population size	
101	208	182		**Activity: Food-Web Connections** • Identify predators and prey within a food web • Model a food web • Recognize interrelationships among organisms in a food web • Compare the model food web with an actual food web	Making and using models Communicating Defining operationally
102–3	209	183		**Exploration: A Tangled Web** • Make a visual representation of a food web • Identify producers, predators, and prey within a food web • Identify animals as herbivores, omnivores, or carnivores *Mankind's God-given dominion*	
104	210–11		123–24	**Answers in Genesis** • Describe relationships among animals and plants in a simple ecosystem • State the sources of food for both people and animals before the Fall • Explain why the kinds of teeth in a skull may not determine the kinds of food an animal eats • Compare and contrast the evolutionary and creationary views of the history of carnivores	
105–6	212–15	184–87	125	• Identify the basic needs of plants and animals • Identify and describe adaptations that help plants survive • Identify and describe adaptations that help animals survive *God's provision for His creation*	
107	216–19	188–91	126–29	• Identify different kinds of symbiosis • Differentiate between instincts and learned behaviors • Give examples of instincts and learned behaviors *God's perfect design* *Consequences of sin* *Christians behavior as showing God's love to others*	
108	220	192	130	**Chapter Review** • Recall concepts and terms from Chapter 7 • Apply knowledge to everyday situations	
109	220			**Chapter 7 Test** • Demonstrate knowledge of concepts taught in Chapter 7	

Lesson	TE pages	ST pages	AM pages	Objectives and Christian Worldview	Process Skills
				Chapter 8: Changes in an Ecosystem	
110	221	193	131	• Recognize ways that people can have dominion over the earth as God has commanded *Mankind's God-given dominion*	
111–12	222–25	194–97	132–34	• Recognize that the earth has many cycles • Identify the seasonal changes that may occur in an ecosystem • Explain the carbon cycle • Differentiate between photosynthesis and respiration *God's orderly design* *God's provision for His creation*	
113–14	226–29	198–201	135–36	• Name two ways that nitrogen is changed into usable compounds • Describe the nitrogen cycle • Identify the parts of the water cycle • Identify and infer some ways that cycles work together in an ecosystem *Interrelationship of the parts of creation* *God's provision for His creation*	
115–16	230–31	202–3	137–38	**Activity: Decomposers at Work** • Recognize that decomposers are a part of many cycles • Identify water as a variable that affects decomposition • Analyze the effects of water on the rate of decomposition	Hypothesizing Experimenting Observing Identifying and controlling variables Recording data
117–18	232–35	204–7	139	• Identify three natural stresses on an ecosystem • Explain how fires and floods can be beneficial to an ecosystem • Identify some effects of a drought • Describe the process of succession • Recognize that sometimes what seems to us like a disaster is actually God's way of maintaining the earth *Consequences of sin* *God's provision for His creation* *God's use of creation for His purpose*	
119–20	236	208	140	**Exploration: Stress Alert** • Research a historical stress, such as a famous fire, flood, or other disaster • Organize and present information about the stress	
121–22	237	209		**Activity: Current Events** • Collect and record information about ecosystems • Organize the information into a notebook for presentation	Classifying Communicating Defining operationally
123	238–39		141–42	**Answers in Genesis** • Explain the water cycle using a model • Relate the cycles of nature to God's care of His creation	
124	240–43	210–13	143	• Identify some manmade stresses • List differing opinions about using natural resources • Differentiate between an extinct species and an endangered species *Mankind's use of God's resources* *Mankind's God-given dominion* *Consequences of sin* *Mankind's responsibility to glorify God*	
125	244	214	144	**Chapter Review** • Recall concepts and terms from Chapter 8 • Apply knowledge to everyday situations	
126	244			**Chapter 8 Test** • Demonstrate knowledge of concepts taught in Chapter 8	

Lesson	TE pages	ST pages	AM pages	Objectives and Christian Worldview	Process Skills
Chapter 9: Sound					
127	249	215–17	145	• Recognize the interrelationship of science concepts • Recognize that technology can be designed to control sound because sound moves in predictable ways *Mankind's God-given dominion* *Mankind's use of wisdom to serve others* *Mankind's responsibility to glorify God*	
128–29	250–53	218–21	146–48	• Define *sound* and *wavelength* • Identify a compression of a sound wave • Differentiate between the frequency and speed of sound waves *Mankind's use of wisdom to serve others*	
130	254–55	222–23	149–50	**Activity: Sound Slide** • Observe how the size of a vibration affects its sound • Change a variable and compare results • Predict the highness or lowness of a sound	Predicting Experimenting Observing Identifying and controlling variables Communicating
131–32	256–59	224–27	151–52	• Define *pitch* and *volume* • Explain how the pitch of a sound wave is related to its frequency • Identify the frequency range of human hearing • Explain how the volume of a sound is related to the intensity of its sound waves • Define and describe *timbre* *God's design for the human body*	
133	260–61	228–29	153–54	**Activity: Shhh, Quiet Please** • Compare the amount of sound absorbed by different materials • Predict which material will absorb the most sound • Rate the loudness of sounds • Identify relationships between materials and their abilities to absorb sound	Hypothesizing Predicting Observing Communicating
134	262–63		155–56	**Answers in Genesis** • Summarize what the Bible says about hearing • Explain why a creationary approach to science is a better approach to solving problems (like hearing loss) than an evolutionary approach	
135	264–68	230–34	157	• Differentiate between *sound* and *noise* • Recognize that a sound fades as its energy is used up • List examples of how echoes are used in nature and technology • Name examples of how an acoustical engineer uses his knowledge of sound *Mankind's imitation of creation* *God's design for the human body* *God's creation for mankind's enjoyment* *Mankind's use of wisdom to serve others* *Christians as faithful witnesses*	
136	269	235		**Exploration: A "Medium" Exploration** • Test the abilities of different mediums to carry sound • Write a paragraph that compares and contrasts the results	
137	270	236	158	**Chapter Review** • Recall concepts and terms from Chapter 9 • Apply knowledge to everyday situations	
138	270			**Chapter 9 Test** • Demonstrate knowledge of concepts taught in Chapter 9	

Lesson	TE pages	ST pages	AM pages	Objectives and Christian Worldview	Process Skills
Chapter 10: Light					
139	271	237	159	• Recognize that God provides for the needs of people *Mankind's use of wisdom to serve others* *Mankind's responsibility to glorify God*	
140	272–75	238–41	160–61	• Identify light as a form of energy • Compare and contrast electromagnetic and mechanical waves • Identify the four properties of waves: wavelength, amplitude, frequency, and speed • Differentiate between the frequency of a wave and the speed of a wave *God's perfect creation*	
141–42	276–79	242–45	162–63	• Differentiate between refraction and reflection • Recognize that the color of an object depends on which colors of light are being reflected • Identify the primary colors of light *God's salvation through Christ* *Faith in the Word of God for guidance* *God's creation for mankind's enjoyment*	
143	280–81	246–47	164	**Activity: Fog Vision** • Test the visibility of colors • Infer which colors are most visible in fog	Hypothesizing Predicting Experimenting Observing Inferring
144–45	282–285	248–51	165	• Explain how light reflects off smooth and rough surfaces • Identify and describe three kinds of mirrors • Identify some technologies that use light • Name some uses for lasers	
146	286–87	252–53	166	**Activity: Angles of Reflection** • Differentiate between the angle of incidence and the angle of reflection • Measure the angle of reflection • Infer the relationship between the angle of reflection and the angle of incidence	Predicting Measuring and using numbers Observing Inferring Defining operationally
147–48	288–92	254–58	167–68	• Identify characteristics of waves found in the electromagnetic spectrum • Name some uses for each type of electromagnetic wave *God's creation of invisible forces* *Mankind's use of wisdom to serve others* *Mankind's responsibility to glorify God*	
149	293–94		169–70	**Answers in Genesis** • Contrast the naturalistic view of the sun's origin with the biblical view • Recognize that the Bible calls Christians to defend their faith	
150	295	259	171	**Exploration: Light at Work** • Identify different ways that light is used in technology • Make a collage that explains how different products use light	
151	296	260	172	**Chapter Review** • Recall concepts and terms from Chapter 10 • Apply knowledge to everyday situations	
152	296			**Chapter 10 Test** • Demonstrate knowledge of concepts taught in Chapter 10	

Lesson	TE pages	ST pages	AM pages	Objectives and Christian Worldview	Process Skills
Chapter 11: Respiratory System					
153	301	261–63	173	• Contrast technology with the marvels found in the human body • Demonstrate how people are being inspired by God's designs to develop new technology *God's perfect design* *Mankind's imitation of creation*	
154	302–4	264–66	174–75	• Identify the respiratory system as the breathing system • Differentiate between involuntary breathing and voluntary breathing • Identify the muscles that help with breathing • Describe the movement of the body and air when inhaling and exhaling *Mankind as God's special creation* *Mankind created in God's image* *God's design for the human body*	
155	305	267	176	**Activity: Breathe In, Breathe Out** • Make a model of a lung • Use the lung model to explain how the diaphragm moves during breathing	Making and using models Inferring Defining operationally
156–57	306–9	268–71	177–78	• Explain how mucus and cilia help keep the respiratory system clean • List the parts of the respiratory system from the nose to the larynx • Describe the function of the epiglottis • Explain how the vocal cords produce sound	
158–59	310–13	272–75	179–80	• Identify and describe the trachea, bronchi, and lungs • Describe the function of the lungs • Explain causes of snoring, hiccupping, coughing, and sneezing	
160	314–15	276–77	181–82	**Activity: How Much Air Is in Your Lungs?** • Calculate the amount of air exhaled • Identify variables that may affect the results	Hypothesizing Measuring and using numbers Collecting, recording, and interpreting data
161	316–17		183–84	**Answers in Genesis** • Describe the unique way God created man • Relate the physical position of Jesus on the cross to His inability to breathe normally, a part of His suffering	
162–63	318–21	278–81	185–87	• Identify some diseases that make it difficult to breathe properly • Describe what happens during an asthma attack • Recognize that allergies are not contagious • Name some reasons why smoking is harmful to your health *God's design for the human body* *Mankind as steward of God's creation*	
164	322–23			**Exploration: Dangers of Smoking** • Explain why it is hard to quit smoking • Identify dangers of smoking • Identify reasons people smoke • List biblical reasons for not smoking *People's responsibility for their actions* *Mankind's responsibility to glorify God* *The human body as God's temple*	
165	324	282	188	**Chapter Review** • Recall concepts and terms from Chapter 11 • Apply knowledge to everyday situations	
166	324			**Chapter 11 Test** • Demonstrate knowledge of concepts taught in Chapter 11	

Lesson	TE pages	ST pages	AM pages	Objectives and Christian Worldview	Process Skills
Chapter 12: Circulatory System					
167	325	283	189	• Illustrate the superiority of God's design over mankind's technology • Glorify God for His wisdom and power *Mankind's imitation of creation* *Faith in the Word of God*	
168–69	326–29	284–87	190–92	• Name the parts of the circulatory system • Describe the path of blood through the heart • Explain the function of the heart's pacemaker *God's design for the human body*	
170	330–31	288–89	193–94	**Activity: How Fast Is the Beat?** • Calculate the heart rate • Calculate how long it takes the heart rate to return to normal • Make a line graph using the heart-rate data	Hypothesizing Measuring and using numbers Collecting and recording data
171–72	332–35	290–93	195	• Identify and describe the three types of blood vessels • Name the largest artery and the largest veins • Differentiate between arteries and veins • Recognize that the exchange of gases takes place in the capillaries • Explain why William Harvey is important as a scientist and a physician *God's immutability* *God as only Creator*	
173–74	336–40	294–98	196	• Identify the contents of blood • Describe platelets, red blood cells, and white blood cells • Name the four main blood types • Describe a blood donation *God's plan for salvation* *God's salvation through Christ*	
175	341	299	197–98	**Activity: Exploring Blood Types** • Demonstrate which blood types can safely mix with each other	Predicting Measuring Making and using models Observing
176	342–43	300–301	199–200	**Activity: Pump and Pour** • Model the heart pumping blood • Compare the model with the function of the heart	Predicting Measuring and using numbers Making and using models Collecting and recording data Defining operationally
177	344–45		201–2	**Answers in Genesis** • Explain why it is important to identify the God of the Bible as the designer of our bodies • Defend from Scripture that Jesus created the world	
178	346–49	302–5	203	• Identify organs that help remove wastes from the body • Recognize that the kidneys help clean the blood • Name three ways to stay healthy • Recognize that no inventions would be possible without God *God's design for the human body* *People's responsibility for their actions* *God as the perfect Creator* *God's love for mankind*	
179	350	306	204	**Chapter Review** • Recall concepts and terms from Chapter 12 • Apply knowledge to everyday situations	
180	350			**Chapter 12 Test** • Demonstrate knowledge of concepts taught in Chapter 12	

1

Student Text pages 1–2
Activity Manual page 1

Objectives

- Define *worldview*
- Recognize that everyone has a worldview
- Identify characteristics of a Christian worldview

Materials

- container

Teacher's Toolkit CD

- Instructional Aid: *Worldviews*

Cut apart the strips on the *Worldviews* page. Fold each and place them in the container. Some of the statements express a Christian worldview, and others do not.

Introduction

- Generate interest and help the students recognize that they already have a worldview.
- Direct a volunteer to draw a strip from the container and read it aloud. Ask the students to vote on whether they agree or disagree with the statement that was read.
- Repeat with the remaining strips.
  Each of these statements expresses a worldview. You might not think that you have a worldview, but you do. If you agreed with a statement, you agree with that worldview.
  In this lesson you will learn about a Christian worldview.

Preparation for Reading

This section includes directives for the students. They should look for the answers as they read silently. The information will be discussed and answered in the lesson.

- Direct the students to look at the headings on pages 1–2. Generate interest in what the students will learn. Direct the students to read silently to find the answers to these questions.
  What is a worldview?
  What global disaster reshaped the surface of the earth?

Teach for Understanding

What is a worldview? a way of viewing the world

What is a Christian worldview based on? the truth of the Bible

What are some ways that God's power is shown in creation? Accept reasonable answers.

- Read aloud Genesis 1:28.
  Who is this verse talking about? Adam, Eve, and all people
  What job did God give them to do? have dominion; to take care of the earth
  What does the word *dominion* mean? the job of taking care of the earth and making it useful

Let's Explore God's World

WORLDVIEW THINKING

Do you know what the word *worldview* means? It means, basically, "a way of viewing the world." We use our worldview to answer questions such as the following:

How did the world begin?

Are people more important than animals?

Is science good or bad?

Answers to questions like these are important. A Christian worldview answers these kinds of questions based on the truth of the Bible. To begin our study, we will review some basics of a Christian worldview in science.

God is unlike anything or anyone.

As you read through this book you will see that God is far more powerful than anything you can imagine. He made the entire universe in just six days! Think of the vastness of outer space. Then think of the tiny parts that make up all the plants and animals and even your body. God made all these things and holds them all together (Genesis 1; Colossians 1:17). Studying science will give you a sense of how great God is.

God put us here to work.

Many adults and children in the world think life is about having fun and not having to work. But God's first command is that the job of all people is to take care of the earth and to make it useful (Genesis 1:28). People in each field of science study the world we live in so that we can use its resources to do better, faster work. Throughout this book you will see examples of machines, materials, and medicines that people have developed. Developing these things helps Christians obey God's first command as well as the command to "love your neighbor as yourself" (Leviticus 19:18; Mark 12:31).

People are important.

Many people in the world think others are important only when it benefits them. It is like a boy who thinks the other players on his soccer team are important only when they help him show off his skills. But the Bible says that other people are important because they are made in God's own image (Genesis 1:26). If you really love God, you will love people made to be like Him. What does this have to do with science? Science helps people care for others. Since people are important, so is science.

Background

Goals for Science

As part of this lesson you may want to discuss the *Goals for SCIENCE 5* found on page v in this book. The goals are written for you as a teacher. You can have the students read some of the Bible verses and give the goals in a way that they will understand.

Weblinks

The BJU Press website offers additional information and links you may find helpful throughout the unit.
bjupress.com/resources/science

The Flood changed everything.
When Adam disobeyed God (Genesis 3), he brought sin and death into the world (Romans 5). God proved His attitude toward sin when He judged the whole earth with a flood (Genesis 6–9). Some people today do not want to think of God's judgment. But the evidence for the Flood is everywhere. Perhaps you wonder where fossils come from or where the dinosaurs went. The answers to all these questions include references to the Flood. This global disaster reshaped the earth's surface, changed the atmosphere, launched an ice age, and reduced all human and animal life to one boat. As you read through this book, remember that a catastrophe of huge proportions occurred during Noah's day.

God is working to redeem this world.
The Bible tells us that we were made to rule over God's creation. But man's sin put this world under a curse (Genesis 3:16–19). Because of His love, God sent His own Son, Jesus, into the world to redeem all who repent of their sin and believe in the gospel (Mark 1:14–15). One day God will make this world perfect again—the way it was before sin (Revelation 21:5). At that future time, God's people will rule over the world with God's Son (Revelation 22:5). Until then, God has called Christians to live lives of good works (Matthew 5:16). This is one reason Christians take science seriously. Science can give people the understanding and tools they need to help others.

Science is about making models.
A model helps explain, describe, or represent something in the world. A scientist often uses models that already exist, but he may make his own. He applies his worldview as he observes and interprets how the model works. The scientist then makes changes to the model as he gathers new information. A goal of scientists is to make existing models better and to create new models as new things are discovered. Because people do not know everything, science changes as the models and ideas of scientists change. But only God really knows and understands what is going on (Ecclesiastes 8:17). A good scientist knows that science is not about getting answers that never change. Science is about making models that work.

These six big ideas will help you think about the science you are learning. Keep these ideas in mind as you read.

1

Explain how this verse relates to science. What we learn in science helps us do our job better.

What makes people different from animals? People are created in God's image.

How can you show God's love to others? by caring for them

What happened in the Garden of Eden that had a negative effect on all people and the world? Adam disobeyed God. His sin brought sin and death to all people and the world.

Why did God send a flood to cover the whole earth? God judged the world for its sin by a worldwide flood.

What are some things that changed as a result of the Flood? The earth's surface was reshaped, the atmosphere changed, and an ice age was launched.

- Ask a volunteer to read aloud Mark 1:14–15.
  What do these verses say that every person needs to do? repent and believe the gospel

  What is the gospel? The good news that salvation comes only in believing that Jesus Christ died and rose again to pay the penalty for sin.
- Discuss how your beliefs about Creation, the Fall, and Redemption through Christ are the basis for your worldview. Tell the students that if they believe that the Bible is true, then they have a Christian (biblical) worldview.
- Discuss how a worldview affects a scientist.
  How might a scientist's worldview about Creation affect his observations? Possible answer: He might think the earth took much longer to form or evolve without God.

  Why do scientists use models? to explain, describe, or represent something in the world

  Once a scientist makes a model, does he keep it the same forever? No, he will make changes as he learns more about what it represents.

  Who knows everything, God or scientists? God

  How does your worldview about who knows everything affect how you study science? Possible answer: I might believe that scientists know everything and that they are always right.
- As time permits, include additional material presented on page vi, *Christian Worldview for* SCIENCE 5.

Activity Manual

Biblical Worldview, page 1

2

Student Text pages 3–5
Activity Manual page 2

Objectives

- Recognize the interrelationship of science concepts
- Distinguish facts and assumptions in the evolution/creation debate
- Evaluate evolutionary assumptions from a Christian worldview

Unit Introduction

- *Science 5* is divided into six units. Grouping chapters together in units helps students understand relationships between science concepts.
- In Unit 1 students will see that the rocks discussed in Chapter 1 may contain the fossils discussed in Chapter 2.
- Chapter 1 focuses on mankind's uses of the minerals and rocks that God created or allowed to form on the earth. Chapter 2 directs attention to the truth that the formation of fossils within the rocks resulted from events and processes controlled by God.
- Use the opportunities that arise to help students connect new information with material that they have already learned.

  Look at the contents page. What are the topics of the chapters we will study in Unit 1? minerals and rocks, dinosaurs and fossils

  Why do you think these chapters are organized into the same unit? Answers will vary. Elicit that the chapters are about rocks and things found in rocks.
- Explain that one way to see the greatness of God is to look at the way all creation fits together. Although the earth is under the curse of sin, God uses all things for His glory. Whether it is a thornbush that provides shelter for an animal or a fire that recycles nutrients into the soil, God controls all of creation.
- *Science 5* emphasizes how mankind uses the resources that God has provided for the earth. As aspects of technology are discussed, point out that mankind's accomplishments are a result of the abilities and gifts that God has given people.

Background

Science Notebook

You may choose for the students to keep science notebooks. The notebook will provide a place to keep Activity Manual pages as well as other papers used in activities, explorations, and other science-related projects. Pages can be organized by chapter or by type, such as reinforcement, review, activity, etc. Keeping a notebook will help the students develop organizational skills. The notebook can also be used in the students' portfolios to demonstrate their progress and ability to think scientifically. Plan to check the content, organization, and neatness of the notebooks at least once per chapter.

Unit Photo

The miniature scientists will be seen throughout the book. Items the scientists use will be shown on their scale. Items that they are interacting within our world will be shown on a scale larger than the scientists. The photo on page 3 shows two different types of geodes.

Weblinks

The BJU Press website offers additional information and links you may find helpful throughout the unit.
bjupress.com/resources/science

2

Look at the picture on page 4.

- Ask volunteers to read different teasers. Discuss the information.

Project Idea

- The project idea presented at the beginning of each unit is designed to incorporate elements of both chapters as well as information gathered from other resources. You may choose to have the students do the project as a culminating activity at the end of the unit or as an ongoing activity while the chapters are being taught.

Unit 1: Geology Museum

- Collect rocks, minerals, and fossils. Identify as many of the items in the collection as possible. Research to find information about each sample in the collection. Record the information from the research on cards that can be kept with the samples. Design a display area for the collection using plastic shoeboxes or other storage containers. The collection may be organized by grouping similar items together in a container. If fossils are not available, research to find the types of rocks that often contain fossils. Include this information with the rocks in the collection. Incorporate information from the chapters as much as possible. Make sure the researchers are sensitive to the use of evolutionary terms and data. Lesson 15 is an activity that helps the student identify resources that contain evolutionary material.

This lesson continues with the Chapter 1 introduction on page 5.

Unit 1 Overview

Lesson	Teacher's Edition	Student Text	Activity Manual	Assessment	Content
Chapter 1: Minerals and Rocks					
2	2–5	3–5	2		• Introduce the book • Worldview thinking • Unit and chapter preview
3	6–9	6–9	3–4	Quiz 1-A	• Layer of the earth • Surface of the earth
4	10–11		5–6		Answers in Genesis • Look at the world with a biblical perspective
5–6	12–17	10–15	7		• Minerals in the earth • Characteristics of minerals
7	18–19		8		Activity: Measuring Mass and Volume
8–9	20–21	16–17	9–10	Rubric	Activity: Salty Crystals • Process skill: observation
10	22–26	18–22	11–13	Quiz 1-B	• Uses of minerals • Finding minerals
11	27	23	14	Rubric	Exploration: Munching Minerals • Research a mineral found in food or beverages
12	28–29		15–18		Study Skill: PQ3R • Reading informational text
13	30–33	24–27	19	Quiz 1-C	• Igneous, sedimentary, and metamorphic rock
14	34–35	28–29	20–21	Rubric	Activity: Rock Hounding • Process skill: classifying
15	36	30	22		**Chapter Review** • Apply knowledge to everyday situations
16	36			Chapter 1 Test	**Chapter 1 Test**
Chapter 2: Fossils and Dinosaurs					
17	37	31	23		• Worldview thinking • Chapter preview
18	38–40	32–34			• Fossils preserved in sediment • Fossils preserved in other materials
19	41	35	24–25	Quiz 2-A	• Two ways of looking at fossils • Evolution and creation
20	42–43	36–37	26		Activity: Fact or Theory? • Process skill: inferring
21–22	44–45	38–39	27–28	Rubric	Activity: Molds and Casts • Process skill: making and using models
23–24	46–49	40–43	29–30	Quiz 2-B	• Evacuating and preparing fossils • Reconstructing and dating fossils
25–26	50–51	44–45	31–32	Rubric	Exploration: Fossil Dig • Model procedures a paleontologist uses
27	52–55	46–49	33		• What dinosaur fossils teach us
28	56–59	50–53	34	Quiz 2-C	• What the Bible teaches us about dinosaurs • Extinction of dinosaurs
29	60–61		35–36		Answers in Genesis • A biblical viewpoint about dinosaurs
30	62	54	37–38		**Chapter Review** • Apply knowledge to everyday situations
31	62			Chapter 2 Test	**Chapter 2 Test**

Minerals and Rocks

1

Dr. Munschall has recently finished fifteen years of study, which he says gives him a better understanding of the journey rocks and minerals have taken in the over 4.5 billion years Earth has existed.

"The stardust from which our solar system started was actually quite simple, consisting of just a few elements," says Dr. Munschall. "Pressures like gravity reshaped this dust into hundreds of different kinds of minerals."

Dr. Munschall believes that volcanoes and water as well as the evolution of plants and animals also had an impact on mineral and rock formation. However, that effect took place over a long period of time through events such as volcanic eruptions and many local floods.

Consider

- Did all minerals come from stardust? How does Dr. Munschall know that?
- Where did rocks come from? Why do they look the way they do now?
- Do Dr. Munschall's ideas agree with what God tells us in the Bible?

5

2

Chapter Introduction

- Students are constantly confronted with evolutionary thinking. If they watch a movie, use the Internet, or have discussions with friends, they will see or hear evolutionary ideas or the ideas that flow from an evolutionary, humanistic worldview. It is critical that students be taught to recognize an unbiblical worldview, evaluate it, and respond intelligently.

Preparation for Reading

- Direct the students to read the introduction to find out what Dr. Munschall believes about the formation of minerals and rocks.

Teach for Understanding

- Engage the students through questions. Help them distinguish between facts and assumptions.

  What is Dr. Munschall's worldview? evolutionary

  Why do you think his worldview is evolutionary? Possible answers: He talks about the earth as being over four billion years old and formed from stardust. He refers to the evolution of plants and animals.

  What is Dr. Munschall trying to figure out? where the minerals came from

  Did Dr. Munschall or anyone else see the formation of rocks? no

  What is Dr. Munschall assuming to be true as he tries to figure out how rocks and minerals were formed? He is assuming that pressure similar to gravity reshaped stardust into different kinds of minerals. He assumes that volcanoes, water, and the evolution of plants and animals had an impact on the formation of minerals and rocks.

- Discuss how Dr. Munschall has left God out of his assumptions about rocks and minerals.

  Who was there when rocks and minerals were first formed? God

  How did the formation take place? God spoke and everything was created in six days.

  How do we know that God created the rocks and minerals? We believe what God says in the Bible about Creation in Genesis 1.

  What are we assuming if we say that God created everything? We are assuming/believing that the Bible is God's Word and that God always tells the truth.

  The Bible describes another event that affected the rocks and minerals we see today. Can you think what this event was? the Flood

Activity Manual

Preview, page 2

The Looking Ahead page is intended to assess the student's prior knowledge before beginning the chapter.

Background

Technology

Technology is any item such as a tool, instrument, or piece of equipment or machinery that people have designed and use. These items may be as simple as a paper clip or as complex as a computer program. Mankind's technology applies scientific principles to solve problems and help his fellow man. Many of mankind's inventions copy and use God's designs in nature.

Chapter Preview

Preview and prereading activities may include using a K-W-L chart, a probe, or an anticipation guide.

Chapter Photo

The photo on page 5 shows small river rocks.

3

Student Text pages 6–9
Activity Manual pages 3–4

Objectives
- Identify and locate the layers of the earth
- Describe features of the core, mantle, and crust
- Explain how weathering and erosion affect sediment
- Define *humus*

Materials
- round apple
- sharp knife

Teacher's Toolkit CD
- Quiz 1-A

Vocabulary
- core
- lava
- mantle
- sediment
- crust
- soil
- magma
- organic

Introduction

Compare the earth to this apple. What is the basic shape of an apple? round, sphere

- Cut the apple in half across the middle, then display the inside of the apple.

How would you describe the inside of the apple? solid with seeds in the middle

What three layers of the apple can you see? the peel, flesh, and core

The earth also has three basic layers.

Preparation for Reading

Preparation for Reading includes directives for the students. They should look for the answers as they read silently. The information will be discussed and answered in the lesson.

- Direct the students to look at the headings on pages 6–9. Generate interest in what the students will learn. Direct the students to read silently to find the answers to the questions.

What are the three layers of the earth?

What is weathering?

Sometimes children try to dig holes that are deep enough to come out on the other side of the earth. But children are not the only ones who find this feat impossible. Even scientists with the most advanced equipment have dug or drilled holes only a few miles deep. Much of what scientists think about the inside of the earth comes from their observations of events near the earth's surface. Events such as volcanoes and earthquakes give scientists a glimpse of what lies hidden beneath the earth's surface.

Layers of the Earth

Geologists are scientists who study the structure, rocks, and minerals of the earth. They divide the inside of the earth into layers. Each layer has characteristics that are different from the other layers. The three main layers of the earth from inside to outside are the core, the mantle, and the crust.

The **core** is the center of the earth. Scientists believe that the core is very hot and under extreme pressure. They think that the inner part of the core is a solid, dense sphere made mostly of iron. Scientists also believe that there is an outer part of the core that is liquid.

Layers of the Earth

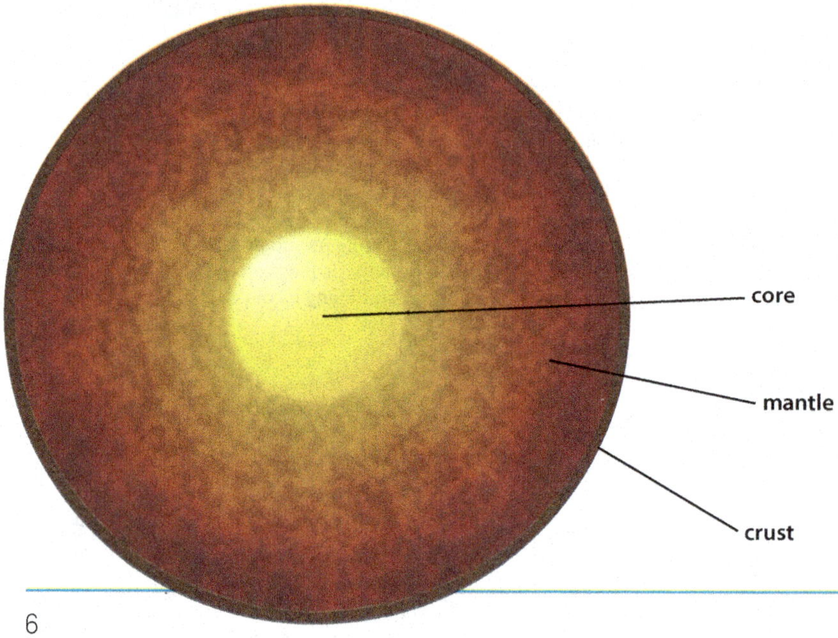

6

Background

Diagrams
Some diagrams from the Student Text are included on the Teacher's Toolkit CD and will be listed in the lesson plans below the Materials list under the Teacher's Toolkit CD heading.

Measurements
Most of the measurements used in *Science 5* are approximations.

Layers or Zones
The layers of the earth are also called *zones*.

Core
The core is believed to be made of iron. Scientists believe that this iron core produces the magnetic field that surrounds the earth.

Theory of Plate Tectonics
The idea that the earth's crust is made up of moving plates that float on the partly melted rock in the earth's mantle is called the theory of plate tectonics.

The next layer located between the core and crust is the **mantle**. This part of the earth's interior surrounds the core and makes up most of the earth's mass. The mantle is made of hot rock. The great pressure in the mantle can cause the rock to flow very slowly over time.

The outer layer of the earth is the **crust**. This thin layer is only about 7 to 45 km (4 to 28 mi) thick. The crust under the oceans is thinner than the crust under the continents.

The earth's crust appears to be made of many pieces rather than one solid sheet. These pieces, called plates, rest on top of the mantle. The slow flow of hot rock in the mantle seems to drag these plates along. The moving plates and their collisions result in earthquakes and also heat the rock in the crust until it melts. Melted rock underground is called **magma**. Magma collects in chambers and can flow toward the earth's surface through deep cracks in the crust. The magma that flows from an opening onto the surface is called **lava**.

• MEET THE SCIENTIST • BENO GUTENBERG

Beno Gutenberg was born in Germany in 1889 to Jewish parents. His father wanted him to work in the soap factory that the family owned. But Gutenberg was more interested in seismology, the study of earthquakes.

As a young man, Gutenberg worked as a scientist in Germany. After World War I, he worked many different jobs. Even with several jobs, he could not earn enough money to live on. Finally, in 1930 he was offered a job in California. There, he was at last able to experience the earthquakes he had spent years studying.

Gutenberg is famous for the methods he used to measure distances and substances inside the earth. In fact, he is known for being the first to determine the size of the earth's core. He also worked with Charles Richter to develop the scale used to measure earthquakes. Gutenberg died in 1960.

7

3

Teach for Understanding

How is the inner part of the core different from the outer part? Scientists think that the inner part is a solid, dense sphere, made mostly of iron. They think that the outer part is made of liquid.

⚙ Why would it be difficult to know the exact composition of the earth's core? Possible answers: Information about the core is based on logical guesses since the core cannot be observed. No equipment has been developed to reach the core.

- On page 7, continue discussing the layers of the earth with the layer in which most of the earth's mass is located, the mantle.

⚙ What is mass? the amount of matter or material

What makes up the mantle? hot melted rock called magma

What happens when magma pushes through cracks in the earth's surface? Volcanoes erupt.

When is magma called lava? when it reaches the earth's surface

⚙ Why can scientists study the mantle more easily than the core? Samples of the mantle come to the surface. The core is very deep within the earth.

What is the thin outer layer of the earth called? the crust

Where is the crust the thinnest? under the oceans

What do scientists call the many pieces of the crust? plates

- Discuss the *Meet the Scientist* information, including what Beno Gutenberg was famous for and what he developed with Charles Richter.

The gear ⚙ indicates a higher-order question. These questions are based on information gathered from the text but require some analysis, synthesis, or evaluation of the text. Supply any prompts or background as needed to guide the students to the answer.

Activity

Model the Layers of the Earth

Materials: equal amounts of red clay and green clay, twice that amount of yellow clay, thread

Roll the red clay into a ball. Cover the red ball with the yellow clay and smooth into a ball. Flatten the green clay into a thin layer and cover the yellow ball. Hold the thread taunt and carefully cut the ball in half.

Identify the layers inside the model earth: red is the core, yellow is the mantle, and green is the crust. The model can be used to aid review of the layers and their features.

Background

Hidden Things

Magma can be used to illustrate a Bible concept. Luke 6:45 tells us, "A good man out of the good treasure of his heart bringeth forth that which is good; and an evil man out of the evil treasure of his heart bringeth forth that which is evil: for of the abundance of the heart his mouth speaketh." Just like a person cannot see deep inside the earth, a person also cannot see the condition of someone's heart. Magma flows through the crust to show what is in the mantle. Hidden things inside a person's heart flow out when he is under pressure. [BAT: 3c Emotional control]

3

God created dry land. What does this land include? rocks, minerals, and soil

- Explain that the Fall of mankind refers to Adam and Eve's sinning against God (Genesis 3:13–24).
- Discuss weathering with the students, including the terms *sediment* and *erosion*.

How does erosion often occur? by wind or water

How might erosion be harmful? by causing fertile land to wash or blow away

What landforms may have been formed as a result of the Flood? canyons, mesas, mountains, and caverns

How is new rock formed today? Possible answers: sediment that settles at the bottom of bodies of water; new rock within the earth that is pushed upward by earthquakes and volcanoes

- Discuss the landforms pictured on the page.

The first picture shows a canyon formed by which river? the San Juan River

Mesa is the Spanish word for "table." Which picture shows mesas—landforms with flat tops like the top of a table? the picture of Monument Valley, Utah

Look at the picture of a cavern in Cheddar Gorge. Do caverns form on the surface or underground? underground

Surface of the Earth

God spoke and the earth was created. Genesis 1:9–10 tells us that God gathered the waters together to form dry land. This land probably included a foundation of rock covered by a thick layer of soil.

Since the Fall of mankind, the earth has been aging and the crust has been wearing away. This wearing away of the earth is called *weathering*. Weathering produces **sediment**, or small bits of weathered rock. This sediment does not usually stay in one place, though. Wind and water often pick up the sediment and move it. The movement of sediment from one place to another is known as *erosion*. Sometimes the sediment settles at the bottom of bodies of water or in dunes.

The Flood that happened in Noah's time caused the entire crust of the earth to change. Some of the rocks and mountains were quickly worn away by the moving water even as other new ones formed. As the floodwaters flowed off the continents into the ocean, they continued to wear away the land. This weathering and erosion caused many landforms such as canyons, mesas, and caverns to form.

Today weathering and erosion continue to wear away the surface of the earth. Pressures within the earth continue to expose magma, which cools to become rock. The new rock is pushed upward by earthquakes and volcanoes.

San Juan River, Utah

Monument Valley, Utah

Cheddar Gorge, Somerset, England

8

Background

Sedimentary Rock

Sedimentary rock forms from compressed, weathered rock. Much of the sedimentary rock on the earth today probably formed as a result of the Flood. Sedimentary rock is discussed in Lesson 10.

Landform Changes

We are not sure how certain landforms were formed. Mankind cannot base how the earth may have changed in the past on how it changes today.

Science Misconceptions

Weathering and Erosion

Weathering and erosion are not the same. Weathering is the process of bits of materials wearing away from the surface of something such as rock. Erosion refers to the movement of those bits of material.

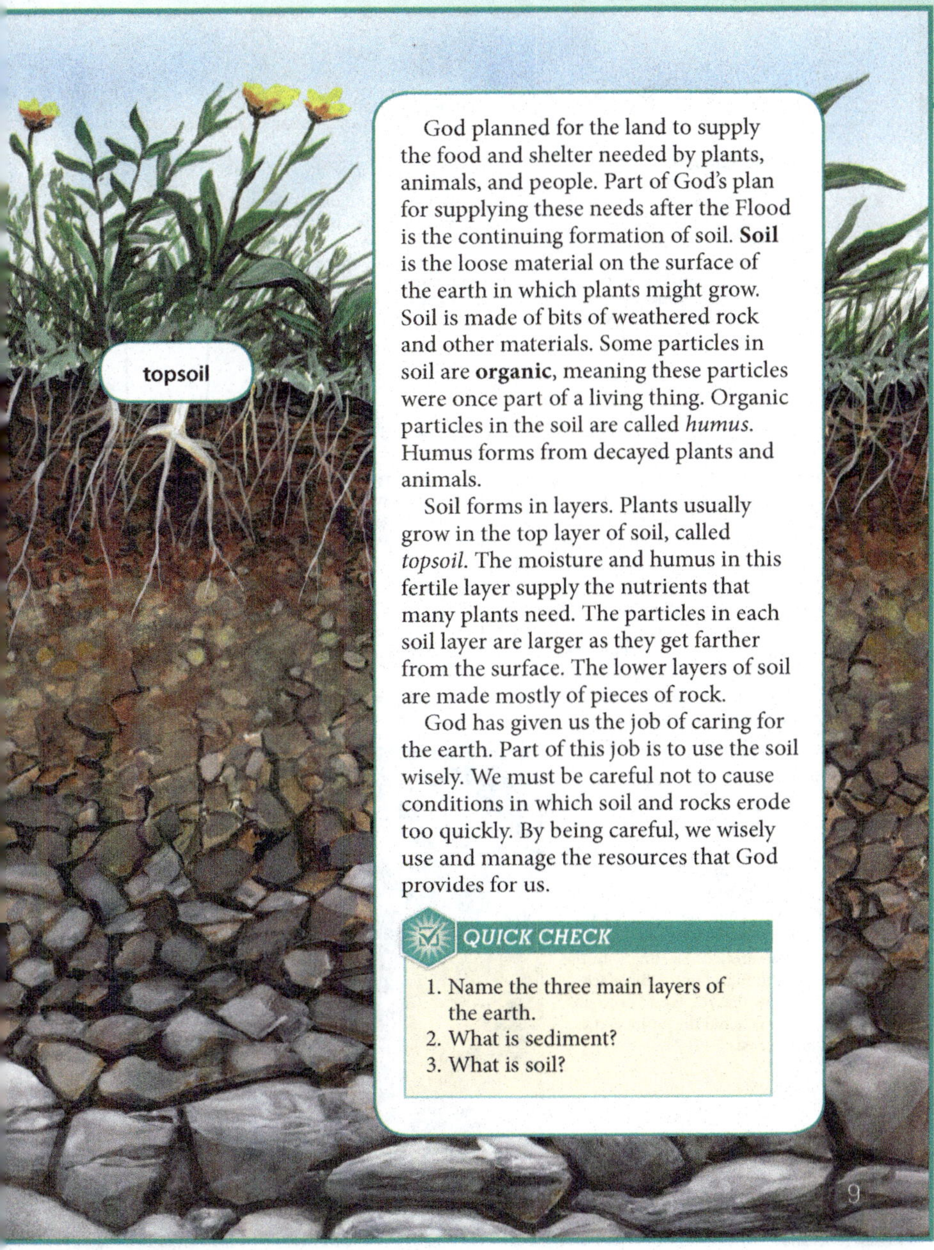

God planned for the land to supply the food and shelter needed by plants, animals, and people. Part of God's plan for supplying these needs after the Flood is the continuing formation of soil. **Soil** is the loose material on the surface of the earth in which plants might grow. Soil is made of bits of weathered rock and other materials. Some particles in soil are **organic**, meaning these particles were once part of a living thing. Organic particles in the soil are called *humus*. Humus forms from decayed plants and animals.

Soil forms in layers. Plants usually grow in the top layer of soil, called *topsoil*. The moisture and humus in this fertile layer supply the nutrients that many plants need. The particles in each soil layer are larger as they get farther from the surface. The lower layers of soil are made mostly of pieces of rock.

God has given us the job of caring for the earth. Part of this job is to use the soil wisely. We must be careful not to cause conditions in which soil and rocks erode too quickly. By being careful, we wisely use and manage the resources that God provides for us.

QUICK CHECK

1. Name the three main layers of the earth.
2. What is sediment?
3. What is soil?

9

3

What is soil? loose material on the surface of the earth that is made of bits of weathered rock and other materials

What does *organic* mean? that the substance was once part of a living thing

What are the organic particles in soil called? humus

What is the source of the organic materials in the soil? decayed plants and animals

Why do you think that a farmer would want soil that is rich in humus for his crops? It would be more fertile, yielding better crops.

- Discuss the layers of the earth, including how particles in the various layers differ in size. They are smaller in the upper layers and larger in the lower layers.

Some areas of the world have little or no topsoil. Do you think that these areas of the world are known for growing crops? no Why? The soil in these areas does not have the nutrients needed to grow crops.

God has given people the job of caring for the earth. What are some ways that we can do this with soil? Possible answers: use soil wisely; avoid erosion; avoid wasting water; dispose of trash properly; avoid causing water and soil pollution

ANSWERS

1. the core, mantle, and crust
2. small bits of weathered rock
3. the loose material on the surface of the earth

Background

Noah's Flood

The account of the Flood of Noah's day is found in Genesis 6–9. The Bible does not tell us many details about the surface of the earth before or after the Flood. Creationists have differing views about how mountains and oceans formed after the Flood. Many creationary scientists think that the earth's surface went through catastrophic changes during the Flood. The changes may have caused land masses to separate and push against one another with tremendous forces, causing mountains to form.

Activity Manual

Reinforcement, page 3

Review, page 4

This page reviews Lesson 3. Point out the Student Text page numbers at the bottom of the page. Explain to the students that these are the pages covered in this Study Guide. After correcting the page, return it to the students to use as a review for the chapter test. Help the students organize a section of their science notebooks for study materials.

Assessment

Quiz 1-A

The quiz may be given any time after completing this lesson.

Several optional quizzes are provided for each chapter on the Teacher's Toolkit CD. The quizzes correlate with the Study Guide pages in the Activity Manual.

Objectives

- Explain why it is necessary to look at the world with a biblical perspective
- Justify from a biblical viewpoint that the layers of the earth did not take millions of years to form

Materials

- clear 1-liter bottle with cap
- 5 different soil samples such as sand, loam, clay, and gravel (100 mL each)
- 5 small plastic bags
- 500 mL of water
- crumpled leaves (optional)
- small twigs (optional)
- grass clippings (optional)
- tool such as a knife or saw to slice the sediment (optional)

Place a sample of each type of soil in a plastic bag.

Introduction

- Display the soil samples in the plastic bags. If you have chosen to use the leaves, twigs, and grass clippings, display those items also.
- Explain that the word *predict* means "to tell about something in advance."
- Invite the students to make a prediction about the activity.
  What would happen if all these soils and items were placed in a bottle with water and shaken and then left to sit for fifteen minutes?
- Direct the students to draw a picture of their predictions in the Prediction bottle on Activity Manual page 5.
- As they are drawing, place each type of soil and water in the bottle. Add bits of leaves, twigs, and grass clippings to the bottle (optional).
- After the students finish drawing their predictions, shake the prepared bottle and display it. The bottle should sit for approximately fifteen minutes. You may continue teaching until fifteen minutes have passed or wait until the end of the lesson before returning to the activity.

Preparation for Reading

- Direct the students to read Activity Manual page 5 silently to find out what Bible glasses are.

A lesson from
Answers in Genesis Name ____________

1:1

Bible Glasses

Making predictions is an important part of scientific inquiry. What do you think would happen after several different earth materials are mixed together in water and shaken in a bottle?

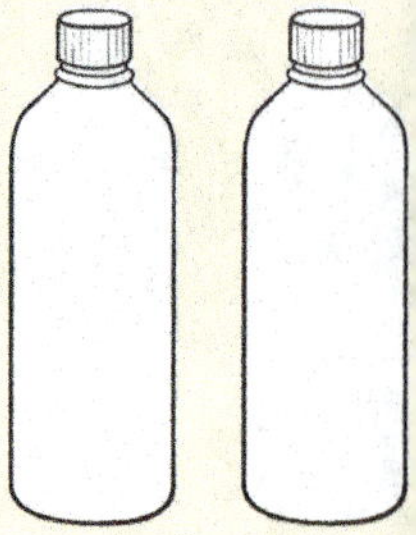

Prediction Observation

Look at the bottle your teacher has. Draw a picture of what you predict the materials in the bottle will look like 15 minutes after shaking.

Studying how earth materials such as these act in water is really interesting and important to a Christian. It helps Christians better understand the structure of soils, especially those deposited by rivers. This knowledge can help make farming and soil management methods more effective. But it also helps Christians understand what the Bible says about the history of the earth.

If your family has taken a trip through hilly country, you may have noticed that highways often cut through hills. Layers of rock are clearly visible. You might ask, "How did those layers get there? How long did it take for them to form? Were the rocks always hard? Why are there different colors and types of materials?"

Some people think they can answer these questions all alone—without God's help. But the wise person knows that all knowledge begins with honoring God and believing His Word, the Bible (Proverbs 1:7). If we try to answer a question without looking to God's Word first, we are acting foolishly. We lack information we need to find the answer. God wants us to depend on and look to Him for understanding—even when we are studying earth materials! We need to think of the Bible as a set of "glasses" that we look through to understand the world.

What do you notice about the layers in the picture? What event in the Bible would help us understand how all these layers could have formed and how the massive canyon could have eroded? Starting with the truth found in the Bible, we know that the Flood of Noah's day covered the entire earth (Genesis 7:17–24).

Rock strata in the Grand Canyon

Science 5 Chapter 1 | Lesson 4 Answers in Genesis 5

Resources

The BJU Press website offers additional links from Answers in Genesis that you may find helpful with this lesson.
bjupress.com/resources/science

Look at the bottle your teacher shook up at the beginning of class. Draw what you observe in the observation bottle next to your prediction.

Complete the section.

1. How did your prediction compare to what actually happened?

2. Compare and contrast your observation with the Grand Canyon picture.
   Answers should include that both have layers. The amounts, sizes, and colors of layers are different.

3. Which layers formed first? Why? These layers would be considered the oldest.
   Possible answer: The bottom layer of larger rocks formed first because the larger, heavier pieces sank to the bottom.

4. Which layers formed last? These layers would be considered the youngest.
   Possible answer: The layers formed by the smaller pieces of earth material would form last as the pieces weigh less.

5. How long did it take the layers in the bottle to form?
   15 minutes

6. How do your answers relate to the layers you see along the highway or in the layers of a canyon? Some scientists believe the earth's layers took millions of years to form. Using your Bible glasses, what would you say?
   Possible answer: The earth layers did not take very long to form. The Bible shows us that the Flood formed the layers of the earth.

Remember to always put on your Bible glasses to answer questions like these!

Teach for Understanding

What are Bible glasses? Answers should include that you use the Bible to study and understand the world we live in.

We all look at the world through "glasses." Christians should look at the world through the glasses of the Bible.

- Explain that when we think about how the rock layers around the world were formed, we should always look to the Bible to see if God has revealed anything that will help us understand how the layers got there. In this case, the Genesis Flood is our obvious starting point.
- Mention that some scientists choose to look at the world with naturalistic glasses and reject the truth in Scripture as a starting point.
- Direct attention to the bottle from the introduction. Instruct the students to draw a picture of what actually happened in the Observation bottle on Activity Manual page 5.
- After the drawing is complete, allow time for the students to answer the first question on Activity Manual page 6.
- Ask a volunteer to read aloud the second question. Remind the students that *compare and contrast* means to show how things are alike and different.
- Discuss the question and then allow the students to answer it.
- Follow a similar procedure to complete Activity Manual page 6.
  We need to always look at the world through our Bible glasses because the fear of the Lord is the beginning of knowledge (Proverbs 1:7).
- You may want to display the bottle for reference throughout the chapter. You may choose to extend the lesson with the demonstration. The bottle or the layers of soil will be used again with the introduction to Chapter 2.

Activity Manual

Answers in Genesis, pages 5–6

Demonstration

Examining Soil Layers

Use the bottle of soils and water from the lesson. Poke holes in the bottom of the bottle and allow the water to drain. After a few days, cut around the bottom of the bottle and lift the bottle off the column of soil layers. Allow the column to dry for several more days.

Direct the students to examine the layers formed in the bottle. Allow them to compare the layers to the strata discussed in the chapter. Use a tool such as a utility knife or a coping saw to cut through the center of the column vertically. Direct the students to observe the layering, especially the position of the plant material and the different degrees of cohesion in the strata.

5
6

Student Text pages 10–15
Activity Manual page 7

Objectives

- Define *mineral*
- Identify crystal structure, luster, hardness, color, and cleavage as characteristics of minerals
- Explain how the Mohs scale is used to determine hardness

Materials

- variety of apples with identifying labels removed
- produce guide

Teacher's Toolkit CD

- Visual 1.1: *Mineral Characteristics*

Vocabulary

- mineral
- hardness
- inorganic
- cleavage
- crystal

Two days have been allotted for this lesson. You may want to teach Lesson 5 the first day and use ideas from Lesson 6 on the second day. You may also incorporate the ideas from Lesson 6 into the text material, making the lesson two days. See page xii for additional information.

Introduction

- Display the apples.

  Are all the apples the same? no

  What are some ways that we could identify the type of each apple? Accept reasonable answers such as by color and shape.
- Guide the students in grouping the apples by color and shape. You may then choose to cut the apples and sort them by inside color, firmness, or taste.
- Display the produce guide and explain that a guide can help identify the apples. Use the guide to identify the apples.

  Just as we used the characteristics of the apples and the guide to help identify each apple, scientists use characteristics and guides to help identify minerals.

Preparation for Reading

- Display the words *pyrite*, *cinnabar*, and *fluorite*. Pronounce the words together. Instruct the students to read silently to find the answers to the questions.

  What is a mineral?

  What are some characteristics that mineralogists use to identify minerals?

Teach for Understanding

What is a mineral? an inorganic substance found naturally in the earth

How is an inorganic substance different from an organic substance? Inorganic substances have never been alive. Organic substances come from plants and animals that were once alive.

Minerals in the Earth

The solid materials in the earth's crust are made of minerals. A **mineral** is an inorganic substance found naturally in the earth. An **inorganic** substance is made of things that have never been alive. Most minerals that make up the earth's crust are solid parts of rock.

Minerals are not just found in the earth. They are also found in you! When God created man, He formed Adam's body from the dust of the ground (Genesis 2:7). God designed your body to use minerals to grow and function properly. Many of the substances that are in your body and the food you eat also make up the minerals in the earth.

Characteristics of Minerals

The earth contains thousands of minerals. Many look very similar to one another. Scientists who study minerals are called *mineralogists*. They work with mineral pieces called *samples*. Mineralogists identify mineral samples by their unique characteristics. Some of the characteristics of minerals are crystal structure, color, luster, hardness, and cleavage.

Crystal Structure

All minerals have a crystal structure. Heat and pressure cause substances within the earth to combine. When this happens, crystals often form. A **crystal** results from the orderly arrangement of the particles that make up the mineral. Some crystals, such as quartz crystals, are very beautiful.

Most mineral crystals have straight sides and sharp corners or edges. However, some crystals, such as the desert rose, form in rounded shapes that resemble a flower. The sizes of crystals vary. Because a mineral hardens as it connects to the parts of a crystal that have already formed, the crystal seems to "grow."

Mineral crystals form when melted rocks or hot gases cool or when dissolved substances separate from water. Crystals from cooled melted rocks

Desert rose

www.iRocks.com

10

Background

Apples and Produce Guides

For the introduction, you may choose to make a list of the names of the apples before displaying them. Some names may be found on packaging or identifying labels.

Produce guides are booklets that can be used to identify apples and other types of produce. They usually include photographs of the produce. You can also find apple information in other resource books or on the Internet. Many state extension offices, orchards, and plant nurseries have information about apple varieties.

Crystalline

Part of the definition of a mineral is that it is crystalline. For this grade level the term *crystalline* has been left out of the definition. The crystalline aspect of minerals is taught as the characteristic of crystal structure.

can be large or small depending on the conditions where they formed. We can find large crystals from dissolved minerals in caves and in spaces left inside rocks when the rocks formed. Small, tightly-packed crystals in cooled melted rocks might indicate the rock cooled quickly. In some rocks, minerals did not form crystals at all.

The substance that forms a crystal determines the number and shape of the sides of the crystal. This means that each mineral has its own crystal structure. Some minerals can have more than one crystal structure depending on how they formed. God has made our world a very orderly place. This order allows people to use these predictable structures to identify minerals.

The shape of the crystals helps identify this mineral as sulfur.

All minerals have a crystal structure, but not all crystals are minerals. Other substances can also form crystals. For example, the sugar that you eat is not a mineral. Sugar is made from plants. Yet if you look carefully at a few grains of sugar, you can see its crystals.

SCIENCE & MATH

When you study mineral crystals, you are studying geology. However, you are also studying geometry. The flat sides, or faces, of crystals are geometric shapes. Different crystals have faces in different geometric shapes. Natural diamonds, for instance, have faces that are triangular in shape. Other crystals have faces that vary in shape. A quartz crystal, for example, looks like a column with six rectangular sides. Most quartz crystals have a point at the upper end.

A mineral crystal keeps growing in its same geometric pattern unless something prevents it. Often, crystals grow into each other. The mineral may look like a jumble of crystals. But if you look at a single crystal, you can see the geometric pattern. The crystals are an example of God's order and design in the earth.

11

Demonstration

Making Rock Candy

Materials: 250 mL water, 1 L sugar, saucepan, spoon, hot plate, candy thermometer, pie plate, wooden skewers, waxed paper

Stir together the water and sugar in the saucepan and heat until 120°C (250°F), or hardball stage as described in a cookbook.

Pour the liquid into the pie plate. Lay the skewers in the liquid so that one end of each skewer rests on the edge of the pan. Make sure that the skewers do not touch each other.

Cover with waxed paper and set aside for at least seven days.

Although the crystals that form are not minerals, the conditions of heat and time are similar to those of mineral formation.

5
6

Teach for Understanding

Where are minerals found? Possible answers: earth, rocks, food, the human body

Why does your body need minerals? to grow and function properly

What is a mineralogist? a scientist who studies minerals

- A chart organizer is found on Activity Manual page 7. You may choose to display the copy of the organizer found on the Teacher's Toolkit CD. The organizer will help the students with the important information about each of the characteristics of minerals. You may guide the students in working through the chart as you teach or allow them to complete the page at the end of the lesson.
- If working on the page during the teaching of the lesson, allow the students to complete the first two questions at this time.

  What characteristics are used to identify mineral samples? crystal structure, color, luster, hardness, and cleavage
- Instruct a volunteer to read aloud the headings of the organizer in Part B on the Activity Manual page.

  Why would it be important to use more than one characteristic in identifying minerals? Some minerals may share some of the same characteristics.
- Direct attention to the first column of the organizer. Guide the students in completing the section as you discuss crystal structure.

  What does a crystal result from? the orderly arrangement of the particles that make up a mineral

  What conditions cause crystals to form? Substances within the earth combine because of heat and pressure.

  Describe the appearance of most crystals. straight sides and sharp corners or edges
- Direct attention to the picture of the desert rose crystal.

  Does the desert rose crystal have straight sides? no How would you describe its crystal structure? rounded, looks like a flower

  What are the sizes of crystals? They vary.

  Do all crystals form from melted minerals that have cooled? no What is another way crystals can form? from dissolved minerals that have separated from water

  Under what conditions do mineral crystals form? when melted rocks or hot gases cool or when dissolved substances separate from water

  What determines the number and shape of the sides of a mineral crystal? the substances from which it is formed

  Are all crystals formed from minerals? no Give an example of a crystal that is not a mineral. Possible answer: sugar
- Direct attention to the picture of the sulfur.

  What helps to identify the mineral as sulfur? the shape of the crystals
- Discuss the *Science & Math* information with the students.

- Discuss the characteristics of minerals and ways that they may be identified.
  What is the color of a mineral? the color that you see
- Ask a volunteer to read the heading of the second column of the organizer. Guide the students in completing the column as you discuss the color of minerals.
  In what colors are minerals found? many colors
  Why do some minerals have a wide range of colors? Impurities, such as dirt or other substances, can cause different colors.
  What is one way that mineralogists can identify minerals that look similar? by conducting streak tests
  What is a streak? the color of the mark made by rubbing the mineral sample on a harder surface
  How does a streak test help identify some minerals? Possible answers: Some minerals leave a streak that is a different color from the color of the mineral. The colors of the streaks of similar-looking samples can be compared.
- Discuss the picture of hematite and its streak.
  What color is the sample of the mineral hematite? black
  What color is its streak? rust red

Amethyst

Smoky quartz

Rose quartz

Color and Streak

A mineral may be identified by its color. The **color** of a mineral is the color that you see. The amounts and variety of elements in a mineral can cause a wide range of colors. For instance, amethyst, smoky quartz, and rose quartz are all types of quartz, but each is a different color.

Some minerals look similar and are hard to tell apart. For example, gold and pyrite are about the same color. In fact, they look so much alike that pyrite is called fool's gold. Mineralogists may use *streak tests* to tell them apart.

The **streak** is the color of the mark that is made when a sample is rubbed on a harder surface. The hard surface causes small bits or powder from the mineral to rub off. A piece of white porcelain or ceramic tile is often used to test the streak color of minerals.

The color of the streak left by a sample may help identify a mineral. Some minerals leave a streak that is a different color from the color of the mineral. Gold and pyrite are similar in color, but their streak tests are quite different. Gold leaves a yellow-gold streak. Pyrite, on the other hand, leaves a greenish-black streak.

The mineral hematite has a black color but a rust red streak.

12

Background

Color

Color is simply the general color or the visual appearance of the whole sample.

Streak

Streak is the color of the powdered form of the mineral. A streak test is one way to produce the powder. Some minerals can also be crushed to produce the powder. A white or black piece of unglazed ceramic is often used as a streak plate.

Luster

The terms used to describe luster may vary. Different mineralogists may use different terms or more than one term to describe the same mineral sample. Most of the terms used imply reference to another substance. *Pearly* literally means that the sample looks like a pearl. *Greasy* means that the sample has an oily appearance. The Student Text does not include all the terms used for luster.

Luster

The **luster** of a mineral is the way that it reflects light. A variety of terms are used to describe the luster of minerals. The mineral cinnabar has a *dull* luster. Very little light reflects from the mineral's surface.

Cinnabar

Both gold and pyrite have a *metallic* luster. A metallic luster is shiny, almost like a piece of aluminum foil.

Pyrite

An opal has a *pearly* luster. This means that the mineral looks similar to the inside of some shells. Colors of light often reflect from the surface of a mineral with a pearly luster.

Opal

A diamond has a *brilliant* luster. Light and colors appear to bounce on the outside and the inside of a carefully cut diamond. The light and colors cause the diamond to sparkle. Minerals that can be cut and polished to a brilliant, or high, luster are often used in jewelry.

Diamond

www.iRocks.com (Diamond)

13

- Direct a volunteer to read the heading of the third column of the organizer. Guide the students in completing the column. Refer to the photos as luster is discussed.

  What is luster? the way that a mineral reflects light

  If a mineral reflects very little light, what kind of luster does it have? dull

  What do minerals with a metallic luster look like? shiny like aluminum foil

  Why do you think minerals like pyrite are described as having a metallic luster? They look like metal.

  What does a pearly luster look like? like the inside of some shells, reflecting different colors

  Why do you think this luster is called pearly? Possible answer: It reflects light and colors as a pearl does.

  What kind of luster does a diamond have? brilliant

  Why would minerals with a brilliant luster be used to make jewelry? Possible answers: The minerals have the ability to reflect light. They have a shiny appearance when polished. They are beautiful.

- Direct a volunteer to read the heading of column four on the organizer. Guide the students in completing the column as you discuss the hardness of minerals.
  What is the hardness of a mineral? how easily it resists scratching
  How is the scale developed by Friedrich Mohs used? to compare the hardness of minerals
  What can be done to determine how hard a mineral is? a scratch test
  What might scratch a soft mineral? a fingernail
- Direct attention to the *Mohs Hardness Scale* diagram as you continue to discuss hardness.
  What number on the Mohs Hardness Scale is the softest? 1
  What is the softest mineral on the scale? talc
  What is the hardest mineral on the scale? diamond
- Instruct volunteers to locate gypsum and quartz on the scale.
  Do you think that gypsum could make a scratch on a piece of quartz? no Why? Gypsum is lower on the scale than quartz, so it is softer and cannot scratch quartz.
- Continue asking similar questions as time allows.

Hardness

The **hardness** of a mineral refers to how easily it resists scratching. Soft minerals may be easily scratched with a fingernail. But some minerals are so hard that they cannot be cut with steel. They must be cut by tools made from even harder materials.

In the early 1800s, a German mineralogist by the name of Friedrich Mohs developed a scale to compare the hardness of minerals. The Mohs scale is still used today. The scale ranks minerals by how hard they are.

A scratch test may be done to determine how hard a mineral is. In a scratch test, the mineral sample is scratched with a substance shown on the Mohs scale. The sample's hardness is decided by which substance can scratch the sample. For example, a sample of fluorite scratches a penny and is scratched by apatite. So you know fluorite's hardness is about 4 on the Mohs scale.

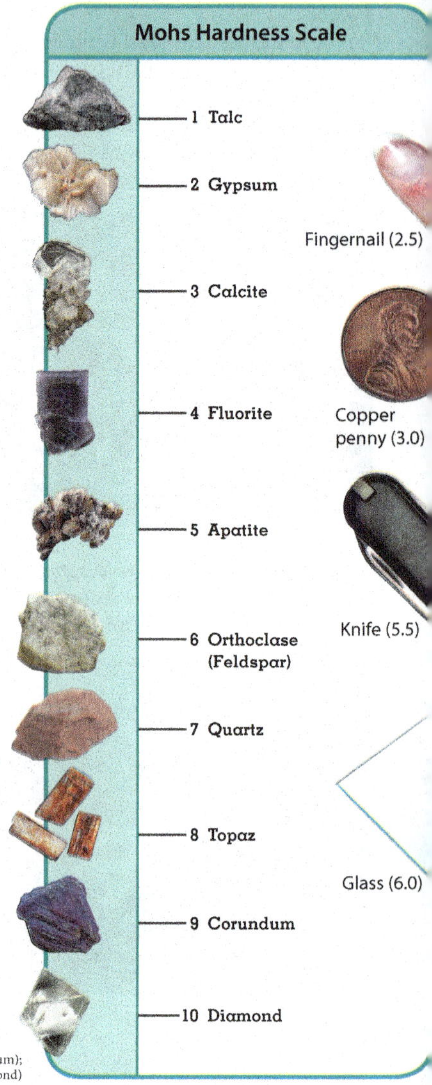

The Mohs scale shows minerals from the softest (1) to the hardest (10). In a scratch test, a harder substance scratches a softer one.

Image by Alan Darbellay, GGGems.com (Corundum); www.iRocks.com (Diamond)

14

Background

Mica

The term *mica* actually refers to a group of about thirty-seven minerals that have the ability to cleave into very thin sheets. Scientists have separated some types of mica into sheets that are only one molecule thick. In the past, thin sheets of transparent white mica, or muscovite, were used for windows. Today these sheets are still used as windows in some cast-iron stoves.

Activity

Observing Cleavage and Fracture

Materials: variety of common rocks and minerals, sturdy plastic bags, hammer, magnifying glass, paper, pencil

Observe and record a basic description of each rock. Work with one rock at a time. Place the rock into a bag. Hit the rock with the hammer until a piece breaks off. Work carefully so that the rock is not completely crumbled by the hammer. Remove the rock from the bag and observe the area of breakage. Record a description of the break(s). Identify the type of break as cleavage or fracture.

This arrangement of sheets of mica is called a book.

Cleavage and Fracture

Different minerals break in different ways. One way that some minerals break is called cleavage. **Cleavage** is the breaking of a mineral along smooth, straight lines or into flat sheets. Many of the minerals used in jewelry have cleavage. People use this characteristic to cut the minerals into beautiful shapes.

Some minerals with cleavage will break into geometric shapes. The smaller, broken pieces of some of these minerals are shaped like the original piece. For example, the crystals of halite, or rock salt, are shaped like cubes. When broken, halite cleaves into smaller cubes that look like the original.

Halite cleaves into small cubes.

Mica has a different type of cleavage. It does not break into cubes or other geometric shapes. Instead, this mineral peels easily into thin layers, or sheets. Some of the sheets are so thin that you can see through them.

Not all minerals have cleavage, however. Some minerals, such as quartz, do not break along straight lines. For example, quartz breaks unevenly in many directions. This type of breaking is called **fracture**. The fracture of some minerals is rough and jagged. Other minerals break along curved lines. Some minerals have fractures that look fuzzy or like thin threads.

The mineral asbestos has thin, fiberlike crystals.

QUICK CHECK

1. What is a mineral?
2. What are five characteristics that mineralogists can use to identify minerals?
3. Scratching a sample with other minerals or substances is a test for which characteristic?

15

5/6

- Ask a volunteer to read aloud the heading of the last column of the organizer. Guide the students in completing the organizer as you discuss cleavage and fracture of minerals.

  What is cleavage? the breaking of a mineral along smooth, straight lines or into flat sheets

  How is mica's cleavage different from many other minerals'? Mica peels into thin layers instead of breaking into geometric shapes.

  What is fracture? a break in some types of minerals that is not in smooth, straight lines or flat sheets

  How do you think cleavage and fracture may affect the way that a jeweler cuts minerals? Possible answer: When a jeweler cuts a mineral, the stone may break or chip along a line of cleavage or fracture and ruin the design that the jeweler wants.

- Discuss the pictures.

  What is the arrangement of mica called? a book

  What kind of cleavage does halite have? It breaks into cubes.

  What shape does asbestos break into? thin, fiberlike crystals

ANSWERS

1. an inorganic substance found naturally in the earth
2. crystal structure, color, luster, hardness, and cleavage
3. hardness

6

- The *Making Rock Candy* demonstration and the *Observing Cleavage and Fracture* activity found at the bottom of the pages in Lessons 5–6 may also be incorporated into the lesson.
- Suggestions for additional activities for this two-day lesson are on the Teacher's Toolkit CD.

Activity Manual

Reinforcement, page 7

Objectives

- Measure mass to the nearest gram
- Measure volume to the nearest milliliter

Materials

Materials listed below and on the student pages are needed for activity lessons.

- sweetened powdered drink mix
- 3 plastic cups with an equal amount of water in each
- 3 spoons
- strainer or colander (optional)
- bucket (optional)

Introduction

- Pour more than the correct amount of drink mix into one cup of water and less than the correct amount into another. Stir. Instruct one student to taste both drinks.
  Does either drink taste the way that you expected? Answers will vary.
  Describe how each tastes. Answers should include that one is too sweet or strong and the other is too watery or weak.
- Add the correct amount of drink mix to the third cup and stir. Be sure the students see you measuring the drink mix. Direct the same student to taste the third drink.
  Does this drink taste the way that you expected? yes
  Measuring is important when preparing food. It is also important in science. Scientists must make careful measurements to have predictable results and accurate, reliable data.

Measuring correctly is an important scientific skill that the students will use in other activities. You may choose to demonstrate both measuring mass and measuring volume before having the students practice measuring. Metric units of measure are used in this lesson and in the activities in *Science 5*.

Teach for Understanding

Mass Measurement

What is meant when we talk about the mass of an object? Guide the students to conclude that mass is the quantity of matter in an object.

- Explain that the size and shape of an object may change. But mass does not change unless some of the matter is added or removed.
- Tell the students that mass is measured using a balance. A balance compares the mass of an object to the known mass of the weights.
  What unit is used to measure mass? the gram

ACTIVITY

Measuring Mass and Volume

Name ______________________

Materials

pebbles
coffee filters (optional)
water
small object, such as a pencil or crayon
balance
metric beakers or measuring cups

Practice measuring mass with a balance. Place the pebbles in a coffee filter as you work. Mark each box after you finish the task.

☐ 1. Measure the mass of one handful of pebbles. ____________ g

☐ 2. Add or remove pebbles until the mass measurement equals 300 g.

☐ 3. Remove the pebbles. Measure the mass of the small object you chose.
Complete the sentence with your information.

The mass of a ____________ is ____________ g.

Practice measuring volume with a beaker or measuring cup. Mark each box after you finish the task.

☐ 4. Pour 300 mL of water into the beaker or measuring cup.

☐ 5. Add a handful of pebbles to the water. Record the level of the water. ____________ mL

☐ 6. Subtract to find the volume of the pebbles. ____________ mL

Background

Measurement

Measuring matter is taught in Chapter 3, *Matter*. Lesson 33 is a measurement lesson that gives students practice in measuring length, temperature, and volume in cubic centimeters.

Science Misconceptions

Mass and Weight

The words *mass* and *weight* are often confused. Mass is the amount of matter in an object. It is measured by comparing an unknown mass with a known mass. Balances are used to measure mass in grams or kilograms. Weight is not the measure of matter. It is the measure of the force of gravity on an object. If the gravity changes, the weight changes. Spring scales are used to measure weight in newtons (1 newton = 0.225 lb).

- Demonstrate the use of the balance. Place a coffee filter on the balance. Place a handful of pebbles in the filter. Adjust the weights and read the measurement. Discuss the procedure.

Volume Measurement

What is meant when we talk about the volume of an object? Guide the students to conclude that volume is the amount of space that an object takes up.

What are some measuring tools used to measure the volume of something such as water? Possible answers: measuring cups, measuring spoons, beakers, graduated cylinders

- Explain that the unit of measure for volume is the liter. A graduated cylinder, metric beaker, or other measuring cup can be used to measure volume.
- Demonstrate measuring 300 mL of water in a beaker. Discuss the procedure.

  You should always place the container on a table and bend to read the markings at eye level. Why should you measure in this way? The measurement is more accurate.

  Since pebbles have irregular shapes, how can you use water and a beaker (measuring cup) to measure the volume of a handful of pebbles? Guide the students to conclude that the volume of the pebbles can be measured by dropping the pebbles into an amount of water and calculating the difference between the water levels. This is known as the displacement method and will be presented in Chapter 3.

- Demonstrate the procedure by adding a handful of pebbles to the water. Read the new level of the water.
- Write the two amounts for display. Guide the students in subtracting to find the volume of the pebbles. Discuss the procedure.

Activity Manual

Activity, page 8

Guide the students as they practice measuring mass and volume.

Monitor the students' measuring accuracy and correct use of each measuring tool. It may be helpful to set up different measurement stations. Then set time limits and let the students move between stations. The materials can stay at each station for reuse. A strainer or colander over a bucket may be used to drain wet pebbles for reuse.

Background

Using an Equal-Arm Balance

Make sure that the balance indicator points to zero. Place the object to be measured in one pan. Add weights to the other pan until the indicator points to zero, showing that the two sides are balanced. Total the amount of the weights to determine the mass of the object.

Meniscus

The curve of the surface of a liquid in a container is called a meniscus. The surface of the water in a container will curve upward at the edges. Look at the bottom of the meniscus (lowest part of the curve) to read the volume of a liquid.

8
9

Student Text pages 16–17
Activity Manual pages 9–10

Objectives

- Follow directions
- Observe the formation of Epsom-salt crystals
- Collect and record observation data

Materials

Materials listed below and on the student pages are needed for activity lessons.

- one sheet of black construction paper
- chalk
- Epsom salt
- table salt
- magnifying glass
- toothpicks

Teacher's Toolkit CD

- Rubrics: *Salty Crystals,* for each student

Introduction

Two lesson days are allotted for this activity. On the first day, introduce and set up the activity. After the observation period is complete, use the second lesson day to discuss student observations and conclusions. See the Activity Helps under Background on this page.

- Make two circles on the construction paper. Label one circle *E*. Place some Epsom-salt granules in the circle. Label the other circle *T* and place some table-salt granules in it. Let the students use the magnifying glass and toothpicks to observe the two types of salt crystals.

  These are pieces, or granules, of Epsom salt and table salt. The pieces are called *crystals.*

  Do you see any differences between the crystals? yes

  Describe the differences that you see. Possible answer: Epsom-salt crystals are long and break easily. Table-salt crystals are small cubes.

Preparation for Reading

- Direct the students to read Student Text pages 16–17 and Activity Manual pages 9–10 before beginning.

Teach for Understanding

Procedure

- Explain that it is not necessary to have exactly 300 g of pebbles but that both bowls in each group should have the same amount.
- Guide the students in coating the pebbles with the salt water. If the pebbles are stirred, direct the students to push the pebbles back into a pile so that some are out of the water.

What shape do you think the crystals will form as they dry? long and thin Why? The crystals should form to look like the original Epsom-salt crystals observed in the introduction.

ACTIVITY

Salty Crystals

Some minerals form into beautiful, large crystals. Others have clusters of tiny crystals or no crystals at all. Some crystals form slowly; others form quickly. Temperature and time are some factors, or conditions, that can affect how crystals form.

In this activity, you will observe how crystals form under different conditions.

Process Skills
- Measuring
- Experimenting
- Observing
- Identifying and controlling variables
- Collecting, recording, and interpreting data

Problem

How do temperature and light affect the size of crystals?

Procedure

1. Complete the hypothesis in your Activity Manual.
2. Measure 100 mL of water and pour it into the saucepan. Add several drops of food coloring.
3. Measure 200 mL of Epsom salts. Stir the salt into the water.
4. Place the saucepan on the hot plate. Wear goggles or other eye protection while heating the liquid. Heat and stir until you cannot see the salt. The water does not have to boil. Remove the saucepan from the heat.
5. Label one bowl with the number *1* and the other bowl with the number *2*. Measure and pile 300 g of pebbles in the center of each bowl.
6. Choose a cool, dark location to place bowl 1. Record this location in your Activity Manual.
7. Choose a warm, bright location to place bowl 2. Record this location in your Activity Manual.
8. Stir the salt water. Measure 100 mL of the salt water into each bowl. Try to wet the pebbles with the water as you pour. Make sure that some wet pebbles remain above the level of the water.

Materials
- 100 mL water
- metric measuring cups
- saucepan
- food coloring
- 200 mL Epsom salts
- large spoon
- hot plate
- goggles
- potholder
- 2 sturdy bowls
- 600 g pebbles
- balance
- Activity Manual

16

Background

Activity Helps

Divide the class into science groups of three or four. Include high, average, and low achievers in each group. Depending on the abilities and personalities of the students, you may need to assign tasks to ensure that each team member participates. If the students use materials that vary from the list printed in the Activity Manual, they should put a line through materials that were not used and then write the materials that were used. This reinforces accurate recording.

If you choose to dissolve all the Epsom salts for the class, stir the mixture frequently while measuring it for the students to use. This prevents the salt from settling too much. The precise mass of the pebbles is not critical, but the same amount must be used in both bowls. The amount should form a pile high enough that several pebbles are above the level of the water.

A minimum of one week is suggested for the observation time. If the observation time is extended, provide more space for recording observations. The water will not completely evaporate within the time of this activity.

The procedures in the Student Text and Activity Manual do not use the terms *mixture, solution, dissolve,* or *evaporation*. You may choose to introduce and use these terms with this activity. The terms are presented in Chapter 3, *Matter*.

9. Place the bowls in the locations that you chose.
10. Observe and record the appearance of the salt water for seven days.

Conclusions

- Did crystals form? If so, which bowl had the larger crystals?
- What characteristics in the different environments do you think affected the formation of the crystals?

Follow-up

- Compare the crystal formation of salt water placed in other locations such as outside or near a fan, an air vent, or an open window.
- Compare the crystal formation of other minerals such as alum or table salt.

17

8
9

Do you think that the water in the bowls should be stirred or touched while we wait for crystals to form? no Why? Guide the students to conclude that the crystals may not form if the water is disturbed.

Process Skill

One process skill will be emphasized in each activity. The process skill emphasized in this activity is observing, which is using the senses to gather information.

What information might you record each day? Possible answers: the level of water in each bowl; any formation of crystals; the size of the crystals; the places in the bowl that crystals form

Why should observations be recorded? Possible answers: to have a record that can be looked at later; to keep information from being forgotten.

Conclusions

- Discuss the formation of the crystals.
  Did the water dry up completely? Answers will vary.
  Is there a difference in the appearance of the crystals on the pebbles and those in the water? Answers will vary.
- The larger crystals tend to form at the bottom. You may choose to have the students scoop some of their crystals from the water and let them dry. Then let them compare their crystals with some of the dry Epsom salts.
  Are there differences in the appearances of your crystals and those from the package? If so, describe the differences. The crystals should appear about the same. However, crystals from the package will be individual and the students' may be stuck together.

Background

Variables

In this activity, temperature and light are the variables being controlled. Another variable that will affect crystal formation is the concentration of the Epsom salts in the water.

Surface Area

The pebbles provide more surface area for the growth of crystals as the water evaporates.

Activity Manual

Activity, pages 9–10

Guide the students as they use these pages to write a hypothesis, observe the crystals, record their observations, and draw conclusions.

Assessment

Rubrics

Select the prepared rubric, or design a rubric to include your chosen criteria.

A rubric is a suggested tool for grading an activity. Guidelines for using the prepared rubrics are included on the Teacher's Toolkit CD. The students should be given a copy of the rubric before beginning the activity. This will allow them to see your established criteria for the activity and for evaluation.

Objectives

- Differentiate between characteristics of precious and semiprecious stones
- List some common uses of minerals
- Recognize that some minerals are metals
- Identify where minerals are found

Materials

- variety of items made from minerals or that use minerals (e.g., jewelry, a bottle of vitamins, toothpaste, fertilizer, a metal can)

Teacher's Toolkit CD

- Quiz 1-B

Vocabulary

- gemstone

Introduction

- Display the items.
  These items have something in common. Do you know what it is? Accept reasonable answers.

Preparation for Reading

- Display the words *amethysts*, *turquoise*, *gypsum*, and *lapidary*. Pronounce these words together.
- Instruct the students to read silently pages 18–22 to find the answers for these questions.
  What is a gemstone?
  What are some ways that minerals are used?
  Where are minerals found?

Teach for Understanding

- Refer to the items used in the introduction as they apply to the material discussed.
  What is a gemstone? a mineral that can be cut and polished
  After a gemstone is cut and polished, what is it called? a gem
  What is a precious stone? a gemstone that is beautiful and rare
  Name some examples of precious stones. Possible answers: diamond, ruby, sapphire, emerald
  Why are some precious stones displayed in museums? Possible answers: They are too valuable to wear. In museums, they are better protected against damage or theft. There they can be seen and enjoyed by many people.
  What is the difference between precious and semiprecious stones? Semiprecious stones are less valuable and are found more easily.

Uses of Minerals

Man has found many uses for the minerals that God placed on and in the earth. Some minerals, such as those used in jewelry, provide beauty and pleasure. The minerals used in glass and building materials help provide shelter. Many minerals, such as those used in lasers and other machinery, help make our lives easier. The minerals in food strengthen our bodies.

Gemstones

Some minerals form beautifully colored crystals that reflect light. A mineral that can be cut and polished for use is called a **gemstone**. Once it is cut and polished, it is called a gem. Gems are often used in jewelry.

Some gemstones are rare, or hard to find. A rare one is worth more money than common ones. A *precious stone* is a gemstone that is beautiful and rare. Some examples of precious stones are diamonds, rubies, sapphires, and emeralds. Large precious stones that do not have impurities are extremely valuable. In fact, many of the larger precious stones are too valuable to wear. These gems are usually placed on display in museums. There they are safe, and people can enjoy their beauty.

Semiprecious stones are more common, or easier to find, than precious stones. Though not worth as much money, semiprecious stones are still very beautiful. Some kinds of semiprecious stones are amethysts, garnets, opals, turquoise, and jade. These stones are often used in jewelry.

For centuries, people have carved stones such as turquoise and jade. These stones have been used in sculptures, ornaments, and even the handles of weapons. Turquoise can be found in the western part of the United States. Turquoise is usually a beautiful blue-green color. Green is the most common color of jade. However, the stone is also found in a variety of other colors. Jade is mostly found in Asian countries.

Jade figures

18

Background

Synthetic

The most notable difference between natural and manmade gems is that synthetic gems have no flaws.

Cubic Zirconia

Cubic zirconia is a synthetic crystal made of zirconium oxide. It is the most common substitute for diamonds. The hardness and luster of cubic zirconia are different from those of a diamond, which is made of carbon. Trained jewelers can easily tell the difference between the two crystals.

Carat

The unit of measurement for the weight of gems and precious metals is the carat. One carat is equal to 200 mg.

Jewel Movement

Some watches and clocks have jewel movements. This term relates to the gear pins of watches and clocks that are wound.

Semi-

The prefix *semi-* means "half of" or "partial." Guide the students in making a list of other words with the prefix *semi-*. Discuss how the prefix affects the words' meanings.

10

SCIENCE & THE BIBLE

Minerals such as precious stones, silver, and gold are very valuable. The Bible reminds us that we were "not redeemed with corruptible things, as silver and gold . . . but with the precious blood of Christ" (1 Peter 1:18–19). Man's sin is so great, he cannot pay for it with money. Silver and gold will be destroyed. Only the life of the perfect Son of God is valuable enough to pay for our sin. And because God loved the world, He gave us His only begotten Son.

Real gemstones form naturally in the earth. But scientists have developed ways to copy this formation in a laboratory. As a result, many gems have a *synthetic*, or manmade, version. Synthetic gems may look very much like the real gems but are not as valuable. Cubic zirconia is a synthetic gem. It looks like a diamond and is sometimes used in jewelry in place of the more expensive diamond.

Diamond

Cubic zirconia

Gemstones are used in more than just jewelry and art, though. Some gemstones are not only beautiful but are also very strong. Diamonds and rubies are the strongest and hardest minerals and can cut many other substances. Chips of these gemstones are added to the drill bits or saw blades of some tools. These kinds of tools are used to cut very hard materials, such as other gemstones or rocks. Diamonds and rubies are also used inside some watches. The gemstones help the gears of the watches move smoothly and last longer.

FANTASTIC FACTS

Rhinestones are used in jewelry and in decorating clothing. In the past, these sparkling stones were cut from rock crystal, a type of clear quartz. Today most rhinestones are made from glass or hard plastic, which are synthetic materials.

19

What does *synthetic* mean? manmade

Name a synthetic stone that resembles a diamond. cubic zirconia

Name some examples of semiprecious stones. Possible answers: amethyst, garnet, opal, jade, turquoise

In addition to jewelry, what are some uses for turquoise and jade? Possible answers: sculptures, ornaments, weapons

Why are diamonds and rubies used in tools and watches? Diamonds and rubies are strong and do not break or wear away easily.

- Discuss *Science & The Bible.*
  Why is the shed blood of Christ more valuable than gold, silver, or precious stones? Only Christ's blood can pay for the sins of man. Salvation is a gift from God that cannot be bought with gold, silver, or precious stones.

Background

Birthstones

Centuries ago, people began associating a different gemstone with each month of the year. Some people believe that the choice of stones came from the Bible. Exodus 28:15–20 identifies the stones that God chose for the high priest to wear. Many of the stones listed are the same as those assigned to the months. Today these monthly stones are called birthstones. Often a birthstone is worn by a person whose birthday is in that month. The pearl is the only birthstone that is not a mineral. Since they are formed by oysters, pearls are organic.

Activity

Finding Gems and Minerals in the Bible

The Bible contains over 1,500 references to gems and minerals. However, many of the names have changed over the years. Guide the students in choosing minerals from the Bible and researching to find their modern or common names. Display or announce the results.

10

What kind of minerals are found in ores? metals

How does the process of smelting work? The ore is crushed and heated. The extreme heat melts the metal, causing it to separate from other materials in the rock.

What are some properties of metals that make them useful to mankind? strong; easy to shape or bend; most conduct electricity

Why are aluminum and titanium used in aircraft and spacecraft? These metals are strong but do not weigh much.

Why would it be important for a spacecraft to be made from lightweight materials? Possible answer: The weight of a spacecraft affects speed, liftoff, and fuel consumption.

What are some common objects that are made from molded metals today? Possible answers: pots and pans, cars, nails, tools

Why are metals sometimes combined? to increase their strength or to change other properties

What is an example of a metal that is actually a mixture of metals? Possible answer: steel

Why do you think that there are different kinds of steel? Possible answer: Steel has different uses. Each kind of steel is made to have different properties, depending on how it will be used.

How are thin sheets of metal used today? Possible answers: jewelry, filing cabinets, shelves, toys

What is gold leaf? thin sheets of gold

How was gold leaf used in Bible times? to cover the wood in many parts of the temple

Metals

Some minerals are also metals. Metals are usually found in rocks called *ores*. Sometimes, to remove the metal from the ore, the ore is crushed and then heated. The extreme heat causes the metal to melt and separate from the other rock materials. This process is called *smelting*.

Metals have several properties that make them very useful to man. Most metals are strong but are still easy to shape or bend. Some soft metals such as gold, silver, and copper are used to make jewelry and sculptures. Most metals also conduct electricity. One metal that conducts electricity especially well is copper. In fact, the wires in your house are probably made of copper.

Copper wire

Titanium and aluminum are two metals that are strong but do not weigh much. Most airplanes and spacecrafts have parts made of titanium. Aluminum is very common. It is often used to make containers, such as those used for canned drinks.

Melted metals can be poured into molds. Early settlers used molds to form musket balls and cannonballs. Even now, many items are formed in this way. Parts of many pots, pans, cars, and tools are made from molded metals.

Sometimes metals are combined to increase their strength or change other properties. Steel is an example of a mixture of metals. Iron is melted with other metals to form different kinds of steel. The different metals determine what kind of properties the steel has. Different kinds of steel are used in things such as buildings and cars.

Many metals can be hammered or flattened into thin sheets. These thin sheets are then used in items such as file cabinets, shelves, and toys.

Since Bible times, craftsmen have skillfully hammered gold into thin sheets. These sheets, known today as gold leaf, can be used to cover wood, stone, or other metals. In 2 Chronicles 3 the Bible tells of how gold covered many wooden parts of the temple.

20

Background

Processing Ore

Smelting is only one way that metals are extracted from ores. Other processes include concentration, leaching, and electrolysis.

Steel

Steel is an alloy of iron ore and other substances. Different substances—such as nickel, chromium, and carbon—are mixed with the iron ore, depending on the type of steel desired. To make stainless steel, iron ore is mixed with chromium.

Alloy

An alloy is a substance composed of the mixture of two or more metals.

Gold Leaf

Gold can be hammered into sheets that are only 0.00001 mm thick.

Quartz

In a quartz watch or clock, the source of power is quartz, not winding. The properties of quartz are used to provide a source of electrical power.

10

Other Minerals

Gemstones and metals are just some of the minerals we use. Each day we use many other minerals. Common minerals such as calcite and sulfur are sometimes added to fertilizers. Gypsum is used in making dry wall. Dry wall sheets form the inside walls of most houses. Fluorite is used in both toothpaste and pottery. Mica has many common uses. It is added to paint and is used as an insulator for electronics. Mica also gives the shine and sparkle to some cosmetics. Silicon is an important part of electronic devices such as cell phones and televisions.

Many plastics, fabrics, and other substances that you use each day contain minerals. Glass is a blend of several minerals. Some foods, drinks, and medicines have minerals added to them. Even some daily vitamin pills include minerals.

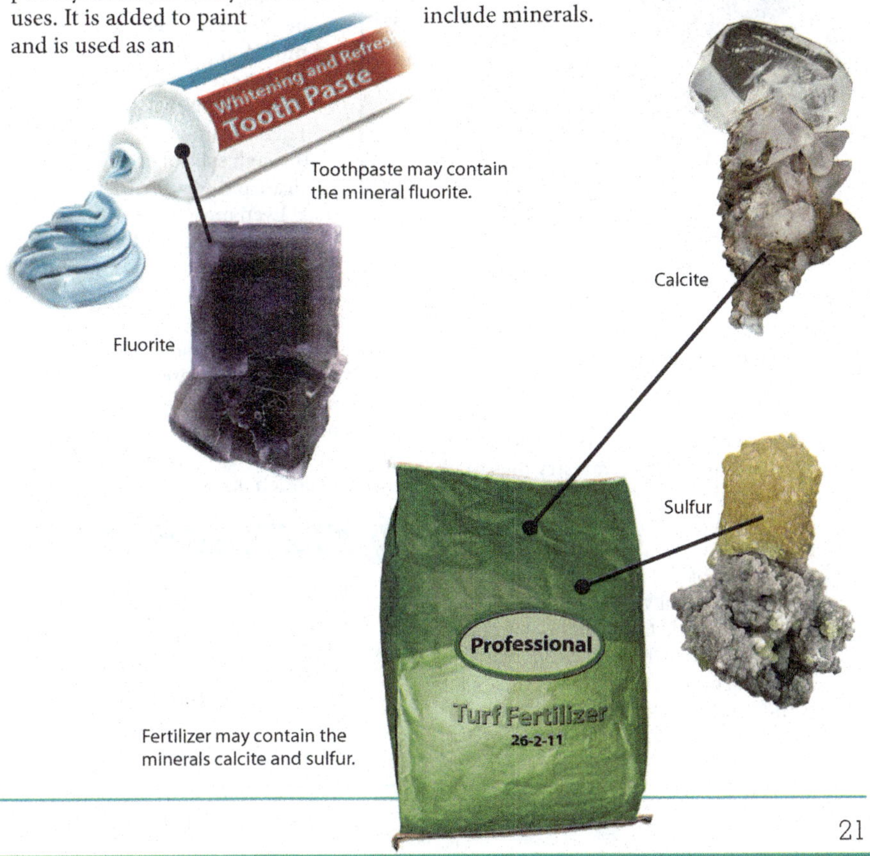

Toothpaste may contain the mineral fluorite.

Fertilizer may contain the minerals calcite and sulfur.

21

What minerals are sometimes used in fertilizers? sulfur and calcite

What do you think would happen to the construction industry if the supply of gypsum were reduced? Possible answers: Since gypsum is used in walls, the industry would have to find other materials to use. Construction costs might get higher.

What mineral is in most toothpastes? fluorite

- Discuss the pictures and the minerals used in each product.

Background

Root Words for Science Occupations

The occupations of people who work with minerals and rocks include geologists, mineralogists, lapidaries, and metallurgists. Help students remember the names by looking at the root for each and connecting the root with the specialty of the person.

A *geologist* is a person who studies geology. *Geology* is the study of the nonliving parts of the earth.

A *mineralogist* is a person who studies minerals. *Minerals* are inorganic substances found in the earth.

A *lapidary* is a person who cuts, polishes, or carves gems. The name comes from *lapid*, the Latin word for "stone."

A *metallurgist* is a person who studies the science of and technology related to metals.

10

Where are most minerals found? buried in the earth's crust

How are minerals removed from the earth? by mining

What is a pocket or strip of minerals within the earth called? a vein

If a vein is located within the shaft of a volcano, what is it called? a pipe

After a mineral is mined from the earth, what must be done before it is manufactured into a product? The mineral must be removed from surrounding rocks and ores, and it must be cleaned and sorted.

How does a dull, rough gemstone become a polished gem? A lapidary carefully cuts and polishes it.

Who is a lapidary? a person who cuts, polishes, or carves gems

ANSWERS

1. a mineral that can be cut and polished for use
2. synthetic
3. a vein

Activity Manual

Review, pages 11–12

These pages review Lessons 5–6 and 10.

Expansion, page 13

Guide the students as they complete this page. It may also be used as independent work for advanced students.

Assessment

Quiz 1-B

The quiz may be given any time after completing this lesson.

Lapidary at work

Finding Minerals

Some minerals are found on the surface of the earth. Most minerals, however, are buried in the earth's crust. The mining industry digs to find and remove minerals and other usable materials from the earth.

The crust of the earth is made up of layers of soil and rock. In these layers mineral veins can form. A *vein* is a pocket or strip of a mineral within the earth. Some minerals also form in vertical cracks through the layers. Vertical veins of minerals can form in volcanic pipes. A pipe is the hardened magma tube that exists inside ancient volcanoes. Pipes often contain diamonds or other minerals that form deep in the layers of the earth's crust. Magma moving toward the surface pushes the diamonds upward through cracks in the earth's crust.

Most minerals cannot be used when they are first mined from the earth. They must be removed from any surrounding rock and ores. Then the minerals are put through special processes to clean and sort them.

Even most gemstones do not sparkle and shine when they are first removed from the earth. Most look like ordinary, dull rocks. Gemstones must be handled with extra care. They are carefully cut by a person called a *lapidary*. These skilled craftsmen know the best way to cut each type of gemstone to reveal the beauty hidden inside.

QUICK CHECK

1. What is a gemstone?
2. What term is used to identify a manmade gem?
3. What is a pocket or strip of a mineral inside rock layers called?

22

Activity

Practice Mining Techniques

Materials: soft chocolate chip cookies, toothpicks, paper towels

Provide each student with a cookie, a toothpick, and some paper towels. Direct the student to use the toothpick to find and remove the "minerals" (chocolate chips) from his "piece of earth" (the cookie). Remind him to dig carefully so the "mineral" is not broken or damaged.

Direct the student to place his "minerals" on a paper towel. He should inspect them for any scratches or broken pieces. After the student has mined all his "minerals," allow him to eat them and his "pieces of earth."

Munching Minerals

EXPLORATION

God designed our bodies to use minerals. Our bodies get these minerals from the things that we eat and drink. Look at the label on a food package. Minerals are often listed on the nutrition label. Nutritionists study the nutrients in foods and how these nutrients affect our bodies.

In this exploration, you are a nutritionist in training. Choose a mineral that has nutritional value. Research to find food and beverage sources of your mineral. Then research the ways that the mineral is beneficial to our bodies. Be prepared to discuss your findings with other nutritionists in training.

What to Do

1. Choose a mineral from the list.
2. Research the mineral that you chose. Your report should include some foods and beverages in which your mineral is found. Also include the ways that God designed our bodies to use this mineral.
3. Display samples of foods or the labels of products that contain your mineral.
4. Prepare a one- to two-minute presentation of your findings.

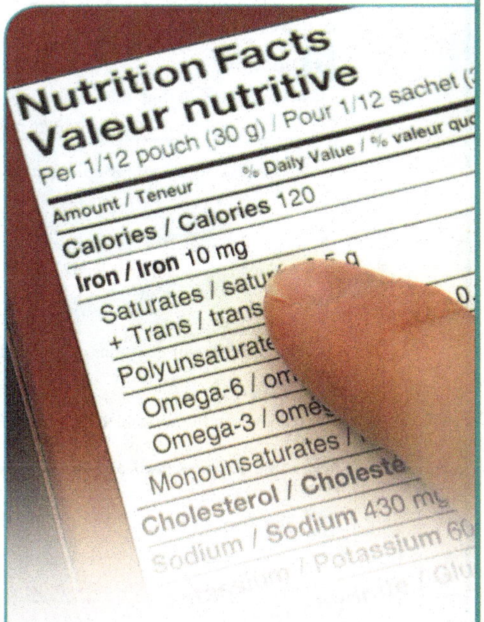

calcium	phosphorus
chromium	potassium
copper	selenium
iron	sodium
magnesium	sulfur
manganese	zinc

23

Background

Presentations

This project provides an opportunity for the students to incorporate various forms of technology such as for researching, preparing the display, and making the presentation.

Objectives

- Research a mineral found in foods or beverages
- Display foods or beverages that contain the mineral
- Prepare an oral presentation

Materials

- assortment of food and beverage nutrition labels

Teacher's Toolkit CD

- Rubrics: *Munching Minerals,* for each student

Introduction

- Display the nutrition labels.
  What do you think is the purpose of these labels on foods and beverages? Elicit that the labels show the amounts of nutrition provided when the items are consumed.
- Explain that some items on a nutrition label are minerals.

The most commonly consumed minerals are listed in the Student Text. You may choose to allow the student to research another mineral. You may also choose to omit the presentation and ask the student to make the display with only written information.

Preparation for Reading

- Direct the student to read silently Student Text page 23 and Activity Manual page 14 before beginning the exploration.

Teach for Understanding

In whose image did God create mankind? His image
Who designed your body to use minerals? God

- Guide the students in choosing minerals to research.
- Review your requirements of the project. Discuss what types of research would be considered reliable.
- Provide a due date for the displays and presentations.

Activity Manual

Exploration, page 14

The page is designed to help the students plan and organize their research.

Assessment

Rubrics

Select the prepared rubric, or design a rubric to include your chosen criteria.

12

Activity Manual pages 15–18

Objectives

- Use the PQ3R method to read informational text

Teacher's Toolkit CD

- Visuals 1.2–1.3: *PQ3R: Reading Skill*

Introduction

- Discuss different purposes for reading.
  What types of books do you read for fun?
  Do you expect to learn new information as you read these books? not usually
  Do you usually have to remember information that you learn while reading for fun? no
  Sometimes you must read to gain new information. What types of books might you read for this purpose? Possible answers: textbooks, encyclopedias, other reference books
  When you read to gain information, such as from your science textbook, do you always understand and remember what you read? Answers will vary.
- In this lesson the student will learn how to use a study skill called PQ3R to help him improve his ability to gain and remember important information as he reads.

Teach for Understanding

- Discuss each step of the PQ3R study skill. You may choose to display the *PQ3R: Reading Skill* visuals as you teach the lesson.
  Look at the Study Skill box on Activity Manual page 15. What do the letters of *PQ3R* stand for? Preview, Question, Read, Recite, and Review
  What does it mean to preview a text? Possible answers: to skim the text; to take a sample look; to get a general idea
- Guide the students in completing the Preview section on Activity Manual page 15 as you teach.
- Explain that in the *Preview* step, the reader looks at the entire selection to get a general idea of what it is about. He **skims, or takes a quick look at, the title, subheadings, illustrations, captions, chart titles, bold words, and italicized words.**
  What do you look at during the *Preview* step? Elicit that you look at the title, subheadings, illustrations, captions, bold words, and italicized words.
  Look at the title on Activity Manual page 17. What do you think this article will be about? the British Crown Jewels
- Explain that after the reader previews the selection, each section of the selection is read separately. The *Question*, *Read*, and *Recite* steps of PQ3R are repeated for each section.

Study Skill: PQ3R

Name ______

Practice the PQ3R study skill as you read "The British Crown Jewels" on pages 17–18. Check each box as you complete the step. Begin by using the *Preview* step.

Study Skill	
Preview	Skim the text.
Question	Ask questions.
Read	Look for answers.
Recite	Answer your questions.
Review	Reread and think about what you have learned.

Preview (Skim)

1. What do you look at during the *Preview* step?
   the title, subheadings, illustrations, captions, bold words, italicized words
2. What do you think this article will be about?
   the British Crown Jewels

Practice the middle three steps of PQ3R (*Question*, *Read*, and *Recite*) for the first section of the article.

Question

3. Make a question from the title of the article, "The British Crown Jewels."
   Possible answers: What are the British Crown Jewels? Where are the British Crown Jewels kept?
4. Make a question from the bold term *monarchs*.
   Possible answer: What is a monarch?
5. Make a question from the italicized term *Tower of London*.
   Possible answer: Where is the Tower of London?

Read

6. Think about your questions as you read the first two paragraphs of the article.

Recite

7. Tell yourself the answers to the questions you wrote in steps 3–5.

Repeat the *Question*, *Read*, and *Recite* steps for the "Jewels for Coronation" section.

Question

8. Make a question from the subheading "Jewels for Coronation."
   Possible answers: Which jewels are famous? What is special about these jewels?
9. Make a question from the bold term *coronation*.
   Possible answer: What is a coronation?

Background

Study Skill

Once PQ3R is learned, remind the students to use it as they read SCIENCE 5 or any other informational material. SQ3R is a study method in which the first step is *Survey*. The *Preview* step in PQ3R is the same as the *Survey* step in SQ3R.

10. Look at the first picture. Make a question from the caption.
Possible answer: What pieces are included in the crown jewels?

11. Make a question from the bold term *orb*.
Possible answer: What is an orb? What does an orb do?

12. Make a question from the italicized term *Westminster Abbey*.
Possible answer: What happened at Westminster Abbey?

R ead

13. Think about your questions as you read the "Jewels for Coronation" section.

R ecite

14. Tell yourself the answers to the questions you wrote in steps 8–10.

Repeat the *Question*, *Read*, and *Recite* steps for the "Protecting the Jewels" section.

Q uestion

15. Make a question from the subheading "Protecting the Jewels."
Possible answers: How are the jewels protected? What are the jewels protected from?

16. Make a question from the italicized term *Jewel House*.
Possible answer: Where is the Jewel House?

17. Make a question from the caption under the picture.
Possible answer: What do the Yeoman Warders wear?

18. Make a question from the bold term *Yeoman Warders*.
Possible answer: What is a Yeoman Warder?

R ead

19. Think about your questions as you read the "Protecting the Jewels" section.

R ecite

20. Tell yourself the answers to the questions you wrote in steps 15–19.

Complete the last step in the PQ3R study method.

R eview

21. Look back at the title, subheadings, captions, bold words, and italicized words. Think about the information you learned.

Background

Monarch

A monarch rules for life and by heredity.

Tower of London

The Tower of London is not just one tower. It is the name given to a group of buildings in the eastern part of London.

Westminster Abbey

The abbey used to be a monastery. It is still used as a church. It is also the final resting place for many of the monarchs. It is located near Buckingham Palace, which is across the city from the Tower of London.

Famous Robberies

Two of the most noted robberies of the crown jewels were the 1303 robbery by Richard de Podlicote and the 1671 robbery attempt by Thomas Blood.

What is the second step of PQ3R? Question

- Explain that in the *Question* step the reader **forms questions from titles, headings, illustrations, captions, charts, bold words, italicized words, or other material he did not understand during the *Preview* step.** The questions may be written.
- Guide the students in completing the Question section on Activity Manual page 15 as you teach.
  Look at the title on Activity Manual page 17. Titles usually tell you what information you will be reading. Make a question from the title on this page. Possible answers: What are the British Crown Jewels? Where are the British Crown Jewels kept?

  Make a question from the bold term *monarchs*. Possible answer: What is a monarch?

  Make a question from the italicized term *Tower of London*. Possible answer: Where is the Tower of London?

  What is the third step of PQ3R? Read
- Explain that after a reader has formed questions about the section, he is to read the section completely. During the *Read* step, he **looks for the answers to his questions**. In addition to the text, he **reads captions and examines any graphic information, such as maps and charts**. He should slow his reading or stop and reread a passage if it is especially difficult.
- Allow time for the student to read the first section on Activity Manual page 17.
  What is the fourth step of PQ3R? Recite
- Explain that the *Recite* step immediately follows each *Read* step. In this step the reader **recites, thinks about, or writes the answers to his questions**. He may also find it helpful to **summarize the section in his own words or take notes**.
- Allow time for the students to read their questions for numbers 3–5 and silently answer them.
- Follow this same procedure for the other two sections of the article. You may choose to work these sections together or independently.
  What is the fifth step of PQ3R? Review
- Explain that after the entire selection is read, the reader reviews what he has learned. In the *Review* step the reader **rereads the titles, headings, captions, and other graphic information and thinks about the answers to his questions**.

Activity Manual

Study Skill, pages 15–18

Guide the student as he applies the PQ3R study skill as he reads for information.

13

Student Text pages 24–27
Activity Manual page 19

Objectives

- Define *rock*
- Identify three types of rock and explain how each is formed
- List examples of igneous rock, sedimentary rock, and metamorphic rock

Materials

- cereal-and-marshmallow treats individually wrapped in plastic, one for each student (See recipe in Activity.)
- sample or photo of sandstone
- magnifying glass

Teacher's Toolkit CD

- Quiz 1-C

Vocabulary

- rock
- sedimentary rock
- igneous rock
- metamorphic rock

Introduction

- Distribute the cereal-and-marshmallow treat squares.
  Keep your treat in the plastic. Examine and describe your square. Possible descriptions: cereal pieces stuck together; square shape; air spaces between cereal pieces
- Circulate the sandstone sample and the magnifying glass for the students to examine.
  Notice that the sandstone is also formed from small pieces.
  Press your treat square tightly with your hands. Describe how your treat changed. Possible descriptions: cereal pieces are smaller and closer together; shape is flatter or rounder
  What caused the changes? Elicit that pressure and possibly heat from their hands caused the changes.
  In a similar way, some sandstone and other rocks change as they are exposed to heat and pressure inside the earth.

Preparation for Reading

- Display the words *igneous*, *sedimentary*, *metamorphic*, and *foliated*. Pronounce the words together.
- Direct the students to mentally make questions using the headings, bold words, and the captions. Instruct them to read silently pages 24–27 looking for the answers to these questions.
  Is a rock the same as a mineral?
  Into what three main groups do geologists classify rocks?

Teach for Understanding

What powerful forces have shaped much of the rock we see? the Flood, earthquakes, and volcanic activity

Rocks

Dr. Munschall thought that local floods and long periods of time caused rocks to look the way they do. However, the Bible tells us that the Flood covered the whole earth with water. This global Flood most likely involved lots of earthquakes and volcanic activity. These powerful forces shaped much of the rock we see and use. Genesis 1 tells us how rocks were originally created, but Genesis 6–8 helps us understand why they look the way they do today.

A **rock** is a hard, natural substance made of one or more minerals. Rocks may also contain organic materials. Since rocks are usually a mixture of several minerals, many types of rocks exist. This makes classifying rocks only by mineral content or appearance very difficult. There would be almost as many categories as there are rocks. So geologists classify rocks according to how they form.

Igneous Rock

Igneous rock forms from cooled magma or volcanic lava. Magma is the very hot, melted rock under the surface of the earth. In Genesis 7:11, the Bible tells us that the fountains of the deep were opened at the beginning of the Flood. These eruptions of magma could have allowed many igneous rocks to form.

Granite forms from magma. The minerals in granite are quartz, feldspar, mica, and hornblende. The arrangement and color of the mineral crystals in the granite determines how it looks. Granite is very hard rock and can be polished to a smooth

Granite

shine. Many statues and monuments are carved from granite. You may even have granite countertops in your kitchen.

Lava is magma that comes out of the earth. Some lava looks like hot foam as it flows along the ground. Foamy lava can also be ejected, or thrown, out of a volcano. As the ejected foam cools, it hardens into a rock called *pumice*. The gases that were in the foam leave holes in the hardened rock. These holes make some pumice rocks so light that they can float on water! Some pumice rocks are the rough "stones" used to produce stone-washed

Pumice

24

Activity

Recipe for Cereal-and-Marshmallow Treats

Materials: 4 cups marshmallows, 3 tablespoons margarine, saucepan, spoon, 6 cups cereal, greased 9 × 13 in. pan.

Combine marshmallows and margarine in saucepan. Heat over medium heat until melted, stirring constantly. Stir in cereal until completely covered. Press in pan. Let cool before cutting into bars.

Background

Types of Igneous Rock

Igneous rock that forms under the surface of the earth is called *intrusive*. Igneous rock that forms on the surface of the earth is called *extrusive.*

Igneous

The word *igneous* comes from the Latin word for "fire," *ignis*. Some other words that come from this same Latin word are *ignite* and *ignition*.

denim. Pumice can also be used in landscaping or a gas grill.

Obsidian is another type of rock that forms from lava. These rocks cool quickly. This, along with other factors, gives little time for crystals to form. As a result the rock looks like shiny, black glass. Obsidian breaks into curved pieces with very sharp edges. For years Native Americans used obsidian to make arrowheads and other tools. Today some surgical instruments are made from sharp pieces of obsidian.

Obsidian

Basalt columns can be clearly seen at the Giant's Causeway in Northern Ireland.

Most igneous rocks are very hard. One of the hardest kinds of igneous rock is *basalt*. It often forms large, wide columns. These columns are tightly packed and usually have six sides. Basalt is often used for construction. Many of the old stone buildings and cobblestone streets in Europe are made from basalt rocks. Even after hundreds of years, they show very few signs of weathering.

SCIENCE & HISTORY

Ballast is any material used to balance the weight in ships. For hundreds of years, heavy rocks were added to or removed from ships to make them heavier or lighter. Ballast rocks left at port cities were often used by the town as building materials. In fact, many of the cobblestone streets in port cities were made from discarded ballast rocks. Today water in special tanks is used as ballast in ships.

25

What is a rock? a hard, natural substance made of one or more minerals and sometimes organic substances

How is a rock different from a mineral? Most rocks are composed of more than one mineral. Some rocks contain organic materials while minerals do not.

Why is classifying rocks by mineral content difficult? Most rocks are a mixture of several minerals.

How do geologists classify rocks? according to how they form

What is igneous rock? rock formed from volcanic magma or lava

What is a possible reason why igneous rock formed according to Genesis 7:11? The fountains of the deep were opened at the beginning of the Flood.

What is a common igneous rock that is used for statues and monuments? granite

Do all pieces of granite look the same? no Why not? The arrangement and color of the mineral crystals can be different.

How is the formation of pumice different from the formation of granite? Pumice forms from lava that has been ejected from a volcano. Granite forms from magma that comes out of the earth and flows along the ground.

What is an unusual characteristic of pumice? It floats.

Look at the picture of obsidian. What are some characteristics of this rock? shiny, black, glassy appearance; breaks into curved pieces with sharp edges

How was obsidian used by Native Americans? for tools and arrowheads

What is one way obsidian is used today? Possible answer: surgical instruments

Why is basalt often used in constructing buildings? As one of the hardest rocks, it shows little weathering even after hundreds of years.

- Discuss *Science & History*.

  In the past, why were rocks often placed in ships? as ballast (to balance weight)

  Why were the cobblestone streets often made of ballast rocks? Possible answers: The discarded rocks were available. They did not need to be dug or gathered.

Background

Stones in Washington, DC

Many government buildings and memorials in Washington, DC, are constructed from rock. Before railroads, rock was quarried near Washington or on the East Coast and brought to Washington by ship. Today rocks from all over the country can be found in the memorials and government buildings. Get a US map and research to find the states that have provided rock for construction in the buildings or memorials in Washington, DC. Mark the map to show the types of rock and the locations where each was quarried.

Stones of Remembrance

Several passages of Scripture record accounts of the Israelites placing or piling stones to remember specific events of God's working in their lives. In 1 Samuel 7:12, Samuel places a stone and names it "Ebenezer." This special stone served as a visual reminder of how the Lord gave them victory over the Philistines. In the hymn "Come Thou Fount," writer Robert Robinson uses words from this verse to challenge Christians to remember God's working in their lives. (It should be noted that some modern versions of the hymn have changed the wording to omit *Ebenezer*.) Christians should raise their own "Ebenezer" stones of remembrance. These stones may include keeping a prayer and praise journal or telling others of God's working.

What is sedimentary rock? rock that forms when layers of sediment and organic material harden

What forces usually move sediments? wind and water

Where does sediment usually settle? at the bottom of bodies of water

Which type of rock is stronger—igneous or sedimentary? igneous

What is an example of a common sedimentary rock? Possible answers: conglomerate, sandstone, limestone, chalk

Describe a conglomerate rock. sedimentary rock that looks like pebbles that are cemented together

When organic material mixes with sedimentary rock, what types of rock can form? coal, limestone, chalk

Which two sedimentary rocks contain the mineral calcite? chalk and limestone

What organic remains are sources of calcite? bones and shells

Layers in sedimentary rock at Twelve Apostles, Australia

Sedimentary Rock

Most of the rocks formed during the Flood were sedimentary rocks. **Sedimentary rock** formed when layers of sediment and organic material hardened. As water or wind moved the particles, some of them settled into layers at the bottom of bodies of water. The layers of sediment were then pressed and cemented into rock. The waters of the Flood would have formed and moved great amounts of sediment that would settle into layers as the water receded. Most sedimentary rocks are not as strong as igneous rocks are.

Sandstone is a common sedimentary rock. The different types of sediment that form it are usually similar in size. Some sedimentary rocks look like pebbles that have been cemented together. This type of rock is called a *conglomerate*.

Conglomerate

Organic material can also settle and compress into the rock. *Coal* formed from decayed plant matter. Perhaps most of the coal we see today was formed by the many plants that died during the Flood. Sea shells contain the mineral *calcite*. Some rocks such as *limestone* and *chalk* are made almost completely of calcite. Calcite mixes easily with water. As a result, limestone and chalk weather easily.

Weathering and erosion have caused large pieces of rock to separate from the chalk cliff at Dover, England.

26

Background

Rock Formation and the Flood

Some creationary scientists believe that most sedimentary rock was formed during the Flood. The erosion caused by so much water would have moved great amounts of sediment. At the time of the Flood, pressure from the water, movements of the earth, or both may have formed the sedimentary rock. Creationary scientists think that it is unlikely that significant amounts of sedimentary rock are still being formed today.

Fossils

Fossils are discussed in more detail in Chapter 2, *Fossils and Dinosaurs.*

Other Metamorphic Rocks

Other rocks mentioned in this chapter may also form types of metamorphic rock. Schist forms from *mica*. Quartzite forms from *sandstone*. Some kinds of gneiss form from *granite*.

13

Metamorphic Rock

Scientists believe that some igneous and sedimentary rocks change while deep below the earth's surface. The changed rocks are called metamorphic rocks. **Metamorphic rock** formed by heat and pressure deep below the earth's crust. The amount of heat or pressure affected the way the rocks changed.

Some metamorphic rocks are *foliated*. These rocks can split into flat sheets that easily break apart. Tremendous pressure squeezed the rocks to form the sheets. One type of foliated metamorphic rock is slate. *Slate* formed from a sedimentary rock called shale. When shale was exposed to additional heat and pressure, it changed to slate. This hard rock easily breaks apart into thin sheets. In the past, sheets of slate were commonly used for chalkboards.

Non-foliated metamorphic rocks do not break into thin sheets. Most of the time, these rocks have tightly packed crystals. An example of this type of rock is *marble*. Heat and pressure formed marble from the sedimentary rock limestone. Marble often contains tiny crystals of various minerals. Sometimes these minerals form thin veins or swirled patterns in the rock. Marble is often used as a building material and for sculptures.

Much can be learned from studying rocks and minerals. But in order to properly understand what you see, you have to start with the Bible. The global Flood described in Genesis reshaped the surface of the earth and left nearly all the rocks we see today. The Bible helps us understand why things look and work the way they do.

Marble

Shale

Slate

QUICK CHECK

1. What is a rock?
2. List the three main groups that geologists use to classify rocks.
3. What term is used to describe metamorphic rocks that can break apart into flat sheets?

27

How is metamorphic rock formed? Heat and pressure deep beneath the earth's surface cause changes in the rock.

How is metamorphic rock different from igneous and sedimentary rock? Metamorphic rock forms from igneous, sedimentary, or other metamorphic rock. Igneous and sedimentary rocks do not form directly from other rock. They form from magma or sediments.

What is the difference between foliated and non-foliated rocks? Foliated rocks can be split into flat sheets that break apart easily, but non-foliated rocks are made of tightly packed crystals and cannot break into sheets.

What type of metamorphic rock was once used for chalkboards? slate

What type of sedimentary rock is slate formed from? shale

What type of sedimentary rock is marble formed from? limestone

What is an unusual characteristic of marble? It sometimes has veins or swirled patterns.

Why do you think that marble is a better choice for building materials and sculptures than limestone is? Possible answer: Limestone weathers easily. Marble is stronger and harder.

ANSWERS

1. a hard, natural substance made of one or more minerals and possibly organic materials
2. igneous, sedimentary, and metamorphic
3. foliated

Activity Manual

Review, page 19

This page reviews Lesson 13.

Assessment

Quiz 1-C

The quiz may be given any time after completing this lesson.

Background

Slate

The term *slate* originally meant "chalkboard." Today the term is used for a variety of flat surfaces associated with writing. The term is also used as a verb associated with scheduling. At one time, scheduling may have been done on a chalkboard because of the ease of erasing and making changes during planning. The term has remained associated with the task even though the type of writing surface has changed.

Activity

Making a Rock Pal

Materials: rocks, glue, assorted craft materials such as paint, markers, felt, yarn, wiggly eyes, pom poms, beads, glitter, chenille wire

Choose a rock and try to identify it. Give the rock a name and personality based on the type of rock that it is. Decorate the rock with a face and other features to indicate its personality. For example, a quartz rock could be named "Quarky" and decorated to resemble a clown.

14

Student Text pages 28–29
Activity Manual pages 20–21

Objectives

- Label rocks in a collection
- Classify rocks according to chosen criteria

Materials

- assortment of colored pencils, colored pens, and colored markers

Correction fluid may pose an inhalation hazard. You may need to use it in a ventilated area. You may want to try a paint marker or other quick-drying paint.

Teacher's Toolkit CD

- Rubrics: *Rock Hounding,* for each student

Introduction

- Combine the pencils, pens, and markers into one pile. Display the pile.
  How could we sort these items into groups? Possible answers: by type of writing instrument, by color
- Choose one method to group the items. Set all the groups aside except for one.
  How could we sort this group into more groups? Possible answers: if the group is all one type of writing instrument, divide it by color; if the group is a color, divide it by types of writing instruments

  In this activity we will be grouping rocks in a similar way.

Preparation for Reading

What do you know about a rock hound? Answers will vary.

- Instruct the students to read Student Text pages 28–29 and Activity Manual pages 20–21.

Teach for Understanding

The emphasis of this activity is to develop the student's ability to set criteria and classify according to those criteria. The actual identification of the rocks is not necessary.

Procedure: Part 1

Choose a rock and examine it. How can you describe your rock? Possible answers: color, surface texture, more than one type of mineral mixed together, shiny, full of holes, crumbly, sharp, smooth

Procedure: Part 2

- Use questions to guide the students as they choose one of their two groups of rocks to sort again.
  Which characteristic do you notice first?

ACTIVITY

Rock Hounding

Process Skills
- Observing
- Classifying
- Communicating
- Defining operationally

No matter where you live, you will find rocks. Rocks are in yards. Rocks are in flower beds and gardens. Rocks are at the edges of roads. Rocks are in parks and at the bottoms of streams. A person who enjoys finding and collecting rocks is called a *rock hound.* A rock hound usually labels his rocks and records information about where each was found. The types of rocks are also identified when possible.

In this activity you are the rock hound who finds, labels, and groups his rocks.

Purpose

Classify rocks.

Materials
10 different rocks
correction fluid
fine-tipped marker
magnifying glass
rock and mineral field guide (optional)
Activity Manual

Procedure: Part 1

1. Use correction fluid to make a white dot about the size of a pencil eraser on each rock. Let the fluid dry completely. Number the rocks by writing on the dots with a fine-tipped marker.
2. Record in your Activity Manual the place where each rock was found.
3. Record the appearance of each rock. Describe its color(s) and features such as layers or stripes of color.
4. Record the texture of each rock.

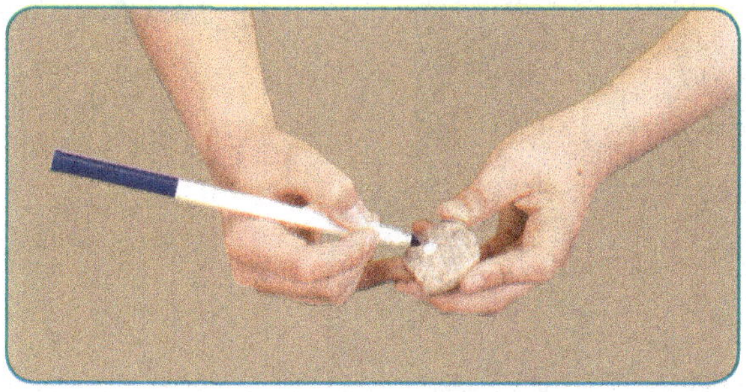

28

Background

Correction Fluid

You may not want to use correction fluid on decorative rocks or rocks from personal collections. You may use dot stickers or pieces of masking tape instead of the correction fluid.

Rock Review

You may want to copy Instructional Aid 1.1: *Which Rock?* for each student. This page reviews the kinds of rocks in a "who am I" type format.

5. Record any other features of the rock such as how the rock breaks.

Procedure: Part 2

6. Sort the rocks into two groups. Each group should share a similar characteristic. Write the characteristic of each group on the chart. List the numbers of the rocks that you placed in each group.
7. Choose the rocks in one of the two groups. Follow the same procedure as in Step 6.

Conclusions

- Did some rocks fit into more than one category? How did you choose where to place them?
- What other characteristics may be used to sort the ten rocks?
- How can you further divide one of the groups of rocks?

Follow-up

- Try to identify the minerals in each rock. Use pages 8–13 of your Student Text or the field guide.

29

14

Look at all of your rocks. Do any of them appear similar? You could group those rocks together.

Look at the descriptions that you recorded for the rocks that you have. Are there any descriptions that are repeated? You could group those rocks together.

- Each group of students does not need to use the same criteria for grouping their rocks. You may choose to incorporate some of the characteristics used to identify minerals.

Process Skills

The process skill emphasized in this activity is classifying, which is the grouping or ordering of objects based on similarities or differences.

- You may choose to write some of the students' descriptions for display as a reference for the students as they work.

Any classification system is based on a set of chosen criteria or standards.

What criteria did you choose to begin your classification system?

Does your chart look the same as the chart from another group? Why or why not?

How would your chart be different if your criteria were based on where the rocks were found?

What do you think a scientist does if he discovers a rock or mineral that fits into more than one category? Elicit that he may consult other scientists, and together they will make the best choice they can.

Sometimes the names or organization of whole categories changes because of new discoveries.

Conclusions

- Discuss the criteria used by each science group. Compare the rocks of different science groups as time permits.

Background

Science Teams

Students should learn to work together cooperatively. They should agree on the choices that are made. Some students may disagree with the choices of others. If this happens, you may need to work out a system for them to vote or draw straws.

Activity Manual

Activity, pages 20–21

Assessment

Rubrics

Select the prepared rubric, or design a rubric to include your chosen criteria.

15

Student Text page 30
Activity Manual page 22

Objectives
- Recall concepts and terms from Chapter 1
- Apply knowledge to everyday situations

Review
- Materials for the Chapter 1 Test will be taken from Student Text page 30 and Activity Manual pages 4, 11–12, 19, and 22. You may review any or all of the material during the lesson. Questions similar to the ones in Thinking It Through, Activity Manual page 22, may appear on the test.

> Information on the Chapter Summary pages reflects the vocabulary and concepts the student should know for the test.

- You may choose to review Chapter 1 by playing the review game or a game from the Game Bank on the Teacher's Toolkit CD.

Activity Manual
Review, page 22
This page requires written responses to application questions.

16

Objective
- Demonstrate knowledge of concepts taught in Chapter 1

Assessment
Chapter 1 Test

CHAPTER SUMMARY

Words to Know

magma
lava
sediment
soil
organic
mineral
inorganic
crystal
gemstone

Key Ideas

Explain what a geologist studies
Identify and describe the layers of the earth
Identify the processes that produce and move sediment
Explain mankind's responsibility to make wise use of God's resources
Explain how the Fall of mankind and the Flood affected the Earth
Explain how crystals form
Identify and describe characteristics of minerals
Explain the purpose of the Mohs scale
Identify ways minerals are used and where they can be found
Describe the relationships between minerals and metals
Compare precious stones and semiprecious stones
Identify some uses of gemstones
Explain how rocks are classified
Describe igneous, sedimentary, and metamorphic rocks and give examples of each

30

Review Game
Precious Gems
Materials: tokens or plastic jewels

Divide the class into two teams. The goal is to "find" the most jewels. Each time a team answers a question correctly, they have "found" a jewel and receive a token or plastic jewel. The game continues until all of the jewels have been "found." The team with more jewels wins.

Variation: Use gold foil-wrapped candy pieces and allow the students to "mine for gold."

Fossils and Dinosaurs 2

While visiting the fossil exhibit at the museum, you hear the tour guide discuss a large fossil of a tree.

"Based on the rock layers, scientists date this particular tree fossil at about 75 million years old," says the guide. "The most unusual feature about this fossil is the way it extends vertically through the limestone and up into this shale deposit. It usually takes thousands of years for water to deposit each of these layers, but because this fossil is standing, it is clear that it must have formed very rapidly."

Then the guide says, "Also notice the top part. Shale deposits are usually formed very slowly. However, here the shale and the limestone must have been put down quickly and at the same time. Perhaps this took place in the flooding of a local river."

Consider

- What viewpoint does the speaker have?
- What is so unusual about this fossil?
- What would have happened to the tree if the shale and limestone had deposited slowly?

31

Background

Worldview

Students are frequently confronted with evolutionary concepts about dinosaurs and fossils. Dinosaurs and fossils are in movies, on the Internet, and especially in educational programs or materials. It is critical that students be trained to recognize an unbiblical worldview, evaluate it, and respond intelligently.

Chapter Preview

Preview and prereading activities may include a K-W-L chart, a probe, or an anticipation guide.

Chapter Photo

The photo on page 31 shows a fossil found in Madagascar. The miniature scientist is holding a small fossilized bone.

Objective

- Evaluate evolutionary assumptions from a Christian worldview

Materials

- bottle with the layers of soil or the cutting of the layers of soil from Lesson 4

Introduction

- Display the layers of soil. Review briefly the concepts from Lesson 4, especially looking through Bible glasses.

Preparation for Reading

- Direct attention to the picture on page 42.
  This is a picture of a tree fossil similar to the one you will read about in the story.
- Direct the students to read page 31 to find out what the tour guide says about the tree fossil.

Teach for Understanding

Is the tour guide looking through Bible glasses? no
Why not? Answers should include that the tour guide does not mention Creation or God. He dates the fossil and the layers of the earth at millions of years. The Bible and the bottle with the soil layers shows how the earth could have formed without taking millions of years.

Could the tree have stood thousands of years while the layers of earth formed around it? No, the tree could not have remained exposed for all that time. It would have rotted or been eroded away.

What does the tour guide conclude about the time it took to bury this fossil? He concludes that the layers formed very rapidly.

What does the tour guide suggest might be the explanation for rapid forming of the layers? a local river flooding

Can the fossil be 75 million years old and have been covered over by a local flood? No, 75 million years would mean there was a slow process of the earth layers forming. A local flood would mean rapid forming of layers and would not take millions of years.

How long did it take to form the layers of soil in the bottle? about 15 minutes

What does a Christian always believe to be true when he is figuring something out? He believes that the Bible is God's Word and that God always tells the truth.

What event does the Bible describe that would rapidly bury the tree in rock layers? the Flood (Genesis 6–9)

Activity Manual

Preview, page 23

The Looking Ahead page is intended to assess the student's prior knowledge before beginning the chapter.

18

Student Text pages 32–34

Objectives

- Define *fossil*
- Compare and describe some types of fossils that form in sediment: petrified fossil, mold, cast, carbon film, trace fossil
- Identify other materials in which fossils are sometimes preserved

Materials

- fossils or photos of fossils

Vocabulary

- fossil
- petrified fossil
- mold
- cast
- carbon film
- trace fossil

Introduction

- Display the fossils or photos.

What are these? fossils or photos of fossils

What do you think caused them to form? Accept reasonable answers.

Preparation for Reading

- Direct the students to mentally form questions from the headings, bold words, and captions on pages 32–34.
- Instruct the students to read the pages silently to find the answers to these questions.

What is a fossil?

What are some types of fossils?

Teach for Understanding

What is a fossil? any part or trace of a living organism that is naturally preserved after it dies

In which type of rock are most fossils found? sedimentary

How are fossils formed? when organisms, or living things, are buried quickly beneath sediment

What parts of an organism usually decompose first? the softer parts

What does ***decompose*** mean? rot away

How might being buried quickly prevent an organism from decomposing completely? Possible answer: Being buried quickly may cause less air and bacteria, which aid in decomposition, to be available.

What do the Bible and the Flood tell us about fossils? Floodwaters and erosion would have created large amounts of sediment in a short time. The sediment probably buried all the creatures that were not in the ark.

Suppose that you are walking along the foot of a cliff one day, and you find some large animal bones lying on the ground. You pick up a bone and study it carefully. It is brown and hard—as hard as rock.

Several questions might come to your mind as you think about these bones. What kind of animal did they belong to? How did they become so hard and stony? What happened to the animal? How long have the bones been there? What can they tell you about life on the earth? These are some of the same questions that scientists seek to answer when they study fossils.

Fossil Formation

A **fossil** is any part or trace of a living organism that is naturally preserved after it dies. Most fossils are found in sedimentary rock. These types of fossils can only form when organisms, or living things, are buried quickly beneath sediment. The softest parts of an organism usually *decompose*, or rot away, first. The harder parts are sometimes preserved as fossils. If not buried quickly, the organism would decompose completely.

The Bible tells us that in the days of Noah God judged the earth with a great Flood. The waters of the Flood caused great erosion. Huge amounts of sediment formed in a short time. It buried all the land and flying creatures that were not on the ark as well as many sea creatures. Trees and other plants were also buried. The fossils that we see today are clear evidence of this worldwide Flood.

Petrified fossil bones

A preserved fossil of a shark's tooth

32

Background

Fossils

It is likely that more fossils are found in sedimentary rock because it weathers more easily than harder types of rocks do. Sedimentary rock is also easier to excavate. It forms on the surface where organisms live, but igneous and metamorphic rocks are more likely to form beneath the surface. Larger fossils are often found below smaller fossils. This contradicts the evolutionary thought that smaller organisms (simpler life forms) lived earlier and so should be found at lower levels. Larger organisms were heavier and probably sank first during the Flood.

Carbon Film

All living things contain carbon. Carbon is discussed in Chapter 8, *Changes in an Ecosystem*.

Activity

Researching Fossils

Fossils are found all over the world. Research and mark on a world map the places where large groups of fossils have been found. Places may be identified as fossil deposits, petrified forests, or quarries.

Fossils Preserved in Sediment

Many fossils are preserved parts of an animal or a plant. These fossils are usually the hardest parts of an animal, such as its bones, teeth, claws, shells, or tusks. Sometimes, as an animal's bone or the wood of a tree decayed, it was slowly replaced by minerals. The minerals hardened over time and formed rock. Such fossils are called **petrified fossils**.

Other fossils also show how an organism used to look, even though the organism has decomposed. This happened when an organism decomposed after the fine sediment covering it hardened to rock. An empty space was left in the rock. This imprint of the organism in the rock is called a **mold**. If fine sediment filled in the empty space of the mold, a **cast** formed. A cast is a copy of the shape of the organism.

Sometimes a plant or animal decayed under the weight of sediment. Parts of the plant or animal turned to carbon forming a dark

Carbon film fossil

image of itself in the rock. This kind of fossil is called a **carbon film**.

Fossils are not always part of a plant or animal itself. **Trace fossils** formed from something that an organism left behind, such as a footprint, a hole where an animal lived, or an animal's droppings. Trace fossils sometimes give scientists clues about how an organism lived. For example, fossilized footprints might show whether an animal walked upright on two feet or walked on all four feet. Fossils of droppings might reveal to scientists what an animal ate.

Cast fossil

Trace fossil

- Refer to the fossils or photos used in the introduction as the types of fossils are discussed.

What kinds of animal parts are usually seen preserved as fossils? the harder parts: bones, teeth, claws, shells, tusks

How are petrified fossils different from other fossils? As an organism decays, it is slowly replaced by minerals that harden into rock.

How does a mold form? An organism is pressed into rock and leaves an imprint.

What is a cast? Sediment fills in a mold, creating a copy of the organism's shape as the sediment hardens.

What is unusual about a carbon film? It is only a carbon image of a plant that has decayed.

What is a trace fossil? a fossil of something, such as a footprint or dropping, that is left behind by the original organism and gives clues about the organism

What kinds of things can scientists learn from trace fossils? Possible answers: how an organism lived; how it walked; what it ate

Why is it important to study fossils? Possible answers: to learn about animals and plants that may be extinct today; to learn about God's creation and His work in Christians' lives; to understand the biblical account of God's judgment on sin during the Flood

Activities

Cast Tiles

Materials: air-dry clay, rolling pin, knife, paper clips, sturdy pieces of plants (such as seed heads, stiff leaves, or tips of pine branches), acrylic paint, water, paintbrush

Knead the clay to soften it. Use the rolling pin to roll the clay until it is 0.5 cm thick. Cut the clay into squares or rectangles. Bend the paper clips into *S* shapes and insert one into the back of each tile to form a loop for hanging. Press a plant piece into the front of each tile. Thin the paint with water and brush a light coat on the front of each tile. Carefully remove the plant piece. Set the tiles aside to harden.

Make a Carbon Film "Fossil"

Materials: fresh leaves, white paper, heavy books

Place the leaves between sheets of paper. Then place several heavy books on top. Wait one to four weeks for the leaves to make marks on the paper. Check periodically during the waiting period. Carbon and other substances in the leaves produce the marks. Encourage the students to record their observations in their journals.

18

What materials, other than sediment, have preserved plants and animals as fossils? amber, ice, permafrost, and tar

What kind of organisms are usually found fossilized in amber? insects and small animals

Why do you think insects are often found in amber? Elicit that, since insects are often on plants, they could easily get trapped in the sap.

What is permafrost? permanently frozen soil

What is one kind of animal that has been found fossilized in permafrost? the woolly mammoth

What are some extinct animals that became fossilized in the Rancho La Brea Tar Pits? ground sloths, ancient bison, saber-toothed cats

How is the formation of fossils in tar pits similar to the formation of fossils in amber? The organisms become trapped in a sticky substance.

- Discuss *Fantastic Facts.*

What was the mammoth found? Siberia

How was Jarkov transported? by helicopter

Why do you think they put Jarkov in a cave rather than a building? Possible answers: The cave was cold. The cave was near the site where Jarkov was found.

What process did scientists use to prevent damage to the mammoth? carefully thawing the ice with a hair dryer

How would using hair dryers prevent damaging the mammoth? Possible answer: A hair dryer would melt ice, whereas tools that could be used to chip away the ice may damage the mammoth.

ANSWERS

1. Organisms were quickly buried in sediment.
2. a cast
3. amber, ice, permafrost, and tar

Fossils Preserved in Other Materials

Not all plant and animal fossils are preserved in sediment. Sometimes plants or animals have been found preserved in amber. *Amber* is sap from a plant buried during the Flood that has hardened into a yellow, gemlike substance. Insects and other small animals are often fossilized in this way.

Scorpion preserved in amber

Fossils have also been found frozen in *ice* and permafrost. *Permafrost* is permanently frozen soil. In the far North, Arctic animals have been found almost fully preserved. The woolly mammoth was a large, hairy animal that looked like an elephant. Woolly mammoths died many years ago. Within the last fifty years, however, scientists have dug mammoth fossils, complete with hair, out of the frozen soil.

Tar can also preserve fossils. In some places thick crude oil oozes out of the earth and forms tar pits. Living things get trapped in the sticky tar and die. The tar then preserves them. Fossils of plants, insects, birds, and mammals have all been recovered from the Rancho La Brea Tar Pits in California. These tar pits even contained many fossils of animals such as the ground sloth, the ancient bison, and the saber-toothed cat.

FANTASTIC FACTS

In 1997 a reindeer herder in Siberia noticed a tusk sticking out of the snowy tundra. Scientists explored the area and guessed that a mammoth's body was buried in the ice. They named the mammoth Jarkov after the man who found it.

In 1999 a helicopter airlifted the huge block of ice containing Jarkov to a cave. Scientists thawed the ice slowly with hair dryers to avoid damaging the mammoth. Although the mammoth's body was not as complete as they had hoped, it provided new information about the animal.

QUICK CHECK

1. How did fossils found in sedimentary rock form?
2. What kind of fossil formed when sediment filled in the space in a rock after the organism decomposed?
3. What materials, other than sediment, have preserved plants and animals as fossils?

34

Background

Preserved Soft Tissue

Soft tissue, such as skin and organs, that is preserved is considered mummified. Mummified organisms are a type of fossil. Many of the frozen fossils and the fossils from the tar pits are mummified. Soft tissue will be discussed in Lesson 29.

Extinction

Dinosaur extinction is discussed in Lesson 21. Extinction is further discussed in Chapter 8, *Changes in an Ecosystem.*

Mammoths

Woolly mammoths are only one type of mammoth. Woolly mammoths were about the same size as the Indian elephants that we have today. Other mammoths include Imperial mammoths and Colombian mammoths. Both were larger than the woolly mammoth.

Science Misconceptions

Fossils and the Flood

Not all fossils formed as a direct result of the Flood. Fossils in amber, ice, permafrost, and tar probably formed sometime after the Flood. Some fossils may also have formed under volcanic ash after an eruption.

Two Ways of Looking at Fossils

Science is based on man's observations. But no one observed how fossils formed. People can look at the same fossils and have very different theories about how the fossils came to be. These theories are based on what the people believe about how the earth began.

Evolution

Most evolutionists believe that the earth came into being by chance. They also claim that life on the earth **evolved**, or developed gradually, over millions of years. They say that rock layers represent *geologic ages*, or long time periods. Books about fossils often use names such as Jurassic, Cretaceous, and Triassic for these ages. According to the theory of evolution, the layer of rock in which a fossil is found reveals how old the fossil is.

Evolutionists say that early forms of life slowly changed into the life forms we see today. They try to support their theory by relating modern plants and animals to their "ancestors" in the fossil record.

Creation

Creationists believe that God's Word in Genesis 1 explains the true origin of the earth. Genesis says that God created the earth and all forms of life in six days by His spoken word. Adam was created during this time. And if we trace human life back to Adam, the earth could be only thousands of years old, not millions.

Creationists also believe what the Bible says in Genesis about the Flood. They believe that this global Flood accounts for most of the fossils we find today. In many places, sediments deposited by the flood waters hardened to become new rock layers. Fossils of many organisms that perished in the Flood were preserved in those rock layers. The layers are evidence that a catastrophe occurred during a short period of time as the Bible says, rather than indicating long periods of time. As scientists continue to study the fossil record, it provides a wealth of evidence that God's Word is true.

QUICK CHECK

1. What determines how a person interprets the fossil record?
2. What event accounts for the formation of most fossils, according to the Bible?

35

19

Objective

- Compare beliefs of evolutionists and Creationists

Materials

- brownies or other baked goods

Teacher's Toolkit CD

- Quiz 2-A

Vocabulary

- evolved

This is a one-page text lesson. You may want to begin the activity in Lesson 20 as part of this lesson.

Introduction

- Share the brownies with the students.

  Do you think that these brownies were made in a factory or at home?

  What color was the mixing bowl that was used?

  How long did they bake?

- Elicit the idea that, although a student may have a guess at how the brownies were made, he does not know all of the details because he was not present when the brownies were baked. Compare the brownies to the beliefs of Creationists and evolutionists. People were not present when the world began. Only God was present. God gave us an accurate account in Genesis of what happened when He created the world.

Preparation for Reading

- Instruct the students to read the page silently to find the answer to the question.

  What are two ways to look at fossils?

Teach for Understanding

What is a theory? an idea or belief about something

What determines how a person looks at the fossil record? his beliefs

What does *evolved* mean? developed gradually

What do evolutionists believe about rock layers and fossils? Rock layers represent geologic ages. Early life forms slowly changed into present-day forms, and the layer of rock in which a fossil is found reveals how old the fossil is.

What do Creationists believe about the origin of the earth? The earth was created by God in six days as recorded in the Bible.

What do Creationists believe about the age of the earth? The earth dates back only to the time of Adam and is no more than thousands of years old.

How do Creationists explain the fossil record? Most fossils are the result of the Flood.

- For many years Creation and evolution have been opposing views of how life began. Many people who do not believe the Bible choose to accept evolution as fact and dismiss Creation as "a religious idea." Science is based on observation, but since neither evolution nor Creation can be proven through observations, both must be accepted by faith. Although the Bible is not a science book, it speaks accurately in all areas of life, including science. Christians trust all of God's Word because God cannot lie (Titus 1:2).

ANSWERS

1. his beliefs
2. the Flood

Activity Manual

Review, pages 24–25

These pages review Lessons 18–19.

Assessment

Quiz 2-A

The quiz may be given any time after completion of this lesson.

20

Student Text pages 36–37
Activity Manual page 26

Objectives

- Identify phrases or statements that indicate a Creationist or evolutionist viewpoint
- Make inferences as to the viewpoint from which literature is written

Materials

- assortment of grade-level books, magazines, articles, or other printed materials that are written from the viewpoints of Creationists and evolutionists (see Background)

This activity is to help the students recognize evolutionary elements in print. You may extend the activity to include information from other media such as science documentary films, computer games, or the Internet.

Introduction

- Discuss characteristics of the Creationist and evolutionist viewpoints in the chart on the page.

  What is a geologic age? a long period of time

  What makes man different from an animal? He is made in God's image.

- Be sure that the students understand that the chart is an example. There are other phrases that might indicate a specific viewpoint. Sometimes none of the phrases on the chart are used.

Preparation for Reading

- Direct the students to read pages 36–37 in the Student Textbook and page 26 in the Activity Manual before beginning the activity.

Teach for Understanding

Parents or teachers will not always be with you as you choose and read books or other materials. You should be able to recognize whether or not a book agrees with your beliefs.

What does a person's viewpoint about fossils usually tell you? whether he believes in Creation or evolution

Why is it important for a Christian to be discerning when reading information or listening to someone's viewpoint? Some people's ideas contradict, or go against, what God teaches in His Word.

How can a Christian young person learn to be discerning? by reading God's Word, listening as it is being taught, and asking parents and others who are knowledgeable about the Bible. Learning to listen carefully and think critically are also important. Much wisdom can also be gained by watching mature believers work through difficult issues.

ACTIVITY

Fact or Theory?

Man makes judgments about the evidence of fossils based on his beliefs. A person who believes God's record of Creation and history will look at fossils in one way. A person who believes in evolution will view fossils in a different way.

When people write about fossils, especially fossils of dinosaurs, they will usually say things that show what they believe. Understanding what view a writer has is important. The Bible says that Christians should be discerning. That means that Christians should understand what is right and wrong.

In this activity you will be given several books or articles to read. You must evaluate whether the writer is writing from an evolutionist viewpoint or a Creationist viewpoint. Use the following chart to help you.

Process Skills
- Inferring
- Collecting and interpreting data
- Communicating
- Defining operationally

Materials
5 books or articles about fossils or dinosaurs
Activity Manual

Creationist Viewpoint	Evolutionist Viewpoint
God created the heavens and the earth.	Earth and space are a result of a sudden explosion.
The earth is a few thousand years old.	The earth is billions of years old.
Fossils are probably a result of the great Flood recorded in the Bible.	Fossils show the great geologic ages of the earth.
God created all the kinds of animals in the beginning.	Different kinds of life have gradually evolved over long periods of time.
Man is God's special creation. He is different from the animals because he is created in God's image.	Man is the highest level of animal.

36

Background

Activity Helps

Try to provide a variety of materials for the students to use. The BJU Press website has additional information on the Resources page.

bjupress.com/resources/science

If the materials are lengthy, you may choose to have the students skim them rather than read the entire book or article.

Bible

The Bible instructs Christians to study God's Word, to recognize truth, and to use discernment (Psalm 119:97–105).

Science Misconceptions

Evolutionist Viewpoint

Not all the material in a book that is written from an evolutionist viewpoint is false.

Purpose

Evaluate viewpoints in books or articles.

Procedure

1. Choose one of the books or articles. Record the title in the chart in your Activity Manual.
2. Skim through the book or article. Look for examples of the author's viewpoint. Record one of these examples in the chart.
3. Repeat steps 1–2 for four more books or articles.

Conclusions

- Which viewpoint did you find more often?
- From which viewpoint is this science textbook written?

Follow-up

- Whenever you read a book or article, keep a record of examples of the writer's viewpoint.

37

20

Process Skill

The process skill emphasized in this activity is inferring, which is drawing conclusions based on incomplete data. Incomplete data would include not having results or not having seen the origin or process that led to the results. Conclusions can also be based on prior knowledge and observations.

What determines how a person views fossil evidence? his beliefs

What a person believes provides the framework for his conclusions. An evolutionist and a Creationist can look at the same fossil and come to very different conclusions based on their beliefs and prior knowledge.

As we read various materials and listen to other people, we can form conclusions about what they believe, based on their viewpoints.

Procedure

- Guide the students as they examine and evaluate the materials.

Conclusions

- Discuss the reasoning behind each student's conclusions.
- Direct the student to also examine his *Science 5* Student Text.

Activity Manual

Activity, page 26

Guide the students as they use this page to evaluate viewpoints in books or articles.

Assessment

An assessment, such as a rubric, is not recommended for this activity.

21
22

Student Text pages 38–39
Activity Manual pages 27–28

Objectives

- Make models of fossils
- Relate models to fossils

Materials

Materials needed for this activity are listed on the Student Text page.

Teacher's Toolkit CD

- Rubrics: *Molds and Casts* for each student

Two lesson days are allotted for this activity. Make the molds and casts on the first day. Use the second day to examine the "fossils" and answer the conclusion questions.

Introduction

Think about stories that you have read about the times of kings and knights. How would a king mark an important decree or document to show that it was official? Elicit that a king would make an imprint of his ring in wax.

How would a person know who made the mark? The mark in the wax showed the details of the original ring.

Scientists study fossils in much the same way. They know that fossils can show details about the original organisms.

Preparation for Reading

- Direct the students to read pages 38–39 in the Student Text and pages 27–28 in the Activity Manual before beginning the activity.

Teach for Understanding

Procedure: Part 1

- If a student uses materials that vary from the list printed in the Activity Manual, he should put a line through each material that was not used, and then write the materials that were used. This reinforces accurate recording.
- Guide the students as they complete their materials lists with the items they are actively using.
- Review the steps aloud as the students work.
- Announce the waiting times for the students.
  Clean your object after removing it from the plaster.
  Wash all the plaster and dust from your hands.
- Identify the place where the students should keep the fossil pieces (both cup parts with plaster and the original object) until the next lesson.

ACTIVITY

Molds and Casts

Process Skills
- Making and using models
- Observing
- Inferring

There are many different types of fossils. One type of fossil looks like an imprint pressed into rock. Another type looks like actual bones or plant parts that have hardened and turned into stone. Details found in and on these and other fossil types can help scientists determine what life may have been like in the past.

In this activity you will make two types of "fossils." Then you will examine them for information.

Purpose

Model fossils.

Materials
- 2 foam cups, 16 oz
- centimeter ruler
- scissors
- foam plate
- modeling clay
- assorted hard objects such as seashells, split wood, nuts, pinecones, or plastic dinosaurs or bugs
- 200 mL plaster of Paris
- 100 mL water
- metric measuring cups
- spoon
- magnifying glass
- Activity Manual

Procedure: Part 1

1. Draw a line around one cup about 5 cm from the top. Cut along the line. Set the bottom part of the cup aside for use in step 6.
2. Place the ring part of the cup on the foam plate. Press a 2 cm layer of clay into the ring. Press the clay tightly against the sides of the ring. Smooth the surface as much as possible.
3. Choose an object. Press the object into the clay and carefully remove it. Wipe any clay from the object.
4. Measure 200 mL of plaster powder and pour it into the other cup.
5. Measure 100 mL of water. Stir the water into the plaster powder. Mix until there are no lumps. It should be thick like pudding. Pour about half the plaster into a 2 cm layer over the clay.
6. Pour the remaining plaster into the bottom part of the first cup. Smooth the surface with the spoon.

38

Background

Activity Helps

Contact with plaster and its dust may cause skin or eye irritation. Make sure hands are washed after working with the plaster. Read the plaster packaging for additional cautions. You may wish to pre-measure the dry plaster into cups to reduce contact with the plaster dust.

The clay must be soft and pliable, but it must be able to make a tight seal in the ring part of the cup. Clays like Play-Doh stretch and produce inaccurate fossils.

Choose objects that will not leave residue in the clay. Brush any loose material from the sticks, rocks, or any other natural objects used.

Adjust the measurements if necessary. The plaster should be the consistency of pudding. If the plaster is too thin, add more dry plaster or extend the waiting times. Use a balance or kitchen scale for any dry measurements.

For step 5, you may choose to have the student make a mark 2 cm above the clay on the inside of the ring part of the cup as a guide for pouring the plaster.

In step 7, make sure the object is not pushed into the plaster so that the plaster closes over it.

7. Wait five minutes and press the object into the wet plaster in the cup bottom. Wait another five minutes and remove the object. The plaster should keep the shape of the object. If the plaster is too soft, press the object back into the plaster and wait one more minute. Make sure to remove the object before the plaster becomes too hard.
8. Let the plaster set overnight.

Procedure: Part 2

9. Peel away the cups and clay from the plaster pieces. Wipe away any clay. The plaster pieces are your fossils.
10. Examine your fossils using the magnifying glass. Identify each as a cast or a mold. Record characteristics and details about each in your Activity Manual.

Conclusions

- How are the fossils alike? How are they different?
- Which fossil provides more information about the original object?

Follow-up

- Try making mold or cast fossils with other materials such as mud or wet sand. Compare the features of these fossils with those made with plaster.
- Make trace fossils of animal footprints.

39

Activity

Making Fossil Paperweights

Materials: plaster fossils from the activity, felt, scissors, glue, paint, paintbrush

Make sure that the plaster fossils are completely dry. For each fossil, cut a circle of felt and glue it to the bottom of the fossil. Decorate the fossil with paint.

21–22

Procedure: Part 2

Work carefully as you peel away the cups and separate the clay from the plaster.

The two plaster pieces are your "fossils."

- Direct attention to Activity Manual page 27.

Which "fossil" is a cast? the one formed when the plaster was poured into the clay

Which "fossil" is a mold? the one formed when the object was placed into the plaster.

As you record characteristics, be sure to include details about each fossil, such as the number and location of grooves or other markings, texture, hair, and number of legs.

Process Skill

The process skill emphasized in this activity is making and using models. Models help us examine and experience things that may otherwise be too small, too large, or too inconvenient to handle.

Why would we use a model to show how some fossils form? Possible answer: The formation of actual fossils occurred long ago.

How were models used in this activity? Possible answer: We made two different types of fossils from the same object. These model fossils helped us distinguish between casts and molds.

Are your model fossils actually fossils? no Why? Possible answers: A fossil is any part or trace of an organism that is naturally preserved after it dies. The model fossils may not have been made from a living organism and were not preserved naturally.

Conclusions

Compare your model fossils with the original object. The 3D cast should look like the original object. The mold is an impression that looks like the reverse of the original object.

Which "fossil" provides you with more information about the original object? the cast

- Instruct the students to complete the Conclusions section in their Activity Manuals.

Activity Manual

Activity, pages 27–28

Assessment

Rubrics

Select the prepared rubric, or design a rubric to include your chosen criteria.

23
24

Student Text pages 40–43
Activity Manual pages 29–30

Objectives

- Define *paleontology*
- Describe how fossils are excavated and reconstructed
- Explain why rock layers do not indicate the age of a buried fossil
- Describe how paleontologists use carbon dating to guess the age of fossils

Teacher's Toolkit CD

- Quiz 2-B

Vocabulary

- paleontology
- excavation

Two days have been allotted for this lesson. You may want to teach Lesson 23 the first day and use ideas from Lesson 24 on the second day. You may also incorporate the ideas from Lesson 24 into the text material, making the lesson two days.

Introduction

Think about your bedroom without any photos of you. Someone that does not know you comes and looks around your room.

What are some things that the person could know about you and how would he know those things? Possible answers: approximate size—from clothing; boy or girl—because of decorations and other items in the room; interests—from books, toys, or other memorabilia

What are some things that the person could not know about you? Possible answers: exact age; facial features; physical strength; skin tone; size of your family; how long you have lived in that house

Some scientists look at fossils in a similar way. They do not have photos or real dinosaurs to look at as they study the fossils. And of course they cannot ask people who were alive with the dinosaurs. Instead scientists must rely on the information that they get from the fossils and the surrounding rock.

Preparation for Reading

- Display the words *paleontology* and *preparators*. Pronounce the words together.
- Direct the students to read silently pages 40–43 to find the answers to these questions.
  What does a preparator do?
  What is an index fossil?

Teach for Understanding

What is paleontology? the scientific study of fossils

What does a paleontologist's job include? studying fossils; finding fossils; transporting fossils to be studied; cleaning or preparing fossils for display

Where do scientists look for fossils? Possible answers: cliffs; beaches; deserts; other areas with sedimentary rock; places where other fossils have been found before

Learning from Fossils

The scientific study of fossils is called **paleontology**. People who study fossils are called *paleontologists*. They do not just study fossils, though. There are many jobs that paleontologists may do. They may find fossils and take them to a place where they can be studied. Or they may clean the fossils and prepare them for display.

Excavating Fossils

Sedimentary rock usually erodes more easily than other types of rocks do. When sedimentary rock erodes, it may reveal buried objects such as fossils. Cliffs, beaches, and deserts are good places to look for fossils. Paleontologists search for fossils in these areas.

Excavation site in East Turkana, Ker

40

Background

Fossil Preparator

A fossil preparator may also help with the excavation and moving of fossils and the reconstruction and reassembly of bones into skeletons.

Fossil Reconstruction

Although some museum guides may state that a model looks exactly like a real dinosaur, scientists cannot know for sure exactly what dinosaurs looked like. The ideas of scientists are based on what living animals look like.

Excavate

The root word for "excavate" is *cave*. Other words with this root are *cavity* and *cavern*. They all come from the Latin word that means "hollow." Excavation is the process or act of digging or hollowing out a solid object. Removing a fossil from rock is one type of excavation.

It is common to find many fossils buried with each other in one place. This place is called a site. When paleontologists find a site, they first take photos and then make a detailed map of it. The location where a fossil was found will be helpful to know later when studying the fossil.

After its location is mapped, the fossil is dug out. Removing a fossil from the rock around it is called **excavation**. This process must be done carefully and may take a very long time. Small fossils can be excavated with simple tools such as dental picks, small chisels, and hammers. Larger fossils may also require cranes and bulldozers.

When a fossil is found, paleontologists usually remove some of the surrounding rock along with the fossil. This extra rock helps to protect the fossil. Scientists number each fossil and then wrap it in foam to keep it from breaking on its way to the laboratory. Fragile fossils are also painted with a protective coating.

Preparing Fossils

When fossils arrive at a laboratory, they need to be cleaned. All the dirt and rock around them must be removed. Paleontologists who clean fossils are called *preparators*. They must work very carefully so the fossils are not damaged. Most fossil preparation is done with small tools and brushes.

Maybe you know someone who had an x-ray taken. The x-ray gave doctors a picture of the inside of the person. Scientists can use the same technology on fossils. X-rays allow scientists to look inside the fossils and learn more about them.

Reconstructing Fossils

Sometimes bones from one animal's skeleton are all found in one place. When this happens, scientists try to put the bones back together to re-create the animal's skeleton. Sometimes all the bones are not available, and scientists must guess about the bones that are missing. Even an almost-complete skeleton, however, cannot show what the animal's muscles or outward appearance was like. So scientists guess. They base their guesses on how modern animals look.

If you visit a natural-history museum, you may see models of animals such as dinosaurs on display. These models were usually made by artists. The artists based the models on scientific data. But they also used their own ideas about what the animals might have looked like.

41

What term is used to identify a place where paleontologists dig for fossils? site

What is excavation? the process of removing fossils from the rock around them

What preparations should be done before excavation? photographing the site and creating a detailed map

Why is it important for paleontologists to work slowly while excavating fossils? The fossils are very valuable and fragile.

Why do scientists usually remove some of the surrounding rock with the fossil? to protect the fossil for its trip to a laboratory

What are other ways that fossils are protected for their trip to a laboratory? They are wrapped in foam or painted with a protective coating.

How are fossils usually marked for identification? They are numbered.

What is the job of a preparator in a lab? Possible answers: clean the fossils with small tools and brushes, x-ray the fossils

Do scientists always know what the original animal or plant looked like? no

- Discuss what some problems are that paleontologists have when trying to reconstruct fossils. Possible answers: missing bones of an animal's skeleton; having to guess about what the animal looked like; relying upon an artist's imagination for a model

What are some things that scientists cannot know for sure from the skeleton of an animal? what the animal's muscles or outward appearance was like

What do some scientists use as a basis for their guesses about what the outward appearance of the animals may have looked like? the appearance of modern animals

Activity

Grid Drawing

Materials: *Grid Drawing* (Instructional Aids: 2.1 on Teacher's Toolkit CD), for each student, pencil

Direct the students to reduce the size of the dinosaur by copying the lines from each square.

Do this activity to prepare the students to draw on a grid in Lessons 25–26.

23
24

What do evolutionists believe about the different layers of rock? The layers represent different time periods.

How do they use that belief to date fossils? They think that the fossil is as old as the layer of rock in which it is buried.

What does evolution teach about the beginning of life on the earth? Life began with simple, one-celled organisms that gradually became more complex until they developed into entirely new life forms.

- Discuss the problems evolutionists have with their way of dating fossils. Fossils are not always found in order from the simplest life forms to the most complex. Some fossils of complex life forms are found in the same layer as fossils of simple ones. Some fossils are in more than one layer.

  What explanation does the Bible give for the problems that evolutionists have with their dating method? The Flood killed and buried all land organisms. The layer of rock where the animal is found probably indicates how fast it was able to move in order to survive longer during the Flood.

Dating Fossils

Evolutionists think fossils formed at different times over millions of years. So they work hard to find out the age of a fossil. They think that the different layers of rock contain fossils from different times in the past. Therefore, a fossil is as old as the rock layer in which it is buried. Rock layers deeper in the earth are considered older than rock layers above them.

Evolution teaches that life on the earth began with simple, one-celled organisms. Over many millions of years, they gradually changed into different, more complex forms of life. Eyes, shells, legs, wings, and skeletons gradually appeared over time where there were none before. According to evolution, the simpler forms of life should be found in lower rock layers and the more advanced forms in the higher layers.

Some scientists believe they can date many kinds of rocks by measuring the substances in them and how they change slowly over time. Paleontologists use this idea to help date fossils. If they find a certain kind of fossil all over the world in rocks dated by geologists, paleontologists give that age to the fossil. Paleontologists call these fossils *index fossils*. They can use an index fossil to date other kinds of fossils found with it in different rocks. So if a scientist finds an index fossil in a rock that has an unknown age, he uses that fossil to date the rock!

But fossils do not always follow the pattern evolutionists expect. For example, complex and simple fossils are sometimes found in the same rock layer. Evolutionists must make exceptions and explanations when they find these kinds of fossils together. Some fossils are buried so they cross more than one sediment layer that geologists say took thousands of years to deposit. If this actually happened, the organism would have decayed long before it could have been completely buried. So evolutionists say that there were many local floods that formed these fossils.

The Bible gives a simpler explanation for these findings. Noah's Flood killed and buried all land organisms not on

These vertical polystrate tree fossils cross several layers of sedimentary rock. The word *polystrate* means "many layers."

Background

Simple Life Forms

Simple life forms are those that have few cells, systems, or functions. Evolutionists believe that complex organisms that have many cells, multiple systems, and greater thinking abilities developed from these less complex, or "simple," forms of life.

CREATION CORNER

In 1990 one of the most complete fossils of a *Tyrannosaurus rex* was found in the Black Hills of South Dakota. Nicknamed Sue after the woman who found it, this dinosaur is now on display at the Field Museum of Chicago. It is the largest of its kind ever found. When Sue was found, its head and hips were close together. The position of the bones suggested that Sue was buried by a quickly moving flood. During the Genesis Flood, only the people and animals in the ark were saved from God's destruction on man's wickedness.

The moving water during the Flood would have been quick and powerful. Even an animal as large as Sue would have been swept away and rapidly buried under the sediment carried by the moving water. When we look at science with a Creationist viewpoint, even dinosaur bones support God's Word.

the ark in sediment during the first forty days. So the age of most fossils is nearly the same. The layer of rock where fossils are found probably indicates how fast animals could move out of harm's way and the order in which they were buried. Small, slow-moving creatures were buried first. The fossils of these creatures are usually found in the deepest layers of rock. Faster-moving sea creatures and land animals were buried later as the water level rose. Their fossils are in higher layers of rock. Humans probably were buried last since they could determine ways to stay alive longer than animals. No human fossils have ever been found in rocks.

The date a scientist gives to a fossil depends on his worldview. Many dating methods assume that evolution is true and that life on earth began millions of years ago. Because these ages do not agree with the biblical account of God's Creation, we know evolution cannot be true.

QUICK CHECK

1. What is a person who studies fossils called?
2. How do paleontologists guess how to reconstruct the fossil of an animal?
3. How do evolutionists date fossils?

43

23–24

Which animals were buried first during the Flood? those that moved slowly

When were the humans buried? probably last, but no human fossils have been found

What do many dating methods assume? Evolution is true. Life on the earth is millions of years old.

When man's ideas conflict, or disagree, with God's Word, what should a Christian do? Possible answer: A Christian's only choices are to trust God or to trust in his own understanding. Christians should put their confidence in a loving Creator.

- Discuss *Creation Corner.*

What did the position of Sue's bones reveal? The position of the bones suggests that she was buried by a quickly moving flood.

What do Creationists believe happened to large dinosaurs like Sue? Creationists believe that large dinosaurs drowned in the Flood with other animals that were not in the ark.

ANSWERS

1. a paleontologist
2. They look at animals that are alive today.
3. The layers of rock represent the different time periods. The fossil is as old as the rock layer it is buried in.

24

- The *Grid Drawing* activity found at the bottom of page 47 may also be incorporated into the lesson. This activity will prepare the students for Lessons 25–26.
- Suggestions for additional activities for this two-day lesson are on the Teacher's Toolkit CD.

Activity Manual

Reinforcement, page 29

Review, page 30

This page reviews Lessons 23–24.

Assessment

Quiz 2-B

The quiz may be given any time after completion of this lesson.

Student Text pages 44–45
Activity Manual pages 31–32

Objectives
- Model the procedures a paleontologist uses while excavating
- Complete a site map

Materials
- square foil pan for each group (see shaded box below)
- soil for each group
- 3–4 different items—such as buttons, broken craft sticks or toothpicks, coins, erasers, or small plastic animals—for each group

Teacher's Toolkit CD
- Rubrics: *Fossil Dig*

The pans for the class should be placed together to form a larger "dig site." Assign each pan a number so that the students can place them back in order. Label the pans using a permanent marker or a type of tag. Types of tags include large dot stickers, masking tape, paper, or signs on toothpicks. Prepare each group fossil pan by filling it with soil and bury three or more objects in the soil. Make sure to bury a broken item in each pan. All the pieces of a broken item do not need to be buried in the same pan but need to be in the same area of the larger dig site. Smooth the soil so that the students will not know the locations of the objects.

Introduction

Why do you think that excavators follow specific procedures as they dig? Possible answer: to make sure that their findings and other information are accurate

In this exploration your team must work together carefully. Accurately record your findings on your *Site Map* and *Fossil Description* pages as you excavate.

Preparation for Reading
- Instruct the students to read silently Student Text pages 44–45 and Activity Manual pages 31–32 before beginning the exploration.

EXPLORATION

Fossil Dig

When paleontologists are digging for fossils, they divide the site into small sections. The sections are then labeled to match sections on a map or drawing of the dig site. As fossils are found, the locations are marked on the map. Paleontologists also record other helpful information about each fossil and its location. This additional information about the fossil may include the depth and position at which it was found, its condition, its color, and the type of soil or rock in which it was found. Accurate records and drawings will help the scientists as they later analyze their notes.

You have been given an area of a site to excavate for "fossils." You and your team of workers are responsible to divide and label your area. As you excavate the site, draw and record information about your findings on a site map.

What to Do
1. Decide the types of information you will record about each fossil.
2. Set the fossil pan carefully in your workspace so that the dirt and buried fossils do not move.
3. Use string to divide the fossil pan into four sections. Label each section to match the *Site Map* in your Activity Manual.
4. Use a spoon to lift some of the soil from one section. Excavate, or dig, slowly and carefully to avoid damaging any buried objects. When you find a fossil, lay the spoon aside and use a toothpick to uncover the fossil. Assign the fossil a number. Before removing the fossil from the soil, draw a picture of the fossil in the matching location on the *Site Map*.

Materials
fossil pan
2 pieces of string, 30 cm long
spoon
toothpick
watercolor paintbrush
Activity Manual

44

Background

Exploration Helps
The grid on the *Site Map* page is smaller than the pan. Doing the *Grid Drawing* activity in Lessons 23–24 may prepare the students for the drawing in this activity.

You may use a deeper container, such as a dishpan, to make two or more layers of strata for the fossils. Use damp sand for the bottom layer and soil for the upper layer.

You may choose to provide centimeter rulers and a balance (mass scale) for the students to measure their "fossils."

5. Carefully lift the fossil out of the pan. If it is broken, be sure that the picture on the map shows the broken part. Use the paintbrush to clean any remaining dirt from the fossil.
6. Examine your fossil. In the *Fossil Log* record any additional information about the fossil's location and a description of the fossil's characteristics.
7. Continue excavating each section of your fossil pan. Remember to clean, map, and describe each fossil.
8. Display your fossils and the information about them.

45

Teach for Understanding

- Why do paleontologists divide excavation sites into small sections? to help them keep accurate records about the fossils; to prevent fossils from being mislabeled
- What kinds of information might a paleontologist record? the fossil's location, including depth and position; the fossil's condition; the fossil's color; the type of soil or rock in which the fossil was found; drawings of the fossil
- Guide the students in deciding what information to record about each fossil. Remind the students to write their descriptions of the items without naming the items. For example, if an eraser is found, it cannot be identified as an eraser. Instead it should be described by its color, shape, and consistency.
- Set a time limit but encourage students to dig carefully.
- Guide the students in discussing possible relationships between the locations of the excavated items.

Activity Manual

Exploration, pages 31–32

Assessment

Rubrics

Select the prepared rubric, or design a rubric to include your chosen criteria.

Background

Local Dig Sites

Use the Internet to search for local fossil or archeological dig sites. Fossil quarries and digs are found in many states. Many of them welcome visitors and volunteers.

27

Student Text pages 46–49
Activity Manual page 33

Objectives

- Recognize that what is known about dinosaurs is based on the observations of fossils
- Name some of the types of information that are known from fossils
- Recognize some of the types of information that can be inferred from fossils

Teacher's Toolkit CD

- Visual 2.1: *Names*

Vocabulary

- dinosaur

Introduction

- Introduce the game "What Am I?" Explain that the students will draw a picture of a creature as you give clues.
- Avoid using words, such as *paws* or *mane,* that will give specific information. Read the clues slowly, giving the students only enough time to make a quick sketch. Use clues such as the following:
  long body
  four strong legs
  large feet
  sharp claws and teeth
  long tail
  furry body
  longer hair on head

  What animal was I describing? a lion

  Compare your drawings. Do all the drawings look like the same animal? no

  Did the clues give you enough information to draw the correct animal? Answers may vary.

  Paleontologists use the clues that they get from fossils to decide what the creatures may have looked like. How accurate do you think that their drawings and models are? Accept reasonable answers. Conclude that the fossils available may not provide complete information and that the scientists may have to make guesses as they work.
- Because of the context of the lesson, some students may think that the creature described in the Introduction will be a type of dinosaur and make their drawings accordingly. If so, discuss how their thoughts affected how they used the information from the clues and that the same thing may happen to scientists.

Preparation for Reading

- Display the *Names* visual and pronounce the names together. If some students appear to be unfamiliar with the names, you may want to leave the visual displayed as the students read.
- Instruct the students to read silently pages 46–49 to find the answers to these questions.

  Were dinosaurs lizards?

  What kinds of information can fossils tell us about dinosaurs?

Dinosaurs

Some people wonder whether dinosaurs really lived on the earth. Fossils indicate that they did. And God's Word describes large land and sea creatures that may have been what people commonly call dinosaurs.

Sir Richard Owen, a British scientist, studied the fossils of these large animals. In 1841 he gave the animals the name *dinosaurs,* which comes from Greek words meaning "terrible lizards." At the time, the dinosaur bones that had been found seemed to be a larger version of lizards. But as more bones were found, scientists decided that dinosaurs were not true lizards. Their skulls were shaped differently. Today most scientists describe **dinosaurs** as an extinct group of land-dwelling, cold-blooded reptiles. This means that the extinct reptiles that flew or lived in the water are not considered to be dinosaurs. However, they are often commonly grouped with dinosaurs.

Dinosaurs might still exist somewhere on the earth, but scientists do not know of any that are still alive today. Whatever is learned about them has to come from the fossils they left behind.

What Their Fossils Teach Us

At one time, there were many different kinds of dinosaurs on the earth. Paleontologists have found fossils of dinosaur bones all over the world. The fossils tell us certain things about the dinosaurs. Some of them were very large, while others were as small as chickens. Some fossils were from reptiles that appear to have been able to fly, and others were equipped to swim and were air breathers.

Brachiosaurus was a type of sauropod. Sauropods were a group of large dinosaurs.

46

Activity

Naming Dinosaurs

Materials: reference books and materials, Internet access

Direct the students to research the root words that form the names of dinosaurs. Compile the information into lists for the students to match the meanings of the names with the characteristics that scientists believe the extinct creatures had.

27

Body Shape and Size

Scientists cannot know exactly what any dinosaur looked like. Paleontologists can look at its skeleton for clues, though. These clues may tell such things as how big the dinosaur was or what its body shape was. From these clues, scientists make guesses.

Some dinosaurs had unusual features. *Stegosaurus* had vertical plates of bone arranged down its spine. Another type of dinosaur, called *Spinosaurus*, had bony spikes on its body. *Tyrannosaurus rex* had very small arms in comparison to the rest of its body. Scientists are not sure how the dinosaurs used these body parts. But we know that God created each creature with the characteristics it needed.

One familiar group of large dinosaurs is the *sauropods*. These plant-eating dinosaurs include the largest animals ever to walk the earth. Fossils of some sauropods indicate they could reach nearly 40 m (130 ft.) long from head to tail. If a *Brachiosaurus*, a type of sauropod, had stood erect with its neck straight, it would have been about 43 ft tall. A male giraffe is only about half this height.

Paleontologists have also found examples of reptiles that had wings. One of these called *Pteranodon* was a large flying creature that had a thin, bony crest on top of its head. Some scientists believe that this crest was used to provide balance during flight.

• MEET THE SCIENTIST • MARY ANNING

Mary Anning (1799–1847) grew up in the coastal town of Lyme Regis, England. Her father taught her how to hunt for fossils along the cliffs by the sea. After her father's death she provided for her family by finding and selling fossils. She is credited with discovering two marine reptiles: *Ichthyosaurus* and *Plesiosaurus.* Anning did not have any formal education. She just read and studied the scientific writings of her day. But she became an expert at identifying the fossils that she found. Her work was recognized by well-known paleontologists, displayed in museums, and discussed in scientific papers.

47

Teach for Understanding

What are two ways that we know that dinosaurs existed? The fossil record proves that they did, and God's Word describes animals that may have been dinosaurs.

The term *dinosaur* comes from two Greek words. What do these two words mean? terrible lizard

Which scientist gave dinosaurs their name? Sir Richard Owen

Why did scientists think that dinosaurs were lizards? At the time, the bones that were found seemed to be larger versions of lizard bones.

What can fossils teach us about dinosaurs? Possible answers: that they existed; their size; that there were different kinds; that they lived all over the world in water and on land; their possible appearance; their diet; their behavior

What part of a dinosaur do paleontologists most often observe to make guesses about the appearance of the dinosaur? its skeleton

- Discuss unusual features of some dinosaurs such as vertical plates of bone arranged down their spines, bony spikes, very small arms, great height, bony crest on the head, and large wings.

  Why is it difficult for scientists to know the function of these unusual body parts? Possible answers: Dinosaurs are extinct, so there are no living species to observe. Fossils are usually limited to skeletons, and there are rarely soft tissues left to observe.

  What kind of dinosaur do scientists believe was the largest creature to live on land? the sauropod

- Discuss *Meet the Scientist.*

  Why was it unusual for Mary Anning to be a noted scientist? She had no formal education. She learned about science by reading and studying the scientific writings of her day.

  In what type of rock did Mary Anning probably find fossils? sedimentary because most fossils are found in sedimentary rock and the cliffs are near an ocean, which would cause sedimentary rock to form

Background

Dinosaurs as Reptiles

True dinosaurs were all land-dwelling vertebrates that walked erect with their legs beneath them. They are usually split into two groups—Ornithischia (bird-hipped) dinosaurs that fed on plants, and Saurischia (lizard-hipped) dinosaurs that included both carnivores and herbivores. Skeletal similarities to living reptiles and the microscopic structure of their bones leads us to conclude that dinosaurs were cold-blooded, but this assumption cannot be proven.

Terrible Lizard

The phrase "terrible lizard" comes from the Greek words *deinos*, meaning "terrible," and *sauros*, meaning "lizard." The term *dinosaur* also refers to creatures other than "lizards."

What did dinosaurs eat before sin entered the world? plants

- Explain that death came after mankind sinned. This means that animals would not have been killed for food before Adam sinned.

What type of teeth do scientists believe plant-eating dinosaurs had? blunt

What could God have designed the sharp teeth of dinosaurs for? The sharp teeth may have originally been designed for tearing bark from trees or cutting tough rinds on fruits.

- Explain that after Adam and Eve sinned, these dinosaurs may have begun to eat meat.

What is the best way to tell what a dinosaur ate? by finding food remains inside of the dinosaur

What can be learned from studying fossils of dinosaur tracks? Possible answers: how the dinosaur walked; whether the dinosaur traveled in a herd

Why did scientists have to revise their ideas about the movements of *Tyrannosaurus rex*? In places where tracks were found, there was no fossil evidence that the dinosaur dragged its tail on the ground.

Why did scientists have the wrong ideas about the movements of *Tyrannosaurus rex*? Because scientists don't have all the facts about the past. They can only make educated guesses based on clues that they find.

Diet and Behavior

Fossils can also give clues about what dinosaurs ate. Genesis 1:30 tells us that before sin entered the world all animals were plant eaters. Only later did some animals become meat eaters.

Dinosaurs with blunt teeth are usually assumed to have eaten only plants. But some dinosaur fossils have very sharp teeth. God could have designed sharp teeth for tearing bark from trees or cutting open fruit with tough rinds.

The best way to tell what a dinosaur ate is to find food remains inside of the dinosaur. For example, *Pteranodon* had a long pouch under its jaw. Fossilized fish bones have been found in this pouch. So scientists believe that this flying reptile held fish in its pouch, much like a pelican does.

Paleontologists can learn many things about a creature's behavior from looking at fossils. Tracks often show whether it was walking or running or whether it was alone or in a herd. You have probably seen models of a *Tyrannosaurus rex*. Its name means "tyrant lizard king." By studying trace fossils, scientists have changed their minds about the way it walked. At one time they thought that it walked upright with its tail dragging the ground. But then trace fossils of *Tyrannosaurus rex*'s tracks were found. These tracks showed footprints but no tail markings. Most scientists now think that *Tyrannosaurus rex* kept its body level and extended its tail as it walked.

Pteranodon

48

Background

Dinosaurs and Birds

According to Scripture, birds were created on Day 5 and the dinosaurs on Day 6 with the other land creatures. Most evolutionists believe that birds evolved from either an early common ancestor of birds and dinosaurs or from one of the more advanced Saurischian dinosaurs similar to *Tyrannosaurus rex*. These dinosaurs have very small front limbs, the opposite of what is needed for flight. Some of the other significant problems of either view include the change from a cold-blooded to warm-blooded metabolism, the radical change from reptile to bird lungs, and the development of complex feathers from simple scales. It is no wonder evolutionists themselves continue to argue about how this occurred.

Classifying Organisms

The Linnaean system of classification was developed to help scientists study organisms more easily. In the system, organisms are grouped according to similarities of physical characteristics. The system was developed by Carolus Linnaeus about 100 years before Charles Darwin published his original works about evolution. Today evolutionary scientists try to use the system to promote evolution.

Oviraptor

Some fossils give us clues about the family life of dinosaurs. We know that some dinosaurs laid eggs because fossilized eggs have been found. One type of dinosaur, called *Oviraptor*, was discovered in the Gobi Desert of Mongolia. The dinosaur's bones were found lying on top of a nest of eggs. At first it was thought that these eggs belonged to another kind of dinosaur. They thought that *Oviraptor* had died in the act of stealing the eggs and so gave this dinosaur a name that means "egg stealer."

But later other fossils of *Oviraptor*s were found on nests as well. Then a fossil of a baby *Oviraptor* was found inside one of the eggs! These nests held *Oviraptor*'s own eggs. The dinosaur was not stealing eggs. It was protecting its own eggs probably from the floodwaters.

Evolutionists would like to find a fossil of an animal that is a cross between a dinosaur and a bird. They think that this would prove that birds evolved from dinosaurs. Even if such a fossil were found, it would not prove evolution to be true. It would only show that God used similar designs in some of His creatures.

In Genesis 1:24–25, the Bible tells us that animals "bring forth . . . after their kind." Babies look like their parents. God planned it this way.

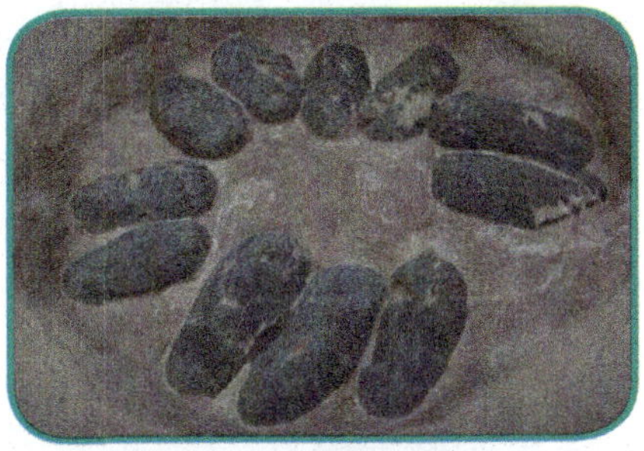

An *Oviraptor* nest with eggs

QUICK CHECK

1. What kinds of clues can scientists gain from observing a dinosaur's skeleton?
2. How could the fossil of a dinosaur give clues about its diet?
3. What characteristics of dinosaur behavior might scientists guess from fossils?

27

How do scientists know that some dinosaurs laid eggs? Fossilized eggs have been found.

Why did scientists give *Oviraptor* a name that means "egg stealer"? The bones of one were found on top of eggs. The scientists thought that the dinosaur died while stealing the eggs.

The conclusions that scientists make as they observe fossils may not be accurate. Explain why scientists had to change their minds about *Oviraptor*. A fossil of a baby *Oviraptor* was found inside an egg.

What does the Bible tell us about babies and their parents? Baby animals are like their parents (Genesis 1:24).

ANSWERS

1. Possible answers: the dinosaur's appearance, its diet, its behavior
2. Possible answers: Remains of food may be found inside the fossil. The shape of the teeth may indicate what a dinosaur ate.
3. Possible answers: how the dinosaur walked; whether it traveled in herds; its family life

Activity Manual

Reinforcement, page 33

28

Student Text pages 50–53
Activity Manual page 34

Objectives

- Realize that man and dinosaurs lived at the same time
- Recognize that some dinosaurs survived the Flood
- Identify biblical animals that may have been dinosaurs
- Name some causes of extinction
- Identify reasons why dinosaurs may have become extinct

Materials

- pictures of dragons

Teacher's Toolkit CD

- Quiz 2-C

Vocabulary

- extinction

Introduction

Many cultures around the world have myths and legends about dragons.

- Display the pictures of dragons.

Do you think that dragons are only from the imaginations of storytellers, or do you think that they may have really lived on the earth with people? Accept reasonable answers.

Preparation for Reading

- Direct the students to mentally make questions using the headings, bold words, and captions.
- Instruct them to read silently pages 50–53 to find the answers for these questions.

Did Noah take dinosaurs onto the ark?

What may have caused the extinction of the dinosaurs?

Teach for Understanding

What do evolutionists believe about dinosaurs and man? They believe that dinosaurs and man never lived on the earth at the same time.

How do we know that dinosaurs and man would have lived on the earth at the same time? The Bible says that God created all the land and sea animals on the fifth and sixth days of Creation. Man was also created on the sixth day, so dinosaurs and man would have lived at the same time.

What does the Bible tell us about dinosaurs on the ark? The Bible says that at least two of every living thing were present on the ark.

What may have been true about the dinosaurs on the ark? Possible answer: They may have been babies or young dinosaurs.

What the Bible Teaches Us

Long before Sir Richard Owen first used the name *dinosaur*, the Bible talked about unusual animals that were very likely dinosaurs.

Evolutionists believe that dinosaurs and man never lived on the earth at the same time. But the Bible says that God created all the land and sea animals during the fifth and sixth days of Creation. And He created man on the sixth day of Creation. So dinosaurs and man would have lived at the same time. God's Word is always accurate. We can trust it to be true even in areas of science.

We also know that God spared some of the dinosaurs during the Flood. Genesis 7 records that Noah took some of every land animal and bird into the ark. This would have included the dinosaurs. So after the Flood, some dinosaurs must have survived for a time.

50

Background

Bible and Kinds

The biblical use of *kind* in Genesis 7:14 is much broader than the use of the term today. For instance, Noah would not have taken every breed of dog that exists today. He probably took a pair of only one kind of dog or canine. The same would be true of other groups of animals, including each kind of dinosaur.

Cultural Stories and the Bible

Many cultures around the world have traditional stories, legends, or myths that parallel the biblical accounts of things such as Creation, the Flood, and the existence of dinosaurs. Some stories are written as children's books. Read one or two of these stories and compare Scripture with the stories that exclude God and the Bible.

Does the story reinforce the truth that the event really occurred?

How is the story different from the Bible?

In Job 40:15–19, the Lord talks about an animal that He called Behemoth. This large, grass-eating animal had a "tail like a cedar." The animal also had bones that were as strong as brass and iron, and it drank great quantities of water. The Bible calls Behemoth "the chief of the ways of God." All these descriptions lead us to think of an animal such as a sauropod. Fossils of sauropods show animals with long necks and long tails. They were plant eaters, and some of them grew to be very large.

In the next chapter of Job, the Lord describes another animal. This one He calls Leviathan. This large sea creature had very strong scales. Job 41:14 says, "Who can open the doors of his face? his teeth are terrible round about." We do not know exactly what kind of animal this was. Some believe it may have been a type of large marine animal known as a *plesiosaur*. This kind of animal also had powerful, interlocking teeth. Fossils of one type of plesiosaur, called a *Kronosaurus*, have been found with 10 ft long skulls!

The book of Job seems to be describing dinosaurs. This means that dinosaurs might have been on the earth during Job's time. Other books of the Bible, such as Psalms and Isaiah, mention dragons and flying serpents. The descriptions of these animals seem to match other kinds of dinosaur fossils that scientists have found.

History has also recorded many stories about dragons. The history of England tells the story of a man named Saint George who killed a dragon. Chinese history includes legends of families who bred and raised dragons. These dragons might have been dinosaurs.

Scientists will probably never be able to tell us exactly what dinosaurs looked like and how they lived. We can be certain, however, that God created these animals. They fit into His perfect plan and fulfilled His purpose.

SCIENCE & THE BIBLE

The book of Psalms mentions sea creatures that it calls "dragons" (Psalms 74:13; 148:7). We think of dragons as creatures only from myths and legends. But "dragons" may have been alive even in Bible times. Over the years, their looks and habits have probably been fictionalized. Now the fossils of dinosaurs may be all that is left of these creatures from long ago.

51

28

What two large animals mentioned in Job could have been dinosaurs? Behemoth and Leviathan

What type of dinosaur might Behemoth have been? a sauropod

How is the Behemoth in Job 40 different from the Leviathan in Job 41? Behemoth lived on land, but Leviathan was a sea creature. Behemoth had a large tail, and Leviathan had strong scales and powerful, interlocking teeth.

What kind of dinosaur might Leviathan have been? a plesiosaur

In what ways are these two creatures the same? They both are very large and strong.

What books of the Bible describe dragons and flying serpents? Possible answer: Psalms and Isaiah

Why do some people believe that these creatures in the Bible may have been dinosaurs? The descriptions seem to match some dinosaur fossils that have been discovered.

In addition to the Bible, where might you find references to dragons? Possible answers: history, literature

What are some things that the Bible teaches us about dinosaurs? They existed. God created them. They lived at the same time as all of God's other creatures and at the same time as man.

- Discuss *Science & the Bible.*

What does *fictionalized* mean? The facts about true events from history may have been changed. Other details may have been added to the story.

Activities

Play a Game of "Telephone"

Play a game in which a sentence is whispered from person to person. Compare the sentence from the last person with the original sentence. Discuss how the results of this game may be similar to the facts of some stories and legends. Both may have changed over time with each retelling.

Dragon Stories

The students may write fictional short stories about a dragon that lives nearby. The time period may be modern or historical.

If dinosaurs are still in existence today, where are they probably located? in remote places

When does extinction occur? when the last of a certain plant or animal dies

What are some things that can cause extinction? disease, lack of food, weather changes

What happened to all of the land animals that were outside the ark? They died in the Flood.

Which kinds of dinosaurs would have survived the Flood? dinosaurs that were on the ark and some marine dinosaurs

Extinction of Dinosaurs

We know that dinosaurs were alive on the earth at one time. Their fossils clearly tell us so. There may in fact still be dinosaurs on the earth. If so, they are very rare, though, and live in places so remote that no one has found them.

Extinction is when the last of a certain plant or animal dies. Extinction can be caused by many things. Disease, lack of food, and weather changes are just a few of the reasons why some kinds of plants and animals are now extinct.

The fossil record points to a time when many animals, including dinosaurs, died all at once. Genesis 7:21–22 tells what happened to the animals that were not in the ark during the Flood: "And all flesh died that moved upon the earth, both of fowl, and of cattle, and of beast, and

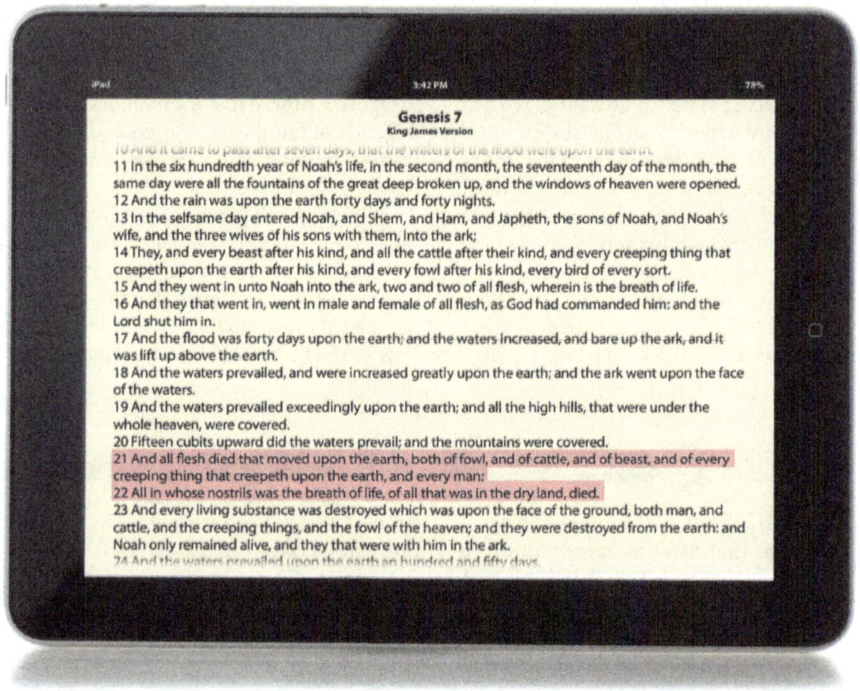

The Bible tells us what happened during the Flood.

52

Background

Causes of Extinction

The specific conditions causing the extinction of fossilized organisms are unknown. Extinction is further discussed in Chapter 8, *Changes in an Ecosystem*.

Changes in Climate

Not all creationary scientists agree on the climatic conditions of the earth before, during, and immediately after the Flood. The specific details were not recorded. Scientists are limited to observing the earth as it exists today. They can only hypothesize, or guess, about past conditions. Models can be made and observed, but whether the models are realistic is still unknown.

of every creeping thing that creepeth upon the earth, and every man: all in whose nostrils was the breath of life, of all that was in the dry land, died." Not all dinosaurs died in the Flood, though. Two of every kind of dinosaur and flying reptile would have been safe in the ark. Also, some of the marine reptiles would have survived.

Evolutionists have formed several theories about how dinosaurs became extinct. One theory says that a meteorite hit the earth and caused major climate changes. The dinosaurs then died from the sudden heat or cold. Another theory says that volcanoes killed the dinosaurs. Less popular theories say that all the dinosaurs died from changes in the air or from diseases. These are ideas that people try to use to explain the extinction of dinosaurs without taking the Flood into account.

The Flood brought many changes to the earth. The Bible is not clear about what the earth's pre-Flood climate was like, but we know it was not like it is now. Some Creationists believe that the earth's pre-Flood climate was very mild, with no extreme temperatures. After the Flood, the earth's climate was very different from what it was before the Flood. It probably was very warm right after the Flood, followed by gradual cooling into an ice age. Some plants and animals, including dinosaurs, may not have been able to adjust to the change.

Dinosaurs may have also had to change their diets. Plants and animals that they had eaten before the Flood may not have been able to thrive in the changed climate. With limited food, different types of animals would have had to compete for survival. God also permitted humans to kill animals for food after the Flood (Genesis 9:2–3). The dinosaurs could have become a major source of food.

No one can prove what happened to the dinosaurs, just as no one can prove how dinosaurs, or any other creature, came to be. No one was there at the beginning to see it—no one, that is, except the Creator Himself. What you believe about dinosaurs and fossils depends on what you believe about the Creator and His Word.

QUICK CHECK

1. What animals mentioned in the Bible and in history records may have been dinosaurs?
2. What is extinction?
3. What are some reasons that Creationists give for the extinction of dinosaurs after the Flood?

53

28

What are some theories that evolutionists use to explain the extinction of dinosaurs? a meteorite hit the earth and caused climate changes; volcanoes; changes in the air; diseases

Why do you think evolutionists have so many different theories? Answers should include that they have no authoritative record of the past, only clues. They reject the account of the Flood in the Bible.

What do some Creationists believe about the earth's climate before the Flood? The climate was mild with no extreme temperatures.

- Read together Genesis 2:6.

What does Genesis 2:6 tell us about the earth's climate before the Flood? that a mist went up from the earth and watered the land

Why do you think it would have been difficult for some plants and animals to survive climate changes after the Flood? Possible answers: Plants and animals could not adjust to the sudden changes. Needed food and nutrients may have changed or been limited.

What are some reasons that dinosaurs may have become extinct? climate changes after the Flood; changes in their diets; limited amounts of food; possibly being hunted by man

Can we prove what happened to dinosaurs? no Why? No record of what happened has been found, and God's Word does not say what happened.

When mankind's ideas conflict, or disagree, with God's Word, what should a Christian do? A Christian's only option is to believe God or to put his trust in his own understanding. A Christian should put his confidence in a loving Creator.

ANSWERS

1. Behemoth, Leviathan, dragons, and flying serpents
2. when the last of a certain plant or animal dies
3. climate changes after the Flood; changes in their diets; limited amounts of food; possibly being hunted by man

Activity Manual

Review, page 34

This page reviews Lessons 27–28.

Assessment

Quiz 2-C

The quiz may be given any time after completion of this lesson.

29

Activity Manual pages 35–36

Objectives

- Justify from a biblical viewpoint that dinosaurs existed and that dinosaurs and people lived together
- Examine scientific evidence to show that dinosaurs are thousands of years old and not millions

Materials

- fossils or pictures of different types of fossils

Introduction

- Ask a volunteer to read aloud the question in the first paragraph. Allow the students to share their experiences with seeing dinosaur skeletons in museums.
- Direct another volunteer to read aloud the remainder of the first paragraph.
- Group the students and encourage them to discuss their responses to the last two questions in the paragraph.
- After their discussion, show the fossils or pictures you brought. Emphasize that fossils come in all shapes and sizes and are formed in various ways.

Preparation for Reading

- Direct the students to read silently the second paragraph on the top of Activity Manual page 35 to find out how they should evaluate museum signs.

Teach for Understanding

What should we believe if a museum sign is not correct according to what the Bible says? We should believe the Bible.

Let's look at the Bible to see what it tells us about dinosaurs living millions of years ago.

- Read together Job 40:14–24. Give the students time to list some characteristics of the animal God is describing to Job on their Activity Manual page.
- Discuss the students' answers. Possible answers: eats grass (verse 15); is strong (verse 16); has powerful muscles (verse 17); has a stiff tail (verse 17); has strong bones (verse 18); the mountains provide him with food (verse 20); lies in the shadow of shady trees (verse 21); does not fear the turbulent river (verse 23)
- Guide the students to conclude that only a sauropod dinosaur fits the description.

God is describing this creature to Job.

Do you think that Job is familiar with the creature? Accept any reasonable answer. Lead the students to understand that Job must be familiar with the creature or Job would not know what God was talking about.

- That Job was familiar with this creature is also supported by the fact that many recognizable creatures are described in Job 38–39.

A lesson from

Answers in Genesis

Name ____________________

1:1

Soft Bones

Have you ever been in a museum and seen a fossil skeleton of a giant dinosaur? Often, signs will tell you what kind of dinosaur you are looking at, where it lived, what it ate, and how many millions of years ago it lived. But wait! How do they know all that stuff? What evidence do they have?

When Christians seek to answer these types of questions, they start in the Bible. Since the Bible is God's Word, everything it says is absolutely true. If a museum sign says one thing and the Bible says the opposite, we must believe the Bible rather than the sign. So, did those dinosaurs live millions of years ago like the museum signs say? Let's see . . .

1. Read Job 40:14–24. As you read, make a list of some of the characteristics of the animal God is describing to Job.
   Possible answers: eats grass, is strong, has powerful muscles, has a stiff tail, has strong bones, is the food of wild animals in the mountains, lies in the shade of the lotus tree, does not fear the turbulent river
2. What kind of creature matches this description? dinosaur (sauropod)
3. If God is describing this creature to Job, had Job ever seen this creature? yes
4. If God is describing a dinosaur that lived with Job, then people lived with dinosaurs. Let's try to confirm this. Read Genesis 1:24–31.
   Do dinosaurs fit in the description of what God created on Day 6? yes
5. On what day was mankind created? Day 6
6. Did people and dinosaurs live together? yes

Science 5 Chapter 2 | Lesson 29 Answers in Genesis 35

Resources

The BJU Press website offers additional links from Answers in Genesis that you may find helpful with this lesson.
bjupress.com/resources/science

When we look at the world through our biblical glasses, it seems clear that man lived with dinosaurs. People who reject the Bible believe dinosaurs died off 65,000,000 years ago and that people appeared fewer than 1,000,000 years ago.

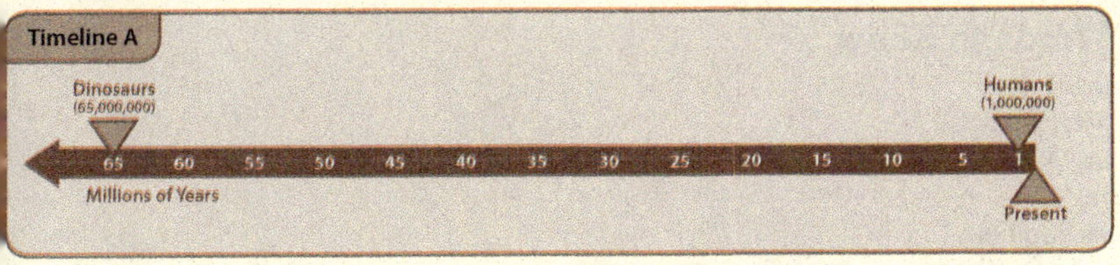

Let's look at some scientific evidence from the fossils. You may know that Egyptian mummies can be thousands of years old. Those mummies can include preserved blood vessels and other materials. Dinosaur fossils are only a few thousand years older than those ancient mummies. If you thought dinosaurs died over 65,000,000 years ago, would you ever expect to find any blood vessels or material that was still soft and stretchy inside of a dinosaur bone? Well, that is exactly what has been found in many fossilized bones! All of the soft tissue like muscle, blood vessels, and blood cells should have rotted away after millions of years.

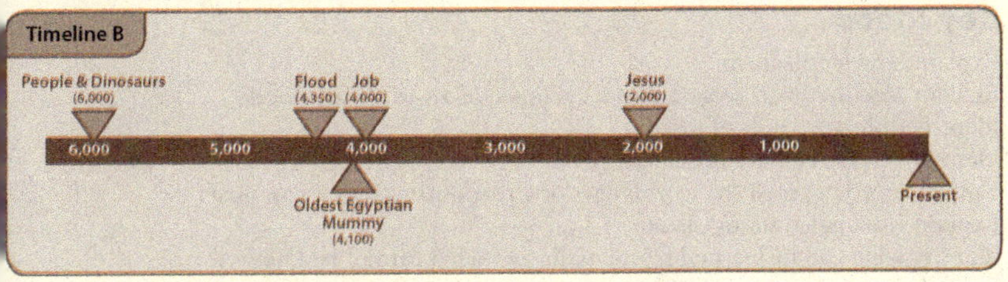

7. So, if there is still soft tissue in some fossils, which of the explanations makes more sense—that dinosaurs lived 65,000,000 years ago or just a few thousand years ago?

____

8. Which of the timelines fits the evidence from the fossils and the evidence from the Bible?

Timeline B

As we discover more about the world God has created, we can see how important it is to trust the Bible to help us understand what has happened in the past.

Background

Soft-Tissue Research

Mary Schweitzer, the paleontologist who first discovered the soft tissue in the *Tyrannosaurus rex* fossil, performed experiments in 2013 to try to replicate the conditions that might have made this preservation possible. Based on her belief that birds evolved from dinosaurs, she soaked some ostrich blood vessels in a solution containing red blood cells and other vessels in plain water. Two years later, the blood vessels in the red blood cell solution were still preserved and pliable, while those in the water had rotted after just a few days. Some evolutionists view the results of Mary Schweitzer's experiment as support for her claim that soft tissue could be preserved for millions of years inside a dinosaur bone. However, her results only prove that ostrich blood vessels can be preserved for at least two years under certain laboratory conditions.

- Ask a volunteer to read aloud question 4 on Activity Manual page 35.
- Read together Genesis 1:24–31. Explain that these verses confirm the idea that people lived alongside dinosaurs.
  Do dinosaurs fit the description of what God created on Day 6? yes
  Dinosaurs would surely be among the land animals God created on Day 6 of the Creation week. They are likely among the "beasts of the earth."
  On what day was mankind created? Day 6
- Conclude that a biblical understanding of the world makes it clear that people lived alongside dinosaurs.
- Ask a volunteer to read aloud the first paragraph on Activity Manual page 36.
- Discuss the times using Timeline A.
  Read the paragraph between the timelines to find out what scientific evidence can be seen in some fossils.
- Discuss the discovery of stretchy soft tissue with amazing detail inside a dinosaur bone. Point out that such discoveries are quite a shock, especially for evolutionists.
- Explain that intact soft tissue has been found in many more fossils. Rather than admit that the soft tissue demonstrates that these fossils cannot be millions of years old, evolutionists are so committed to their worldview (the evolutionary glasses they look at the world through) that they must insist there is some mechanism that allows this preservation to happen. They claim it happens because evolution is true!
  The Bible has a much more reasonable answer. Looking at this evidence through biblical glasses, we might expect to find intact soft tissue if the creatures had been buried in the Flood only about 4,350 years ago.
- Share the Background information with the students.
- Use the timelines on Activity Manual page 36 to help the students understand how the Bible helps explain the soft tissue in these bones.
- Reinforce that the Bible always gives us a proper place to begin our thinking because it is the true account of the history of the universe.
  When we examine fossils, we can only learn the truth if we start our thinking from the Bible. Fossils are not millions of years old but only thousands.

Activity Manual

Answers in Genesis, pages 35–36

30

Student Text page 54
Activity Manual pages 37–38

Objectives
- Recall concepts and terms from Chapter 2
- Apply knowledge to everyday situations

Review
- Materials for the Chapter 2 Test will be taken from Student Text page 54 and Activity Manual pages 24–25, 30, 34, and 38. You may review any or all of the material during the lesson. Questions similar to the ones in Thinking It Through, Activity Manual page 38, may appear on the test.
- You may choose to review Chapter 2 by playing the review game or a game from the Game Bank on the Teacher's Toolkit CD.

Activity Manual
Bible Connection, page 37
This page uses the Bible to answer some statements made by evolutionists about dinosaurs.

Review, page 38
This page requires written responses to application questions.

31

Objective
- Demonstrate knowledge of concepts taught in Chapter 2

Assessment
Chapter 2 Test

CHAPTER SUMMARY

Words to Know
fossil
petrified fossil
mold
cast
carbon film
trace fossil
evolve
paleontology
excavation
dinosaur
extinction

Key Ideas
Explain how fossils form
Explain how the fossil record shows evidence of a worldwide flood
Identify different kinds of fossils
Identify ways fossils are preserved
Compare and contrast the worldviews of Creationists and evolutionists
Explain what paleontologists do
Identify what can be learned from fossils and what cannot be known
Describe ways fossils are dated
Identify how dinosaurs got their name
Explain how we know man and dinosaurs lived at the same time
Explain what often causes extinction for plants and animals
Identify reasons why dinosaurs may have become extinct

54

Review Game
Build a Dinosaur
Materials: a picture of a dinosaur skeleton for each team

Draw the same number of grid lines on each picture and cut the picture apart along the lines.

Divide the class into teams. For each correct answer, the team member gets one piece of his team's skeleton. The first team to reassemble its picture wins.

Teacher
Notes

32

Student Text pages 55–57
Activity Manual page 39

Objectives

- Recognize that God created different kinds of matter to melt at different temperatures
- Recognize that learning about matter and how it works is important to glorify God and serve others
- Give an example of how God's design of the properties of matter benefits people

Unit Introduction

- The purpose of this unit is to show the relationships between matter and energy. As you teach this unit, direct the students' attention to how they experience matter and energy in everyday life.
- Chapter 3 discusses the states of matter, the physical properties of matter, and mixtures and solutions.
- Chapter 4 explores different forms of energy, specifically thermal energy. Gaining or losing thermal energy can cause matter to change states.

  Look through Unit 2. What kinds of topics do you think you will be studying in this unit? Possible answers: matter, energy, heat

  Why do you think these chapters are organized into the same unit? Answers will vary. Elicit that the energy discussed in Chapter 4 affects the matter mentioned in Chapter 3.

Background

Unit Photo

The photo on page 55 shows the sunrise over a lake landscape.

Weblinks

The BJU Press website offers additional information and links you may find helpful throughout the unit.
bjupress.com/resources/science

Look at the picture on page 56.

- Discuss the picture with the students. Allow volunteers to read aloud each of the teasers.

Project Idea

- The project idea presented at the beginning of each unit is designed to incorporate elements of both chapters as well as information gathered from other resources. You may choose to have the students do the project as a culminating activity at the end of the unit or as an ongoing activity while the chapters are being taught.

Unit 2: Keep It Cool

- Provide a half-gallon juice or milk carton for each student or group. Prepare the cartons by cutting a rectangular door of the same size on one side of each.
- Each student will then design a "cooler" that will slow the time that it takes for a freezer pop to melt. Each "cooler" must be made from a carton that will be able to contain a freezer pop standing in a cup. The student provides the materials to complete his design. Limit the placement of additional materials to the inside of the carton.

- Test the "coolers" by placing a frozen freezer pop in a cup inside each cooler. Place another freezer pop in a cup outside the cartons to use for comparison. Record and compare the melting times.
- Discuss relationships between the melting times and the materials used.
- Extend the project by changing conditions. You may choose to change the location or blow identical hair-dryers on each.

This lesson continues with the Chapter 3 introduction on page 66.

Unit 2 Overview

Lesson	Teacher's Edition	Student Text	Activity Manual	Assessment	Content
Chapter 3: Matter					
32	64–67	55–57	39		• Worldview thinking • Unit and chapter preview
33	68–71	58–61	40		• Volume of a liquid and a solid • Mass, weight, and density
34	72–73		41		**Activity: Measuring Length, Volume, and Temperature**
35–36	74–79	62–67	42–44	Quiz 3-A	• States of matter: solid, liquid, gas • Changing states of matter
37	80–81		45–46		**Activity: A Science Experiment** • Using a scientific method
38	82–83	68–69	47–48	Quiz 3-B	• Classifying matter • Atoms and molecules
39	84–85	70–71	49–50	Rubric	**Activity: Separating a Mixture** • Process skill: experimenting
40	86–89	72–75	51		• Mixtures • Separating substances in mixtures
41	90–93	76–79	52–54	Quiz 3-C	• Solutions • Using matter
42	94–95		55–56		**Answers in Genesis** • God created matter
43	96–97	80–81	57–58	Rubric	**Activity: A Disappearing Act** • Process skill: hypothesizing
44–45	98–99	82–83	59	Rubric	**Exploration: Float a Boat** • Demonstrate buoyancy
46	100	84	60		**Chapter Review** • Apply knowledge to everyday situations
47	100			Chapter 3 Test	**Chapter 3 Test**

Lesson	Teacher's Edition	Student Text	Activity Manual	Assessment	Content
Chapter 4: Energy and Heat					
48	101	85	61		• Worldview thinking • Chapter preview
49–50	102–5	86–89	62		• Potential, kinetic, and thermal energy • Temperature
51	106–7	90–91	63–64	Rubric	**Activity: Rock Heaters** • Process skill: experimenting
52	108–10	92–94	65–66	Quiz 4-A	• Effects of thermal energy on matter • Measuring thermal energy
53	111	95	67–68	Rubric	**Exploration: Energy for Your Body** • Track Calorie consumption
54–55	112–15	96–99	69–70		• Heat • Conduction, convection, and radiation
56	116–17	100–101	71–72	Rubric	**Activity: Keeping Warm** • Process skill: measuring and using numbers
57	118–20	102–4	73–75	Quiz 4-B	• Fuel sources and uses • Unwanted heat
58	121–24	105–8	76	Quiz 4-C	• Space capsules and space shuttle • International Space Station and modern spacecraft
59	125–26		77–78		**Answers in Genesis** • Exercise biblical dominion through operational science
60	127	109	79	Rubric	**Exploration: Moon Station** • Design a piece of equipment for a moon station
61	128	110	80		**Chapter Review** • Apply knowledge to everyday situations
62	128			Chapter 4 Test	**Chapter 4 Test**

Matter 3

Andrea's neighbor broke her ankle recently. Wanting to be helpful, Andrea and her family have been doing yard work and running errands for her. Today Andrea decided to encourage her neighbor by making her brownies. Andrea needs to melt some chocolate for the frosting.

But what if all kinds of matter melted at the same temperature? How would that change things for Andrea? The pan, spatula, and chocolate would all melt at the same time. These types of problems would be unending. However, God created different kinds of matter to melt at different temperatures. Learning about matter and how it works is important for using it to glorify God. God's provision of matter offers many ways to serve others.

Consider

- What does Andrea need to know about matter in order to make brownies?
- Is it important how Andrea makes brownies for her neighbor?

57

Background

Chapter Preview

Preview and prereading activities may include using a K-W-L chart, a probe, or an anticipation guide.

Chapter Photo

The photo on page 57 is the Cooper River Bridge in Charleston, South Carolina. The bridge connects the city of Charleston with the town of Mount Pleasant. It is North America's longest cable stay span. It features the latest in seismic considerations with eight lanes of north/south-bound traffic and a shared bicycle/pedestrian lane overlooking the Charleston Harbor.

Chapter Introduction

Can you think of a Bible verse that teaches us to show love to other people? Answers will vary.

- If the students cannot come up with a command or a reference, ask a student to look up a passage and read it. Suggested verses are Leviticus 19:18; Matthew 5:43–44; 19:19; 22:39; Mark 12:31–33; Luke 10:27; Romans 13:9, 10; Galatians 5:14; and James 2:8. Each of these verses contains the command to love your neighbor as yourself.
- Help the students understand that loving your neighbor is not a suggestion but something God commands us to do.

Preparation for Reading

- Direct the students to read the introduction to find out a benefit from materials melting at different temperatures.

Teach for Understanding

Why are Andrea and her family helping their neighbor? The neighbor broke her ankle. They wanted to be encouraging and show God's love to her.

What is Andrea making for her neighbor? brownies

What does Andrea need to know about matter in order to make brownies? at what temperature chocolate melts

- Allow the students to share what they think it would be like if all kinds of matter melted at the same temperature.

How is this problem taken care of for us? God created different kinds of matter to melt at different temperatures.

How did this benefit Andrea? only the chocolate melted

Is it important to learn about matter and how it works? yes Why? It is important so that we can use matter correctly to glorify God and to serve others.

Is it important how Andrea makes the brownies? yes Why? If she left an ingredient out or baked the brownies at the wrong temperature, the brownies may not taste good. Doing all to the glory of God means making the best brownies you can.

Activity Manual

Preview, page 39

The Looking Ahead page is intended to assess the student's prior knowledge before beginning the chapter.

Objectives
- Define *matter*
- Explain how to find the volume of a solid and of a liquid
- Differentiate between mass and weight
- Recognize that volume, mass, and weight are ways by which matter can be measured
- Explain how density is related to mass and volume

Teacher's Toolkit CD
- Visuals 3.1–3.2: *Mass*; *Weight*

Vocabulary
- matter
- volume
- mass
- weight
- density

Lesson 7 in Chapter 1 teaches measuring mass and measuring volume using milliliters. Lesson 34 in Chapter 3 teaches measuring temperatures, as well as measuring length and volume using cubic centimeters. Refer to the Background notes in these lessons for additional information about each type of measurement.

Introduction
- Discuss varieties of measurement.

  What unit would you use to find the distance from your bed to the bathroom? meter, foot, or yard

  How does the post office determine the heaviness of a box? Possible answers: weigh the box, use a scale

  How can you find how many glasses of milk are in a gallon of milk? Possible answers: Measure the size of the glass and divide one gallon size by the glass size. Count the number of glasses of water it takes to fill an empty container.

  What are we doing in each of these situations? measuring

  Are we measuring the same thing each time? no

  Different measurements measure different characteristics of substances.

Preparation for Reading
- Direct the students to mentally make questions with the headings, the bold and italicized words, and the captions.
- Instruct the students to read pages 58–61 to find the answers to these questions.

  What are three ways to find the volume of a substance?

  How is weight different from mass?

Measuring Matter
Serving others often involves using objects. Andrea used ingredients and utensils to make her brownies. Although objects are different in many ways, they are all alike in at least one way. Each is made of matter. **Matter** is anything that has volume and has mass. Volume and mass are characteristics of matter that we can often measure. Sometimes we use the term *substance* to refer to matter.

Volume of a Liquid
The **volume** of a substance is the amount of space that it takes up. The standard unit of metric measurement for volume is the liter (L). Many soft drinks come in two-liter bottles. A smaller unit for volume is the milliliter (mL). Liquid medicine often uses this unit. A teaspoon of medicine is about 5 mL.

We measure the volume of a liquid by using a graduated container. *Graduated* means that the container is divided into equally marked parts. A graduated container has the units of measurement marked on its side. A liquid is poured into the container. The level of the liquid is then compared with the numbers on the side of the container. The numbers show the volume of the liquid. Scientists use containers called graduated cylinders, but even the measuring cups in a kitchen are a type of graduated container.

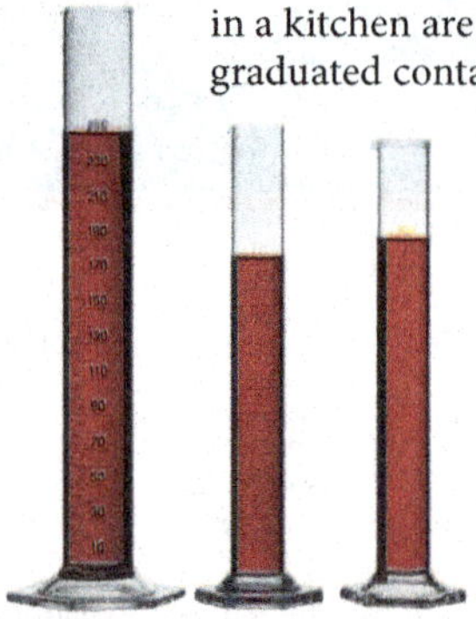
Graduated cylinders

SCIENCE & THE BIBLE

In 2 Kings 4 the Bible tells us about a widow who owed money. Her sons were going to be taken as slaves unless she could pay her debts. All that the widow had was a pot that held a small volume of oil. The prophet Elisha told the widow to borrow vessels, or large containers, from her neighbors. She started pouring her little bit of oil into the vessels and kept pouring until all of the vessels were filled. Elisha told her to sell the oil to pay her debts. God caused the small volume of oil to increase to fill many large pots. In this way, He miraculously provided for the widow's needs.

58

Background
Mass, Gravity, and Weight
Every object in the universe that has mass attracts every other object that has mass. This attraction is called *gravity*. The amount of the attraction depends on the mass of each object and how far apart they are. The gravity of an object with greater mass will pull an object with lesser mass toward it. When you stand on a scale, you are measuring the pull of gravity of the earth (the object with the greater mass) on you (the object with the lesser mass).

Cubic Measurement
Cubic units of measure are used to measure three-dimensional objects. The three dimensions of these objects are length, width, and height. The measurement is named *cubic* because the cube is a basic shape with these dimensions. Cubic measurements are often written with a ^{3}. This represents the three dimensions that were factored into the number. There are other formulas for finding the volumes of cylinders, spheres, cones, pyramids, and other regular geometric solids.

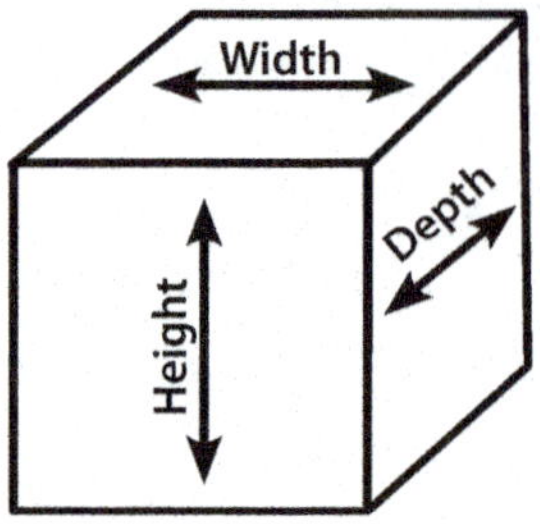

Volume of a Solid

There are two ways to measure the volume of solid objects. To measure the volume of regular shapes, such as cubes and rectangular solids, multiply the measurements of the length, width, and height of the object. The volume is written as a cubic measurement. An object that is 15 centimeters (cm) long, 10 cm wide, and 10 cm high has a volume of 1,500 cubic centimeters. A cubic centimeter is written as *cm^3* or *cc*.

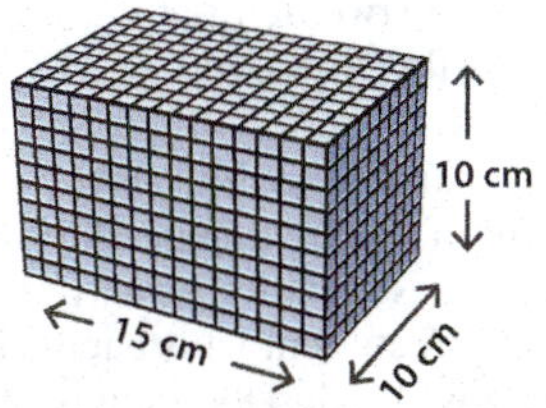

length × width × height = volume
15 cm × 10 cm × 10 cm = 1500 cm^3

However, we cannot always measure a solid object by using its length, width, and height. For instance, it would be difficult to measure a rock or a marble in this way. The volume of irregularly shaped objects can be measured using a method called *water displacement*.

You can see water displacement when you add marbles to the water in a graduated cylinder. The volume of the marbles is the difference between the beginning level of the water and the level after the marbles were added. The volume can be written in milliliters or liters. For example, suppose a cylinder contains 250 mL of water. After the marbles are added, the level of the water is at 300 mL. The volume of the marbles is 50 mL.

300 mL − 250 mL = 50 mL

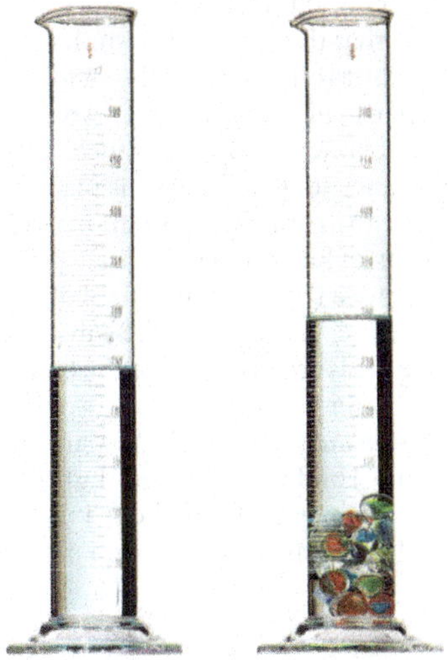

250 mL 300 mL

In the metric system one cubic centimeter is equal to one milliliter. Scientists usually use milliliters to describe the volume of a liquid. They use cubic centimeters to describe the volume of a solid. So the volume of the marbles in the example is also equal to 50 cm^3.

59

Teach for Understanding

What is matter? anything that has volume and mass

What is another word for matter? substance

What is volume? the amount of space that a substance takes up

How can you measure the volume of a liquid? Pour the liquid into a graduated container. Read the number on the side of the container at the level of the liquid.

What is a graduated container? a container with evenly divided markings that is used to measure the volume of liquids

Give some examples of graduated containers and their uses. Possible answers: measuring cups for cooking, graduated cylinders in laboratories, droppers for liquid medicines, beakers, syringes

How can you measure the volume of a rectangular-shaped solid? Measure the sides of the object and multiply its length, width, and height measurements.

How is volume written? as a cubic measurement

- Write the measurements of the rectangular solid and work the problem with the students.

If an object that is a rectangular solid has measurements of 20 cm long, 10 cm wide, and 20 cm high, what is its volume? 20 cm × 10 cm × 20 cm = 4000 cm^3

Can the volume of all solids be measured by multiplying the length, width, and height measurements? no Why not? Some solids are round or have irregular shapes.

When would you use water displacement to measure the volume of an object? if the object has an irregular shape or if it is difficult to measure the height, length, or width

What do we mean by "irregularly shaped object"? The object's length, width, and height cannot be measured.

- Write the displacement of the irregularly shaped solid.

If an irregularly shaped solid displaces 125 mL of water, what is its volume? 125 mL How do you know? The amount of water that is displaced equals the volume.

If you measure the volume of an object by water displacement, what unit of measurement can you use? the milliliter or the liter

- Write *1 mL*.

What is 1 mL equal to in the metric system? 1 cubic centimeter

- Write = *1 cm^3* after 1 mL.

What are two ways to write the unit for cubic centimeters? cm^3 or cc

- Write the two measurements.
- Discuss *Science & The Bible* on page 58.

What miracle did God work with the widow's small volume of oil? Even though she had only a little bit of oil, she was able to fill many large vessels. That way she could sell the oil and pay her debts. God provided for her needs by increasing her volume of oil.

Science Misconceptions

Properties of Matter

Although volume, mass, and weight are often referred to as properties of matter, they are not properties but are actually ways in which matter can be measured. *Physical properties* of matter are the characteristics of matter that distinguish it from other matter. These properties include color, shape, texture, odor, taste, electrical conductivity, magnetism, and density. *Chemical properties* of matter are characteristics that matter displays as it interacts with other matter. For example, combustion is a chemical property of gasoline as it reacts with oxygen.

Activity

Make a Graduated Container

Materials: clear cup, smaller container (at least one-fourth the size of the cup), water, marker or tape

Fill the smaller container with water. Pour the water into the cup. Mark the top of the water level with the marker or tape. Repeat until the cup is full of water. The cup now has even graduations, or markings. It is a graduated container.

If the smaller container is not already a standard unit of measure, you may want to think of a name for this new unit.

33

What is mass? the amount of material in a substance

What instrument is usually used to measure mass? a balance

What are some metric units used to measure mass? gram and kilogram

- You may want to do the *Air Has Mass* demonstration at this time.

What does weight measure? the amount of force that gravity places on a substance

Are mass and weight the same thing? no How are they different? Mass is the amount of material in a substance. Weight measures the amount of force that gravity places on the matter.

What instrument do we use to measure weight? a scale

When does the mass of an object change? when the amount of matter in the object changes

- Display the *Mass* visual as you discuss mass on the earth and the moon.

What may cause an object to have different weights in different locations? if the force of gravity is different

- Display the *Weight* visual as you discuss weight on the moon.

Why does a person weigh less on the moon than on the earth? There is less gravity on the moon than on the earth.

Why do you think that a person weighs more on Jupiter? Jupiter has greater gravity than the moon or the earth does.

Mass and Weight

Mass is another way that matter is measured. **Mass** is the amount of material in a substance. It is usually measured by comparing an unknown mass with a known mass. For example, a student with an unknown mass can sit on one end of a seesaw. A substance with a known mass, such as a sack of sand, can be placed on the other end. If the seesaw balances, we know that the mass of the student is the same as the mass of the sand.

We usually measure mass with an instrument called a balance. The substance to be measured is placed on the instrument. Standard masses, objects of known masses, are used to balance the unknown mass of the substance. The mass of the substance to be measured is equal to the sum of the standard masses.

Some metric units of mass are the gram (g) and the kilogram (kg). The mass of a paper clip is about 1 g. A one-liter bottle of water has a mass of about 1 kg.

It is usually easy to determine which of two objects has the greater mass. A large dog has more matter than a small cat does and, therefore, has more mass. The dog also weighs more than the cat does. But the dog's mass and the dog's weight are not actually the same.

Mass and weight are related, but they measure different things. Mass measures the amount of matter. **Weight** measures the amount of force gravity exerts on the matter. If we measure two items at the same location, the object with greater mass weighs more.

Mass can change only as the amount of matter changes, but weight can vary based on the gravity at a location. A student whose mass is 40 kg would weigh 88 pounds (lb) on the earth. But that student would weigh about 220 lb on Jupiter and only 15 lb on the moon. The difference occurs because the gravity of the moon and the gravity of Jupiter are different from the gravity of the earth.

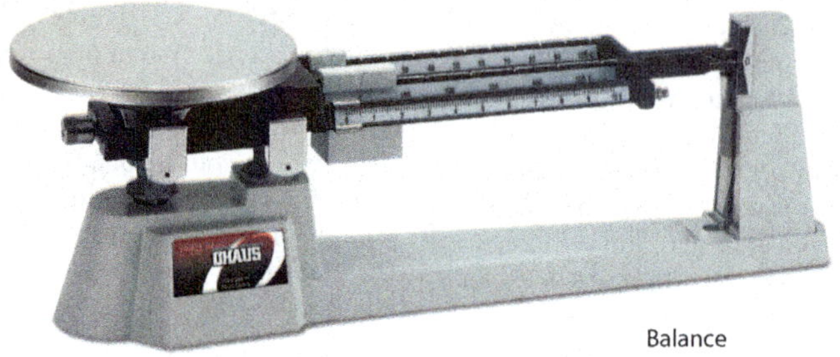

Balance

60

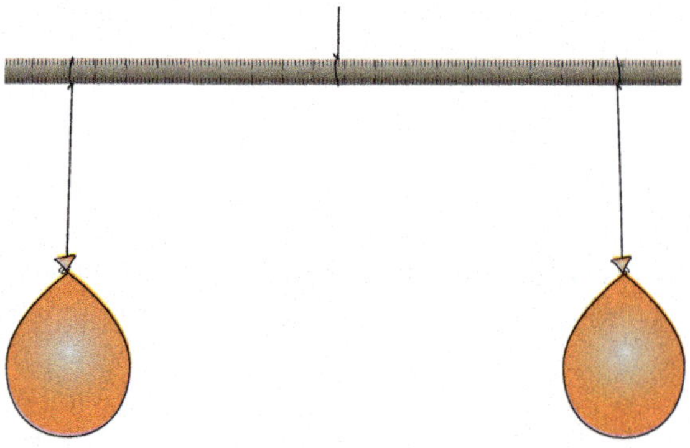

Demonstration

Air Has Mass

Materials: 2 same-sized balloons, string, meter stick, safety pin

Inflate both balloons to the same size. Use string to attach one balloon to each end of the stick. Attach a piece of string to the middle of the stick. Hold the string so that the stick and balloons hang like a balance. Adjust the placement of the string so that the stick is level.

Matter has mass. Do you think that air has mass? Answers will vary.

What do you think will happen if one balloon is popped? Answers will vary.

Pop one balloon with the pin, leaving the deflated balloon attached.

What difference is there between the balloons? The inflated one still has air.

What about the balance changed after the balloon popped? The side with the inflated balloon sank lower.

What does the position of the balance tell us about the air in the inflated balloon? Possible answers: The air has mass. The air has enough mass to pull the balance down.

33

Density

Density is the amount of matter in a certain volume. Suppose you are given the task of carrying one of two boxes in a race. The two boxes are the same size and look the same on the outside. However, one box is filled with cotton. The other box is filled with rocks. You would probably choose the box with the cotton inside. The two boxes have the same volume, but the box with cotton has less mass than the box with rocks has. This difference in mass makes the box of cotton less dense than the box of rocks is.

The density of an object is calculated by dividing its mass by its volume. The formula is mass ÷ volume = density. An object that has 12 g of mass and takes up 3 mL of volume has a density of 4 grams per milliliter. Notice this density is measured in grams per milliliter. It is written as *g/mL*.

CREATION CORNER

Most substances have a greater density in a solid form than in a liquid form. One exception is water. During the winter, ice floats on top of lakes and rivers instead of sinking. If the ice sank, it would continue to displace the water. The water on top would freeze and sink. All the living things in the water would die. However, God designed ice to be less dense than water. In this way, He provided a top layer to preserve life under the ice in the winter.

Object	Density
gold	19.3 g/mL
lead	11.4 g/mL
silver	10.5 g/mL
water	1.0 g/mL
ice	0.92 g/mL
wood (pine)	0.7 g/mL
wood (oak)	0.4 g/mL
cork	0.2 g/mL

mass ÷ volume = density
12 g ÷ 3 mL = 4 g/mL

QUICK CHECK

1. What is matter?
2. How is mass different from weight?
3. What is density?

61

What is density? the amount of matter in a certain volume

Look at the definition for density. What term could we substitute for "amount of matter"? mass

What mathematical formula can you use to figure out the density of an object? mass ÷ volume = density

If you have two objects that take up the same amount of space but one is heavier than the other, what can you say about the heavier object? The heavier object is denser.

- Write *mass of 100 g* and *volume of 25 mL* as you read the problem. Allow a volunteer to write the measurements in the formula for density. Work the problem together.

Use the formula for density to find the density of an object with a mass of 100 g and a volume of 25 mL. 100 g ÷ 25 mL = 4 g/mL

- Discuss the chart of densities.

What is the density of water? 1.0 g/mL

Which object in the chart is the most dense? gold, 19.3 g/mL

- Discuss *Creation Corner*.

Which are usually more dense, solids or liquids? solids

What substance is more dense as a liquid than as a solid? water

How did God create ice to float to benefit living things? The layer of ice keeps the water underneath from freezing. If all the water froze, the living things in it would die.

ANSWERS

1. anything that has volume and mass
2. Mass is the amount of material in matter. Weight measures the force of gravity on matter.
3. the amount of matter in a certain volume

Activity

Measuring Types of Matter

Materials: measurement instruments, various types of matter

Place materials and measurement instruments at centers in the classroom to give each student the opportunity to practice measuring various types of matter.

Provide problems for the students to solve using measurement. For example, place a balance and several books at one center, asking the question, "What is the difference between the mass of the science book and the mass of the reading book?"

The volume of some items is difficult to measure. To practice density, you may choose to label an item with its volume so that the students only have to measure its mass before doing the calculations.

Activity Manual

Reinforcement, page 40

Objectives

- Measure length to the nearest millimeter
- Measure volume using cubic centimeters
- Measure temperature to the nearest degree

Materials

- meter stick
- thermometer

Metric units of measure are used in this lesson and in the activities in *SCIENCE* 5. Calculating the volume of cylindrical objects is not covered in *SCIENCE* 5.

Introduction

We rely on tools and instruments to help us measure matter around us. Most families listen to a weather report regularly. Part of the report is the current temperature. This measurement tells us the temperature of the air at the time of the report. Learning to accurately measure is an important part of science.

Teach for Understanding

Linear Measurement

What basic unit of measure is used for length and distance? the meter

What tools are used to measure length? Possible answers: meter stick, ruler, tape measure

- Display the meter stick and ruler.
  Which tool would be more appropriate to measure the length of a room? the meter stick
  Which tool would be more appropriate to measure the length of your pencil? the ruler
- Review how to read the markings on a ruler as needed.
  What do the numbered markings on the ruler indicate? centimeters
  What do the smaller markings indicate? millimeters
  How many millimeters are in 1 cm? 10
- Demonstrate measuring the length, width, and height of the cardboard box. Round the measurements to the nearest whole centimeter.
- Write the measurements for display to use in the Volume Measurement part of this lesson.

Measuring Length, Volume, and Temperature

Name ______________________________

ACTIVITY

Materials

SCIENCE 5 Student Text
centimeter ruler
calculator
box (Use a box large enough to hold textbooks.)
container of room-temperature water, labeled *A*
container of cool water, labeled *B*
container of hot water, labeled *C*
3 thermometers

Practice measuring length with a centimeter ruler. Measure your Student Text. Round your measurements to the nearest centimeter.

1. Measure the length of your book. ______________ cm
2. Measure the width of your book. ______________ cm
3. Measure the height of your book. ______________ cm

Practice calculating volume.

4. The volume of my book is ______________ cm^3.
5. The volume of the box is ______________ cm^3.
6. I estimate that ______________ textbooks will fit into the box.

Practice measuring temperature with a thermometer.

7. The temperature of the water in container *A* is ______________ °C.
8. The temperature of the water in container *B* is ______________ °C.
9. The temperature of the water in container *C* is ______________ °C.

Background

Activity Arrangement

You may set up one or more temperature stations. Set time limits and have the students take turns at the stations. The students not at a station could practice linear measurement and volume calculations at their desks.

Volume Measurement

- Review that volume can be measured in liters.
  Volume can also be measured in other ways. How can we find the volume of the box since it cannot fit into a beaker or measuring cup? The formula for the volume of regular-shaped objects is length × width × height = volume.
- Multiply the measurements of the box to find its volume in cubic centimeters. Review math procedures as needed. Calculators may be used.
- Display the volume of the cardboard box for the students to use on Activity Manual page 41.

Temperature Measurement

> Thermometers vary. Compare several thermometers. The starting temperature, or room temperature, may be different on each. With this in mind, most of the activities in *Science 5* ask that a starting temperature be recorded. These activities usually compare the differences between readings rather than the actual readings. Also, if several thermometers are used in the same activity, do not rearrange them during the activity. Keep the same thermometer with the same container or substance.

- Display a container of water.
  What instrument can be used to measure temperature? a thermometer
- Place the thermometer in the water.
  What is the metric unit of measure for temperature? Celsius (centigrade)
- Read the thermometer in the water. Review how to read a thermometer as needed.
- Some Celsius thermometers are marked for every degree, others for every two degrees. Review how to read your specific thermometers.

Background

Equivalent Measurements

One milliliter is equal to one cubic centimeter. Because these measurements have this relationship, they can often be used interchangeably.

Activity Manual

Activity, page 41

Guide the students as they practice measuring length, volume, and temperature.

You may choose to have the students place their Student Texts into the cardboard box to test their volume calculations.

35
36

Student Text pages 62–67
Activity Manual pages 42–44

Objectives

- Identify and describe the three states of matter
- List examples of solids, liquids, and gases
- Define *physical change*
- Recognize that a change of state is a physical change
- Differentiate among melting, freezing, vaporization, and condensation

Materials

- variety of objects to change appearances (e.g., sunglasses, a hat, a wig)
- photos of yourself when you were younger (optional)
- wet paper towel

Teacher's Toolkit CD

- Quiz 3-A

Vocabulary

- physical property
- solid
- liquid
- gas
- physical change
- melting
- freezing
- vaporization
- condensation

Two days have been allotted for this lesson. You may want to teach Lesson 35 the first day and use ideas from Lesson 36 on the second day. You may also incorporate the ideas from Lesson 36 into the text material, making the lesson two days.

Introduction

- Put the sunglasses on or ask a volunteer to put them on.
  What has changed? You put on sunglasses.
  Did my identity change? no
  Am I still the same person? yes
- Repeat with several other items. Continue to reinforce that your identity has not changed, only your physical properties have.
- Display photos of yourself when you were younger.
  Did my identity change because I was a different size? no
  My mass and volume changed as I grew older, but my identity did not change. I am still the same person. Physical properties can change many things about a substance, but the identity of the substance does not change.

Preparation for Reading

- Instruct the students to mentally make questions using the headings, the bold and italicized words, and the captions.
- Direct them to read silently pages 62–67 to answer these questions.
  What are the states of matter?
  What process changes a liquid to a solid?

Physical Properties

When God created matter, He gave it many characteristics. Some of these are called physical properties. A **physical property** is a characteristic of a substance that can be observed without changing the identity of the substance. Some physical properties are easy for us to observe with our senses. The color, texture, and smell of a substance can usually be readily observed. Other properties can be harder to observe. These properties include things like a substance's density or its ability to conduct electricity. Whether easy or hard to observe, these characteristics are all important because they help us distinguish one substance from another.

Rose

Spruce

States of Matter

One physical property of matter is whether it is a solid, liquid, or gas. These properties are called the states of matter. Almost all matter on the earth is in one of these states.

All matter is composed of tiny particles that are always in motion. These particles attract each other. This attraction tends to hold the particles together. The motion of the tiny particles and the strength of their attraction determine a substance's state of matter.

62

Background

Phases or States

Some sources use the term *phase of matter* rather than *state of matter*. The terms are interchangeable.

Conservation of Matter

This basic principle of science states that matter cannot be created or destroyed. When matter changes states, it is not created or destroyed. The mass of the matter stays the same.

Activity

Using Physical Properties

Put an assortment of coins in a coin-sorting bank and observe how it sorts.

What physical property does this bank use to separate the coins? size

Inquire at a bank, vending company, or possibly the school office to observe a commercial machine that sorts coins.

Currency sorters use the physical property of appearance (optical) as they sort and count paper money. Many of these machines can also identify counterfeit currency by using magnetism, infrared light, or ultraviolet light in addition to the optical examination.

Solid

A desk, a pencil, and a book are all solids. **Solid** is the state of matter in which a substance has a definite volume and shape. If you put a pencil in a glass, the pencil stays in its own shape. It does not take on the shape of the glass. The particles that make up a solid do vibrate, but they do not move very much. They have a strong attraction to each other. This strong attraction keeps the particles close together and allows solids to keep their shape.

Solid

Liquid

Water, juice, and soft drinks are all liquids. **Liquid** is the state of matter in which a substance has a definite volume but not a definite shape. A liquid takes the shape of the container that it is in. The same amount of liquid can be poured from a tall, thin container into a short, wide container. The liquid's volume stays the same, even though the shape of the liquid changes.

The particles in liquids move more than the particles in solids do. Yet the attraction between the particles is still strong enough to keep the matter from completely spreading apart.

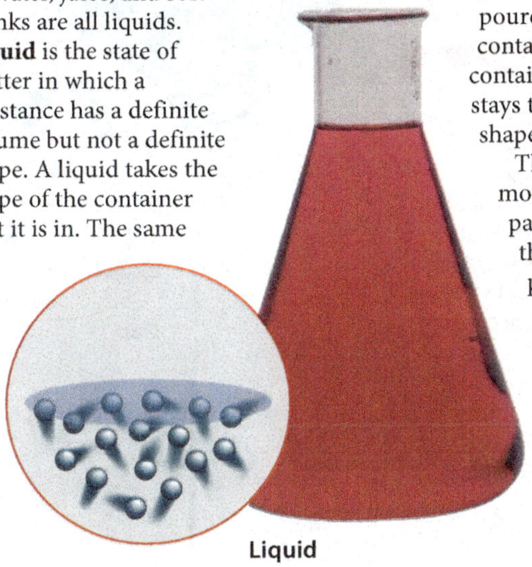

Liquid

63

Activity

Demonstrate Different Strengths of Attraction

Materials: three or more magnets of different strengths, iron nail

Guide the students as they test the attraction between the nail and the magnets. Discuss the differences of attraction and identify the strongest.

How much did you have to pull to separate the nail from each magnet?

Was it the same for each magnet?

Which magnet required you to pull hardest?

Arrange the magnets in order according to strength.

Put the nail and the strongest magnet together. Hold the nail and let the magnet hang. Gently shake the nail. The magnet should not fall.

How does the strength of the attraction of the strongest magnet compare to the attraction of the particles within a solid? The attraction is strong and holds the particles together.

Hang each of the other magnets, one at a time, from the nail and gently shake. The movement of the magnets should appear looser than for the strongest magnet, and the magnets may fall.

Teach for Understanding

What is a physical property of a substance? a characteristic of a substance that can be observed without changing the identity of the substance

Name some physical properties that are easy to observe. Possible answers: color, texture, smell

Name some properties that are harder to observe. Possible answers: density, the ability to conduct electricity

Look at the picture of the rose. What are some physical properties of a rose that you could observe? Possible answers: color, softness, smell

Look at the picture of the spruce branch. What are some physical properties of a spruce branch that you could observe? Possible answers: color, smell, roughness

What are the states of matter? solid, liquid, and gas

What determines the state of matter of a substance? the amount of attraction between its particles

What does *attract* mean? pull toward oneself

- Choose a volunteer to read the headings across the top of the chart on Activity Manual page 42.
- Ask another student to read the headings across the left side of the chart.
- As you discuss a solid, guide the students in completing the first row of the organizer on the Activity Manual page.

  Name the two characteristics of a solid. a definite shape and a definite volume

  Describe the motion and attraction of the particles in a solid. The particles do not move much. They have a strong attraction to each other.

  What does *strong attraction* mean? Possible answer: Each particle pulls other particles tightly toward it.

  What would happen to the shape of a solid if you put it in a bowl? It would stay the same shape.
- As you discuss a liquid, guide the students in completing the second row of the organizer on Activity Manual page 42.

  Name the two characteristics of a liquid. no definite shape (or takes the shape of its container) but a definite volume

  Describe the motion and attraction of the particles in a liquid. They move more than solids do, and attraction between particles is still strong enough to keep the matter from completely spreading apart.

  What would happen to a liquid if you put it in a bowl? It would take the shape of the bowl.
- You may want to do the *Demonstrate Different Strengths of Attraction* activity at this time.

- As you discuss a gas, guide the students in completing the third row of the organizer on Activity Manual page 42.

  Name two characteristics of a gas. no definite shape (or takes the shape of its container) and no definite volume (or expands to fill container)

  How are gases like liquids? take the shape of their containers

  How are gases different from solids and liquids? Gases can expand to fill the volume of their containers and can be compressed.

  What does *compressed* mean? pressed together or into less space or volume

  Describe the motion and attraction of the particles in a gas. They are very active and have a very weak attraction to one another. This allows them to spread apart until they meet a barrier that contains them.

- You may want to do the *Compression* demonstration at this time.
- Discuss *Fantastic Facts*.

  What is a fourth type of matter? plasma

  What source of energy is associated with all three plasma examples mentioned in your reading? light

Gas

Air in a tire and helium in a balloon are examples of gases. **Gas** is the state of matter in which a substance does not have a definite shape or a definite volume. Like liquids, gases take the shape of their containers. But unlike solids or liquids, gases can also expand to fill the volume of their containers.

Gases are also different in that they can be easily compressed. Perhaps you have seen a balloon being filled from a tank of helium gas. The gas in the tank is compressed. Only a small amount of helium gas is released into each balloon, but the gas expands to fill and inflate the balloon.

FANTASTIC FACTS

We often talk about the "three states of matter," but most of the matter in the universe is made up of a fourth type called plasma. Plasma occurs only at very high temperatures such as those found on the sun. On the earth, plasma naturally occurs in a lightning bolt. Manmade plasma causes a fluorescent light to shine. Some scientists think plasma may have great potential as a future energy source.

The particles in gases are very active. They have only a weak attraction to each other. Because of this, gas particles spread apart until they meet a barrier that contains them.

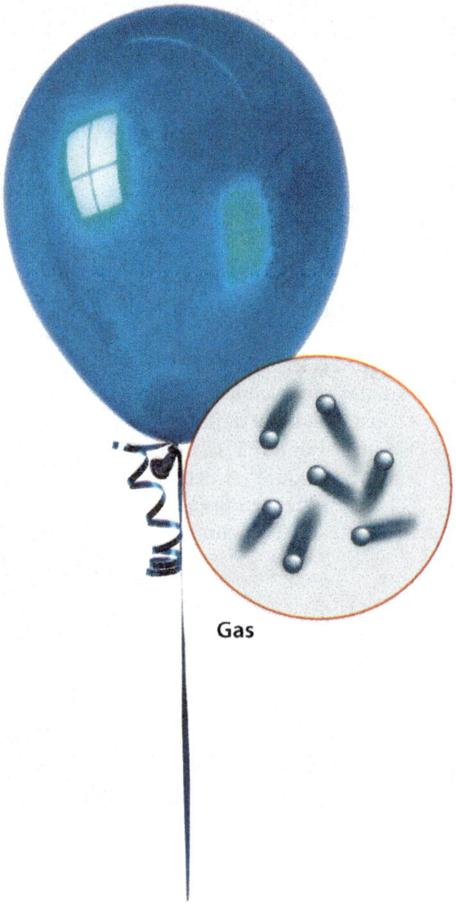

64

Demonstration

Compression

Materials: 2 slices of fresh bread

Display the slices of bread.

Although it is difficult to see how a gas compresses, we can easily see how bread compresses.

Both of these slices of bread take up about the same volume. What will happen if I compress one slice of bread?

Compress one slice of bread.

Does the bread still have the same volume? no

Has the mass of the bread changed? no How do you know? No matter was added or removed.

Although the second slice of bread takes up much less space, it has not changed mass. Gases are somewhat like this. They can compress or expand to fill the available space.

State of Matter		
State	**Shape**	**Volume**
solid	definite shape	definite volume
liquid	no definite shape	definite volume
gas	no definite shape	no definite volume

Changing States of Matter

People cannot create or destroy matter. Only God can. In John 1:3, the Bible reminds us that all things were made by God. However, matter can change in different ways. If you use a hammer to break a rock into smaller pieces, the rock changes in size, but the pieces are still rock. The rock has had a physical change. A **physical change** is a change in matter that does not form a new substance.

A physical change also occurs when matter changes from one state to another. Ice cream is usually a solid. It has a definite shape and a definite mass. But if you leave it out of the freezer for a while, it will change to a liquid. It has the same mass, but it will take the shape of whatever container you put it in. The ice cream has not changed into another substance, nor has it lost or gained matter. The ice cream just changed its form, or state of matter.

The temperature at which a substance changes from one state of matter to another is one of its physical properties. This physical property can help us describe and identify different kinds of matter.

An ice cube is the solid state of water. An ice cube has its own shape and volume. If you take the ice cube out of the freezer, though, the ice starts to change to a liquid. This process of a solid changing to a liquid is called **melting**. The temperature at which a solid changes to a liquid is called its *melting point*. Different solid substances melt at different temperatures. Ice melts at 0 °C (32 °F), but lead melts at about 328 °C (622 °F). Most metals have very high melting points. Knowing information about how a solid melts allows us to use that solid in ways that can help others.

65

- Discuss the *State of Matter* chart.
  Which state of matter has a definite shape? solid
  Which states of matter have a definite volume? solid and liquid
  Which state of matter does not have a definite shape or a definite volume? gas
  What is a physical change? a change in matter that does not form a new substance
  When a solid melts and changes to a liquid, is that a physical change? yes Why? It is still the same substance. It has not lost or gained matter.
  When matter changes forms, what happens to the amount of matter in the substance? The amount stays the same.
  Think back to the way we introduced this lesson. What kinds of physical changes did we have? Answers will vary.
  During a physical change, the substance itself does not change. It may look different or feel different, but it is still the same substance.
  What is happening to a substance when it melts? The solid substance is changing to a liquid substance.
  What do scientists call the temperature at which a solid changes to a liquid? the melting point
  Do all substances change states at the same temperatures? no
  How does this fact help scientists? Possible answer: The temperatures at which a substance changes states is a property that can help identify the substance.
  What are the melting points of ice and lead? Ice melts at 0 °C (32 °F). Lead melts at 328 °C (622 °F).
  Which melts at a lower temperature, ice or lead? ice
- Refer the students to the chapter introduction on Student Text page 57. Discuss the knowledge Andrea needed to know about melting chocolate for the brownies.
- Explain that nothing in God's creation is by accident. The fact that substances melt at different temperatures so that we can use them for our benefit is an example of God's lovingkindness toward us.

Background

Other States of Matter

Solids, liquids, and gases are the states of matter found under normal conditions on the earth. However, matter can also exist under extreme conditions. The fourth and fifth states of matter are plasma and Bose-Einstein condensate. Matter in the plasma state exists under extremely hot conditions, such as those found in a lightning bolt or on the surface of the sun. Bose-Einstein condensate exists under extremely cold conditions. This fifth state was discovered in the 1990s. Details about its properties are still being determined.

Activity

Writing a Fictional Story

Direct the students to write a fictional story about a place in which matter has different states and properties from those that actually exist. For example, the place may have metals that melt at very low temperatures and wood that almost never burns. Under these conditions, cooking over a campfire would be very different. Include an illustration of at least one scene from the story.

35
36

What do scientists call the process of a liquid changing to a solid? freezing or solidifying

Does a substance have to be cold to freeze? no Explain. Freezing means that a substance is changing to a solid. It does not mean that the substance is cold. Our perception of cold is usually based on the temperature at which water freezes. However, all substances do not freeze at that temperature.

If a substance solidifies, what is happening? It is becoming a solid.

What do scientists call the temperature at which a substance changes from a liquid to a solid? the freezing point

What is the relationship between the freezing point and the melting point of a substance? They are the same temperature.

- Direct attention to question number 1 on Activity Manual page 42. Guide the students as they list the processes.

What is vaporization? the process of a liquid changing to a gas

What is another word for vaporization? boiling

When does boiling occur? when vaporization is occurring throughout a liquid and bubbles of gas rise and break away from the surface of the liquid

What is a boiling point? the temperature at which boiling occurs

At what temperature does water boil? 100 °C (212 °F)

Think about a pan of boiling water on a stove. Do you think that the pan has a higher or lower boiling point than the water in the pan does? higher Why? Possible answer: The pan has a higher boiling point because it is a solid. Water is a liquid and boils before the pan does.

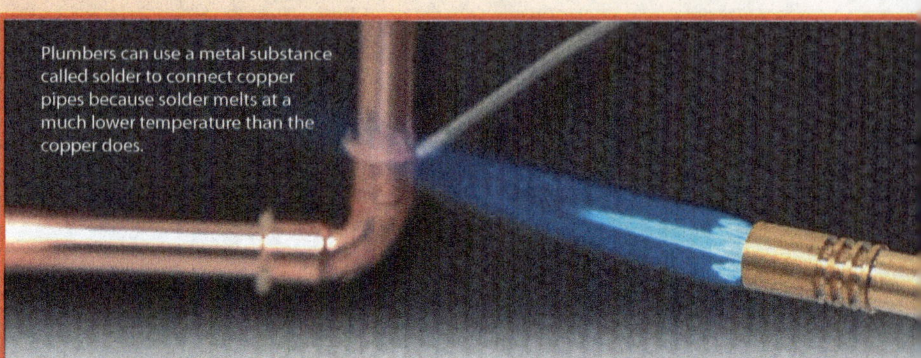

Plumbers can use a metal substance called solder to connect copper pipes because solder melts at a much lower temperature than the copper does.

Matter can also change from a liquid to a solid. If you put an ice cube tray of water into the freezer, the water freezes into solid ice cubes. The process of a liquid changing to a solid is called **freezing**.

We usually think of freezing as meaning something is cold. But in science a substance does not have to be cold to be frozen. Any substance that is changing from a liquid to a solid is freezing. Another word that we could use to describe this is *solidifying*. Think about a piece of solid chocolate. The chocolate can be melted to use for cooking. But if melted chocolate is left at a cool room temperature it will change back to a solid. It does not have to be in a freezer to freeze, or solidify. The temperature at which a substance changes to a solid is its *freezing point*. The melting point and the freezing point of a substance are the same temperature.

Another way matter can change is from a liquid to a gas. This can be harder for us to see, but it happens all around us. The process of a liquid changing into a gas is called **vaporization**. A liquid is boiling when vaporization is occurring throughout the liquid and bubbles of gas rise and break away from the liquid's surface. The temperature at which this occurs is called the *boiling point*. Different substances have different boiling points, just as they have different melting points. Water boils at 100 °C (212 °F).

Boiling water is an example of vaporization.

66

Background

Scientists Apply States of Matter

Many types of scientists use the basic principles about matter and how it changes states. For example, biochemists must know the boiling and melting points of the substances that are used to develop new medications. A physicist uses knowledge about the masses and densities of substances as he studies the interactions of matter. A chemist studies atoms and elements and how they combine to form solids, liquids, and gases.

Freezing and Boiling Points

Although you may choose to have the student memorize the freezing and boiling points of water, emphasize that these temperatures are not the same for all types of matter.

Vapor and Evaporate

Both words are from a Latin word that means "steam." The prefix *e-* has the same meaning as *ex-*, so it gives the idea of "outside" or "away." As water evaporates, it forms vapor that moves away from the liquid.

Evaporation is another form of vaporization. A substance evaporates when it changes from a liquid to a gas at a temperature below its boiling point. Evaporation occurs only at the surface of a substance. You may have noticed a puddle of water that gradually dried up during the day. The water did not boil, but it did change to a gas through evaporation.

A substance can also change from a gas to a liquid. This process is called **condensation**. Perhaps you have set a cold glass of liquid on a table. After a little while, you saw drops of water on the outside of the glass. Water vapor in the air had cooled and changed back to a liquid. The *condensation point* is the temperature at which a gas turns to a liquid.

We often see water changing states. But in certain conditions almost all matter changes states. Even iron can get hot enough to change to a vapor. The condition at which a substance changes states is a physical property that helps scientists identify the substance.

Water vapor in the air condenses on the outside of a cold glass.

QUICK CHECK

1. What two things determine a substance's state of matter?
2. What are two ways gases differ from liquids and solids?
3. What is freezing?
4. What is vaporization?

67

35–36

- Use a wet paper towel to wet the chalkboard before discussing evaporation. Allow time for the chalkboard to dry.

We say that the chalkboard has dried, but what happened to the water that was on the chalkboard? It evaporated.

Was the water boiling? no

What process occurs when a liquid changes to a gas at a temperature below its boiling point? evaporation

How is evaporation different from boiling? Evaporation is vaporization that occurs only at the surface of a liquid. Boiling is vaporization occurring throughout a liquid.

What is the process of a gas changing to a liquid called? condensation

Give an example of a gas condensing. Possible answer: drops on the outside of a cold drink

Does the process of changing states of matter happen only with water? no Explain. Almost all matter can change states.

- Direct attention to question number 2 on Activity Manual page 42. Guide the students as they list the processes.

ANSWERS

1. the motion of the particles and the strength of their attraction
2. no definite volume and can be easily compressed
3. the process of changing from a liquid to a solid
4. the process of changing from a liquid to a gas

36

- The *Compression* demonstration and the *Using Physical Properties, Demonstrate Different Strengths of Attraction*, and *Writing a Fictional Story* activities found at the bottom of the pages in Lessons 35–36 may also be incorporated into the lesson.
- Suggestions for additional activities for this two-day lesson are on the Teacher's Toolkit CD.

Activity

Classifying Matter

You may want to copy Instructional Aid 3.1: *What Is the Matter?* for each student. This page allows the students to find different types of matter in a Bible passage and then classify that matter into one of its three states.

Activity Manual

Study Skill and Reinforcement, page 42

Review, pages 43–44

These pages review Lessons 33 and 35–36.

Assessment

Quiz 3-A

The quiz may be given any time after completion of this lesson.

Objectives

- Use a scientific method

Materials

- clear cup (approximately 8 cm in diameter)
- water
- plate
- roll of pennies (More pennies may be needed if a larger cup is used.)

Vocabulary

- problem
- materials
- hypothesis
- procedure
- conclusion
- follow-up

This experiment is intended as a teacher demonstration. The emphasis is to explain the steps of a scientific method and practice some of the process skills that will be used in other activities. The students will complete Activity Manual pages 45–46 as the lesson is taught.

Introduction

- Explain that a scientific method is a systematic approach to problem solving.
- Explain that although the specific steps of the scientific method may vary, they should include using a hypothesis to identify a problem, experimenting to test the hypothesis, and drawing conclusions to test the accuracy of the hypothesis.

Teach for Understanding

- Ask a student to read aloud the blue box under *Problem*.
  What is the problem of an experiment? the question that the experiment will answer
- Explain that the problem is worded as a question and is specific rather than general.
- Display the problem for this experiment and ask a volunteer to read it aloud.
  Write the problem on Activity Manual page 45.
- Choose a volunteer to read aloud the green box under *Materials*.
- Display the materials: a cup, a plate, water, and a roll of pennies.
  List the materials you will use.
- Ask a volunteer to read aloud the yellow box under *Hypothesis*.
- Discuss the definition of *hypothesis*.
  The hypothesis is written as a statement rather than as a question. Complete the hypothesis sentence.
- Share the information about how God encourages Christians to discern what is true.

A Science Experiment

Name ____________________

ACTIVITY

Problem

Each experiment begins with a **problem**, a *question* that needs to be answered.

Write the problem.

How many pennies can be added to a full cup of water before the water spills out of the cup?

Materials

The list of **materials** includes all the materials and equipment you will use to conduct the experiment. In some experiments, all of the materials will be listed for you. In others, you will choose some of the materials.

List the materials you will use.

1 clear cup
plate
water
1 roll of pennies

Hypothesis

A **hypothesis** is a scientist's testable idea of the answer to a problem. As you conduct experiments, your hypothesis is a *statement* of your idea of how the problem will be answered through that experiment.

Complete the hypothesis sentence.

____________ pennies can be added to a cup of water before the water spills out.

Procedure

The **procedure** is the steps of an experiment. The steps must be followed exactly to ensure the accuracy of the results. Scientists **record** their observations and results throughout an experiment.

Follow the procedure.

1. Place the cup on the center of the plate.
2. Fill the cup with water so that the water level is even with the top edge of the cup but no water overflows.
3. Carefully slide the pennies into the water, one at a time. Count the pennies as you put them in.

Background

Discerning What Is True

Understanding and using the scientific method can also be very useful for someone's spiritual life because it develops critical thinking. God made people's minds and wants them to think critically and ask questions about all of the world around them. God encourages Christians to test and examine ideas that are presented to them. This even includes religious ideas. In Acts 17:11, the Bereans were commended because they did not just accept what someone said but did their own research in the Scriptures to find out if what they were hearing was true. Critical thinking is most useful when it begins with the truth of God's Word. [BAT: 6a Bible study]

Continue adding pennies until the number reaches the number in your hypothesis or until water spills out of the cup.

Record your observations and results.

If water did not spill, form a new hypothesis about the number of pennies that can be added.

Continue the procedure until water spills.

umber of pennies in your hypothesis	Did water spill?

nclusions

ıe **conclusions** evaluate what was learned from the experiment. ey relate and apply the information to other areas.

wer the questions.

How many pennies did it take for water to spill?

Did your results support your hypothesis? If not, was your prediction too high or too low?

llow-up

ıe **follow-up** suggests ways to vary the experiment to answer new estions. This involves changing the materials or conditions of the periment.

mple: Use a liquid other than water.

te your own follow-up to this experiment.

ssible answers: Use marbles instead of pennies. Change the size of the cup.

- Direct a student to read aloud the purple box under *Procedure*. Explain that the steps in a procedure are the steps that a scientist follows as he tests his hypothesis.
  What is the procedure of an experiment? the steps that will be followed
  Why is it important to follow the steps exactly? to ensure that the results are accurate
- Explain that scientists usually repeat tests before drawing final conclusions. A scientist must follow his procedure precisely to make his information valid.

Use a document camera (or an overhead projector) to view the cup of water in profile. This will help a group of students to see any drops of water that fall. Fill the cup as full as possible. The water should bulge slightly above the rim of the cup. Avoid touching the water with your fingers.

- Ask a volunteer to read aloud the procedure steps before beginning.
- Set up and demonstrate the experiment.
  Record your observations on the chart on Activity Manual page 46. Write the number of pennies that were added before the experiment stopped. Write *yes* or *no* to answer the question in the second column.
- Discuss whether the hypothesis was correct. If not, guide the student in adjusting his hypothesis, repeating the activity, and recording his observations.
  Having an incorrect hypothesis is not bad. Scientists often have to adjust and rewrite hypotheses as they test a problem.
- If water has not spilled after three tries, continue adding pennies until it does.
- Ask a volunteer to read aloud the tan box under *Conclusions*. Help the students draw conclusions based on the experiment. Generally, conclusions are inferences based on what the students have seen.
  Was your hypothesis correct? If not, was your prediction too low or too high?
- Ask another student to read aloud the yellow box under *Follow-up*.
  What details or conditions about this experiment could we vary or change? Possible answers: the kind of liquid; the size of the cup; the temperature of the water; the type of coins; the way the penny is dropped

Background

Surface Tension

The water is able to bulge above the rim of the cup because of surface tension. The positive and negative properties of water molecules cause the molecules to attract one another and form a "skin." This attraction is called *cohesion*. Because of surface tension, creatures such as water striders can walk on the surface of water. Test the surface tension of the water in the cup by carefully placing a paper clip on the water. The paper clip should float on the surface.

Activity Manual

Activity, pages 45–46

These pages are designed to be used during the discussion.

38

Student Text pages 68–69
Activity Manual pages 47–48

Objectives

- Identify atoms as small particles of matter
- Differentiate between elements and compounds
- Contrast chemical changes and physical changes

Materials

- assortment of colored interlocking bricks
- something built from interlocking bricks, such as a truck

Teacher's Toolkit CD

- Visual 3.3: *Periodic Table of Elements*
- Quiz 3-B

Vocabulary

- atom
- element
- molecule
- compound
- chemical change

Introduction

- Display the interlocking bricks and the object.
  Have you ever built things from interlocking bricks? Answers will vary.
  This truck is made up of many pieces. The pieces are different colors and shapes.
  Like this truck, all matter is made up of very small particles. In today's lesson we will look at some of the ways these particles combine to make matter.

Preparation for Reading

- Direct the students to mentally make questions using the heading, the bold words, and the captions.
- Instruct them to read silently pages 68–69 to find the answers to these questions.
  How are compounds different from elements?
  What is a chemical change?

Teach for Understanding

What small particles make up all matter? atoms

What is an element? matter that is made up of only one kind of atom

Is most matter made up of just one kind of atom or many kinds of atoms? many kinds

- Build and display a small structure of interlocking bricks using pieces of only one color.
  This object represents an element. It is made of only one kind of "atom."
- Display the *Periodic Table of Elements* visual.
  You may have seen a table like this one. It shows the more than 100 different elements that scientists have named.
  What are molecules? particles formed by atoms joining with other atoms

Classifying Matter

Atoms and Molecules

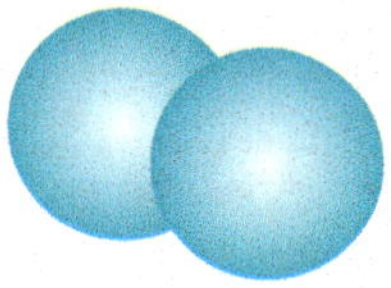

O_2 is a molecule made up of 2 oxygen atoms.

All matter is made up of small particles called **atoms**. These atoms are much too small for you to see. Most matter contains more than one kind of atom. The paper used to make this page is made of different kinds of atoms. Your chair is made from a variety of atoms. You are even made of dozens of kinds of atoms.

Some matter contains only one kind of atom. This type of matter is called an **element**. Gold, silver, and oxygen are all elements. Scientists have named more than 100 different elements.

Atoms may join with other atoms to form particles called **molecules**. Air contains molecules of oxygen and molecules of water. Each molecule of oxygen contains only the element oxygen. However, each molecule of water contains two elements: hydrogen and oxygen. Substances such as water are called compounds. A **compound** is formed by combining atoms of different elements.

A compound is a new substance. It is different from the elements that formed it. Water is different from either hydrogen or oxygen. Both hydrogen and oxygen are usually found as gases. Hydrogen is an explosive gas that burns easily. Oxygen is needed to help many other things to burn. But water does not burn. It can be used to put out fires.

2 atoms of hydrogen | 1 atom of oxygen | 1 molecule of water

68

Background

Atom

An atom is the smallest particle of an element. Atoms can exist alone or in combination with other atoms. The structure and behavior of atoms and molecules are discussed more thoroughly in *Science 6*.

Element

An element is a substance that is made of only one type of atom. The *Periodic Table of the Elements* classifies elements according to their physical and chemical properties.

Molecule

A molecule is the smallest particle of a substance that retains all the properties of that substance. A molecule forms when two or more atoms join together.

Compound

A compound is a substance that is formed from the chemical combination of two or more different elements. These elements are always found in the same proportion throughout the substance. A compound cannot be separated through physical means.

Every molecule of a compound is chemically identical. For example, every molecule of water is the same. Each molecule of water always has two atoms of hydrogen and one atom of oxygen.

Compounds form through a chemical change. A **chemical change** occurs when the atoms in an element or compound combine to form a new substance. Unlike a physical change, a chemical change causes the elements or compounds to lose their individual properties. They change into a new substance with different properties. For example, the elements sodium and chlorine are both poisonous to people. When the atoms in these elements join, however, they form the compound known as table salt.

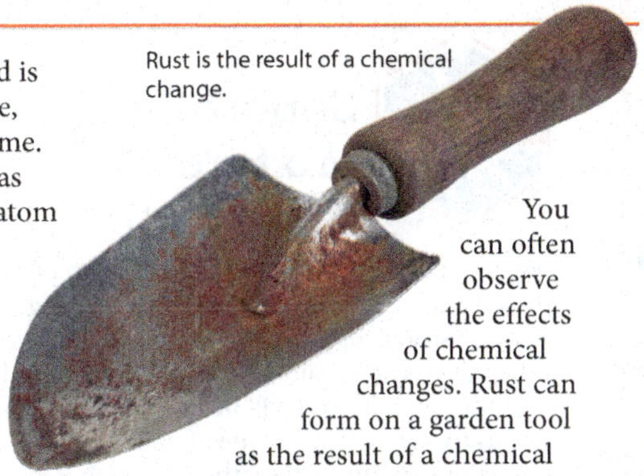

Rust is the result of a chemical change.

You can often observe the effects of chemical changes. Rust can form on a garden tool as the result of a chemical change. This change happens when the oxygen in water and the iron in the tool combine. They form a new substance that we call rust. Chemical changes can also occur in the kitchen. Brownies are the result of chemical change. The properties of the flour, eggs, and other ingredients change into a tasty substance—brownies.

QUICK CHECK

1. What is all matter made of?
2. What is an element?
3. What is a chemical change?

69

Do all the atoms in a molecule need to be the same kind? no

- Connect five interlocking bricks of one color into a stack. Connect another five pieces of more than one color into a second stack.

These stacks represent molecules. Molecules can be made from one kind of atom or from different kinds.

What type of molecule is made of atoms from different elements? a compound

Which of these "molecules" represents a compound? the one with different colors How do you know? A compound is made of atoms of different elements, and the colors represent different atoms of different elements.

Does a compound have the same characteristics as the elements that make up the compound? no Explain. The compound is a new substance. It is different from the elements that formed it.

What is a chemical change? the process of two or more elements or compounds combining to form a new substance

What happens to the individual elements in a chemical change? They lose their original properties and take on new properties.

Name some examples of compounds. Possible answers: water, table salt, rust, cake, brownies

- You may want to do the demonstration at this time.

What are two types of changes you have learned about? physical and chemical

ANSWERS

1. atoms
2. matter that is made up of only one kind of atom
3. the process of two or more elements or compounds combining to form a new substance

Demonstration

Demonstrate a Chemical Change and a Physical Change

Materials: candle, candle holder, match, piece of foil

Light the candle.

A chemical reaction is occurring as the wick burns. How do we know this is a chemical change? The wick changes to ash. It is no longer a piece of string.

Tip the candle so that some wax drips onto the foil.

What kind of change is the wax going through? a physical change

What physical changes have occurred in the wax? It changed from a solid to a liquid, then back to a solid.

Activity Manual

Review, pages 47–48

These pages review Lesson 38.

Assessment

Quiz 3-B

The quiz may be given any time after completion of this lesson.

Objectives

- Plan a procedure for separating the parts of a mixture
- Apply the physical properties of the items that make up a mixture
- Experiment to test predictions
- Infer how to physically remove a dissolved item from water

Materials

- paper towels
- resealable plastic sandwich bag for each group
- 10 mL iron filings for each group (see Background)
- 30 mL rice for each group
- 30 mL sand for each group
- 30 mL sugar for each group

Teacher's Toolkit CD

- Rubrics: *Separating a Mixture,* for each student

The iron filings, rice, sand, and sugar should all be placed in the resealable sandwich bags. You may choose to have the students measure the items for their mixtures or you may place the items in the bags yourself.

Introduction

- Display a bag of the prepared mixture.

  This bag contains iron filings, rice, sand, and sugar. It is a mixture because each item can be separated from the others based on at least one physical property.

  In this activity you will plan how to separate each item from this mixture and then test your predictions.

Preparation for Reading

- Direct the students to read silently Student Text pages 70–71 and Activity Manual pages 49–50 before beginning the activity.

Teach for Understanding

Process Skills

The process skill emphasized in this activity is experimenting, which is setting up a procedure to test a hypothesis.

- Explain what experimenting is.

  What do you need in order to experiment? materials and the necessary conditions for testing the hypothesis

  Would a scientist stop testing a hypothesis if one experiment did not prove it? probably not

ACTIVITY

Separating a Mixture

Process Skills
- Predicting
- Experimenting
- Observing
- Inferring
- Communicating

Matter has many properties. Size, shape, and mass are some physical properties that we can observe and measure. The ability of matter to dissolve in water or to be attracted to a magnet are also physical properties. By using these properties and others, you can separate a mixture into its individual parts.

In this activity you will be given a mixture of four items. Your task is to separate each type of item from the mixture. You may need to use a different method for each item. You may also need to use more than one method to isolate a specific item. Plan your procedure carefully. One method of separation may affect another.

Materials
- mixture of iron filings, rice, sand, and sugar
- 4 containers
- paper plate
- cheesecloth
- funnel
- magnet
- spoons
- strainer or sieve
- tweezers
- water
- Activity Manual

Problem

How do you separate a mixture into its individual substances?

Procedure

1. Get your mixture from your teacher.
2. Plan the order in which you will separate each substance. You may use any of the materials included in the materials list. Record your plan in your Activity Manual.
3. Choose the material that you will use to separate the first substance. Record the name of the substance, the method that you will use to separate it, and the property of matter that the method uses.
4. Test your method for the first substance. Record the results. If needed, try other methods until most of the substance is removed.
5. Following your plan, repeat steps 3–4 for the other substances. Remember to record your information at each step.

70

Background

Activity Helps

Cheesecloth may be purchased at most grocery stores.

Use purchased iron filings. Cut-up pieces of steel wool pads can be used in place of iron filings, but the pieces can cut or puncture skin. Purchased iron filings are safer. See information on the BJU Press website for places to find iron filings.

bjupress.com/resources/science

Place the magnets in plastic bags to make the cleanup easier. Instruct students to keep the magnet in the plastic bag as they work.

Discourage several groups from using the water first but do not tell them why. Water causes the items to stick together, which may hinder the other methods from working easily.

Conclusions

- Did your plan work?
- Was there a better method or order that you could have used?

Follow-up

- Use other materials or methods to separate the substances.
- Separate a mixture containing other substances.

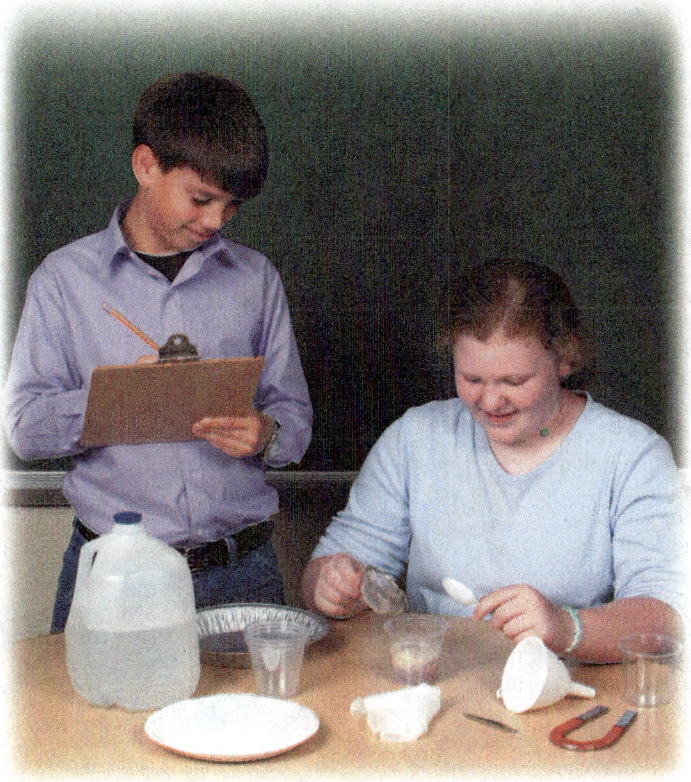

71

39

What could you do if your results do not prove your hypothesis? Possible answer: rethink the hypothesis and experiment again

Why is experimenting important to scientific discovery? Elicit that, without experimentation, ideas are only the opinions of people and are not proven facts.

Procedure

- There is not one "correct" procedure for this activity. The students are to plan their methods based on their own reasoning.
- This is an inductive lesson in which the students should use their own background knowledge as they make choices. Try not to lead the students to your own conclusions. You may not think a method is correct or best; however, the students should test each method, making adjustments as needed, until each item has been successfully separated from the mixture.
- If needed, discuss the property that each of the available materials uses. For example, the tweezers can be used to separate an item according to its appearance. (Given enough time, the entire mixture could be separated with tweezers!)

Conclusions

- Choose one student from each group to explain what methods his group used to separate the items. Discuss methods that were similar or different from one group to another.

Activity Manual

Activity, pages 49–50

Assessment

Rubrics

Select the prepared rubric, or design a rubric to include your chosen criteria.

40

Student Text pages 72–75
Activity Manual page 51

Objectives
- Define *mixture*
- Explain the difference between a mixture and a compound
- Give some examples of mixtures
- Identify some ways that substances in a mixture can be separated using physical properties

Materials
- snack mix, enough for each student to have a small amount

Be alert to students with nut allergies when choosing the snack mix.

Vocabulary
- mixture

Introduction
- Display the snack mix.

  What ingredients make up this snack mix? Answers will vary depending on the mix that you have.

  What are some ways that you could separate this snack mix into its individual ingredients? Possible answers: by shaking, by sorting

  Today we will look at some types of mixtures and different ways to separate them.

Preparation for Reading
- Display the word *aerosol*. Pronounce the word together.
- Direct the students to mentally make questions using the headings, the bold word, and the captions.
- Instruct them to read silently pages 72–75 to find the answers to these questions.

  What are some examples of mixtures?

  How can you separate salt from sand?

Teach for Understanding

What is a mixture? two or more substances that are physically combined

How is a mixture different from a compound? Each particle in a mixture keeps its identity. A compound forms a new substance, so the particles in a compound lose their identities and take on new ones.

Mixtures

Sometimes different substances combine without going through chemical changes. A **mixture** consists of two or more substances that are physically combined. Unlike the molecules in a compound, the particles of each substance in a mixture keep their identity. Think of a snack mix. A snack mix is a mixture that can include pretzels, peanuts, and raisins. The pretzels, peanuts, and raisins did not change to other substances when they were added to the mixture. They each kept their own identities and properties.

The amount of each substance in a mixture may vary. If your family likes lots of raisins, you can add more raisins to the snack mix. You also could add other substances such as crackers and candy pieces.

Each substance in a mixture can be separated from the mixture. Since each substance keeps its properties, it may be identified and removed. You may like to pick all the peanuts from the snack mix and eat them first. The peanuts look and taste different from the pretzels and the raisins.

The matter in a mixture may sometimes undergo a physical change. Physical changes include changes of physical properties such as the state of matter, color, or texture. Strawberries and ice cream go through a physical change when they are blended to form a milkshake. Combining the strawberries and the ice cream does not form a new substance, though. You can still see the strawberries in the ice cream.

In some mixtures the individual substances are not spread evenly throughout the mixture, as in the snack mix. Another example of this kind of mixture is granite. If you look closely at a piece of granite, you will see that several kinds of minerals have "mixed" to form it. However, the minerals are not mixed evenly throughout the granite. Each piece of granite may have a different mixture of minerals in it.

A snack mix is an example of a mixture.

72

Background
Types of Mixtures
In some mixtures, the individual substances are not spread evenly throughout the mixture. These mixtures are called heterogeneous mixtures. In other mixtures, the substances are evenly distributed throughout the mixture. This kind of mixture is called a homogeneous mixture, or a solution. Solutions are discussed in Lesson 41.

Activity
Identify Mixtures
Direct the students to make a list of any mixtures they may have used that day. For example, their lists might include orange juice, cereal and milk, hair spray, snack mix, or iced tea. Discuss their lists and verify that each thing listed is a mixture.

Granite

It is easy to identify the substances in some mixtures as being evenly mixed or not. You can easily see that vegetable soup is not evenly mixed. Likewise, if you look at the sand on a beach, you can see a variety of rocks and shells in it. No two handfuls of sand have exactly the same substances in them.

Some mixtures separate if they sit for a while. Orange juice, some liquid medicines, and paint are mixtures that separate. These mixtures may have labels that say "Shake well before using." Shaking these mixtures combines the substances again.

The shaving cream that many men use each morning is also a type of mixture that is not evenly mixed. Shaving cream mixes a gas with a liquid to form a mixture called foam. An aerosol is also a mixture of liquid in a gas. In some aerosols, you may even be able to see the droplets of liquid. Some hair sprays come as aerosols.

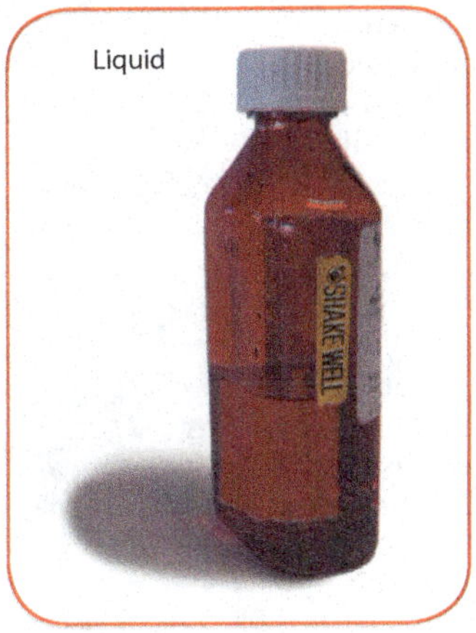

Some mixtures separate and must be combined again by shaking.

73

- Distribute some snack mix to each student and instruct them to begin eating.
  Did you pick out one ingredient to eat first? Did you pick out an ingredient that you do not like or cannot eat? Answers will vary.
  How were you able to identify and remove those ingredients? Answers will vary, but elicit that this was possible beca-use each ingredient kept its own physical properties.
  What happens to the physical properties of substances when the items in a mixture undergo a physical change? The properties change.
  What are some examples of physical properties that might change? Possible answers: the state of matter, the color, the texture, or the taste
  The substances in some mixtures are not spread evenly throughout the mixture. What is one type of rock that is an example of this kind of mixture? granite
  Many foods that we eat are mixtures in which the individual substances are not evenly distributed. What are some foods and beverages that are mixtures like this? Possible answers: trail mix, vegetable soup, orange juice, tossed salad, cereal and milk, banana split
  What are some examples of mixtures that are not food? Possible answers: shaving cream, aerosols, hair spray, paint, any item that needs to be stirred or shaken before use
- Discuss the pictures.

Background

Aerosols

Aerosol is a general term. It often refers to either a solid or a liquid mixed with air or another gas. Some examples of aerosols in the air are smoke, haze, and dust. The term *aerosol* is also used to refer to an aerosol can. In an aerosol can, a liquid product, such as hair spray or shaving cream, mixes with a gas as it is released from the can. In some products, the gas makes bubbles, forming a foam. In other products, such as hair spray, the gas causes the product to become tiny droplets. Different gases are used with different products.

Many products are dispensed in aerosol cans. For example, some types of whipped cream, cooking spray, deodorant, insect repellent, spray paint, disinfectant, and cleaner come as aerosols.

The first aerosol spray can was invented in the 1920s by Erik Rotheim, a Norwegian inventor and engineer. During World War II, American scientists invented the first aerosol insect repellent that soldiers could use. Although those first aerosol containers were large, the easy-to-use insect repellent was a valuable resource for soldiers fighting in the Pacific. Today aerosol containers come in all sizes and are used for thousands of products.

40

Why is it possible to separate substances in a mixture? They are not held together chemically.

What kind of properties do you use to separate a mixture? physical properties

What are some ways to separate mixtures? Possible answers: by appearance, by size, by shaking, by density

If you were asked to separate a mixture of coins, what physical properties would you use to separate the mixture? Possible answers: size, weight, color

- Discuss the pictures at the top of the page.

The types of matter that are included in the mixture are rice and beans.

Why did the beans come to the top of the rice after the mixture was shaken? Items often settle into layers because of their sizes. The beans are larger than the rice, and larger items are usually pushed to the top.

What is density? the amount of matter in a certain space

Look at the picture of the oil and vinegar. What causes the layers seen in oil-and-vinegar salad dressing? Vinegar is denser than oil. The oil rises above the vinegar.

- You may choose to use oil, vinegar, and a tall, clear container to show how liquids can separate into layers based on their densities.

Before shaking

After shaking

Separating Substances in Mixtures

Substances in mixtures are not held together chemically. They can be separated using their physical properties. Some mixtures are easy to separate based on how the parts look. You could look at a snack mix and easily remove each ingredient. You could take out all the pretzels and put them in one pile. Another pile could then be made of all the peanuts, and so on for each food item.

Suppose you had a jar full of coins. You could separate them by how they look. But if you had a coin sorter, you could separate the coins more quickly. The coin sorter separates the coins based on each coin's size.

Sometimes you can separate dry items by gently shaking them. The smaller items fill in the gaps at the bottom of the container. The larger items are pushed to the top. The items often settle into layers because of their sizes.

When each of the liquids in a mixture has a different density, the mixture may separate into layers. The order of the layers depends on how dense each liquid is. The densest liquid will be at the bottom. The least dense liquid will be at the top. Oil-and-vinegar salad dressing separates in this way. Oil is not as dense as vinegar and goes to the top. Until you shake the mixture, the two liquids will stay in different layers. If you are careful, you can separate the mixture by pouring off most of the oil.

74

Background

Water Purification

Most water is not pure water; it is a mixture. The substances mixed with the water can include salts, other minerals, chemicals, sediments and debris, or parasites and other disease-carrying substances.

Many methods have been developed to purify water for drinking and other uses. The most common, and probably the oldest, method is boiling. Other methods—such as filtering, adding chemicals, or using ultraviolet light—are also used to purify water.

Distillation is one method of boiling water to separate its components. This method captures the steam and condenses it back to a liquid for use. Some bottled water is purified by this method.

Most city water-treatment plants use a combination of methods. Some sediments separate and settle to the bottom while the water is in special holding tanks. The water is then pumped through a variety of filters that are specially designed to remove or absorb pollutants. Finally, chemicals are added. Of course, adding chemicals makes the water a mixture again, but the chemicals disinfect the water by removing substances that can cause disease.

SCIENCE & HISTORY

One method of separating mixtures had a great impact on the history of the United States. This method called gold panning was once used to separate gold from the mixture it was in. In 1848 James Marshall found gold in a river in California while building a sawmill for John Sutter. This began the great California gold rush.

Many men rushed to try their hand at panning for gold. To pan for gold, a man would put a small amount of dirt, gravel, and water from a stream into a shallow pan. He would then slowly shake the pan. The movement caused the materials to separate based on how dense each material was. Gold has a high density and would sink to the bottom. Much of the other material could be carefully poured off with the water. Gold and some other dense rocks would remain at the bottom of the pan. The shiny gold flakes and nuggets would then be picked out of the pan.

Solids may also separate based on density. Perhaps you have seen how sediment in a stream forms. The densest substances settle out first. Then other substances settle out according to how dense each is.

Not all mixtures separate easily, though. Suppose you had a mixture of salt and sand. The pieces of salt and the pieces of sand would be nearly impossible to separate from each other. However, you could use some of the physical properties of salt and sand to make them easier to separate. One of the substances dissolves in water and one does not. If you added water to the mixture of salt and sand, the salt would dissolve in the water. The sand would settle to the bottom of the water. After pouring out the salt water, the sand would be separated through its physical properties.

However, the mixture would not yet be completely separated. The salt would still be combined with the water. A way to separate the salt and water would be to boil the salt water. Boiling causes the water to change from a liquid to a gas. The salt would remain in the container.

QUICK CHECK

1. What is a mixture?
2. How is granite an example of a mixture that is not evenly mixed?
3. What are three ways that you might separate a mixture?

75

What physical properties of sand and salt make it possible for us to separate them? Salt dissolves in water, but sand does not.

How could you separate the salt out of the water? by boiling the salt water

How does vaporization aid in the separation of mixtures? by causing the liquid in a liquid-solid mixture to evaporate, leaving the solid behind

- Discuss *Science & History*.

How was separating mixtures important in the United States in the 1840s? People panned for gold in the California gold rush.

How does panning work? A person scoops water, dirt, and gravel into a pan and shakes the mixture so that the gold pieces, which are denser, sink to the bottom.

ANSWERS

1. two or more substances that are physically combined
2. Granite is made of many kinds of minerals, and the different minerals are scattered throughout the granite.
3. Possible answers: by appearance, by size, by shaking it, by density

Activity Manual

Expansion, page 51

This page presents additional information about classifying mixtures.

Activity

Separating Ink Colors

Materials: clear plastic cups, water, black felt-tip pens (each made by a different manufacturer), 2 cm wide strips of coffee filter, paper clips, pencils or craft sticks

Follow the same procedure for each pen. Pour about 3 cm of water into a cup. Balance a pencil or craft stick across the top of the cup. Make a large dot about 5 cm from one end of a strip of filter. Bend a paper clip into an S shape. Attach one end of a paper clip to the end of the filter. Hang the paper clip on the pencil so that the end of the strip touches the bottom but not the sides of the cup.

Observe any changes to the strip and the dot as the water moves up the strip. The dot should separate into different colors, revealing the colors that were mixed to make the color of ink. Repeat with the other black markers.

The process used in this activity is a type of chromatography, or a technique for separating some mixtures. The strip with separated colors is called a chromatogram. Chromatography has many uses. It can be used to identify different solutes dissolved in the same solvent or to detect the level of pollutants in the water. It is also used by pharmaceutical companies, hospitals, law-enforcement agencies, and manufacturers.

Objectives
- Identify a solution as a type of mixture
- Identify the parts of a solution
- Define *concentration*
- Explain ways to increase the rate of dissolving

Materials
- 2 cups
- water
- powdered drink mix
- spoon

Teacher's Toolkit CD
- Quiz 3-C

Vocabulary
- solution
- solute
- solvent
- concentration
- solubility

Introduction
- Mix a weak mixture of powdered drink mix and water in one cup and a regular-strength mixture in the other cup.
- Choose a student to taste both mixtures.

  Do both mixtures taste the same? no

  How would you describe the strength of the flavor of each? Elicit that one tastes stronger than the other.

  Both mixtures contain the same ingredients, but they taste different.

Preparation for Reading
- Display the words *solute*, *saturated*, and *solubility*. Pronounce the words together.
- Instruct the students to read silently pages 76–79 to find the answers to these questions.

  What is the difference between a solute and a solvent?

  What can affect the rate of dissolving?

Teach for Understanding
What is happening to a substance when its particles separate and spread evenly through a mixture? It is dissolving.

What is a solution? a mixture in which all the substances are spread evenly throughout

- Direct attention to the mixtures tasted in the introduction.

  Were the powdered-drink particles spread evenly throughout the water? yes

  What is another name for this kind of mixture? a solution

Solutions
Salt and sand do not mix evenly. But water added to the mixture does mix evenly with the salt. Salt and water form a solution. A **solution** is a mixture in which all the substances are spread evenly throughout. The salt dissolves in the water and seems to disappear. *Dissolving* is the process in which the particles of one substance separate and spread evenly throughout another substance.

Other substances can dissolve and form solutions as well. Suppose you mix sugar into iced tea. You cannot observe the grains of sugar in the tea because some physical properties of the sugar have changed. However, if you taste the tea, you know that the sugar is there. The sugar has undergone a physical change to form a solution, but it has not become another substance.

Scientists have special terms they use to describe the parts of a solution. A **solute** is the substance that is dissolved. A **solvent** is the substance that dissolves the other substance. In salt water the salt is the solute, and the water is the solvent.

Water is the most common solvent on the earth. It is often called the "universal solvent" because so many things dissolve in it. Water's ability to dissolve many substances is very important to life on the earth.

Solute—sugar Solvent—water Solution—sugar water

76

Background
Solutions
The solvent in a solution is usually a liquid. Occasionally the solvent is a gas or a solid. If the solvent is a gas, the solute must be another gas, because solid or liquid particles are too large to remain in solution in a gas. Below are some key properties of solutions.

- Solutions are homogeneous. The particles are spread evenly throughout the solution (see Activity Manual page 51).
- Solutions do not separate while standing.
- The entire solution can pass through a filter.
- Solutions are usually clear, not cloudy.

Names of Solutions
Many solutions are so common that they have their own names. For example, a solution of water and sugar is called a syrup. Table salt mixed with water forms brine. A sterile solution of salt and water is called a saline solution. Seltzer water is a solution of carbon dioxide and water.

You might think that all solutions are formed from a solid being dissolved in a liquid. That is not the case, however. The solute and the solvent of a solution can actually be a solid, liquid, or gas. This means that there are different types of solutions. Air is a solution of gases. The oxygen you breathe and the carbon dioxide plants need are two of the most common gases in the air solution.

A soft drink is a solution of a gas and other substances in a liquid. The bubbles you see are from the carbon dioxide gas that has dissolved in the liquid. The carbon dioxide in the soft drink is the solute, and the liquid is the solvent.

You use many metal products that are formed from a mixture of two solids. These mixtures of two or more metals are called *alloys*. A nickel coin is actually an alloy of the solid metals nickel and copper. Even fourteen-carat gold jewelry is gold that has other metals added to it.

A soft drink is a solution of a gas mixed in a liquid. The gas is the solute, and the liquid is the solvent.

A nickel is made from a type of solution called an alloy.

- Do the substances in a solution experience a physical change or a chemical change? a physical change How do you know? The physical properties changed, but the substances did not change into another substance.
- Refer to the pictures at the bottom of page 76 as you discuss solutes, solvents, and solutions.

  What are the two parts of a solution? the solute and the solvent

  What is a solute? a substance that is dissolved

  What is a solvent? a substance that dissolves another substance

- In our powdered-drink-mix mixture, which is the solute and which is the solvent? The drink mix powder is the solute, and the water is the solvent. How do you know? The drink mix is the solute because it dissolves in the water. The water is the solvent because it dissolves the powder.

  Why is water often called the universal solvent? Many things can dissolve in it.

  Are all solutions formed from solids and liquids? no Which states of matter can be used as the solute or solvent? solid, liquid, and gas (but not every combination is possible)

  What are some examples of different solutions? Possible answers: air, soft drinks, tea, nickels, some jewelry

  What is an alloy? a mixture of two or more metals

  List some common alloys. Possible answers: nickel coins, steel, fourteen-carat gold

  What is the difference between a mixture and a solution? A mixture is two or more substances combined in a physical process. A solution is a type of mixture in which the substances are evenly distributed.

Background

Alloys

Alloys are mixtures of two or more metals. The solid metals are heated together until they melt. The particles of the molten metals mix together. As the mixture cools, the particles are permanently mixed together. For example, brass is a solution of copper and zinc. The copper is the solvent, and the zinc is the solute.

Activity

Identify Common Solutions in Which Water is the Solvent

Direct the students to collect labels or containers of products that are dissolved or can be dissolved in water. Organize the labels into categories according to how they are used.

If some labels or containers are difficult to obtain, you may choose to photocopy the label from a product or print product information from the Internet.

41

What does the term *concentration* refer to? the amount of solute that is dissolved in the solvent

If you tasted a solution that is concentrated, what word might you use to describe it? strong

If you tasted a solution that is diluted, what word might you use to describe it? weak

How could you increase the strength of a diluted solution? add more solute to the solvent

Why might you need to change the concentration of a cleaning fluid? Possible answer: Dirtier areas may need a stronger solution.

What does it mean for a solution to be saturated? The solvent is holding all of the solute that it can dissolve.

What are three things that we can do to speed up the rate of dissolving? Increase the surface area of the solute, stir the solution, and increase the temperature of the solution.

Concentration and Saturation

The **concentration** of a solution is the amount of solute that is dissolved in the solvent. Suppose you mix a powdered fruit-drink mix with water. The powder contains flavoring and sweetener. When you mix the powder with water, the powder is the solute, and the water is the solvent. By altering the amount of solute, the concentration of the fruit drink can be changed to suit your taste.

If you put a lot of the powdered mix into the water, the solution will taste strong. It will be concentrated. *Concentrated* means that the solvent is holding more than a normal amount of solute.

Concentrated

If you put just a little of the powdered fruit mix into the water, the solution will taste weak. It will be diluted. A solution is *diluted* when there is less than a normal amount of solute in the solvent.

Sometimes the concentration of a solution can be adjusted to make the solution more useful. Often a cleaner is mixed with water to form a cleaning

Diluted

solution. To wipe counters and sinks, you might use a diluted solution with only a weak concentration of cleaner. To clean a very dirty floor, though, you would use a more concentrated solution.

If you continue to add more and more solute into a solvent, the solution will become *saturated*. This means that the solvent is holding all of the solute that it can dissolve.

Rate of Dissolving

Sometimes a solute can take a long time to completely dissolve in a solvent. There are ways to speed up the process, however. One way is to increase the surface area of the solute. This can be done by crushing,

78

Background

Rate of Dissolving

A rate is the comparison, or relationship, of one measurement to another. Rate is also related to time. The rate of dissolving, also called the rate of solution, is a measurement of how fast a solute dissolves in a solvent. This rate varies depending on a variety of conditions. These conditions include the total surface area of the solute, any movement of the solution, the amount of solute already dissolved in the solvent, and the temperature of the solute or solvent.

Solubility

Solubility refers to how much solute a solvent is able to hold. Three main factors affect the solubility of a solute: the type of solute and type of solvent used in the solution, the temperature, and the pressure. An increase in temperature usually increases the solubility of solutions. However, increasing the temperature of a gas solution decreases its solubility.

Also, increasing and decreasing pressure usually affects only gas solutions. Increasing the pressure increases the solubility of a gas solution, and decreasing the pressure decreases the solubility. You can observe the results of decreasing pressure by opening a soft-drink bottle. The bubbles indicate that some gas particles can no longer be dissolved in the liquid and gas is escaping. This process is called effervescence.

41

crumbling, or cutting it into small pieces. The more solute that touches the solvent, the faster the solute dissolves. That is why rock salt dissolves more slowly than a teaspoon of salt does.

A substance with greater surface area will dissolve more quickly than one with less surface area.

Another way to speed up the rate of dissolving is to stir the solution. Stirring brings more fresh solvent into contact with the solute. It helps spread the solute faster. Perhaps you have seen someone sweeten coffee with sugar. Usually he does not just put the sugar in the coffee. He also stirs the coffee. This causes the sugar to dissolve more quickly.

Temperature also affects how fast a solute can dissolve. This is because heat increases solubility. **Solubility** is the ability of a solvent to dissolve a certain amount of a solute. The particles in a heated solvent move faster than when the solvent is cooler. This increases the rate of contact with the solute and speeds up the dissolving process. Faster moving molecules also spread farther apart. That leaves more room for the particles of the added solute to dissolve.

Suppose that you want to add sugar to tea. The sugar will dissolve more quickly if the tea is hot. The sugar does not dissolve as easily in cold tea. Instead the sugar tends to settle to the bottom of the glass.

Using Matter

Often our love is shown by what we do with physical objects. You have learned about matter and some of its properties. But there is much more to learn about matter and how it works. Jesus said that the second great commandment is to love your neighbor as yourself (Matthew 22:39). Measuring and mixing matter can help you obey that command—even if it is by baking brownies.

QUICK CHECK

1. What is a solution?
2. In a solution of salt water, what is the solvent and what is the solute?
3. What are three ways to speed up the rate at which a solute dissolves in a solution?

79

- Direct attention to the picture of the sugar cubes. Point out how there is more surface area once the cubes have been crushed.

  What is solubility? the ability of a solvent to dissolve a certain amount of solute

  Why does heating a solution usually increase its solubility? Molecules in a heated solution move faster than when the solvent is cooler.

- Direct attention to the chapter introduction on Student Text page 57.

  What example of God's care is given in the introduction? A pot on a stove and a spatula do not melt at the same temperature that chocolate melts.

  What state of matter are the pot and the spatula? solid

  Do you usually think about the pot remaining solid when you use it on a stove? probably not

  When God created the world, He cared about you. Name ways that God's design of matter shows His care and provision for us. Answers will vary.

  What is the second greatest commandment? to love your neighbor as yourself

  We can glorify God by using matter to obey this commandment.

ANSWERS

1. a mixture in which all the substances are spread evenly throughout
2. Salt is the solute. Water is the solvent.
3. increasing the surface area of the solute, stirring, and increasing the temperature

Activity Manual

Expansion, page 52

This page presents descriptions of substances that the student identifies as either diluted or concentrated.

Review, pages 53–54

These pages review Lessons 40–41.

Assessment

Quiz 3-C

The quiz may be given any time after completion of this lesson.

Objectives

- Recognize that God created the matter in the universe from nothing
- Provide examples from Scripture of how the universe was created
- Identify the object of faith for materialists (matter) and Christians (God and the Bible)

Preparation

- 3 paper bags
- 3 resealable plastic bags
- sand or a wooden block
- water

Mark each paper bag with a number (1–3). Prepare the three resealable plastic bags by putting sand or a wooden block in one, water in another, and leaving one empty, flattening out all the air. Place a prepared plastic bag in each of the paper bags. Place the paper bags around the room.

Introduction

- Instruct the students to walk by each paper bag.
- Tell them to reach in without looking and determine whether each plastic bag contains a solid, a liquid, or a gas.
- Direct the students to record their answers on Activity Manual page 55.
- After they have recorded their observations, discuss the answers.

  What was in the first two bags? solid (sand) and liquid (water)

  How did you know what was in the bags? Elicit that their hands were acting as scientific instruments to gather data that they interpreted to come to their conclusion.

  What was in the third bag? Elicit that nothing was in the bag. Some students may believe there was air in the bag. Explain that if there were air in the bag, it would have been puffy, not flat, so there was nothing in the bag.

- Point out that many people believe that matter and energy are all that exist. We call this *materialism*.

  Can you think of a synonym for the word *matter*? Possible answers: substance, stuff, material, solid

  What is the door made of? wood

  Is wood a type of matter? yes

  Can it be measured? yes

  People who look at the world through lenses of materialism put their faith in what can be detected and measured, such as matter and energy.

  Materialism is based in the faith that people can determine truth only with the physical parts of the world.

  Does a materialistic worldview agree with a biblical worldview? no

---

A lesson from
Answers in Genesis Name ______________

1:1

Seeing the Invisible

Identify what is in each bag as *solid, liquid,* or *gas.*

Bag 1 ____________ Bag 2 ____________ Bag 3 ____________

If you look around the room you will see many things. All these things are made up of matter. You can touch them, measure them, throw them, and determine what color they are.

But scientific study by measurement and observation can only answer certain kinds of questions. One question this kind of study cannot answer is where matter came from in the first place. A different source is needed to discover the truth about the beginning of matter.

Read Hebrews 11:1–3. Use this text to answer the following questions.

1. According to verse 3, where did the world *not* come from?
   It did not come from the things that we can see.
2. According to verse 3, where did the world come from?
   It came from the word of God. He made it from nothing (things not visible).
3. According to verse 1, what is the only way for an individual to relate to something that is invisible or still in the future?
   by faith or by believing
4. Give some examples of things that are real but invisible or still in the future.
   Possible answers include love, hate, friendship, gravity, sunrise, and summer vacation.
5. What is the only way to know about where matter came from in the first place?
   by faith or believing in the Bible

Science 5 Chapter 3 | Lesson 42 Answers in Genesis 55

Resources

The BJU Press website offers additional links from Answers in Genesis that you may find helpful with this lesson.

bjupress.com/resources/science

There are other places in Scripture that talk about how God created matter. We call these cross references.

6. Read John 1:1–5. Who is the "Word" and the "Him" who created all things? (Hint: Look at verse 14 if you need help.)
   This is a reference to Jesus who is referred to as the Word.
7. Look at John 1:1–2. What verse in the Old Testament is being referred to in these verses? (Hint: Look at the first three words of verse 1.)
   The first three words of verse 1 are a reference to Genesis 1:1. Jesus, as a member of the Trinity, was present at Creation with the Father and the Spirit.

Some people called materialists reject the Bible and *believe* that only matter and energy exist. Rather than looking at the world through faith in God—through biblical glasses—they look at the world through faith in glasses labeled "materialism."

Christians place their faith in God and what He has revealed in the Bible so they can understand the world around them. Materialists place their faith in their own abilities and science to understand the world. Both views are based on faith, but Christians put their faith in God and not in their own abilities.

8. Respond to this statement: "It is fine to believe in the Bible if you want to be a kind and loving person. But if you want to understand science and know about where the world came from, you cannot take the Bible seriously."
   This statement demonstrates faith in science but not in the Bible. It fails to accept the claim that the Bible is God's Word. God's Word tells us about being kind and loving, and it also tell us where everything came from. If a person does not take the Bible seriously about where everything came from, he will not take it seriously about being kind and loving.

What are you putting your faith in—the God who says He created everything out of nothing or your own abilities to see the world correctly? Just as a person must accept by faith that Jesus is the Creator, he must also accept by faith that Jesus took on flesh to save mankind from their sins. Without accepting this truth by faith, nothing in science really matters.

Can something be made of nothing?

- Leave this question unanswered as the students begin to look at the biblical passages.

Teach for Understanding

- Ask a volunteer to read aloud the paragraphs on Activity Manual page 55.
- Read aloud Hebrews 11:1–3.
- Work through numbers 1–5 about the verses on Activity Manual page 55 together.
  How do people understand the world around them? Christians believe what God has revealed so that they can understand the world around them. Materialists have faith that science can explain everything about the world and matter.
  The Bible clearly points us to believe that God has created the universe as He has revealed in Scripture so that we can understand His creation better.
- Direct the students to read silently the first two sentences on Activity Manual page 56.
  What is a cross-reference? a Scripture reference that talks about the same subject as the original reference
- Read aloud John 1:1–5.
- Explain that these verses point to the work of Jesus in Creation. Christians accept Creation by faith so that they can understand the world God has created.
- Discuss and guide the students in answering numbers 6 and 7.
- Ask a volunteer to read aloud the paragraphs above and below the glasses.
- Discuss the paragraphs.
- Discuss the statement in number 8 and guide the students in answering it.
- Read aloud Colossians 1:9–18. Emphasize the role of Jesus as both Creator and Savior.
- Remind the students that they can be delivered from the kingdom of darkness into the kingdom of Christ through faith in what Jesus has done in becoming human and dying as our substitute.
- Direct the students to read silently the last paragraph on Activity Manual page 56.

Activity Manual

Answers in Genesis, pages 55–56

43

Student Text pages 80–81
Activity Manual pages 57–58

Objectives
- Predict how surface area will affect the rate of dissolving
- Relate results to other situations

Materials
- hammer or coffee grinder
- plastic bag or paper towels

Teacher's Toolkit CD
- Rubrics: *A Disappearing Act* for each student

Crush the peppermints individually before the activity. Place a peppermint into a plastic bag or wrap it in a paper towel. Use a hammer or coffee grinder to crush the candy until all the pieces are about the same size. Clear cups work best to let the students see when the candy is dissolved.

Introduction

What are some ways that we can decrease the amount of time that it takes to dissolve a substance? increasing surface area of the solute, stirring, increasing temperature

- Display a whole peppermint.

  Can we increase the surface area of this peppermint? yes How? by crushing it

  Today we will test to see how differences in surface area affect the time that it takes peppermint candies to dissolve.

Preparation for Reading

- Direct the students to read silently Student Text pages 80–81 and Activity Manual pages 57–58 before beginning the activity.

Teach for Understanding

Process Skill

The process skill emphasized in this activity is hypothesizing, which is writing a testable statement that tells what a person thinks will happen.

Is the hypothesis written as a statement or as a question? as a statement

If the results of an experiment indicate that your hypothesis is wrong, what might you do? Possible answers: Repeat the experiment to see if you get the same results. Change your hypothesis, then repeat the experiment.

ACTIVITY

A Disappearing Act

When a chef needs melted chocolate for a recipe, he could melt an entire block of chocolate whole, but he usually does not. Instead, he cuts the chocolate into small pieces before he melts it. If he needs a large amount of melted butter, he probably cuts the butter into smaller pieces, too. Why would he do that? Does the surface area of the substance make a difference?

In this activity you will use peppermints to find out how the surface area of a solute affects its ability to dissolve.

Process Skills
- Hypothesizing
- Experimenting
- Observing
- Inferring
- Defining operationally

Problem
Which will dissolve first—a whole peppermint or a crushed peppermint?

Materials
2 clear plastic cups, 9 oz
100 mL hot water
metric measuring cups
1 whole peppermint
1 crushed peppermint
stopwatch
2 spoons
Activity Manual

Procedure
1. Complete the hypothesis in your Activity Manual.
2. Label one cup *whole* and the other cup *crushed*.
3. Pour 50 mL of hot water into each cup.
4. Add a peppermint to each cup at the same time. Be sure that the label of the cup and the type of peppermint that you add match.

80

Background

Grouping Suggestion
If more than one group is doing the activity, you may choose to use different temperatures of water for each group. Supply thermometers for each group to measure the temperature of their water. Record the temperature next to the hot water listed in the materials. Compare the results to reinforce the information on Student Text page 79.

5. Begin timing the peppermints as they dissolve. Stir each cup until a peppermint is dissolved. Stop timing when one of the peppermints has completely dissolved.
6. Record your observations.

Conclusions

- Did your results support your hypothesis?
- Which peppermint dissolved first?
- How did the surface area of a peppermint affect the time it took to dissolve?

Follow-up

- Use a different type of candy. Compare the results.
- Use a larger amount of water. Compare the results.

81

Procedure

- Monitor the rate of stirring for each cup. Students in the same group should stir at the same rate.
- The student who is stirring should pause periodically to check if the candy has dissolved.

Conclusions

- Ask questions such as the following to help the students decide how to answer question 4.
  How many lollipops would you need to test your prediction?
  Would it matter if the lollipops were different kinds or shapes?
  Would one person test both lollipops or would you use two people?
  What kind of data would you record?
  How could you determine if your hypothesis was correct?
  Would you need to repeat your tests to be sure that your data was accurate?

Activity Manual

Activity, pages 57–58

Assessment

Rubrics

Select the prepared rubric, or design a rubric to include your chosen criteria.

Student Text pages 82–83
Activity Manual page 59

Objectives
- Design a clay boat that will float
- Demonstrate buoyancy

Materials
- pictures of boats and ships made from various materials (e.g., a barge, an ocean liner, a sailboat, a canoe)
- axe head
- pail of water (large enough to put the axe head in)

The weight of clay given is only a suggested amount. Some clay is packaged in blocks that are about the size that you need. An equal length of heavy duty aluminum foil may be substituted for the clay.

Teacher's Toolkit CD
- Rubrics: *Float a Boat* for each student

Vocabulary
- buoyancy
- buoyant force

Introduction
- Display the pictures of boats and ships.

  What do these boats and ships all have in common? Elicit that they all float.

  What is different about each? Possible answers: its size; the materials it is constructed from; its use; the way it is powered

  These boats and ships are made from a variety of materials, but they can all float. We will be learning how this is possible. Then you will design and make your own boat and test its ability to float.

Preparation for Reading
- Display the word *buoyancy*. Pronounce the word together.
- Instruct the students to read silently Student Text pages 82–83 and Activity Manual page 59 before beginning the exploration.

Teach for Understanding
What is buoyancy? the ability of an object to float

What determines the buoyancy of an object? its volume and mass

How does gravity affect the buoyant force of an object? The object pushes down on the water due to gravity, and the water pushes back on the object.

EXPLORATION

Float a Boat

Buoyancy

If you were to build a ship, what materials would you use? Perhaps you would use wood to make a ship like those used by the early explorers. Most modern ships are huge vessels made of thousands of tons of steel. During World Wars I and II, supplies of steel were limited, so some ships were made of concrete!

The ability of an object to float is called its *buoyancy*. Wood is known for its buoyancy. Until it becomes saturated with water, wood naturally floats. Steel and concrete, however, are not known for their buoyancy. These substances are usually expected to sink. Yet huge floating vessels are made of these materials. How do the vessels stay afloat?

Whether an object floats or sinks is based on its volume and mass. An object placed on water pushes down on the water because of gravity. When an object pushes down, it displaces, or moves the water. The water also pushes back on the object. This push against the object is known as *buoyant force*.

Several factors affect volume and mass. The shape of the object affects how much water it displaces. The surface area of an object is also a factor. A ball of clay will sink. But if you flatten the clay into a boat shape, it can float. The mass of the clay stayed the same, but its surface area changed.

An object's ability to float depends on its density. Density is the amount of matter in a certain space. For an object to float, its density must be less than the liquid it is floating in. Increasing the mass or volume of an object increases its density. Because ice is less dense than water is, ice floats on the top of water. An object such as an iron anchor is very dense, so it will sink.

You may wonder how a huge ship can float. Although the ship's hull is very heavy, the open spaces within its hull are filled with air. The air helps make the overall density of the ship less than the density of the water so that the ship will float.

82

Background

Concrete Boats
Throughout history boats have been made from materials that naturally float. However, sometimes nonbuoyant materials are used because of other characteristics, such as strength or flexibility. One very heavy material that has been used to make boats is cement. The first cement boats were built by Joseph Louis Lambot in 1848. Since then, cement has become a common material used in boat construction. Often small bits of buoyant materials are mixed with the cement or concrete to help it float better.

Today the engineering departments of many colleges and universities compete in annual concrete canoe contests. You may want to have the students research to find out more about the design and construction of these heavy boats that float.

A Contest

You have been invited to take part in a boat-building contest. Using only the given materials, you will build a boat. The goal is to make the boat that will hold the most pennies and remain afloat. Because the pennies increase the density of your boat and may cause it to sink, think carefully about your design!

What to Do

1. Plan the design of your boat. Sketch the design in your Activity Manual.
2. Use the clay to build a boat.
3. Test your boat in the dishpan of water to make sure that it floats. Make adjustments as necessary before the contest.
4. Float the boat and add pennies one at a time until the boat sinks. Record the number of pennies that your boat held before it sank.

Materials
- 100 g modeling clay
- large dishpan or similar container
- water
- pennies
- Activity Manual

83

How does the displacement of water affect an object's ability to float? For an object to float, the weight of the object must be less than or equal to the weight of the water that it displaces.

Why do you think that anchors are made of iron instead of wood? Possible answer: The shape of the anchor and its density cause it to sink. Iron sinks because it is denser than water. Wood floats unless it is saturated with water.

Explain how a large ship is able to float. It has open spaces in its heavy hull that are filled with air.

- Display the axe head.

Why do you think an iron ship can float? Possible answers: Open spaces within its hull are filled with air. The air helps make the density of the ship less than the density of the water. The weight of the ship is spread over a large area. The weight of the water that the ship displaces is greater than its own weight.

Can an axe head float? Answers will vary.

- Put the axe head in the pail of water. Read and discuss 2 Kings 6:1–7.

Why do you think the axe head did not float? Possible answers: It does not have air in it. The weight is compacted together.

The axe head's floating was a miracle. Only God could put aside the laws of nature and cause this miracle to happen.

What to Do

- Provide time for the students to design and build their boats. Allow them to test their boats without pennies before the contest begins.

Activity Manual

Exploration, page 59

Assessment

Rubrics

Select the prepared rubric, or design a rubric to include your chosen criteria.

46

Student Text page 84
Activity Manual page 60

Objectives

- Recall concepts and terms from Chapter 3
- Apply knowledge to everyday situations

Review

- Materials for the Chapter 3 Test will be taken from Student Text page 84 and Activity Manual pages 43–44, 47–48, 53–54, and 60. You may review any or all of the material during the lesson. Questions similar to the ones in Thinking It Through, Activity Manual page 60, may appear on the test.
- You may choose to review Chapter 3 by playing the review game or a game from the Game Bank on the Teacher's Toolkit CD.

Activity Manual

Review, page 60

This page requires written responses to application questions.

47

Objective

- Demonstrate knowledge of concepts taught in Chapter 3

Assessment

Chapter 3 Test

CHAPTER SUMMARY

Words to Know

matter	freezing	solution
volume	vaporization	solute
mass	condensation	solvent
density	atom	concentration
physical property	element	solubility
melting	mixture	

Key Ideas

Identify tools and methods used to measure mass and volume
Explain the difference between mass and weight
Identify the states of matter and their characteristics
Describe the processes of melting, freezing, vaporization, and condensation
Explain what is happening when matter solidifies
Explain the relationship between evaporation and vaporization
Compare and contrast a physical change with a chemical change
Compare and contrast a molecule and a compound
List several characteristics of a mixture
Identify examples of mixtures
Identify ways to separate substances in a mixture
Identify examples of solutions
Describe the process of dissolving
Explain why water is called the universal solvent
Describe a solution that is concentrated, saturated, or diluted
Identify ways to increase the rate of dissolving

84

Review Game

Changing States

Materials: 2 thermometer shapes drawn for display

Mark increments on the thermometers. Choose a substance, such as water, and label its melting/freezing point and its vaporization/condensation point. Possible substances:

Water: 0 °C; 100 °C

Diamond: 3,500 °C; 3,930 °C

Gold: 1,064 °C; 2,856 °C

Oxygen: –218 °C; –183 °C

For each correct answer to a review question, the team colors in a section on the thermometer, starting at the bulb of the thermometer. The first team to reach "vaporization" (or to have their substance become a gas) wins. You may choose to have the team erase a colored section for each incorrect answer.

Energy and Heat 4

Jeff was helping his father repair the refrigerator in the church kitchen. As they worked, his dad explained a principle to Jeff.

"When God created our world, He established certain principles that always work the same way. One of those is the fact that heat flows from something warm to something cool. Inside the walls of the refrigerator are tubes with liquid in them. This liquid is cooler than the inside of the refrigerator. The liquid absorbs heat from the refrigerator as it flows through the tubes. As it warms it turns into a gas. A pump compresses the warm gas and it flows through tubes on the outside of the refrigerator. There it releases extra heat into the kitchen and becomes a cool liquid again. This process happens over and over. Knowing how heat moves and being able to control it has allowed people to develop tools such as refrigerators and air conditioners."

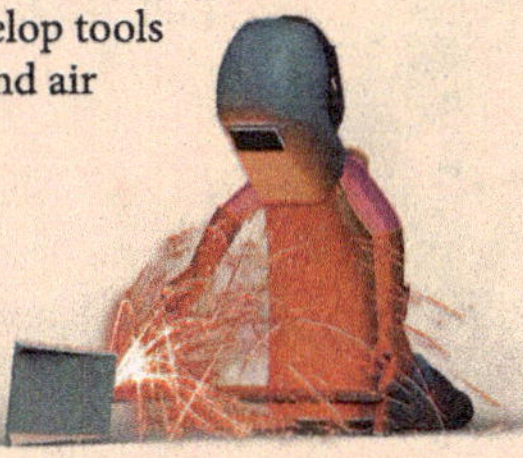

Consider

- What happens to the liquid in the tubes as heat is added or removed?
- How is Jeff's dad using his knowledge to serve others?

85

Objective

- Explain the importance of energy and heat in designing useful technology

Introduction

Look at the picture at the top of the page. What do you think is happening? Answers should include that food is being cooked on a campfire.

What is the chapter title? Energy and Heat

Why do you think this picture was chosen to represent the chapter title? Answers will vary.

Preparation for Reading

- Instruct the students to read silently page 85 to find out why it is important to know how heat moves.

Teach for Understanding

- Explain that God provides and faithfully maintains all things needed for life. His natural laws, or principles, are always the same.

What principle was Jeff's father explaining to him? Heat flows from something warm to something cool.

Describe how this principle works in a refrigerator. The liquid in the refrigerator tubes absorbs heat from the refrigerator. As the liquid warms, it turns into a gas. A pump compresses the warm gas. It flows through the tubes on the outside of the refrigerator where it releases the heat.

What did people who developed the refrigerators and air conditioners need to know? how heat moves and how to control it

People can observe the consistent ways that God designed for energy to change forms and for heat to move. By studying these natural patterns, people can design and invent technology, such as the refrigerator, that is helpful and useful.

Activity Manual

Preview, page 61

The Looking Ahead page is intended to assess the student's prior knowledge before beginning the chapter.

Background

Weblinks

The BJU Press website offers additional information and links you may find helpful throughout the unit.
bjupress.com/resources/science

Chapter Preview

Preview and prereading activities may include using a K-W-L chart, a probe, or an anticipation guide.

Chapter Photo

The photo on page 85 shows food being prepared over a campfire. The miniature scientist is welding.

Refrigeration

The processes used in refrigeration are much more complex than the information given in the introduction. Refrigeration involves the compression and expansion of gases. However, the basic properties described, as well as the additional properties of matter and heat used in refrigeration, are examples of the consistency of the world that God created.

49 50

Student Text pages 86–89
Activity Manual page 62

Objectives

- Define *energy*
- Differentiate between potential energy and kinetic energy
- Recognize that energy is often classified as either potential or kinetic
- Recognize that the amount of thermal energy depends on the temperature and mass of a substance
- Differentiate between thermal energy and temperature

Vocabulary

- energy
- potential energy
- kinetic energy
- thermal energy
- temperature
- thermometer

Two days have been allotted for this lesson. You may want to teach Lesson 49 the first day and use ideas from Lesson 50 on the second day. You may also incorporate the ideas from Lesson 50 into the text material, making the lesson two days.

Introduction

- Recall the story of "Goldilocks and the Three Bears."

  Why did the bears go for a walk? They thought their porridge was too hot.

  When Goldilocks tried their porridge, did she think that the porridge in all the bowls had the same temperature? no

  Describe her responses. Papa Bear's porridge was too hot, Mama Bear's porridge was too cold, and Baby Bear's porridge was "just right."

Preparation for Reading

- Display the word *Celsius*. Pronounce the word together.
- Instruct the students to mentally make questions using the headings, bold and italicized words, and captions.
- Direct them to read silently pages 86–89 to find the answers to these questions.

  What are two ways that scientists classify energy?

  What is thermal energy?

  How is temperature different from thermal energy?

Teach for Understanding

The term *temperature* is defined on Student Text page 88. The student's basic understanding of temperature should be adequate for him to understand the material presented on pages 86–87.

What is energy? the ability to do work

What are some different forms of energy? Possible answers: light, sound, chemical, electricity, mechanical, thermal

Does energy always stay in the same form? No, it can change from one form to another.

Energy

Think about some of the things that you do each day. Maybe you have a refrigerator in your kitchen. You may have turned on a light. Hopefully you ate a good breakfast. Perhaps a car or bus took you to school. Someone adjusted the temperature in your classroom to make you comfortable. While these are very different activities, they all involve energy.

Energy is the ability to do work. Some forms of energy are light, sound, and electricity. Machines, such as cars, use a type of energy known as mechanical energy. Thermal energy causes your room to be the temperature it is. Even the food that you eat contains a form of energy. The food provides chemical energy for your body's needs.

You can see that energy comes in many forms. But energy would not be very useful if it could not change from one form to another. A flashlight works because of converted energy. The chemical energy in the batteries changes to electrical energy. This energy then causes the filament of the bulb to glow, releasing light energy.

Scientists sometimes classify energy as potential energy or as kinetic energy. **Potential energy**, or stored energy, is the energy that an object or substance has because of its position or condition. A roller coaster at the top of the track has potential energy. As the roller coaster starts down the track, some of its potential energy changes into kinetic energy. **Kinetic energy** is the energy that an object or substance has because of its motion. A flying jet and flowing water both have kinetic energy. They are both moving.

God is the Creator of all matter and energy. But people use matter and energy as tools to meet their needs. Sometimes they change energy from one form to another. At other times they use matter to produce energy.

A flashlight changes energy from one form to another.

Potential energy

Kinetic energy

86

Background

Types of Energy

Scientists usually classify energy as either potential (stored) energy or kinetic (moving) energy.

Forms of Energy

Energy can occur in many forms. Some forms of kinetic energy are thermal, electrical, sound, and radiant (light) energy. Some forms of potential energy are chemical, stored mechanical, gravitational, and nuclear energy.

Thermal Energy

Thermal energy, sometimes called heat energy, is the *total* kinetic energy in a substance. Thermal energy is measured in calories.

Science Misconceptions

Misuse of Terms

The terms *heat* and *heat energy* are sometimes used incorrectly to describe the concept of thermal energy.

Thermal Energy

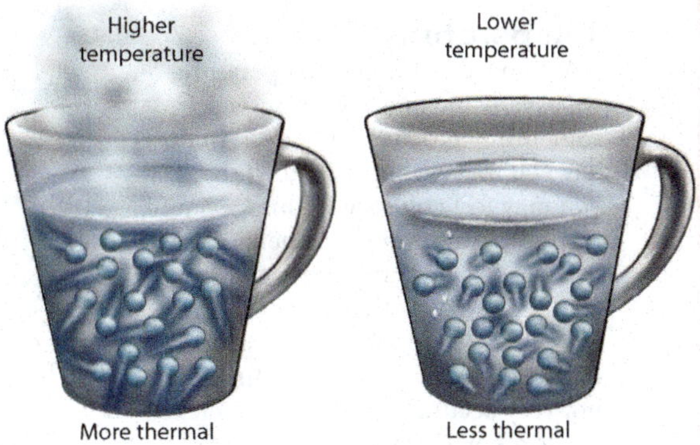

All matter is made up of tiny particles called atoms and molecules. They are always in motion. This means they have kinetic energy. One kind of kinetic energy is thermal energy. **Thermal energy** is the total kinetic energy of the particles in a substance.

The thermal energy of a substance depends on the temperature and mass of that substance. As the temperature of a substance increases, the particles move faster. The faster the particles move, the more thermal energy the substance has. For example, the particles in a cup of hot water move faster than the particles in a cup of cold water do. Therefore, hot water has more thermal energy than the same amount of cold water does.

The mass of a substance also affects the thermal energy of that substance. If two substances are at the same temperature, the substance with the greater mass will have more thermal energy. More mass means that more particles are present and moving. The more particles that are moving, the more thermal energy a substance has. For example, a large pitcher of water at room temperature has more thermal energy than a small glass of water at room temperature does.

87

What are two ways that energy is classified? potential energy and kinetic energy

What is potential energy? stored energy; the energy that an object or substance has because of its position

What is kinetic energy? the energy that an object or substance has because of its motion

- You may want to do the *Differences Between Kinetic and Potential Energy* demonstration at this time.
- Direct attention to the pictures of the roller coasters on page 86.

Why is the first picture an example of potential energy? The coaster is not moving but is in position to roll down.

Why is the second picture an example of kinetic energy? The coaster is in motion.

Who created energy? God

What are two things that people can do with energy? change energy from one form to another and use matter to produce energy

Why do atoms and molecules have kinetic energy? They are always in motion.

What is one kind of kinetic energy? thermal energy

What is thermal energy? the total kinetic energy of the particles in a substance

What two things does the thermal energy of a substance depend on? the temperature and mass of that substance

- Direct attention to the pictures of the cups at the top of page 87.

Are the particles in both cups moving? yes

What is different about the movement of the particles in the two cups? The particles are moving faster in the cup with more thermal energy than in the cup with less thermal energy.

Which would have more thermal energy—a stick of butter in the freezer or a stick of butter at room temperature? the butter at room temperature Why? It is warmer.

- Direct attention to the pictures of the pitcher and the glass.

The liquids in the pitcher and in the glass are both the same temperature. Why does the liquid in the pitcher have more thermal energy? It has more mass.

Which would have more thermal energy—one pint of ice cream or one quart of ice cream? one quart Why? One quart of ice cream has more mass than one pint has.

Demonstration

Differences Between Kinetic and Potential Energy

Materials: book or other object

Place a book near the edge of a desk or table.

What kind of energy does the book have? potential energy

Gently bump the book so that it falls to the floor.

What kind of energy did the book have while it was falling? kinetic

Why are the terms *hot* and *cold* not accurate terms to use for science experiments? These terms may have different meanings to different people; one person might feel hot while another person in the same room feels cold.

Does your skin give you an accurate measure of temperature? no Why not? Our bodies adjust to temperature changes. Different people have different sensitivities to temperature.

- You may want to do the *Touch as an Inaccurate Indicator of Temperature* demonstration at this time.

What is temperature? the measure of the average kinetic energy of the particles in a substance

In your own words, what does *average* mean? Possible answers: about the middle amount; has some higher values and some lower values

Why is an average used to measure temperature? Not all the particles in a substance move at the same speed.

How is temperature different from thermal energy? Temperature measures the average kinetic energy. Thermal energy is the total kinetic energy in a substance. A substance that has great mass might have a great deal of thermal energy, but the average speed of its particles may be slow.

What instrument is used to measure temperature? a thermometer

- Share as much of the following information about prefixes and suffixes as you would like.
- Display the prefixes *therm-* and *thermo-*. Explain that both mean "heat."

What are some words that begin with these prefixes? Possible answers: thermal, thermometer, thermostat

- Display the suffix *-meter*.

The suffix *-meter* means "a measuring tool." Using the meanings of the prefix and suffix, how would you define *thermometer*? a tool that measures heat

- Display the suffix *-stat*.

The suffix *-stat* comes from a Greek word meaning "to stand still" or "to keep something stationary." How would you define *thermostat*? a device or instrument that keeps the heat the same

- You may want to do the *Finding Averages* activity at this time.

Temperature

When we talk about temperature, we usually mean how hot or cold something feels to us. On a summer day you may feel hot. On a winter day you may feel cold. Sometimes, even in the same conditions, one person may feel comfortable, another may feel cold, and still another may feel warm. *Hot* and *cold* are words that we use to describe how we feel, but they are not very accurate or useful for science experiments. Scientists need terms that have more exact meanings.

One word scientists use in an exact way is *temperature*. **Temperature** is the measure of the *average* kinetic energy of the particles in a substance. Particles are always moving. However, not all the particles in a substance move at the same speed. Some particles move faster than average. Others move slower. So, finding the average speed is more accurate than saying how hot or cold a substance feels.

Temperature is not the same as thermal energy. A substance with great mass has a lot of thermal energy. However, the average speed of its particles may be slow. The substance may still feel cold. In this case, it has a lot of thermal energy but a low temperature.

How can temperature be measured exactly, though? God designed your body to adjust to different temperatures. For this reason, you cannot measure temperature by how something "feels." If you go swimming in a pool, the water may feel cold at first. But after a while, the water no longer feels cold to you. The water has not gotten warmer. Your body has adjusted to it. So, to get an accurate measurement of temperature, an instrument that does not adapt to its surroundings must be used.

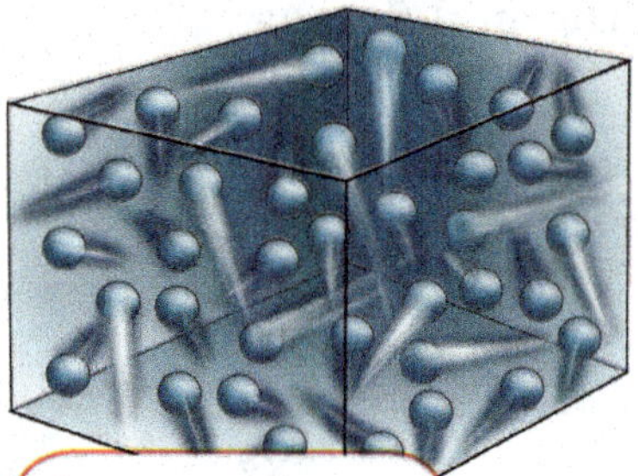

Temperature is the measure of the average kinetic energy of all the particles in a substance.

A **thermometer** is an instrument that measures temperature. One type of thermometer is made of a glass tube containing red-colored alcohol. Numbered markings on the tube allow you to measure temperature in units of degrees. You have probably had your temperature taken with a digital thermometer. The temperature shows as a number on a small screen.

Temperature is usually measured in degrees Fahrenheit or degrees

88

Demonstration

Touch as an Inaccurate Indicator of Temperature

Materials: 3 large containers, hot water, ice water, room-temperature water

Note: Be sure the hot water will not burn the students.

Put equal amounts of a different temperature of water into each container. Direct a student to place one hand in the hot water and one hand in the ice water. He should leave his hands in the water for at least one minute.

Direct the student to take his hands out of the hot water and ice water and place both hands into the room-temperature water.

Describe how the room-temperature water feels to each hand. The hand that had been in the hot water senses the water to be cold, and the hand that had been in the ice water senses the water to be hot.

Discuss how the sense of touch is not an accurate measure of temperature.

Celsius. Some people use the Fahrenheit scale. The nurse probably takes your temperature using the Fahrenheit scale. The weatherman usually tells the temperature outside using this scale. On the Fahrenheit scale water freezes at 32 °F and boils at 212 °F.

Scientists, however, usually use the Celsius scale to measure temperature. The Celsius scale is based on the freezing point and boiling point of water. On the Celsius scale water freezes at 0 °C and boils at 100 °C.

Thermometer

QUICK CHECK

1. What is energy?
2. How is potential energy different from kinetic energy?
3. How is thermal energy different from temperature?

MEET THE SCIENTIST • LORD KELVIN

Fahrenheit and Celsius are not the only scales used to measure temperature. The Kelvin scale was developed by William Thomson. Born in 1824 in Belfast, Ireland, Thomson studied thermodynamics, the science of heat and energy. His temperature scale is based on the movement of molecules. Instead of degrees, his scale uses kelvins. He placed zero at the point where he believed molecules no longer move or have kinetic energy. Absolute zero, written as 0 K, on his scale is equal to about −273 °C.

Thomson also had a great impact on communication technology. His inventions allowed the first telegraph cable to be laid across the Atlantic. In 1866 he was knighted by Queen Victoria of England. Later, his title became Lord Kelvin.

Although Lord Kelvin was a highly honored scientist, he was also a humble Christian. He said, "If you think strongly enough, you will be forced by science to the belief in God." He continued to study science until his death in 1907.

89

Background

Celsius Scale

Scientists use the Celsius scale for measuring temperature. Most countries also use the Celsius scale. The United States is one of the few countries that uses the Fahrenheit scale.

Scientist Daniel Fahrenheit

The Fahrenheit scale was developed by Daniel Gabriel Fahrenheit (1686–1736), a German scientific instrument maker. He used three temperatures to calibrate his thermometer: the freezing point of a mixture of water, salt, and ice; the freezing point of pure water; and the average temperature of a normal human body. He divided this scale into equal units that he called degrees Fahrenheit.

Scientist Anders Celsius

The Celsius, or centigrade, scale was devised by Anders Celsius, a Swedish astronomer (1701–44). He divided the temperature range between the freezing and boiling points of pure water into 100 equal intervals. He called the boiling point of water 0° and the freezing point of water 100°. The Celsius scale that we use today is actually an inverted form of his temperature scale. The numbers on the scale were reversed after his death.

49
50

- Explain that a scale is a series of evenly spaced marks along a line. Scales are sometimes used to measure substances.

  What scales are usually used to measure temperature? Fahrenheit and Celsius

  What is the boiling point of water on the Fahrenheit scale? 212 °F the freezing point of water? 32 °F

  What is the Celsius scale based on? the freezing point and boiling point of water

  What is the boiling point of water on the Celsius scale? 100 °C the freezing point? 0 °C

- Discuss *Meet the Scientist.*

  What temperature scale was developed by William Thomson? the Kelvin scale

  What did he choose to use as the start of his scale? the temperature at which he believed molecules no longer moved or had kinetic energy

  What did he call this point? absolute zero

  How many degrees Celsius is equal to absolute zero? −273 °C

  Why do you think Thomson might have made his scale as an alternative to the Celsius scale already in use by scientists? Accept any reasonable answer. It kept scientists from having to work with negative numbers in many of their lab calculations.

  Who knighted Thomson? Queen Victoria

  What title did Thomson receive later in his life? Lord Kelvin

  What did Lord Kelvin think that science would cause a person to do? to believe in God

ANSWERS

1. the ability to do work
2. Potential energy is the energy that an object or substance has because of its position or condition. Kinetic energy is the energy that an object or substance has because of its motion.
3. Thermal energy is the total kinetic energy of the particles in a substance. Temperature is the measure of the average kinetic energy in a substance.

50

- The *Differences Between Kinetic and Potential Energy* demonstration and the *Touch as an Inaccurate Indicator of Temperature* demonstration activity found at the bottom of the pages in Lessons 49–50 may also be incorporated into the lesson.
- Suggestions for additional activities for this two-day lesson are on the Teacher's Toolkit CD.

Activity Manual

Reinforcement, page 62

Objectives

- Predict how the mass of a substance affects the amount of thermal energy it can transfer
- Experiment to test a hypothesis

Materials

- brick or large stone
- large potato

Teacher's Toolkit CD

- Rubrics: *Rock Heaters*, for each student

> If needed, one pot of boiling water could hold the bags of rocks for several groups. You may choose to have the water already boiling before the rocks are put in.

Introduction

- Display the brick and potato.

  How would you like to sleep with either of these near your feet? Answers will vary.

  Before the times of central heating, people often did just that. Today we will experiment to see why they may have used items such as these to keep themselves warm.

Preparation for Reading

- Instruct the students to read silently Student Text pages 90–91 and Activity Manual pages 63–64 before beginning the activity.

Teach for Understanding

Decide which mass of rocks would be able to transfer the most thermal energy. Write your answer as your hypothesis.

Process Skill

> The process skill emphasized in this activity is experimenting, which is simply setting up a procedure to test a hypothesis.

How do you test a hypothesis? by experimenting

What do you need in order to experiment? materials and necessary conditions for testing your hypothesis

As you experiment, what should you do with any information that you gather? Record it. Why? Possible answers: to have a record of what happened during the experiment; to help form conclusions

Should you repeat experiments? yes Why? Possible answers: to make sure that the results are accurate; to determine the accuracy of the hypothesis; to test any changes in variables or the hypothesis

ACTIVITY

Rock Heaters

Process Skills
- Hypothesizing
- Measuring and using numbers
- Collecting and recording data
- Defining operationally

People have benefitted from thermal energy for many years. In times past, houses usually did not have a source of heat in the bedrooms. In cold weather people would often heat a brick by placing it in or close to the main fireplace. At bedtime they would cover the brick with cloth and place it at the foot of the bed. The warmth from the brick would help keep the bed warm.

Do you think that the size of the brick affected how long it stayed warm? In this activity you will use rocks to test how the mass of a substance affects its thermal energy.

Problem

How does the mass of a substance affect the amount of thermal energy the substance transfers to water?

Materials
- 2 squares of cheesecloth
- 150 g aquarium rocks
- balance
- 2 pieces of string
- goggles
- hot plate
- saucepan
- water
- tongs
- 2 containers
- metric measuring cups
- 2 thermometers
- Activity Manual

Procedure

1. Place a piece of cheesecloth on a balance and measure 50 g of aquarium rocks. Remove the cheesecloth and place the other piece on the balance. Measure 100 g of aquarium rocks.
2. Gather the corners of each piece of cheesecloth together so that it forms a bag with the rocks inside. Tie a string around the gathered top of each bag to hold it shut.
3. Predict which mass of rocks will transfer more thermal energy to the water. Complete the hypothesis in your Activity Manual.
4. Add water to the saucepan and place it on the hot plate. Put the bags of rocks into the saucepan. Wear goggles or other eye protection while heating the rocks and water.
5. Heat the water until boiling. Boil the rocks for 10 minutes.

90

6. Label one container *50 g* and the other container *100 g*. Pour 150 mL of room-temperature water into each container.
7. Measure the temperature of the water in the containers. Record the temperatures.
8. Place each bag of heated rocks into the appropriate container.
9. Measure the temperature of the water again after 5 minutes. Record the temperatures.

Conclusions

- Did your results support your hypothesis?
- How was the mass of the rocks related to the change in temperature of the water?

Follow-up

- Use two different substances that have the same mass.
- Make the bags identical in both substance and mass but use different amounts of water.

91

Procedure

- If needed, help the students measure the rocks and tie the cheesecloth bags.

If the squares of cheesecloth have approximately the same mass, the students could place the cheesecloth on the balance and then measure their rocks directly on their cheesecloth. Or you may choose to place a coffee filter on the balance to keep the rocks in place.

- Supervise the students' handling the boiling water to prevent accidents.
- Announce the times for steps 5 and 9.

Thermometers vary. When comparing several thermometers, the starting temperature, or room temperature, may be different on each. With this in mind, the activity asks for a starting temperature, and then compares the differences between readings rather than the actual readings.

Conclusions

- Guide the students as they use their charts to complete the questions.

Why do you think that people used items such as a brick or potato rather than bags of small rocks? Elicit that a larger item, or an item with more mass, holds heat longer.

Imagine that you have a brick and a potato that are the same size and warmed to the same temperature. Which would stay warm longer? the brick Why? A brick has more mass than a potato does.

Predict what volume of water could be put in the container receiving the 100 g of rocks if you wanted both containers to warm the same number of degrees. Accept reasonable answers. Since the larger bag of rocks contains twice the thermal energy, it could raise 300 mL of water (twice the volume) the same temperature as the smaller bag warmed 150 mL.

Activity Manual

Activity, pages 63–64

Assessment

Rubrics

Select the prepared rubric, or design a rubric to include your chosen criteria.

52

Student Text pages 92–94
Activity Manual pages 65–66

Objectives

- Recognize that increasing or decreasing thermal energy can cause matter to change to a different state
- Explain what happens during thermal expansion
- Define *calorie*
- Recognize that substances differ in their ability to store thermal energy

Materials

- saucepan with water, about 5 cm deep
- hot plate
- empty soup can, clean and with label removed
- old crayons, broken into very small pieces
- 1 spoon
- potholders
- small foil candy cups, one for each student

Teacher's Toolkit CD

- Quiz 4-A

Vocabulary

- thermal expansion
- calorie

Introduction

- Boil water on the hot plate. Display the broken crayons.
  In what state of matter are the crayons? solid
- Add the crayon pieces to the can. Carefully place the can in the boiling water.
- Stir the crayons until they melt. Carefully pour a few drops of melted crayon into each candy cup. **Note:** Hot wax can cause burns.
  What happened to the crayons? They melted. Elicit that they changed states.
- Distribute one cup to each student.
  Do not touch the melted crayons.
  What state are the crayons changing to now? solid

Preparation for Reading

- Direct the students to mentally make questions using the headings, bold words, and captions.
- Display the word *buckle*. Explain that when buckle is used as a verb it can mean "to bend or crumple."
- Instruct them to read silently pages 92–94 to find the answers to these questions.
  What could happen to a liquid if thermal energy is added to it?
  What unit is used to measure thermal energy?

Teach for Understanding

What are some ways that thermal energy affects the things around you? Possible answers: Hot food cools. Ice cream melts. Fire warms the air.

What determines the amount of attraction between the particles in a substance? the speed at which the particles move

Matter and Thermal Energy

You already know of ways that changes in thermal energy affect you and the things around you. If you wait too long to eat hot food, it cools off. If you do not eat your ice cream quickly enough, it melts. A fire burning in a fireplace warms the air in the room. All of these situations involve the flow of thermal energy.

State Changes

The moving particles in matter attract each other much as magnets attract each other. The amount of attraction can vary, though. It depends on the speed at which the particles move. As the temperature of a substance increases, the speed of its particles increases. As the particles move faster, there is less attraction between them.

Particles in a solid do not move very much. They just vibrate back and forth. Because their small vibrations do not force the particles very far apart, the attraction between the particles is great. In a liquid the particles move a little faster and there is less attraction. Particles in a gas move rapidly and cover great distances. There is little attraction between the particles.

As a substance is heated, it gains thermal energy. This additional energy causes the particles to move faster and to move farther apart. When farther apart, they have less attraction. A solid may become a liquid, or a liquid may become a gas.

If a substance is cooled, the opposite occurs. It loses thermal energy and may change to a state of matter in which the particles move less and are attracted to each other more. For example, when the gas in refrigerator tubes cools, it changes to a liquid.

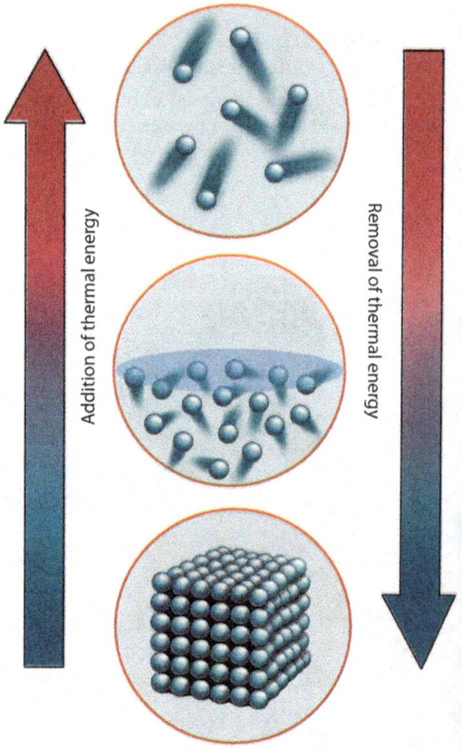

92

Spaces between sections of a bridge allow the sections to expand and contract.

The reverse is also true. When a substance is cooled, the particles move less rapidly and occupy less space. In this case, the substance shrinks, or contracts. Almost every substance expands when it is heated and contracts when it is cooled.

When the temperature of most substances changes only a little, they expand or contract only a small amount. However, even a small amount can cause great problems. Engineers must allow room for these small changes in size as they plan structures. If you look at a bridge, sidewalk, or highway, you may see spaces between the sections. These spaces allow the structures to expand and contract without buckling or cracking.

Thermal Expansion

Sometimes, when the temperature change is not very great, a substance does not change its state. The substance's particles may only move faster and occupy a larger space. This property is known as **thermal expansion**.

A liquid thermometer uses thermal expansion to measure temperatures. When thermal energy is added to the liquid in the thermometer, it increases the speed of the liquid's particles. The liquid in the thermometer expands and takes up more space. This increased volume causes the liquid to show an increased temperature on the thermometer's scale.

TRY IT YOURSELF

You can observe how thermal expansion works. Blow up two balloons about halfway. Place one balloon in the refrigerator or freezer. As the thermal energy is removed from the balloon, the balloon shrinks. Place the other balloon in a very warm place. Observe what happens. The increased thermal energy causes the gas in the balloon to expand and take up more space.

93

Background

Liquids in Thermometers

The thermometer described on Student Text page 93 is a bulb thermometer. These thermometers contain alcohol if the liquid is red or mercury if the liquid is silver. Both alcohol and mercury expand and contract with changes in temperature. Although a mercury thermometer is more accurate, it should not be used in the classroom due to the risk of mercury contamination if it is dropped and broken.

As you discuss state changes, remind the student of terms discussed in Chapter 3, *Matter,* such as *melting point*, *freezing point*, *condensation point*, and *boiling point*.

52

Do particles move faster in a solid or in a gas? in a gas

Which state of matter has the strongest attraction between its particles? solid

How does a substance gain thermal energy? by being heated

What does the additional thermal energy do to the particles? It causes them to move faster and to move farther apart.

If a liquid is heated enough to change to a different state of matter, will it become a solid or a gas? a gas

- Direct two or more students to stand side by side with their arms touching. Ask the students to walk in place slowly. They should walk slowly enough to stay in the same place.

  How do these students represent the particles of a solid? They are moving slowly and staying close to each other in about the same place.

  Suppose that we heat the particles. What will they do? move faster

- Direct the students to jog in place.

  What type of substance do the students represent now? a liquid How are they like a liquid? As they increase speed, it is harder for them to stay close and in the same place.

  What is needed for a liquid to become a gas? an increase in temperature

- Direct the students to run in place as fast as they can.

  As the students move faster, how are they like the particles in a gas? It is harder for them to stay close and in the same place.

  Do substances always change states when they are heated or cooled? no

  What happens when the particles of a substance move faster and occupy a larger space without changing to another state? thermal expansion

What is expansion? spreading out or apart, becoming larger

When a substance is heated, does it usually expand or contract? expand

How does a thermometer demonstrate thermal expansion? The liquid takes up more space when it is heated and shows an increased temperature reading.

What does it mean to contract? to become smaller or to shrink

When do most substances contract? when they are cooled

When a substance expands, does it usually expand a huge amount or a little? a little

Why is even a small amount of expansion or contraction a problem? Possible answer: Structures such as sidewalks, bridges, and highways could crack or buckle from even a small amount of movement.

52

What is temperature? the measure of the average kinetic energy of the particles in a substance

What unit is used to measure temperature? the degree

What is thermal energy? the total amount of kinetic energy in a substance

What unit is used to measure thermal energy? the calorie

What is a calorie? the amount of thermal energy needed to raise the temperature of 1 g of water 1 °C

Do all substances have the same ability to store thermal energy? no

Why do most metals feel hotter than plastics feel? Most metals transfer thermal energy more quickly than plastics.

ANSWERS

1. the speed of the particles of the substance increases
2. to allow for thermal expansion
3. the calorie

Activity Manual

Review, pages 65–66

These pages review Lessons 49–50 and 52.

Assessment

Quiz 4-A

The quiz may be given any time after completion of this lesson.

Measuring Thermal Energy

Since thermal energy and temperature are not the same, they are not measured in the same way. One of the units used for measuring thermal energy is the calorie. A **calorie** (cal) is the amount of thermal energy needed to raise the temperature of 1 g of water 1 °C.

Raising the temperature of 10 g of water 1 °C takes 10 cal of thermal energy. It also takes 10 cal to raise 1 g of water 10 °C. The amount of thermal energy required depends on both the mass of the substance and its change in temperature.

The plastic grips on the handlebars protect your hands from the hot metal on a warm summer day.

Stored Thermal Energy

Some substances heat or cool more quickly than others do. This means that the substances have different abilities to transfer thermal energy. Perhaps you have noticed that in the summer the metal parts of a bicycle feel hotter than the plastic parts do. Most metals can store a lot of thermal energy quickly. Plastics transfer thermal energy less quickly than most metals.

QUICK CHECK

1. What happens to particles of a substance as it is heated?
2. Why do engineers put spaces between the sections of a bridge?
3. What is one unit of measurement used to measure thermal energy?

94

Background

Calorie

Although calories are often associated with food, they are actually a unit used to measure energy. Even gasoline has calories! Another unit sometimes used to measure energy is the joule. One calorie is equal to about four joules. The calorie associated with food can be called the kilocalorie, which is equal to about 4,000 joules.

Student Text page 95
Activity Manual pages 67–68

53

Energy for Your Body

EXPLORATION

You may have heard someone say, "That food has a lot of calories in it." He was probably thinking about his possible weight gain. But calories in food are actually a measure of the energy available from that food. A food Calorie is 1,000 times bigger than the calorie a scientist uses. A food Calorie is labeled as a *kcal* (kilocalorie) or *Cal*. However, it is usually written simply as *calorie*.

Your body converts the Calories from the food that you eat into energy for doing work. If the amount of Calories that a person eats is the same as the energy that he uses, his weight stays about the same. If he eats more Calories than he uses, he tends to gain weight. If he eats less, he loses weight. Other things, such as growth or illness, can also affect how many Calories a body needs.

Most of the Calories that the body uses are for basic functions such as breathing and maintaining body temperature. The energy needed for these functions is called the resting metabolic rate (RMR). A person's RMR is dependent on factors such as age, weight, height, and gender. This rate can be adjusted, based on average daily activity, to show about how many calories a person needs each day.

What to Do

1. Calculate your resting metabolic rate (RMR) and adjust it based on your average daily activity.
2. For three days record everything that you eat and drink.
3. Research to find an estimate of the number of calories in each of the foods and drinks that you consumed. Add the results to find a total calorie count for each day. Average the totals to find an average daily amount.
4. Compare your average daily calories with your adjusted RMR.

95

Objectives

- Recognize that a food Calorie is also called a kilocalorie
- Calculate the resting metabolic rate
- Track Calorie consumption for three days

Materials

- reference materials such as cookbooks or other calorie charts
- bathroom scale
- yard stick or tape measure
- calculator for each student
- see Background for additional materials and helps

Teacher's Toolkit CD

- Instructional Aid 4.1: *Calorie Chart*, for each student
- Rubrics: *Energy for Your Body*, for each student

Introduction

What is your favorite food? Answers will vary.

Some types of foods provide nutrients and minerals that the body needs. Other foods do not have many of those nutrients. But all foods have Calories.

Preparation for Reading

- Direct the students to read silently Student Text page 95 and Activity Manual pages 67–68 before beginning the exploration.

Teach for Understanding

- Discuss the paragraphs at the top of the page, answering any questions the students may have.
- Guide the students as they complete their own weight and height information and convert the measurements to metric on the Activity Manual page.
- Instruct them to write those measurements in either the boys' or the girls' equation to calculate their RMRs. Remind the students to complete the calculations in parentheses before adding.
- Remind the students to write down what they eat or drink for each meal or snack on the Activity Manual page. They should include foods such as butter, sugar on their cereal, and gravy. Encourage the students to be as accurate as possible.

Activity Manual

Exploration, pages 67–68

Assessment

Rubrics

Select the prepared rubric, or design a rubric to include your chosen criteria.

Background

Additional Materials and Helps

You may want to place a bathroom scale somewhere in the room where students can weigh themselves privately.

Many Internet sites include Calorie charts. You may choose to do an Internet search for "Calorie charts" and allow the students to look up the Calorie information for the food that they ate.

Display an 8 oz glass to help the students estimate the amount of beverages that they drink. You may choose to use other objects to help the students estimate the serving sizes of common foods.

You may choose to have the students also record the approximate amount of food, such as 2 rolls, 1 scoop of mashed potatoes, or 8 oz of water.

The students may record their food choices for only one day instead of three days.

54
55

Student Text pages 96–99
Activity Manual pages 69–70

Objectives

- Define *heat*
- Recognize that heat always flows from a warmer substance to a cooler substance
- Identify and describe three ways that heat occurs
- Differentiate between conductors and insulators

Materials

- beanbag

Vocabulary

- heat
- conduction
- conductor
- insulator
- convection
- radiation

Two days have been allotted for this lesson. You may want to teach Lesson 54 the first day and use ideas from Lesson 55 on the second day. You may also incorporate the ideas from Lesson 55 into the text material, making the lesson two days.

Introduction

- Choose two students. Direct one student to stand in the front of the room and the other to stand in the back of the room. Give the beanbag to the student in the front.
  What are some ways that the student in the front could get the beanbag to the student in the back? Possible answers: tossing it; walking up to the other person and handing it to him; having other people help pass the beanbag
- Allow the student to choose one of the ways mentioned to transfer the beanbag to the other student.
  Just as there were several ways that the beanbag could get from one person to the other, there are different ways that energy can be transferred from one substance to another.

Preparation for Reading

- Direct the students to mentally make questions using the headings, bold words, and captions.
- Instruct them to read silently pages 96–99 to find the answers to these questions.
  What are the three ways that matter can be heated?
  Which type of heat does not require any matter?

Teach for Understanding

What is thermal energy? the total kinetic energy of particles in a substance

How is heat related to thermal energy? Heat is the transfer of thermal energy from one substance to another.

What happens to a substance when it gains thermal energy? It is heated. when it loses thermal energy? It is cooled.

Heat

Thermal energy is the total kinetic energy of the particles in a substance. But thermal energy of the particles does not usually stay in one place. It moves, or transfers. **Heat** is the transfer of thermal energy from one substance to another.

Heat always flows from a warmer substance to a cooler one. When this happens, we say the cooler substance has been heated. The cooler substance has gained thermal energy. The warmer substance has lost thermal energy, or cooled. Heat can flow in several ways: conduction, convection, or radiation.

Conduction

Conduction is heat transfer that occurs when particles bump into each other but do not change location. This often happens in solids. The particles in a solid are tightly packed together. Being tightly packed together allows the energy to flow easily between the particles.

When part of a solid is heated, the increased thermal energy of the heated particles causes them to vibrate faster.

As the particles vibrate faster, they cause the particles around them to vibrate faster as well. These particles then cause even more particles to vibrate faster. Eventually all the particles of the solid are vibrating faster. This process allows thermal energy to flow through the entire object, making it warmer.

Conduction does not just occur within a substance. It can also occur between substances. For example, a metal spoon in a cup of hot chocolate heats up through conduction. At first, the hot chocolate has more thermal energy than the spoon does. The hot chocolate's particles move faster than the spoon's particles do. As the faster hot chocolate particles bump into the slower spoon particles, the spoon is heated. The hot chocolate particles lose some energy and get cooler. The spoon particles gain some energy and get warmer. In time, the thermal energy will become equal, and the two substances will have the same temperature.

The hot chocolate causes the spoon to heat up by conduction.

96

Background

Conductors and Insulators

In a solid, the particles are closely packed together. However, the amount of space between the particles varies for different substances. The closer together the particles are, the faster the transfer of heat can take place. Metals are good conductors. They have little space between their particles. Wood and glass are effective insulators since they have greater space between their particles. Another factor increasing conduction is the free movement of electrons. Conductors often have loosely held electrons while the electrons of insulators are less mobile.

First Refrigerator

Thomas Moore, a Quaker from Maryland, is known for inventing the first refrigerator in 1803. He wrote an essay describing a type of icebox that he had patented. In his essay and patent, he used the term *refrigerator* to describe his icebox. This was the first known use of the word. His essay explains his understanding of conductors and insulators. His refrigerator was a covered cedar tub with a large tin container inside. The space between the tub and container could be filled with ice. The entire tub was wrapped in rabbit fur and cloth. In his essay, Moore identifies the tin as a conductor and the cedar, rabbit fur, and cloth as nonconductors.

Most cooking on a stove top uses conduction. A stove burner heats a pan. The pan heats the food. You would also feel the results of conduction if you touched the hot pan. Your hand would be cooler than the hot pan, and the pan would quickly heat your hand. The quick transfer of energy would cause pain and might burn your hand.

Conductors are substances that allow heat to move through them easily. The best conductors are metals. For this reason metals such as silver, copper, and aluminum often are used when rapid heating is desired. Many kinds of cookware are made of copper or aluminum.

Insulators are substances that do not allow heat to move through them easily. Wood, brick, glass, and plastic are good insulators. Cookware may have a wooden or plastic handle that acts as an insulator so your hands do not get burned.

Building Insulation

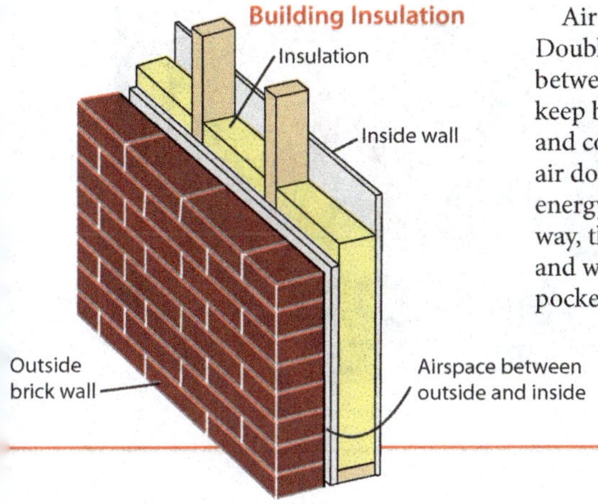

A bird fluffs out its feathers to help insulate itself from the cold.

You might be surprised to learn that air is also a good insulator. Birds fluff out their feathers on cold days to keep warm. Air between a bird's feathers and its body helps insulate the bird from the cold. You use the same principle when you wear several layers of clothing to help keep warm in the winter.

Air also helps insulate buildings. Double-paned windows trap air between the glass panes. This helps to keep buildings warmer in the winter and cooler in the summer. The trapped air does not easily transfer thermal energy by conduction. In a similar way, the insulation used in the ceilings and walls of buildings often has many pockets of air to reduce heat loss.

97

Does heat move from a warm object to a cold object or from a cold object to a warm object? from a warm object to a cold object

In the winter, if someone tells you to close the door because you are letting all the cold air in, is that an accurate description of how heat is moving? no Explain. Since heat flows from higher temperatures to lower temperatures, the heat is actually moving from inside to outside. The room feels cooler because it loses thermal energy.

What are three ways that heat flows from one substance to another? conduction, convection, and radiation

What is conduction? heat transfer that occurs when particles bump into each other but do not change location

Which state of matter is usually heated through conduction? solid Why? The particles in a solid are tightly packed together.

Can two different substances transfer energy through conduction if they are not touching each other? no Explain. For conduction to happen, the substances must be able to touch each other.

Name an example of something being heated by conduction. Possible answers: a metal spoon in a cup of hot chocolate; a pan on a hot stove; a hand on a hot pan

How is a conductor different from an insulator? A conductor allows heat to move through it easily, but an insulator does not.

- You may want to do the *Heat and the Process of Conduction* demonstration at this time.

Name some substances that are good conductors. Possible answer: metals, such as silver, copper, and aluminum

Name some substances that are good insulators. Possible answers: wood, brick, glass, plastic

Air is a good insulator. Based on what you know about air and conduction, why is air a good insulator? Air is a gas, so its particles are far apart. Since few particles touch, not much energy is transferred.

- Explain that the term *insulate* comes from the Latin word *insulatus*, which means "to make into an island."

Why do you think this term is used in science? Elicit the idea that an island is separate from other land forms. Insulators separate heat from an object, or prevent heat from flowing through an object.

Demonstration

Heat and the Process of Conduction

Materials: saucepan, water, hot plate, all-metal spoon with a long handle, wooden spoon with a long handle, cold butter, stopwatch

Pour some water into the saucepan and heat the water to boiling. Press equal amounts of cold butter onto the handles of both spoons. Make sure that the butter on both spoons is the same distance from the tip of the bowl of the spoon. Place the spoons in the boiling water so that the butter is outside of the pan, facing downward. Time how long it takes for the butter to slide or fall off the spoons. Compare the results.

54
55

How is convection different from conduction? In convection, the particles move from one place to another. In conduction, the particles do not move from one place to another.

In which states of matter does convection usually occur? in liquids and gases

- Explain that in the picture of the boiling water the color blue means *cold* or *cooler* and the red means *hot*.

Describe how water particles in a pan of water move during convection. Use the picture to help you. The water in the bottom of the pan is heated by conduction. The heated particles move faster and move upward, while the cooler particles fall downward. The particles at the bottom are heated, and the whole process occurs again.

What term describes this movement of particles? a convection current

- Explain that a convection current does not occur in steps. As parts of it rise, other parts sink. The process is continuous.

Name an example of a convection current. Possible answers: boiling water in a pan; air currents in thunderstorm clouds; magma under the earth's crust

Convection

In conduction the particles transfer energy but do not move from one place to another. But not all heat transfer works this way. **Convection** is heat transfer that occurs when particles carry thermal energy as they move from one place to another.

Convection usually occurs in liquids and gases. For example, as you heat water, only some of the water particles touch the bottom of the pan. These particles are heated by conduction. As they get hotter, they move faster and spread out. This spreading out makes the water at the bottom become less dense. The particles in the hotter water are pushed up as the particles in the cooler, denser water above sink. The new, cool particles at the bottom are then heated, and the entire process starts again. This process is continuously repeated. The movement of particles is called a *convection current*. The current moves particles that carry thermal energy from one place to another.

You have probably seen the convection currents in a pan of boiling water. There are also other kinds of convection currents. Huge thunderstorm clouds are caused by convection currents in the air. Most geologists think that the hot rock under the earth's crust flows in this type of current as well.

Convection currents

98

Background

Convection Currents

Convection currents can be found in places such as the atmosphere, oceans, hot chocolate, and houses. These currents occur in a variety of substances known as fluids. Any substance that flows—such as water, honey, air, or shampoo—is a fluid and can be a medium for a convection current.

Radiation

Thermal radiation is also called electromagnetic radiation. Electromagnetic radiation includes visible light (light that can be seen) and the other electromagnetic waves. The electromagnetic spectrum is discussed in Chapter 10, *Light*. Like other electromagnetic waves, thermal radiation does not need matter in order to transfer heat. However, this radiant energy can pass through transparent materials, such as air and glass.

Radiation

Radiation is heat transfer that occurs when there are no particles to carry the energy. Both conduction and convection need matter for heating. Radiation, however, happens without particles directly touching or particles moving. The transfer of thermal energy is indirect. Radiation can occur even through empty space.

Solar radiation, or energy from the sun, is an important source of both heat and light for the earth. The sun heats the earth even though a great distance separates them. There is no matter that can be used to transfer the energy of the sun to the earth. Yet energy from the sun still reaches the earth by radiation. Much of this solar radiation is absorbed by the earth and converted to thermal energy.

Not all substances absorb the same amount of radiation. The amount depends on the type of surface the substance has. Dark colors absorb radiation better than light colors do. This is why you feel warmer in the sunshine when you wear dark clothing than when you wear light-colored clothing.

The sun is not the only source of radiation, though. The glowing coils of a toaster use radiation to brown your bread. You also can feel the warmth of radiation from a glowing light bulb.

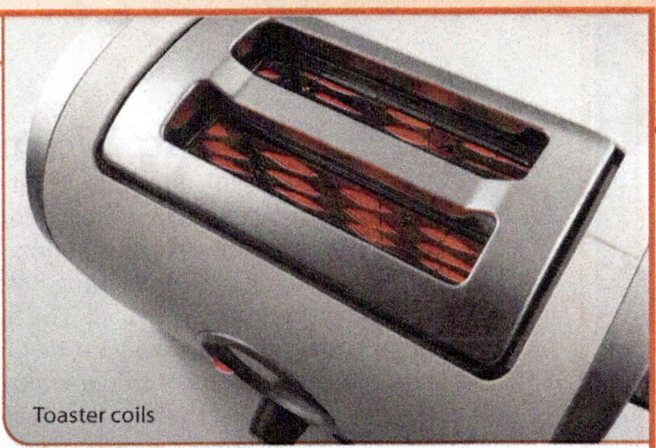
Toaster coils

All three forms of heat can be observed in a campfire. *Conduction* heats the air particles that directly touch the fire. As the thermal energy of the particles increases, their temperature also increases. The particles become less dense. They rise, and cooler air particles sink and replace them. This causes *convection* and heats the air above the fire. Several feet to the side of the fire, you can sense the heat mostly by *radiation* from the flames. Knowing how heat moves and being able to control it allows people to use heat in ways that benefit others.

QUICK CHECK

1. Which type of heating requires particles to touch each other?
2. What is the difference between an insulator and a conductor?
3. In which type of heating does matter move from one place to another?
4. Which type of heating transfers energy without using particles?

99

54
55

What is radiation? heat transfer that occurs when there are no particles to carry the energy

How is radiation different from conduction and convection? Radiation can happen with or without matter. The particles of matter do not need to be moving or touching for radiation to happen. However, both conduction and convection need matter in order to transfer energy.

What is the energy from the sun called? solar radiation

Why do you feel warmer in the sunshine when you wear dark-colored clothing than when you wear light-colored clothing? Dark colors absorb the sun's radiation better than light colors do.

What are some sources of radiation? Possible answers: the sun, a toaster, a glowing light bulb

Explain how a campfire uses all three forms of heat. Air particles directly in contact with the fire are heated by conduction. As those particles get warmer, they move faster. The hot particles move upward, and cooler air particles move downward, starting a convection current. If you are standing several feet to the side of the fire, you sense the heat mostly by radiation.

- You may want to do the *Movement of Particles for Heat Transfer* activity at this time.

ANSWERS

1. conduction
2. An insulator does not allow heat to move through it easily. A conductor allows heat to move through it easily.
3. convection
4. radiation

55

- The *Heat and the Process of Conduction* demonstration and *Movement of Particles for Heat Transfer* activity found at the bottom of the pages in Lessons 54–55 may also be incorporated into the lesson.
- Suggestions for additional activities for this two-day lesson are on the Teacher's Toolkit CD.

Activity Manual

Reinforcement, page 69

Reinforcement, page 70

This page helps students identify ways heat is transferred in a scene.

Activity

Movement of Particles for Heat Transfer

Materials: 3 small beanbags, several students, 2 chairs

Conduction: Use one beanbag and four or five students. Direct the students to stand in a straight line with their hands touching. Explain that the beanbag represents the heat that is being transferred. Give the beanbag to the first student. Direct him to pass the beanbag down the line, with each student placing the beanbag into the hand of the next student.

Conduction requires a direct contact between particles.

Convection: Use three beanbags and three students. Place two chairs near each other. Direct two students to stand on the chairs and one student to stand between the chairs. Give one beanbag to each student. Direct the students to pass the beanbags in a circular path.

In a convection current, the heated particles move upward as the unheated particles move downward.

Radiation: Use one beanbag and two students. Direct the students to stand apart from each other. Give the beanbag to one student and ask him to throw the beanbag to the other student. The second student should catch the beanbag.

Radiation is the heating that occurs even when there is no matter to carry the thermal energy.

56

Student Text pages 100–101
Activity Manual pages 71–72

Objectives

- Predict which type of insulation will best keep hot water warm
- Test different types of insulation to determine which is the most effective
- Measure and use numbers in an activity

Materials

Materials needed for this lesson are listed on the Student Text page.

Teacher's Toolkit CD

- Rubrics: *Keeping Warm*, for each student

Cut 12 cm squares of cotton batting, craft foam, bubble wrap, and aluminum foil for each group. Some 3 oz cups have a wider bottom than others do. Cups with a wider bottom will be less likely to tip over when wrapped with insulation.

Introduction

What are some ways that people keep picnic foods or drinks cold? Possible answers: put them in a cooler or thermos; freeze them and let them thaw; pack them in a box with ice

What are some ways that people can keep food warm for an extended time? Possible answers: cover it; place it under a heat lamp; keep it in an oven on low heat; put it in a thermos

Preparation for Reading

- Direct the students to read silently Student Text pages 100–101 and Activity Manual pages 71–72 before beginning the activity.

Teach for Understanding

Process Skill

The process skill emphasized in this activity is measuring and using numbers, which is using standard measuring devices and techniques to quantify information.

What kinds of measurements are being used in this experiment? milliliters, degrees

Why is it important to measure the amount of water that we put in each cup? If the amount of water is not the same in each cup, the temperature measurements will not show which insulation worked best.

Why is it important to measure the temperature? Without measuring, we are simply guessing which insulation works best.

ACTIVITY

Keeping Warm

One of the most important reasons to study heat is to protect people. Often that means using insulation. Insulation keeps something from getting hot. Many cooking pots have insulated handles. The handles stay cool and keep you from burning your hand.

Insulation can also keep hot things warm, though. For example, an insulated thermos can keep hot foods warm or cold foods cool. Many animals, such as polar bears, have air-filled spaces in their fur. The air helps the animals keep warm during the winter. Likewise, insulation in the ceilings or walls of buildings keeps heat inside during cold weather.

In this activity you will test several kinds of insulation to determine which is the most effective.

Process Skills
- Hypothesizing
- Predicting
- Inferring
- Collecting and recording data
- Communicating

Materials
- 5 plastic cups, 3 oz
- cotton batting
- rubber bands
- craft foam
- bubble wrap
- aluminum foil
- 250 mL hot water
- metric measuring cups
- 5 thermometers
- plastic wrap
- Activity Manual

Problem

Which kind of insulation will keep hot water warm the best?

Procedure

1. Wrap cotton batting around one of the cups. Be sure to cover the bottom and the sides of the cup. Use a rubber band to secure the cotton batting against the cup.
2. Prepare three more cups: one wrapped with craft foam, one with bubble wrap, and one with a double thickness of aluminum foil. Use rubber bands to secure each material. Do not wrap anything around the fifth cup.
3. Predict which cup will best keep the hot water warm. Complete the hypothesis in your Activity Manual.
4. Pour 50 mL of hot water into each cup. Place a thermometer in each cup.
5. Cover the top of each cup with plastic wrap. Do not cover the thermometer with plastic wrap. You should be able to read the temperature without removing the plastic wrap.

100

6. Measure and record the starting temperature for each cup.
7. Leave the cups undisturbed for five minutes. Then measure and record the water temperature in each cup. Measure and record the temperatures again after another five minutes.
8. Calculate the difference between the starting and ending temperatures for each cup.

Conclusions
- Did your results support your hypothesis?
- Which cup had the greatest change in temperature? Why?
- Which type of insulated cup would you choose to hold hot chocolate? Why?

Follow-up
- Test other materials as insulation.
- Use ice cubes instead of hot water. Determine which insulation is the best for keeping ice cubes from melting.

101

- Emphasize the importance of measuring accurately. If we fail to measure accurately, how useful is the information that we gain from the experiment? The information is unreliable because it has no specific data that can be quantified.

Procedure
- If needed, help the students prepare the cups. It is easier to cover a cup with a square if the cup is first placed upside down on a table. The cup wrapped with craft foam will be less likely to tip over if it is wrapped similar to the way you would wrap a gift. If necessary, the students can prop the cups against other objects for support.

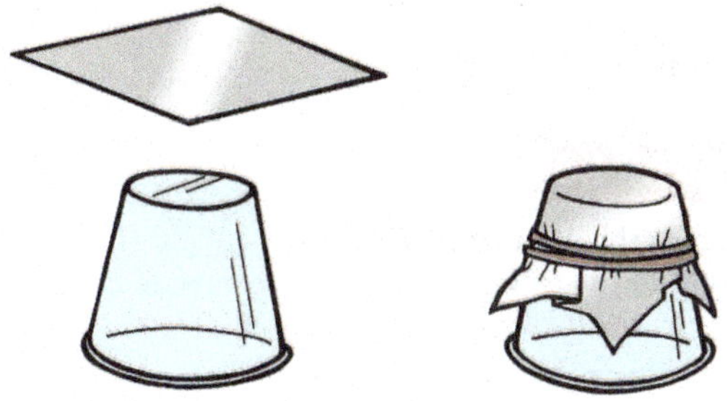

- The students should prepare all the cups before adding the hot water. This should keep the starting temperatures about the same.
- Remind the students not to cover the thermometer with the plastic wrap. They should be able to read the temperature on the thermometers without disturbing the plastic wrap.
- Direct the students to check the temperatures after 5 minutes and then after 10 minutes.

Conclusions
Which cup had the greatest change in temperature?

Was your hypothesis correct?

Which type of insulated cup would you choose to hold hot chocolate? Why?

Activity Manual
Activity, pages 71–72

Assessment
Rubrics
Select the prepared rubric, or design a rubric to include your chosen criteria.

57

Student Text pages 102–4
Activity Manual pages 73–75

Objectives

- Identify some common fuels
- Distinguish between renewable and nonrenewable resources
- Name some ways fuel is used
- Give examples of unwanted heat

Materials

- pictures of a car or another type of vehicle, a factory, and a campfire

Teacher's Toolkit CD

- Quiz 4-B

Vocabulary

- fuel
- fossil fuel

Introduction

- Display the pictures.

  What do the things pictured here have in common? Answers will vary, but elicit that all need fuel.

  Many things use fuel in order to release energy to do something, to move something, or to make something. In this lesson we will discuss some different kinds of fuels and the energy they release.

Preparation for Reading

- Direct the students to mentally make questions using the headings, bold and italicized words, and captions.
- Instruct them to read silently pages 102–4 to find the answers to these questions.

  What are fossil fuels?

  What is a byproduct?

Teach for Understanding

- Direct attention to the captions of the pictures.

  What are all these called? fuels

  What is fuel? any substance that releases energy when it burns

  Which of the pictures on the page are common fuels? coal, natural gas, and firewood

  What are fossil fuels? fuels that formed when the remains of plants and animals were buried quickly

  When were most fossil fuels probably formed? during the Genesis Flood

  Which of the pictures is a fossil fuel? coal, crude oil, and natural gas

  What is a resource? anything that can meet the need of a living thing

  What is a nonrenewable resource? a resource that is no longer forming today

  Which type of fuel is nonrenewable? fossil fuels

Heat at Work

If you tried to burn a piece of granite, you would find it very difficult. It would take a tremendous amount of heat to do so. You also would have to use much more energy than you could gain from the hot rock. Some kinds of matter are good sources of thermal energy, though. These kinds of matter are often called fuels.

Firew

Fuel Sources

A **fuel** is any substance that releases energy when it burns. Some common fuels are wood, oil, gasoline, and natural gas. Each contains matter that releases thermal energy by a chemical change. As the fuel burns, some of it changes into thermal energy.

Some of the fuels that we use are known as fossil fuels. **Fossil fuels** were formed when the remains of plants and animals were buried quickly during the Genesis Flood. Fossil fuels include coal, crude oil, and natural gas. The worldwide Flood of Noah's day, recorded in Genesis 7–8, can account for the abundance of fossil fuels that we find in the world today.

There are still large quantities of fossil fuels around the world. But these fuels are often described as *nonrenewable*. A nonrenewable resource is one that is no longer forming today.

Another form of fuel is wood. This fuel has been used throughout history. Today many people use fossil fuels instead of wood. However, there are still places in the world where wood is the only available fuel. Unlike fossil fuels, wood is a *renewable* resource, which means it can be replaced. New trees can grow naturally, or people can plant new ones.

Coal

Natural gas

Crude oil

102

Fuel Uses

Releasing thermal energy is usually the purpose for burning a fuel. Many people burn fuel to keep warm in cold weather. As fuels burn, they can heat homes and other buildings. Natural gas, heating oil, coal, and wood are all common fuels used for heating.

However, fuels are not just used to heat homes. Some people use natural gas to cook food, dry clothes, or heat water. Sometimes wood is used to cook food as well.

Power plants use fuel to produce electricity. The burning fuel gives off thermal energy. This energy heats water and produces steam. The steam is used to turn a propeller-like device called a turbine. The turbine then turns a generator to produce the electricity.

Fuel is important for transportation.

Many other industries also burn fuels. Without heat from fuels, many products could not be made. For example, iron ore, a rock-like mineral used to make steel, must be melted to remove any impurities. However, the melting point of iron is very high, about 1,535 °C. Fossil fuels produce the very high temperatures needed for this process. Once the impurities are removed from the liquid iron, it is mixed with other substances and poured into molds. As the metal cools and solidifies, it forms many of the steel products that you use.

Steel plant

Most forms of transportation involve the burning of fuels made from crude oil. Almost all cars burn gasoline made from crude oil. Large trucks use diesel fuel, which is also made from crude oil. Even jet fuel is a product of crude oil.

103

What are renewable resources? resources that can be replaced or grown again

What is one fuel that is renewable? wood

What is the purpose for burning a fuel? to release thermal energy

What are some ways that fuel is used? Possible answers: for warmth, for cooking, for drying clothes, for heating water, for generating electricity, for manufacturing, for transportation

What are some common fuels used for heating? natural gas, heating oil, coal, and wood

Describe how power plants use fuel to produce electricity. The thermal energy heats water and produces steam. The steam turns a turbine. The turbine turns the generator, producing electricity.

What type of fuels can produce the high temperature needed to melt iron ore? fossil fuels

- Direct attention to the picture of the steel plant. Discuss the process of making many of the steel products that we use.

What kind of fuel is often used in transportation? fuels made from crude oil

- Explain that crude oil, or petroleum, is not the same oil that is put into automobiles or on door hinges. Crude oil is the oil as it comes out of the ground. It is processed, or refined, and eventually made into a usable form, such as oil, gasoline, or kerosene.

Activity

Fuel Collage

Collect pictures of different ways that fuels are used. Make a collage or mobile using the pictures.

57

What is often produced when energy changes from one form to another? thermal energy

Is this thermal energy, or heat, always the desired result? no

What name is given to the unwanted heat or waste product? byproduct

What does *byproduct* mean? a waste product or something unwanted

What is an example of a byproduct? Possible answers: heat given off by a light bulb; heat from an engine or another machine; heat given off by the human body

What could happen if a machine produces too much heat as a byproduct? The machine could become damaged.

What helps remove excess heat from a car engine? fluid in a radiator and a fan

What is another type of machine that uses a fan to remove excess heat? a computer

What kind of cooling system did God create for our bodies? sweat

ANSWERS

1. when the remains of plants and animals were buried quickly
2. Possible answers: heating, cooking, drying clothes, generating electricity, making industrial products, transportation
3. Possible answers: removing heat from a car engine or cooking hotdogs over a campfire.

Activity Manual

Review, pages 73–74

These pages review Lessons 54–55 and 57.

Expansion, page 75

This page uses a diagram to show the uses of fuel.

Assessment

Quiz 4-B

The quiz may be given any time after completion of this lesson.

Unwanted Heat

Often when energy changes from one form to another, thermal energy is also produced. Sometimes the heat is the desired result. At other times, though, the heat is not wanted. In these cases, it is a waste product, or *byproduct*. The heat from a light bulb is an example of this. The electrical energy in the light bulb produces light. Light is the desired form of energy, but the light bulb also gets very hot. The energy that produces the light also creates heat. This heat is a byproduct.

Unwanted heat can sometimes damage equipment. For example, when a car engine burns fuel, the engine produces mechanical energy. This energy moves the car, but it also produces heat. Some of this heat is a byproduct. If not removed, it will cause the engine to get too hot. To prevent that, a car radiator uses fluid to transport heat away from the engine. A fan near the engine also helps move away extra heat.

Other machines also use fans to remove unwanted heat. A computer produces a lot of heat as electricity moves through its parts. This heat could damage sensitive parts of the computer. So a fan in the computer moves heat away from these parts.

Sweating helps the body remove extra heat.

Even your body has its own type of cooling system. God designed the human body to maintain a certain temperature range. When you exercise, though, your body converts energy from one form to another. This produces extra heat. The blood in your body carries this heat close to your skin's surface so it can be released. Your body may also produce sweat. The sweat evaporates on your skin and helps cool your body. Because of this loss of water, though, you should drink plenty of fluids when you exercise. Replacing the fluids helps maintain your body's cooling system.

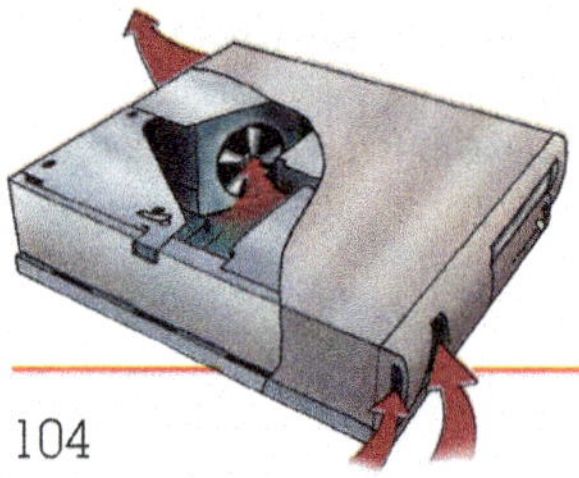

A fan moves air through a computer to help remove unwanted heat.

QUICK CHECK

1. How were fossil fuels formed?
2. What are three uses of fuels?
3. Give an example of how thermal energy affects everyday life.

104

Background

Sweat

Sweat evaporating from the skin is also part of the excretory system. The organs involved with this system are discussed in Chapter 12, *Circulatory System*.

Heat in Space

People have designed many kinds of technology to control heat. Sometimes the purpose is to make equipment run better. Other times it is to make people more comfortable. During space travel, though, controlling heat is not just a matter of efficiency or comfort. It is a matter of survival. The lives of the space crew depend on it.

Space is a very cold place, but the sun's radiation causes objects in space to heat up. The earth is protected by the atmosphere. Without the protective layer of the earth's atmosphere, objects in space get very hot if not cooled in some way. As a spacecraft travels, it can move from an area of intense heat to areas of intense cold within seconds. Scientists and engineers must plan carefully to keep the spacecraft and everything inside it within a specific temperature range.

The bottom of a space capsule is curved.

Space Capsules

At the beginning of the space program, rockets designed for manned space travel had a capsule at the top of the rocket. Only the space capsule, with the astronauts inside, returned to the earth. This capsule had a slightly curved bottom. While entering the earth's atmosphere on its return, the bottom of the capsule faced toward the earth. This design slowed the capsule down.

Although the design slowed the capsule down, the curved bottom also produced much friction. The heat from this friction caused the capsule to get very hot. The exterior could reach temperatures higher than 1,000 °C (1,832 °F) as it dropped through the earth's atmosphere. No one inside the capsule would have been able to live through it. So scientists and engineers had to find a way to remove some of the heat.

To solve the problem, scientists used an idea similar to how sweat cools your skin. They put a material on the capsule's outside that melted and vaporized as the capsule heated up. As the material vaporized, it cooled the capsule. Sweat keeps us cool in much the same way.

105

Objectives

- Explain why controlling heat is necessary
- Explain how scientists controlled heat for the reentry of space capsules
- Name two types of insulation used on space shuttles
- Name some ways that thermal energy is part of our everyday lives

Teacher's Toolkit CD

- Quiz 4-C

Introduction

What are some reasons we use technology to control heat? Possible answers: to be comfortable, to keep equipment from breaking, to cook or prepare food

These reasons are not limited to heat on the earth.

Preparation for Reading

- Instruct the students to mentally make questions using the headings, italicized words, and captions.
- Direct them to read silently pages 105–8 to find the answers to these questions.

How did scientists cool a space capsule?

Why would the solution for controlling heat on a space capsule not work for space shuttles?

What are the tiles on a space shuttle used for?

Teach for Understanding

Why is controlling heat during space travel important? for survival and for keeping the space crews safe

Which type of spacecraft was used at the beginning of the space program—the space shuttle or the space capsule? the space capsule

When a spacecraft returns to the earth, how hot can its exterior become? more than 1,000 °C (1,832 °F)

Why were capsules designed with a slightly curved bottom? The curved bottom would help slow the capsule down when it reentered the earth's atmosphere.

What was a problem with that design? It produced a great deal of friction and would cause the capsule to become very hot.

How did scientists solve the problem? They put a material on the capsule that would melt and vaporize. As the material vaporized, it helped cool the capsule.

Background

United States Space Program

NASA, the National Aeronautics and Space Administration, began on July 27, 1958, while Dwight D. Eisenhower was president of the United States. In April 1961 a Russian astronaut became the first man in space. Less than a month later, Alan Shepard made the first American space flight.

The early space program used rockets, with a space capsule attached to the top of the rocket. *Freedom 7*, the space capsule used by Alan Shepard, was only a little more than 3 m (10 ft) high and 2 m (6.6 ft) wide. This launch was part of Project Mercury.

The first space shuttle, *Columbia*, was launched in April 1981. Unlike the space capsules, the space shuttles could be reused for other space flights.

Could the cooling method used on a space capsule work on a space shuttle? no Why? Space shuttles were intended to be used again, but a space capsule was used only once.

Why did a space shuttle experience more friction and higher temperatures than a space capsule? It was larger and had more surface area.

What does *surface area* refer to? the entire outside surface of an object

Why do you think that more surface area causes more friction? Possible answer: More surface area provides more places to produce friction.

Which part of a shuttle became the hottest when it was returning to Earth? the bottom Why? It produced most of the friction that slowed the shuttle down.

What types of insulation were used on space shuttles? ceramic tiles and blanket-like waterproof glass cloth

Which type of insulation was used to cover the bottom of a shuttle? special tiles designed to protect against very high temperatures

What probably caused the explosion of the space shuttle *Columbia*? tiles that were broken or damaged

The tiles on the bottom of a shuttle were made to withstand very high temperatures.

Space Shuttle

As scientists and engineers developed new spacecraft, new problems had to be solved. The cooling method used on a space capsule would not work on a space shuttle. Space shuttles were meant to be used again. They could not use material that would melt and vaporize.

A space shuttle was also larger than a space capsule. More size means more surface area to cause friction. It also means that more surface area gets hot. As a shuttle returned to the earth, the shuttle's exterior could reach 1,650 °C (3,000 °F). This would cause the shuttle's aluminum frame to melt if not protected.

To protect the shuttle and its crew, several kinds of insulation were used. One kind is like a heavy blanket. It was made with waterproof glass cloth and was quilted. This kind of insulation protected areas of the shuttle that do not get the hottest.

A shuttle also used ceramic tiles as insulation. A shuttle returning to the earth glided like an airplane. In doing this, the bottom of the space shuttle produced most of the friction that slows the shuttle down. However, the friction also caused the bottom of the shuttle to heat up more than any other part of the shuttle did. For this reason, special tiles designed to protect against very high temperatures covered the bottom of the shuttle.

Faulty tiles can create serious problems. Scientists believe that tile damage caused the breakup of the space shuttle *Columbia* in 2003. The tiles may have been broken or damaged during liftoff. When the shuttle returned to the earth, the tiles could not protect the shuttle from the intense heat. The shuttle broke apart as it flew back into the earth's atmosphere.

The ceramic tiles on space shuttles were carefully checked.

106

Science Misconceptions

Space Shuttle Orbit

The space shuttle did not orbit in outer space. Along with the International Space Station, it orbited within the earth's atmosphere, about 350 km (217 mi) above the surface of the earth.

Background

Columbia Explosion

The space shuttle *Columbia* exploded over the southern United States on February 1, 2003, just minutes before its scheduled landing. All seven astronauts were killed.

Shuttle Tiles

A space shuttle may have had more than 33,000 tiles on it. Each tile was unique and could not be switched with any other tile. Computer numbers helped identify the location where each tile should be placed. Some of the tiles were black and were used to protect areas of the shuttle that experienced higher temperatures. Other tiles were white and were placed on areas of the shuttle that experienced lower temperatures.

International Space Station

The *International Space Station* (ISS) is a large spacecraft that is also a science lab and a space habitat. People from several countries have occupied it since 2000. The ISS orbits the earth about every ninety minutes. When it is facing the sun, the temperature is about 121 °C (250 °F). However, when it is in Earth's shadow, away from the sun, the temperature drops to −157 °C (−250 °F).

International Space Station

Similar to the shuttle, the ISS is insulated by a special blanket made of mylar sheets and dacron fabric. This blanket has layers that stay separated to keep heat from being conducted through the layers. It covers most of the space station.

The insulation on the ISS keeps heat outside the station, but it also keeps heat inside. Computers and other electronic equipment generate a lot of heat as they work. Just imagine how much heat is produced by all the equipment astronauts use!

The ISS also has a system, similar to an air conditioning unit, that cools the inside of the station. Pipes inside the station circulate water through devices that help absorb and remove the excess heat. However, it is so cold outside the station that water would freeze quickly. So the heat is transferred to pipes filled with ammonia. The heated ammonia circulates through large radiators on the outside of the station. There the heat is released into space. The ammonia does not freeze because it circulates back inside the station before it loses enough heat to freeze.

107

58

How is insulation used on the International Space Station? It keeps harmful heat outside the shuttle but keeps some heat inside the shuttle to protect the crew from the coldness of space.

Why is keeping too much heat inside a shuttle a problem? Possible answers: The equipment on the shuttle produces a great deal of heat. Temperatures would get too hot for people. Heat could damage equipment.

Describe how the International Space Station keeps the inside cool. Pipes circulate water through devices that help absorb and remove excess heat. The heat is transferred to pipes filled with ammonia. The heated ammonia circulates on the outside of the station where the heat is released into space.

Why is it important to avoid exposure to the extreme temperatures that exist in space? They are a threat to human life.

Activity

Space Exploration

Materials: books and other resources

Encourage the students to decide whether they think that the benefits of space travel are worth the expense and danger involved. Divide the students into two groups. Let one group represent the position for continuing space travel and exploration and the other represent the position for canceling the program. Provide resources for each group to research and find facts to support its position. Choose several students to represent each side in a panel-discussion forum.

Point out to the students that technological advances often cause differing views. Scientists, engineers, and those providing money must make decisions based on many factors.

58

What is an unmanned research spacecraft called? a probe

What kind of probe is the *Curiosity*? a rover

Where is the *Curiosity*? on the planet Mars

What methods to control heat are used on the *Curiosity*? Some parts are painted with a special gold paint. There is a special layer of insulation called aerogel. Fluid-filled tubing allows unwanted heat to be released out into space. Switches automatically turn heaters on or off.

What are some ways that thermal energy affects our everyday lives? Possible answers: choosing clothing; using freezers to prevent ice cream from melting; choosing spoons to stir boiling liquids

How can a Christian glorify God by studying about thermal energy? He can protect human life from the byproducts of thermal energy.

ANSWERS

1. Since God values life, it is important for the safety of the crew and the protection of the equipment
2. friction as the space vehicles move through the air
3. electronic equipment, such as computers

Activity Manual

Review, page 76

This page reviews Lesson 58.

Assessment

Quiz 4-C

The quiz may be given any time after completion of this lesson.

Modern Spacecraft

Radiators and heaters have often been used to control heat in space. But those are heavy and take up a lot of room in the spacecraft. New spacecraft and equipment are often lighter, smaller, and more efficient than previous ones. As we learn more, scientists and engineers make changes and add new technology.

The space capsules and shuttles were manned missions. People traveled in those spacecraft. However, some spacecraft are robotic. An unmanned research spacecraft is called a probe. Probes that can land on a planet or other space objects and move around are called rovers. The rover *Curiosity* arrived on Mars in 2012.

Scientists are using *Curiosity* to study Mars. Mars is colder than Earth, especially at night. The rover uses a combination of methods to control heat. Some parts of it are painted with a special gold paint that helps radiate heat outwards away from the rover's body. There is also a special layer of insulation called aerogel that traps heat inside the rover's body. Fluid-filled tubing allows unwanted heat to circulate through the rover's system and be released out into space. Switches automatically turn heaters on or off to maintain electronics and battery temperature during night on Mars.

Through the space program, we have learned many things about our universe. All our knowledge could not have been gained, however, if we were not able to control heat. We are affected by thermal energy in many ways every day. We dress in clothing that is appropriate for how warm or cool the day is. We put ice cream in the freezer so that it will not melt. We use a wooden or plastic spoon, instead of a metal one, to stir boiling liquids. Each of these decisions is based on our knowledge of how thermal energy and heat affect us.

Thermal energy is an amazing part of God's creation. It can be used as a powerful tool in many different areas. Everything from refrigerators to electricity requires understanding thermal energy. Byproducts of thermal energy can sometimes be dangerous. Protecting human life also means learning about thermal energy. A Christian can glorify God by studying how to use thermal energy.

QUICK CHECK

1. Why is controlling heat while in space so important?
2. What causes the heat that occurs when space vehicles take off or land?
3. What technology causes a lot of heat inside the ISS?

108

A lesson from
Answers in Genesis Name ____________________ 1:1

Serving through Science

When someone makes an invention to make his life better, he also helps others in the process. You are probably grateful that someone invented things such as cars, eyeglasses, refrigerators, and medicines. How does a scientist or engineer decide how to solve a problem? This activity will help you understand how to solve a practical problem.

Materials
ceramic mug
metal teaspoon
plastic spoon
tape
boiling water
2 ice cubes (the same size)
stopwatch

Procedure

1. Place the spoons on the cup as shown and tape the handles in place.
2. Add the water to the cup so that it comes just below the level of the spoons near the rim.
3. Place an ice cube in each spoon. Record a description of the appearance of the ice on each spoon. Touch the handle of each spoon where it touches the rim of the mug and record your observation. Observe the ice and touch the spoons every 2 minutes for 10 minutes, recording your observations each time.

	2 min	4 min	6 min	8 min	10 min
Ice on metal spoon					
Touch of metal spoon					
Ice on plastic spoon					
Touch of plastic spoon					

Objectives

- Show how Christian scientists can do operational science in order to exercise biblical dominion
- Give examples of discoveries that show that operational science does not need to refer to evolutionary principles to be successful
- Explain why biomimicry is an example of exercising dominion to love our neighbor and to glorify God

Materials

- ice-cream maker

Preparation for Reading

- Direct the students to read the first paragraph on Activity Manual page 77 to find out why they will be doing the activity.

Teach for Understanding

Why will you be doing the activity? to help understand how to solve a practical problem

Procedure

- Guide the students as they set up and work the activity.
- Remind them to record their observations in the chart at the bottom of Activity Manual page 77.

Conclusions

How does the thermal energy in the hot water compare to the thermal energy in the ice cube? Lead the students to understand that the molecules in the hot water have more kinetic energy than the molecules in the ice.

Which spoon was the best conductor of heat? The metal spoon was the more efficient conductor of heat.

What observations allowed you to conclude this? Possible answers: The metal spoon was warmer to the touch than the plastic spoon. The ice melted faster in the metal spoon than in the plastic spoon.

Describe the heat transfer that caused the ice on the metal spoon to melt. The thermal energy from the water was transferred by conduction to the ceramic cup, then to the spoon, and finally to the ice, causing the ice to change from a solid to a liquid (or to melt). There was also transfer of energy through convection as the air moved energy from the hot water to the spoon as well as through conduction with the air around the ice. However, these would be at a much smaller scale than the heat from the water.

Which material would be best to use as a container to keep ice cold the longest, metal or plastic? The plastic would be a better choice because it transfers heat more slowly, less efficiently, than the metal.

59

- Display the ice-cream maker. Show how different materials are used in the various parts. The outside container is usually made of plastic or wood as an insulator to keep the ice/salt mixture cold. The inner container is typically made of aluminum, stainless steel, or galvanized steel to allow the heat of the cream mixture to be efficiently transferred to the ice/salt mixture, cooling the mixture inside.
- Direct the students to read silently the remainder of Activity Manual page 78 to find out what Dr. Raymond Damadian invented.

  What did Dr. Damadian invent? the MRI

  How does his invention help others? allows doctors to look inside the body without cutting it open

  Scientific applications can provide many benefits to make our lives easier and also to lessen the effects of the Fall.
- Discuss God's original perfect creation referring to Genesis 1:26–31.
- Instruct the students to turn to page 1 in their Student Texts.

  What is the second big idea about a Christian worldview? God put us here to work.
- Direct the students to skim the paragraph.

  What is the job God gave all people? to take care of the earth and to make it useful
- Read aloud Mark 12:30–31.

  What two commandments are obeyed when someone develops machines, materials, and medicines to help others? the two greatest commandments: loving God with all your heart, soul, mind, and strength and loving your neighbor as yourself

  As Christians exercise dominion in a fallen world, serving others and reversing the effects of the Fall are ways that we can honor God and love others.
- Explain that despite the claims of some people, Christians can be real scientists. An evolutionary worldview is not necessary to do scientific inquiry. There are many biblical creationists who reject evolutionary views and do marvelous things in science.

Activity Manual

Answers in Genesis, pages 77–78

Conclusions

1. How does the thermal energy in the hot water compare to the thermal energy in the ice cube?

   ______

   ______

2. Which spoon was the better conductor of heat?

   ______

3. What observations allowed you to conclude this?

   ______

   ______

4. Describe the heat transfer that caused the ice on the metal spoon to melt.

   ______

   ______

5. Which material would be better to use as a container to keep ice cold longer, metal or plastic?

   ______

A scientist who develops materials to use as insulators or heat conductors is working to benefit other people. Christians can work to make products and develop ideas with the goal of honoring God by serving their fellow humans. Serving others in this way is an example of what Jesus taught in Mark 12:30–31.

God created the world to be perfect, but our sin has broken God's perfect creation. Now the world is full of diseases and pain. As cures are found for diseases and medicines are developed to help relieve suffering, scientists help to lessen the effects of the Fall.

There are many Christian scientists who are working to make people's lives better. Some people claim that a person cannot be a good scientist if he believes in God. They think that the only true scientists are those who believe in evolution. But is their thinking correct? How much evolution did you use to decide which material was the better conductor of heat? Absolutely none!

Dr. Raymond Damadian is a Christian who believes that God created the world in six days. He used the talents that God gave him to help his fellow humans who were suffering under the effects of the Fall. He invented the MRI machine that allows doctors to look inside the body without cutting it open. This machine has helped doctors heal many people. We should give thanks to God for people like Dr. Damadian who serve others through science.

If you were a scientist, what would you want to develop to serve others?

Resources

The BJU Press website offers additional links from Answers in Genesis that you may find helpful with this lesson.
bjupress.com/resources/science

Moon Station

EXPLORATION

The earth's atmosphere traps heat from the sun. This helps keep the earth from experiencing extreme changes in temperature. But the earth's moon has no atmosphere. The moon's daytime temperature can reach around 120 °C (250 °F). At night the temperature can drop to −173 °C (−279.4 °F). The length of time from one sunrise to another is also longer than that on the earth. It takes over twenty-seven Earth days for the moon to complete one rotation, or day.

Imagine that scientists have asked you to help design equipment for a station on the moon. You will need to consider the extreme temperature changes on the moon as you design your equipment. You will also need to keep in mind that the extreme temperatures last for many Earth days.

What to Do

1. Decide what your equipment will do. Will it be used for clothing, housing, communication, transportation, or something else?
2. Fill in your Activity Manual and draw a sketch of your piece of equipment.
3. Be able to explain how your invention is designed to withstand the temperatures on the moon.

109

Objectives

- Design a piece of equipment for a moon station
- Research equipment developed for the space program

Materials

- various trade books or other resources to research equipment used in the space program

Teacher's Toolkit CD

- Rubrics: *Moon Station*, for each student

Introduction

In our town, how high do temperatures get in the summer? Answers will vary.

How low do the temperatures get in the winter? Answers will vary.

The temperatures on the moon are much hotter and much colder than on the earth. What kind of protection would you need if you had to experience exceedingly cold temperatures or exceedingly hot temperatures for several days? Answers will vary.

Preparation for Reading

- Direct the students to read silently Student Text page 109 and Activity Manual page 79 before beginning the exploration.

Teach for Understanding

What protects the earth from the extreme temperature changes of the sun? the earth's atmosphere

How long are the days on the moon? twenty-seven Earth days

- Guide the students as they plan and design the equipment. Help them correctly use the scientific concepts presented in this chapter. The designs do not have to be realistic. Encourage the students to be creative in their designs.
- Determine the type of presentation that you want from the students. The presentations may be oral or written but should explain the elements shown in the diagrams.

Activity Manual

Exploration, page 79

Assessment

Rubrics

Select the prepared rubric, or design a rubric to include your chosen criteria.

61

Student Text page 110
Activity Manual page 80

Objectives

- Recall concepts and terms from Chapter 4
- Apply knowledge to everyday situations

Review

- Materials for the Chapter 4 Test will be taken from Student Text page 110 and Activity Manual pages 65–66, 73–74, 76, and 80. You may review any or all of the material during the lesson. Questions similar to the ones in Thinking It Through, Activity Manual page 80, may appear on the test.
- You may choose to review Chapter 4 by playing the review game or a game from the Game Bank on the Teacher's Toolkit CD.

Activity Manual

Review, page 80

This page requires written responses to application questions.

CHAPTER SUMMARY

Words to Know

energy	thermometer	conductor
potential energy	thermal expansion	insulator
kinetic energy	calorie	fuel
thermal energy	heat	fossil fuel
temperature		

Key Ideas

Explain what factors determine the thermal energy of a substance
Identify and compare scales used to measure temperature
Give an example of how thermal expansion affects everyday life
Explain how thermal energy is involved with matter changing states
Identify the calorie as the unit used for measuring thermal energy
Recognize that heat flows from a warmer substance to a cooler one
Compare how heat moves by conduction, convection, and radiation
Identify examples of conduction, convection, and radiation
Identify materials that are good conductors and insulators
Explain what causes a convection current
Identify the sun as an important source of radiation
Identify uses of fuels
Give examples of how unwanted heat can be removed
Summarize ways that scientists and engineers have controlled heat while developing equipment for space travel

110

Review Game

A Heated Challenge

Materials: two sets of vocabulary-word cards

Prepare two sets of cards for each vocabulary word by writing each letter of a word on a separate card. Use a different color ink or card for each word. This will enable you to see which letters go with which word. Vocabulary words can be found on Student Text page 110 under Words to Know.

Divide the class into two teams. Put the letter cards for one of the vocabulary words face down in front of each team. For each correct answer to a review question, allow a team member to draw one letter. The letters on the other team's cards should not be visible. When a team has drawn all the letter cards for their word, the team members should unscramble and identify their team's vocabulary word. If the answer is correct, add a point to that team's score. Give additional points if the team can also define or explain the meaning of the word. Give the teams another word and continue asking review questions as time allows.

62

Objective

- Demonstrate knowledge of concepts taught in Chapter 4

Assessment

Chapter 4 Test

Teacher Notes

63

Student Text pages 111–13
Activity Manual page 81

Objectives

- Recognize the interrelationship of science concepts
- Recognize, from a Christian worldview, reasons for studying climate
- Understand the role of meteorology in preserving human life
- Apply the biblical teaching on the value of human life to everyday situations

Teacher's Toolkit CD

- Visual 5.1: *Southeastern United States*

Unit Introduction

- The purpose of this unit is to explain the influence of climates on all people and guide the students in applying this knowledge in ways that glorify God.
- Chapter 5 discusses the atmosphere, weather patterns and conditions, and severe weather. Chapter 6 explores how weather conditions affect the land and create unique biomes for living things. These biomes have a variety of plants and animals that are characteristic of those areas. Chapter 6 also discusses water biomes and wetlands.

  Look through Unit 3. What kinds of topics do you think you will be studying in this unit? Possible answers: weather, biomes

  Why do you think that these chapters are organized into the same unit? Answers will vary. Elicit that the chapters are about climates and their effects on the earth.

  Look at the picture. It is the Orkhon Falls in Mongolia. What do you think the weather was like when the picture was taken? Possible answers: spring or summer day; warm and sunny; with the gray clouds a possibility of rain

  Look at the miniature scientist at the top of the page. Is he in the same kind of climate as the waterfall? no What kind of climate is the scientist experiencing? Possible answers: cold, snowy, and windy

Because of the Climate

Background

Unit Photo

The Orkhon Falls are located in the Orkhon Valley of Mongolia. The water from the Ulaan (meaning "red") River falls from a height of 24 m into the Orkhon River. The local people call the waterfall Ulaan Tsutgalan, which means "red flows."

Weblinks

The BJU Press website offers additional information and links you may find helpful throughout the unit.
bjupress.com/resources/science

- Ask volunteers to read the teasers on page 112. Discuss the information.

Project Idea

- The project idea presented at the beginning of each unit is designed to incorporate elements of both chapters as well as information gathered from other resources. You may choose to have the students do the project as a culminating activity at the end of the unit or as an ongoing activity while the chapters are being taught.

Unit 3: Tracking Weather in Different Biomes

- Choose six or more locations around the world. The locations should represent the different biomes described in Chapter 6. While discussing Chapter 5, the students should track the weather for each location. This information may be found in newspapers and on the Internet. Provide a separate chart for recording the weather information for each location. Then, while discussing Chapter 6, the students should use the weather information to help them decide the type of biome for each location. Research to find additional information about each biome. Display the charts and biome information on a world map.

This lesson continues with the Chapter 5 introduction on page 133.

Unit 3 Overview

Lesson	Teacher's Edition	Student Text	Activity Manual	Assessment	Content
Chapter 5: Weather					
63	130–31	111–13	81		• Worldview thinking • Unit and chapter preview
64–65	134–37	114–17	82–83	Quiz 5-A	• Contents and layers of the atmosphere • Air pressure
66–67	138–43	118–23	84–86	Quiz 5-B	• Air temperature, air masses, and fronts • Global and local winds
68	144–45	124–25	87	Rubric	Activity: Temperature Changes • Process skill: identifying and controlling variables
69–70	146–51	126–31	88	Quiz 5-C	• Precipitation and humidity • Clouds
71	152–54	132–34			• Severe weather • Storm safety
72	155	135		Rubric	Exploration: Dangerous Extremes • Make and present a severe weather safety poster
73	156–57	136–37	89–90	Quiz 5-D	• Forecasting the weather and using weather maps • Job and tools of a meteorologist
74–75	158–59	138–39	91–92	Rubric	Activity: Weather Observatory • Process skill: collect, record, and interpret data
76	160–61		93–94		Answers in Genesis • A biblical viewpoint about the Ice Age
77	162	140	95		**Chapter Review** • Apply knowledge to everyday situations
78	162			Chapter 5 Test	**Chapter 5 Test**

Lesson	Teacher's Edition	Student Text	Activity Manual	Assessment	Content
Chapter 6: Biomes					
79	163	141	97		• Worldview thinking • Chapter preview
80	164–67	142–45	98		• Introduction to land biomes • Tundra
81–82	168–71	146–49	98–100	Quiz 6-A	• Coniferous forest • Deciduous forest
83–84	172–75	150–53	101–2	Quiz 6-B	• Grasslands • Deserts
85	176–77	154–55	103–4	Rubric	Activity: Help Prevent Water Loss! • Process skill: predicting
86–87	178–180	156–58	101, 105–7	Quiz 6-C	• Tropical rain forests • Mountains
88–89	181	159		Rubric	Exploration: Build a Biome • Research and make a model of a biome
90–91	182–87	160–65	108–10	Quiz 6-D	• Aquatic biomes and wetlands • Man's stewardship
92	188–89		111–12		Answers in Genesis • A biblical viewpoint of the location of the Garden of Eden and climate change after the Flood
93	190–91	166–67	113	Rubric	Activity: From Dirty to Clean • Process skill: inferring
94	192	168	114		**Chapter Review** • Apply knowledge to everyday situations
95	192			Chapter 6 Test	**Chapter 6 Test**

Weather 5

In September 1900 there was a large storm in the Gulf of Mexico. During this time Isaac Cline was working at the weather station of Galveston, Texas. According to telegraph messages from other weather stations, the storm was expected to move north into Florida. But no one knew for certain since they could not easily track the storm's movements.

It was Friday and Mr. Cline could tell that a storm was coming. He raised storm flags over the weather station. On Saturday morning, he watched the wind and the waves, but he still did not know how severe the storm was. By that afternoon, the storm struck hard. The wind blew more than 120 mi/h. Flood water rose more than 20 ft in some places. Finally about midnight, the storm was over. All the homes between the beach and downtown Galveston were completely destroyed. About 8,000 people died, including Mr. Cline's wife.

Consider

- Why was it important for Mr. Cline to warn others?
- What technology do we have today that would make it easier to warn others?

113

Chapter Introduction

Have you seen or heard anything on the news recently about a major storm?

- Discuss the impact of that storm on people's lives.

Preparation for Reading

- Direct the students to read silently page 113 to find out what happened during the storm of 1900.

Teach for Understanding

Do you know what type of storm took place in 1900? a hurricane

- Display the *Southeastern United States* map. Point out the different locations during the discussion.

In what body of water was the hurricane? Gulf of Mexico

Where was the hurricane expected to go? into Florida

Why was the direction of the hurricane not known for sure? The equipment to track storms did not exist at that time.

In what city did the hurricane come ashore? Galveston

Look at the map. In what state is Galveston located? Texas

How did Mr. Cline let the people of Galveston know that a storm was coming? He raised storm flags over the weather station.

What made this hurricane such a terrible disaster? All the homes between the beach and downtown Galveston were destroyed. About 8,000 people died.

What do you think are some reasons that the storm caused so much damage and loss of life? Possible answers: Equipment was not available for early warnings or to know the strength of the hurricane. The construction of homes was not designed to withstand the storm. The area did not have any protection against storm flooding.

- Explain that although people cannot control the weather, the weather does follow certain patterns. Scientists study the patterns and regularities that they see in creation. As scientists learn more about the way God's creation works, they can use what they learn of the weather to design better technology that improves people's surroundings and can save their lives.

Activity

Research Hurricanes

Divide the students into groups. Direct each group to compare what happened with the storm of 1900 and the advances made with hurricanes today. Assign each group to research weather prediction about the storm, warning others about the storm, or survival of the storm.

Background

Chapter Preview

Preview and prereading activities may include using a K-W-L chart, a probe, or an anticipation guide.

Chapter Photo

The photo on page 113 is of Hurricane Catarina over the Atlantic Ocean in March 2004.

Activity Manual

Preview, page 81

The Looking Ahead page is intended to assess the student's prior knowledge before beginning the chapter.

64
65

Student Text pages 114–17
Activity Manual pages 82–83

Objectives

- Describe the atmosphere
- Define *air pressure*
- Recognize that gravity pulls the atmosphere toward the earth
- Name an instrument that measures air pressure
- Identify and describe the two lower layers of the atmosphere

Materials

- winter jacket

Teacher's Toolkit CD

- Visual 5.2: *Layers of the Atmosphere Organizer*
- Quiz 5-A

Vocabulary

- atmosphere
- air
- air pressure
- altitude
- weather
- ozone

Two days have been allotted for this lesson. You may want to teach Lesson 64 the first day and use ideas from Lesson 65 on the second day. You may also incorporate the ideas from Lesson 65 into the text material, making the lesson two days.

Introduction

- Display the jacket.

  What is the purpose of wearing a jacket? Elicit that a jacket is a covering worn for warmth and protection from weather.

  Although the earth does not wear a jacket, it has a covering that provides warmth and protection. God designed this covering so that life can exist on the earth.

Preparation for Reading

- Direct the students to mentally make questions using the headings, bold and italicized words, and captions.
- Instruct them to read silently pages 114–17 to find the answers to these questions.

  Is air pressure greater at the surface of the earth or at higher altitudes?

  Which layer of the atmosphere protects the earth from harmful rays from the sun?

Teach for Understanding

For ease in reading and discussing Chapter 5, *air* is used many places in which *atmosphere* would be more scientifically correct. See Science Misconceptions.

What is the thin blanket of gases and particles that surrounds the earth called? the atmosphere

There are many parts of weather other than storms. These parts are important as well. God has given people the ability to observe weather and learn about it. The weather that we observe takes place in a protective layer of air that surrounds the earth.

Atmosphere

In the beginning God created an **atmosphere**, a thin blanket of gases to surround the earth. If the earth were the size of an apple, the atmosphere would be similar to the apple's skin. The atmosphere is very important to life on Earth, though. It lets in the right amount of energy from the sun and provides the air we breathe. It also prevents extreme weather differences on the earth.

Contents of the Atmosphere

One part of the atmosphere is air. But what is air? **Air** is a mixture of nitrogen, oxygen, carbon dioxide, and other gases in the atmosphere. We cannot see or feel these gases. God provided the kinds and amounts of atmospheric gases to support the kinds of life He created. Plants need the nitrogen and carbon dioxide to produce food. People and animals need oxygen to survive. Providing an atmosphere with this blend of gases is one way God meets the needs of living things.

Another part of the atmosphere is dust particles. When you clean your bedroom, you probably stir up dust. Dust particles may seem annoying when you are cleaning. However, they are useful in the atmosphere. Dust helps water droplets and ice crystals in the clouds form rain, snow, and other types of precipitation.

Gases in the Atmosphere

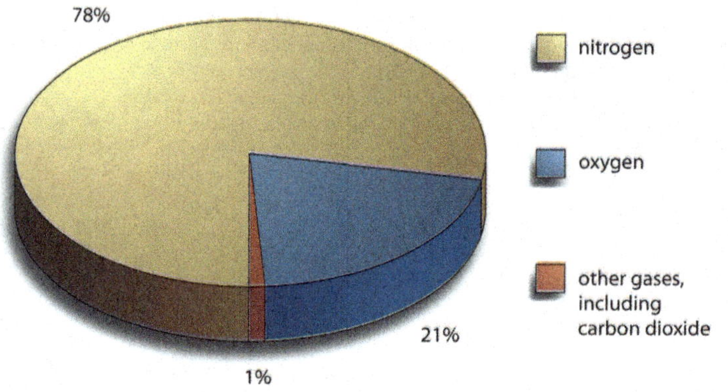

114

Background

Weight

Weight is the measure of the force of gravity on an object. If the gravity changes, the weight changes.

Measuring Air Pressure

Air pressure is measured using a barometer. The unit of measure is a millibar. Some barometers measure air pressure using millimeters or inches of mercury. The average air pressure at sea level is 1,013.25 millibars, or 29.92 inches of mercury.

Sea Level

Sea level is the average level of the surface of the ocean between high and low tides.

Science Misconceptions

Although the terms *atmosphere* and *air* are often used interchangeably, they are not actually the same. The atmosphere, or gaseous layer that surrounds the earth, is composed of air and other particles such as dust. Air is the mixture of gases found in the atmosphere.

Air Pressure

Gravity pulls the atmosphere close to the surface of the earth. At any moment there are thousands of pounds of air pressing on you. The weight of the air is called **air pressure**. Near the surface of the earth the atmosphere presses down on each square inch with 14.7 pounds of weight. It does not seem that air would weigh very much, but there is a huge amount of air above the earth.

As the air is pulled toward the earth by gravity, it increases the pressure. Air pressure is greatest at the surface of the earth and decreases as you move away from the surface of the earth, or gain altitude. **Altitude** is the measurement of the distance above sea level. The altitude affects air pressure. There is less air pressure on a high mountain than at the seashore. However, altitude is not the only factor that affects air pressure. Other factors, such as changes in temperature, also affect air pressure.

One of the tools Isaac Cline used at his weather station was a barometer. A *barometer* is an instrument used to measure air pressure. Knowing the air pressure is important when forecasting the weather. Changes in air pressure may mean that a storm is coming. Pilots are also alert to changes in air pressure because the changes can affect flying conditions.

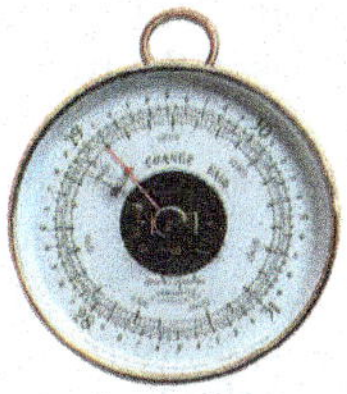

Barometer

CREATION CORNER

Some places in our bodies, such as our lungs and middle ears, are filled with air. When we breathe in and out, the air pressure in our lungs changes to equalize the outside air pressure. Our inner ears are often closed, however. This means the air pressure inside our ears may not be the same as the pressure outside. The inequality may cause our ears to hurt as we travel up and down mountains or fly in a plane. To help prevent this, God has designed our bodies with a tube that goes from our middle ears to our throat. When we swallow, we open the tube and allow the air pressure inside and outside our ears to equalize.

115

What force holds the atmosphere close to the earth? gravity

What are two ways that the atmosphere is important to us? It provides the air that we breathe and prevents extreme weather differences from occurring on the earth.

What gases make up air? nitrogen, oxygen, carbon dioxide, and other gases

- Discuss the *Gases in the Atmosphere* circle graph on page 114.

Which two gases make up most of the air we breathe? nitrogen and oxygen

What percentage of the air is oxygen? 21%

What are some ways that the gases in air are used? Plants use the nitrogen and carbon dioxide for producing food. People and animals need oxygen to survive.

How is dust helpful in the atmosphere? Dust helps form rain, snow, and other types of precipitation.

What is air pressure? the weight of air

What is the air pressure on 1 sq. in.? 14.7 lb

Can you feel the weight of the air? no Why do you think that is? Accept reasonable answers. Elicit that God designed our bodies to handle the air pressure on and around us.

Where in the atmosphere is air pressure the strongest? at the surface of the earth

- You may want to do the *Calculating Air Pressure* activity at this time.

What is altitude? the measurement of the distance above sea level

How does the air pressure at a lower altitude compare with the air pressure at a higher altitude? Air pressure is greater at lower altitudes than at higher altitudes.

Is altitude the only factor that affects air pressure? no What else can affect air pressure? temperature changes

What instrument is used to measure air pressure? a barometer

Name some situations in which it would be helpful to know the amount of air pressure. Possible answers: when forecasting storms, when flying an airplane

- You may want to do the *Make a Barometer* activity at this time.
- Discuss *Creation Corner*.

Why do our ears sometimes hurt when we travel up or down mountains or fly in a plane? The air pressure inside our ears is not equal to the air pressure outside our ears.

How can you keep your ears from hurting when you travel up or down a mountain? Possible answers: by swallowing, by chewing gum, by gently blowing your nose

Activity

Calculating Air Pressure

Materials: ruler, newspaper, calculator

Lay the ruler on the top of a desk so that one-third of it extends over the edge. Open two double-pages of newspaper and lay them over the ruler. Smooth the paper flat against the table.

Predict how the newspaper will move when the ruler is hit.

Direct a student to hit the end of the ruler that is not covered with newspaper.

Did the paper move as predicted? Answers will vary.

What prevented the newspaper from catapulting away from the ruler? Elicit that the weight of the air, or air pressure, held the paper against the table.

What is the pressure of air on 1 sq. in.? 14.7 lb

How would you find out the amount of air pressure that is pushing on the newspaper? Measure to find the area of the newspaper in square inches. Multiply the number of square inches by 14.7 lb.

For example, if the newspaper is 22 × 25 in., the area of the newspaper is 550 sq. in. The pressure on the newspaper is 8,085 lb. The weight of an average female elephant is about 8,000 lb!

64–65

Is the atmosphere thick or thin? Compared with the size of the earth, it is thin.

- Display the *Layers of the Atmosphere Organizer*. Guide the students as they take notes in the organizer on Activity Manual page 82 during the discussion.
- Explain that the discussion will begin with the layer of the atmosphere closest to the earth or the bottom of the organizer.

  Which layer of the atmosphere is closest to the earth? the troposphere
- Allow the students to write the name of the layer of the atmosphere closest to the earth in the box at the bottom of the organizer and use the space to take notes.

  Is the tallest mountain, Mt. Everest, higher than the troposphere or lower? Since the troposphere continues above Mt. Everest, it is lower.

  How much of the air in the atmosphere is in the troposphere? about half

  What is air? the gases found in the atmosphere

  How does the troposphere help keep the earth warm? The troposphere holds in the radiant heat from the sun.

  How does the temperature of the air in the troposphere change at higher altitudes? The air gets colder. Why? The thinner air cannot hold as much thermal energy as denser air can.

  What is weather? the condition of the atmosphere at any moment in time

  In which layer of the atmosphere does weather occur? the troposphere

  What does the troposphere contain that affects the weather? particles and water droplets

  What layer of the atmosphere is above the troposphere? the stratosphere
- Direct the students to label the stratosphere on the organizer and use the space to take notes.

  How are the air temperatures in the stratosphere different from the air temperatures in the troposphere? In the stratosphere, the temperatures increase with altitude instead of decreasing as they do in the troposphere.

  What is located in the upper part of the stratosphere? the ozone layer

  What is ozone? a form of oxygen

  What does the ozone layer do? It protects the surface of the earth from harmful rays of sunlight.

  What is the next layer of the atmosphere? the mesosphere

  What happens in the mesosphere? Meteors burn up.
- Give the students time to label and take notes in the mesosphere section on the organizer.

Layers of the Atmosphere

God designed the atmosphere to have several layers. Each layer is different. Together, though, these layers protect the earth and give it a controlled climate.

The *troposphere* is the layer of the atmosphere that is closest to the earth. But even the tallest mountain on the earth, Mt. Everest, reaches only partway into the troposphere. About half of the air in the atmosphere is in the troposphere. The troposphere is also where weather occurs.

The air in the troposphere holds in the radiant heat from the sun. This helps keep the surface of the earth warm. At higher altitudes, the air in the troposphere gets thinner and colder. Just as a thin sweater cannot keep you as warm as a thick coat can, the thinner air at higher altitudes cannot hold as much thermal energy as the denser air near the surface can.

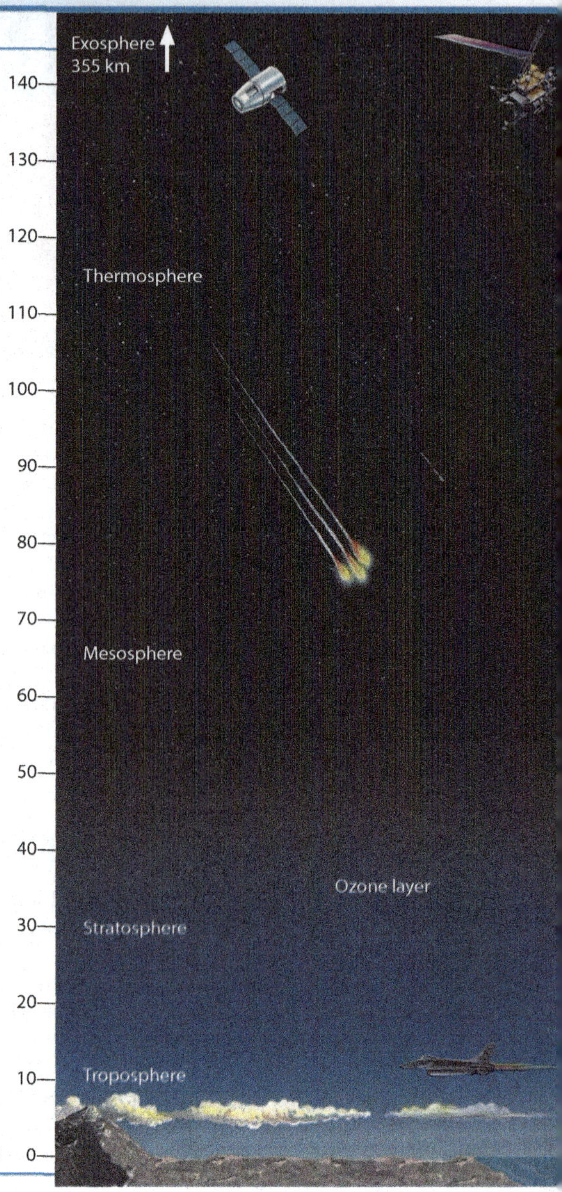

Altitude in km

116

Background

Layers of the Atmosphere

Scientists divide the atmosphere into layers by various criteria. The method used in *Science 5* divides the atmosphere into layers according to temperature.

Boundaries of the Layers

The layers of the atmosphere do not have specific boundaries. The measurements in the *Layers of the Atmosphere* diagram are approximate.

Stratosphere

The temperatures in the stratosphere increase as the altitude increases because of the ozone layer in the stratosphere. Ozone absorbs many of the ultraviolet rays from the sun. The ultraviolet rays cause the ozone molecules to heat up. This, in turn, causes the stratosphere to get warmer as the altitude increases.

Exosphere

The exosphere is the layer of the atmosphere farthest from the surface of the earth. The exosphere extends beyond the *Layers of the Atmosphere* diagram. This layer has very little air pressure. The outer edge of the exosphere gradually fades into space.

Weather is the condition of the atmosphere at any moment in time. The troposphere contains many particles and water droplets that affect the weather. The particles reflect light, allowing us to see colorful sunrises and sunsets.

The *stratosphere* is the next layer above the troposphere. The air in the stratosphere continues to get thinner as you travel away from the surface of the earth. Unlike the troposphere, the stratosphere increases in temperature as the altitude increases.

God made the stratosphere to protect the earth. During the day, the stratosphere shields the earth from the harmful effects of the sun. At night, the stratosphere helps hold in some of the heat that the sun provided during the day.

The ozone layer is in the upper part of the stratosphere. **Ozone** is a form of oxygen. Molecules of ozone have three atoms instead of two as in normal oxygen. The layer of this gas is one way the stratosphere protects the surface of the earth from the sun's harmful rays.

The higher layers of the atmosphere also help protect the earth. Meteors heading toward the earth burn up in the *mesosphere*. These burning meteors are the "shooting stars" we sometimes see. Many of our weather and communication satellites are in the *thermosphere*. The *exosphere* is the layer of the atmosphere farthest from the earth. This layer gradually fades into space.

Some people compare the atmosphere to a greenhouse. A greenhouse is a special building, often made of glass. It completely surrounds and protects the plants inside from extreme heat or cold. The sunlight, temperature, and moisture inside a greenhouse are carefully adjusted to provide the ideal environment for the plants. In a similar way, God designed the atmosphere to completely surround the earth. He planned for the atmosphere to protect the earth and provide the proper weather needed for each area of the earth.

QUICK CHECK

1. What is the atmosphere?
2. What is air pressure?
3. What is the relationship between altitude and air pressure?
4. In which layer of the atmosphere is the ozone layer found?

117

64–65

In which layer of the atmosphere are weather and communication satellites? the thermosphere

The International Space Station is located in the thermosphere. It orbits at around 400 km (250 mi) above the earth.

- Give the students time to label and take notes in the thermosphere section on the organizer.

  What layer of the atmosphere is farthest from the earth? the exosphere

- Give the students time to label and take notes in the exosphere section on the organizer.

  What is a greenhouse? a building that provides an ideal environment for plants and protects the plants from extreme heat or cold

  How do some people compare the atmosphere to a greenhouse? In the same way that a greenhouse protects plants, the atmosphere protects the earth and provides the weather the earth needs.

- Explain that God created the world and the atmosphere around it to meet the needs of the plants, animals, and people on the earth. As stewards of God's creation, people must use wisdom as they use and care for the resources God has given.

ANSWERS

1. the thin blanket of gases and dust particles that surrounds the earth
2. the weight of air
3. As altitude increases, air pressure decreases.
4. the stratosphere

65

- The *Calculating Air Pressure* activity, *Writing about an Air-Balloon Trip* activity, and *Make a Barometer* activity found at the bottom of the pages in this lesson may also be incorporated into the lesson.
- Suggestions for additional activities for this two-day lesson are on the Teacher's Toolkit CD.

Activity Manual

Study Skill and Reinforcement, page 82

This page is an organizer for the layers of the atmosphere.

Review, page 83

This page reviews Lessons 64–65.

Assessment

Quiz 5-A

The quiz may be given any time after completing this lesson.

Activities

Writing About an Air-Balloon Trip

Write a short story about an imaginary air-balloon trip through the layers of the atmosphere. Include details of the temperature changes, things that may be seen in each layer, and other conditions that may be faced.

Make a Barometer

Materials: large balloon, glass jar, rubber band, 8 cm drinking straw, straight pin, tape, cardboard

Cut the balloon, stretch it over the mouth of the jar, and secure it with the rubber band. Tape the pin to the end of the straw to make a pointer. Tape the straw to the balloon as in the illustration. Cut a strip of cardboard 5 cm wide and 5 cm taller than the jar. Starting at the top, draw ten lines 1 cm apart down the strip. Beginning with 10 at the top, number the lines down to 1. Stand the strip so that the pin points at 5. For best results, place the barometer near a window.

Read and record daily measurements. Discuss comparisons of the readings and the weather conditions. When possible, read the measurements before, during, and after a rainstorm.

Student Text pages 118–23
Activity Manual pages 84–86

Objectives
- Compare and contrast high-pressure air masses and low-pressure air masses
- Define *front* and describe three types
- Explain how temperature affects wind
- Differentiate between global winds and local winds
- Name examples of global winds and local winds

Materials
- stiff paper
- tape
- thread
- lamp with an incandescent bulb

Cut a circle from the stiff paper about 20 cm in diameter. Cut a spiral as illustrated. Insert a 30 cm length of thread through the center of the spiral and secure with tape.

Teacher's Toolkit CD
- Quiz 5-B

Vocabulary
- wind
- air mass
- front
- global winds
- local winds

Two days have been allotted for this lesson. You may want to teach Lesson 66 the first day and use ideas from Lesson 67 on the second day. You may also incorporate the ideas from Lesson 67 into the text material, making the lesson two days.

Introduction

- Hold the spiral above the lamp that is turned off.
  What effect does the turned-off lamp have on the spiral? none
  What difference in temperature is there between a light bulb that is turned on and one that is not? The light bulb that is on gives off heat.
- Turn on the lamp and hold the spiral above it.
  What effect does the lamp have on the spiral now? The spiral turns.
  A light bulb that is on heats the air around it. Air moves as its temperature changes.

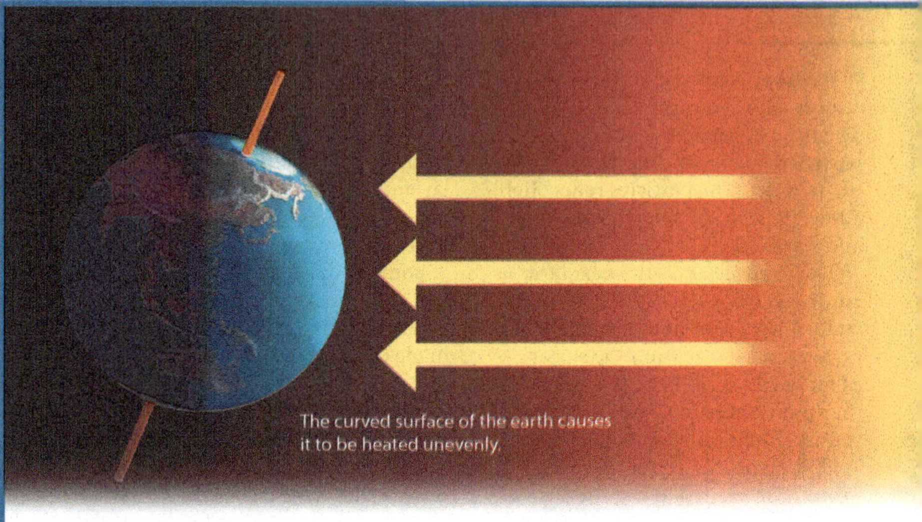

The curved surface of the earth causes it to be heated unevenly.

Moving Air

You are surrounded by air that is usually moving. Moving air is called **wind**. Sometimes wind moves so slowly that the air seems perfectly still. Other times it blows violently and destroys everything in its path.

We can use and enjoy wind in many ways. A light breeze may cool us off after a ball game. A stronger wind may allow us to fly a kite high in the air. In times past, great winds pushed the ships of explorers across the oceans. Today some people exercise dominion by using wind to help produce electricity.

Air Temperature

Although the heat from the sun comes through the atmosphere, the sun does not warm the air much. Instead, the sun warms the earth's surface, which then warms the air above it. Air temperature is a result of the heating and cooling of the earth's surface.

The curved surface of the earth affects temperatures. Sunlight hits some places from overhead and other places at a slant. The places that receive overhead sunlight are warmer than those that receive sunlight at a slant. Also, the part of the earth facing the sun receives heat and is warmer than the part facing away from the sun.

Since the earth's surface does not heat evenly, the air in the atmosphere does not heat evenly either. This is why days are usually warmer than nights.

118

Activity

Movement from Cold to Warm

Materials: 4 foam cups, very hot water, ice, loaf pan, room-temperature water, food coloring

Fill one cup with ice and one with hot water. Arrange the four cups as illustrated.

Place the loaf pan so that one end is over the cup of ice and the other end is over the cup of hot water. Fill the pan about three-fourths full of room-temperature water. Let the water sit for several minutes so any movement from the pouring has stopped.

How do you think the ice and water in the cups are affecting the water in the pan? The hot water is warming the pan and water above it, and the ice is cooling the pan and water above it.

Remind the student how heat moves by convection. This is discussed in Chapter 4, *Heat*.

The warm water represents warm air, and the cool water represents cool air.

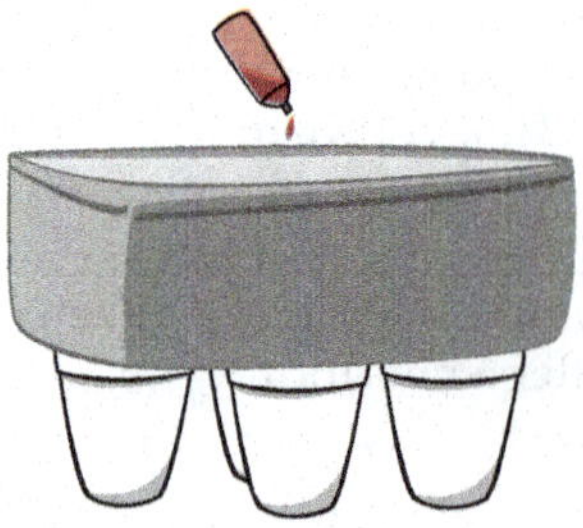

Add one drop of coloring into the center of the pan. (Only the teacher should handle the coloring.)

Direct the students to look at the coloring.

66
67

Air Masses

A large body of air that has about the same temperature and moisture is called an **air mass**. An air mass gains its characteristics as it moves over the earth's surface. For example, an air mass over cool land or water becomes cooler. An air mass over an ocean will also collect moisture. As an air mass moves over a hot desert it becomes warmer and drier.

Air masses also vary in density and air pressure. The gas particles in cool air are close together, causing cool air masses to be denser than warm ones. This is why cool air often has greater pressure. The zone of highest pressure in a cool air mass is called a *high pressure center*, represented by a capital *H* on a weather map. Because cool air cannot hold as much moisture as warm air, cool air masses usually have cool, dry air.

A cool air mass is cooler than the land under it. Over several days, the warm surface of the earth heats the air mass. Warming allows the air to hold more moisture and causes the gas particles to spread out. As the air becomes less dense, the air pressure of the warmer air mass becomes lower. The area of lowest pressure in a warm air mass is a *low pressure center*. It is shown by a capital *L* on a weather map.

SCIENCE & HISTORY

Hot air balloons use changes in air density to move up and down. Heating the air in a balloon moves the air particles farther apart. As the air inside the balloon becomes less dense than the outside air, the balloon rises. As the air inside the balloon cools, the balloon drops.

Hot air balloons have been used by the military since the Civil War. The balloons were used to observe the location and size of enemy armies. Today some balloons are still used for surveillance along our country's borders.

119

What do you notice about the food coloring? It moves along the bottom of the pan toward the warm water, then rises and moves toward the cool water, and then falls.

Air currents move in much the same way. Cooler air moves toward warmer air, warms and rises, and then falls as it cools again.

You may choose to place a thermometer at each end of the pan to compare temperatures. A clear glass pan may show the circulation better, but the thickness of the glass may prevent the changes of water temperature needed to produce good circulation.

Preparation for Reading

- Display and pronounce the word *anemometer.*
- Instruct the student to read silently pages 118–23 to find the answers to these questions.
  What type of winds did Christopher Columbus use to sail to the New World?
  What instrument is used to measure wind speed?

Teach for Understanding

The heating of air is basic to the behavior of weather. The relationship between matter and heat is discussed in Chapter 4, *Heat.* Reinforce concepts about heat as they relate to the behavior of air throughout this chapter.

What is wind? moving air

How do some people exercise dominion by using wind? to help produce electricity

How does the way that sunlight hits the earth affect the temperature? Areas of direct sunlight are warmer than areas of slanted sunlight.

Why would the air temperatures be warmer at the equator than in Europe? The equator receives more direct sunlight than Europe does.

What is an air mass? a large body of air that has about the same temperature and moisture throughout

What often determines the characteristics of an air mass? the surface of the earth that it is over

Why is a cool air mass denser than a warm air mass? The gas particles are closer together.

Describe a high-pressure air mass. It has cool, dry air and often moves toward the earth's surface.

What happens when the warm surface of the earth heats the air? The air becomes less dense and lighter because the gas particles spread apart.

Describe a low-pressure air mass. It has warm, moist air that usually moves above cooler air.

Why would a low-pressure air mass sometimes bring rainfall? It can hold more moisture than a high-pressure air mass can.

Why would a high-pressure air mass move toward the earth's surface and a low-pressure air mass move away from the earth's surface? Cool air falls, and warm air rises.

- Help the students remember the density of an air mass and its type of air pressure. Notice that *high pressure* and *heavy air mass* both begin with an *H*. Notice that *low pressure* and *light air mass* both begin with an *L*.
- Discuss *Science & History.*
  Describe how a hot air balloon operates. Heating the air inside the balloon causes the air particles to move farther apart and become less dense than the outside air. The difference in density causes the balloon to rise. Cooling the inside air causes the balloon to drop.
- You may want to do the *Movement from Cold to Warm* activity at this time.

66
67

What is a front? the boundary of two unlike air masses; the place where two unlike air masses meet

What are three kinds of fronts? a cold front, a warm front, and a stationary front

Does a cool air mass usually push under or over a warm air mass? under

What kind of front would this be? a cold front

What type of weather does a cold front bring? rain, possibly with thunderstorms and hail, followed by colder, drier weather

What results from a warm air mass moving over a slower cool air mass? a warm front

What type of weather does a warm front bring? several days of light rain or fog, followed by warm, moist weather

Which moves faster, a cold front or a warm front? a cold front

How are stationary fronts different from warm fronts and cold fronts? In a stationary front, the air masses push against each other but stay in place.

Is the weather of a stationary front more like the weather of a cold front or that of a warm front? a warm front

- Explain that a cold front can be compared to a collision between two cars. Both happen suddenly and may be very severe. A warm front is similar to one car passing another on the highway. The front moves more slowly and is less intense.
- Discuss the *Cold Front*, *Warm Front*, and *Stationary Front* diagrams.

Which kind of front shows a warm air mass moving over a cold air mass? a warm front

Why does the stationary front diagram not show one air mass moving over another air mass? because in a stationary front the air masses push against each other but stay in place

Is it possible to have a cold front in the summer? yes Explain. A cold front occurs when cooler air moves under warmer air. This may cause thunderstorms and sometimes hail.

- You may want to do the *Formation of Fronts* demonstration at this time.

Fronts

When two unlike air masses meet, the boundary between the masses is called a **front**. A cool air mass may push itself under a weaker warm air mass. This type of front is called a *cold front*. It often brings stormy weather that may include thunderstorms and hail. A cold front usually moves quickly through an area. After the cold front passes, the weather is usually cooler and drier than before.

A *warm front* occurs when a warm air mass moves over a slower cool air mass. A warm front moves slowly and can take several days to leave an area. As a result, a warm front often brings several days of light rain or foggy weather. When a warm front passes, the following days may have warm, moist weather.

Sometimes air masses push against each other but do not move. When this happens, the boundary is called a *stationary front*. A stationary front may stay in the same place for several days. It usually causes cloudy skies and rainy conditions.

Cold Front

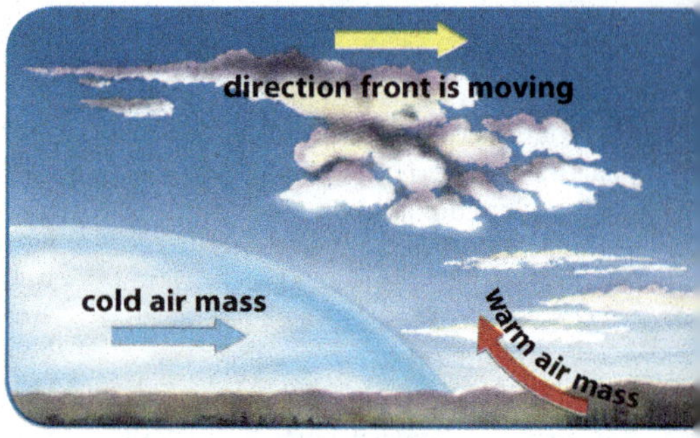

Warm Front

Stationary Front

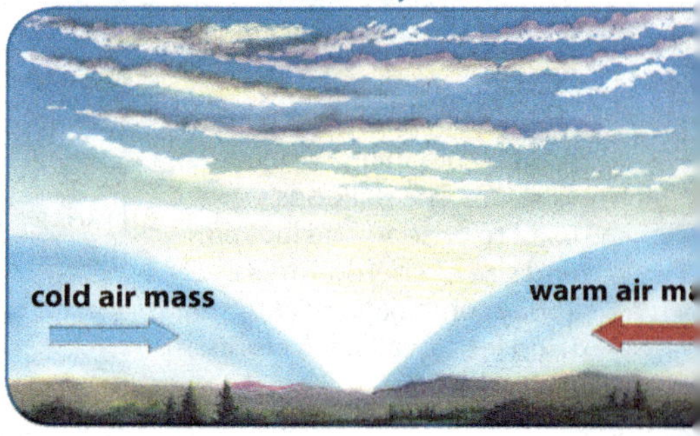

120

Demonstration

Formation of Fronts

Materials: several different-sized sponges

Demonstrate the movement of a *cold front* by pushing a large, thick sponge under a thin sponge. The large sponge represents a cool air mass, and the thin sponge represents a warm air mass.

Demonstrate the movement of a *warm front* by using two sponges about the same thickness to represent the air masses. Move both sponges in the same direction. Gradually have the second sponge slide over the first and push it along.

Demonstrate a *stationary front* by using two sponges about the same thickness to represent the air masses. Push the sponges against each other without passing.

If needed, label each sponge according to air mass, temperature, or air pressure.

66
67

Kinds of Winds

Air varies in temperature. Many factors, such as location or cloud cover, affect the air temperature and can make it warmer or cooler. For instance, air near the surface of the earth is warmer than air at higher altitudes. Air is also warmer near the equator and cooler at the poles of the earth. Changes in temperature cause wind. Warmer air rises. Cooler air moves toward the earth to replace the warmer, rising air. These and other air patterns influence the movement of winds.

Global Winds

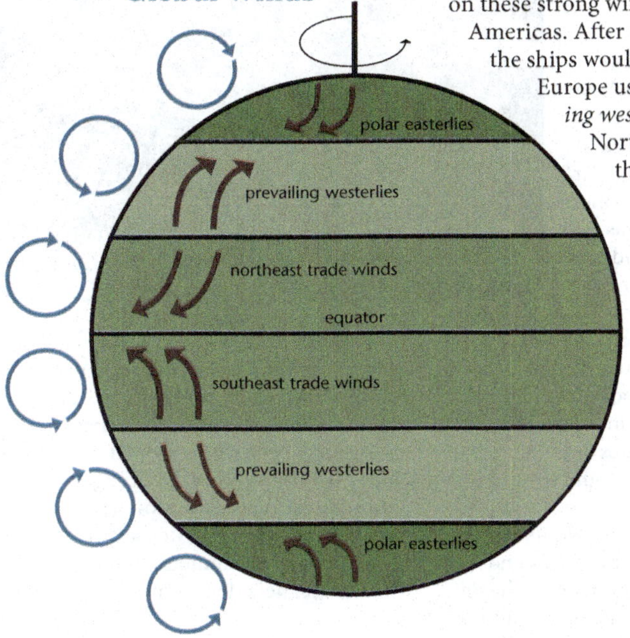

Global Winds

The uneven heating of the earth causes global winds. **Global winds** occur in large, circular belts around the earth. Global winds include trade winds, the prevailing westerlies, and the polar easterlies. Winds are identified by the direction from which they blow.

The prevailing easterlies, or *trade winds*, blow from the east toward the equator. Christopher Columbus used the prevailing easterlies when he sailed to the New World. These winds became known as the trade winds because trade ships relied on these strong winds to sail to the Americas. After trading their cargo, the ships would then return to Europe using the *prevailing westerlies*. In the Northern Hemisphere, these winds blow northward from the west. Near the North Pole and the South Pole, the *polar easterlies* blow from the east and move away from each pole.

121

What causes winds? changes in temperature

How does air move when its temperature changes? Warm air rises. Cool air moves toward the earth to replace the warm air.

Where are air temperatures the warmest? near the earth's surface

What happens to the temperature of air as it rises? It cools.

What influences the movement of winds? the patterns of warm air and cool air

- Refer to the *Global Winds* diagram during the discussion.

What are global winds? winds that move in large, circular belts around the earth

How are winds identified? by the direction from which they blow

What are some examples of global winds? trade winds, prevailing westerlies, polar easterlies

What is another name for trade winds? prevailing easterlies

How did trade winds help Christopher Columbus in his explorations? These winds blow from the east to the equator. His ships were powered by sails. Trade winds moved the ships west from Europe to the Americas.

How did global winds help colonization in the New World? Explorers and settlers used the trade winds to arrive in the Americas and used the prevailing westerlies to ship goods back to Europe.

In which direction do polar easterlies blow? from the east, away from each pole

- Provide time for students to observe weather maps of your area for several days. Direct them to make generalizations about the movements of air masses and fronts and relate the movements to global winds.
- You may want to do the *Rotation of the Earth and Winds* activity at this time.

Activity

Rotation of the Earth and Winds

Materials: paper circle, pin, pencil

Mark the center point of the circle. The center point is one of the poles. The outer edge of the circle is the equator.

One student holds the circle in place with the pin in the center mark.

A second student places the tip of the pencil against the pin.

At the same time, the first student will turn the circle and the second student will draw a line from the pin straight toward himself.

Does the line look straight? no Why? The line curved as the circle moved.

Global winds blow in northerly or southerly directions. The rotation of the earth causes these winds to blow slightly to the east or west.

What are local winds? winds that are influenced by temperature changes in a small area or place

How is a local wind different from a global wind? A local wind affects only a small area or place, but a global wind goes around the world.

What causes land and water to heat? the sun

What causes land and water to cool? the lack of sunlight

Which warms and cools faster, water or land? land

What are some examples of local winds? sea breezes, land breezes, mountain winds, and valley winds

- Refer to the *Sea Breeze* and *Land Breeze* diagrams during the discussion.

How is the movement of sea breezes different from that of land breezes? Sea breezes blow inland from the water all day. Land breezes blow toward the sea at night.

Why do the breezes blow in from the water during the day? Land is warmer than water. The air over the land rises, and the cooler air moves in to replace it.

Why do the breezes switch at night? At night, the air over the water is warmer than the air over the land, so the air moves toward the water.

How is the movement of sea breezes and land breezes similar? They both blow toward the warmer place.

- Explain that both global winds and local winds are named for the direction from which they blow.

Local Winds

Other winds, known as **local winds**, are influenced by temperature changes in a small area or place. During the day, both land and water absorb heat from the sunlight. At night the land and water cool as heat moves toward the atmosphere. Land and water absorb heat and cool off at different rates, or speeds. Water warms and cools more slowly than soil and rocks do.

Places where water and land meet often have local winds called *sea breezes* and *land breezes*. Winds form as cooler air moves to replace warmer, rising air. During the day, the air over land is warmer than the air over large bodies of water. This causes sea breezes to blow inland from the cooler water. At night, the water and the air above it cool down slowly and stay warm longer than the land does. This causes land breezes to blow from the cooler land toward the water.

Valley winds and *mountain winds* are other types of local winds and form in a similar way. During the day, the sun warms the side of a mountain and the air around it. As the air warms, it becomes less dense. Cooler, denser air from the valley moves up the mountain to replace the warmer, rising air. These winds are called valley winds. At night, air at higher altitudes cools quickly. Mountain winds occur as cooler, denser air flows down the mountain to replace the warmer, less dense valley air.

122

Background

Wind Vane

A wind vane is sometimes called a weather vane.

Rear Admiral Sir Francis Beaufort

Francis Beaufort was born in Ireland in 1774. He joined the Royal Navy at the age of thirteen. In 1805 he became the commander of the HMS *Woolwich*. While commanding this ship, Beaufort kept a daily journal and developed a numerical scale for recording wind conditions. This early scale was based on his observations of the reactions of his ship's sails to the force of the wind. As the scale became more widely used, it was modified for general use. Today the Beaufort scale includes wind velocity (speed), as well as verbal descriptions of the wind's effects on the surroundings. The scale is used by both meteorologists and mariners. Many versions of the scale exist and are used in different contexts.

the wind is blowing. If the winds become too strong, an anemometer may become damaged. The one at Isaac Cline's weather station was blown off the building by the strong winds.

Wind vane

Anemometer

A *wind vane* is an instrument that shows the direction wind is blowing from. It shows the direction of the wind by pointing into the wind.

The speed of wind is measured using an *anemometer*. An anemometer looks like several cups attached to a post. As the wind blows into the cups, they spin around the post. A dial that looks similar to the speedometer in a car shows how fast

QUICK CHECK

1. What is wind?
2. Does a cool air mass have high pressure or low pressure?
3. How does a warm front form?
4. Does warm air generally rise or fall?

123

66–67

Is air less dense in a valley or on a mountain? on a mountain

Which warms and cools faster, mountain air or valley air? mountain air

- Refer to the *Valley Wind* and *Mountain Wind* diagrams during the discussion.

Look at both diagrams. In which direction is the wind blowing? up the mountain during the day and down the mountain at night

What instrument is used to measure wind direction? a wind vane

How is wind speed measured? with an anemometer

How does an anemometer work? Wind causes the cups attached to the post to spin, and the speed is shown on a dial.

ANSWERS

1. moving air
2. high pressure
3. A warm air mass moves over a cool air mass.
4. rise

67

- The *Movement from Cold to Warm* activity, *Rotation of the Earth and Winds* activity, *Local Wind Speed* activity, and *Formation of Fronts* demonstration found at the bottom of the pages in Lessons 66–67 may also be incorporated into the lesson.
- Suggestions for additional activities for this two-day lesson are on the Teacher's Toolkit CD.

Activity

Local Wind Speed

Materials: copy of Instructional Aid 5.1: *Beaufort Scale*, for each student; paper to record wind observations

An anemometer is not the only way to determine wind speed. This can be done visually using the Beaufort scale.

Give each student a copy of the *Beaufort Scale* page.

Skim the *Wind Effects on Land* column. What scale number is used when the wind raises dust and loose paper into the air? 4

About how fast would this wind be moving? 13–18 mi/h

How would you describe this wind? a moderate breeze

What happens to trees during a strong breeze? The large branches begin to move.

What is the speed of the wind during a strong breeze? 25–31 mi/h

Direct the students to observe the effects of the wind at certain times for two days. They should record their observations and compare them to the Beaufort scale. Use the Beaufort scale to estimate the wind speed.

You may choose to have the students compare their estimated wind speeds with the wind speed given in local daily weather reports.

Activity Manual

Reinforcement, page 84

Review, pages 85–86

These pages review Lessons 66–67.

Assessment

Quiz 5-B

The quiz may be given any time after completing this lesson.

68

Student Text pages 124–25
Activity Manual page 87

Objectives

- Predict whether water and soil will warm or cool at the same rate
- Identify and control variables
- Measure and record temperatures
- Relate temperature changes to the ability of each substance to hold and give off heat

Materials

Materials needed for this activity are listed on the Student Text page.

Teacher's Toolkit CD

- Rubrics: *Temperature Changes*, for each student

Several hours or the night before the activity, set aside the soil and water so that both will be at room temperature. The pair of containers for each group must be the same to avoid introducing another variable.

Introduction

What is the main cause of land and sea breezes? differences in temperature between the land and the water

What type of experiment do you think we could set up to find out if this is true? Accept reasonable answers.

- Display the empty containers.

Do you think that the temperatures of small amounts of soil and water in these containers will change in the same way as the large land masses and large bodies of water that cause land and sea breezes? Answers will vary.

Preparation for Reading

- Direct the students to read silently Student Text pages 124–25 and Activity Manual page 87 before beginning the activity.

Teach for Understanding

Process Skill

The process skill emphasized is identifying and controlling variables.

What is a variable? a changeable part of an experiment

Variables can include the length of time or the amount of material used.

How many variables can you change while working on an experiment? A scientist changes only one variable at a time and compares the results.

Why is only one variable changed at a time? If more than one variable is changed, a scientist cannot tell which variable caused the results that he is recording.

ACTIVITY

Temperature Changes

Process Skills
- Measuring
- Observing
- Inferring
- Recording data

When you go swimming, the water temperature does not always feel the same as the air temperature. On a hot day the water may seem cooler. On a summer night the water may feel warmer. Are there really differences in the temperatures, or are your senses fooling you?

Problem

Which substance warms and cools faster—water or soil?

Procedure

1. Complete the hypothesis in your Activity Manual.
2. Put 4 cm of soil into one container. Fill the other container with 4 cm of water.
3. Place a thermometer into each container. Make sure that the bulbs of the thermometers are under the surface of the soil or water.
4. Measure the temperature of each container and record the starting temperatures.
5. Place the containers outside in a sunny place that is not exposed to wind. Measure and record the temperature of each container after 30 minutes and then after 60 minutes.
6. Move the containers to a shady location that is not exposed to wind. Measure and record the temperature of each container after 30 minutes and then after 60 minutes.

Materials
- 2 containers
- potting soil
- centimeter ruler
- water
- 2 thermometers
- Activity Manual

124

Conclusions
- Draw conclusions about the substances that increased and decreased temperatures the most.
- Use your results to explain the causes of land and sea breezes.

Follow-up
- Compare the temperatures after longer periods of time.
- Test pairs of containers in other temperature situations.
- Test the temperature changes of other materials such as sand or gravel.

125

What will be the same about each container throughout this activity? Possible answers: both containers are the same size, shape, and material; both have thermometers; both have the same depth of material in them; both were in the sun and in the shade for the same lengths of time

What variable will we be changing? the material in the container—one has soil, the other has water

Procedure
- Emphasize the importance of measuring carefully. If both of your containers did not contain the same depth of soil and water, how do you think that would affect the results? Accept reasonable answers. Would you be able to make an accurate comparison? no
- Emphasize the importance of reading the thermometers accurately.
- Instruct the students not to remove the thermometers from the containers. The reading on the thermometer begins changing immediately when it is removed from the soil or water. Instead, the students should lift each container to eye level when reading the thermometer. How might removing the thermometer from the soil or water affect the activity? The thermometer reading would probably change as the thermometer begins taking the temperature of the air.
- If the student has trouble reading the exact temperature, suggest that a strip of paper or other straight edge be used to align the level of the liquid in the thermometer with the degree line markings.
- You may need to guide the students in recognizing which substance had the greatest change in temperature. The students should not need to calculate the differences, but you may choose to have them subtract to find the temperature differences at each time interval.

Conclusions
How can we compare our results with the causes of land breezes and sea breezes? The soil warmed and cooled faster than the water. This is the same factor that causes sea and land breezes.

Activity Manual
Activity, page 87

Assessment
Rubrics
Select the prepared rubric, or design a rubric to include your chosen criteria.

69
70

Student Text pages 126–31
Activity Manual page 88

Objectives
- Define *precipitation*
- Differentiate among rain, sleet, snow, and hail
- Define *humidity*
- Identify and describe three basic shapes of clouds

Teacher's Toolkit CD
- Quiz 5-C

Vocabulary
- precipitation
- humidity
- stratus cloud
- cumulus cloud
- cirrus cloud

Two days have been allotted for this lesson. You may want to teach Lesson 69 the first day and use ideas from Lesson 70 on the second day. You may also incorporate the ideas from Lesson 70 into the text material, making the lesson two days.

Introduction

Have you ever thought that the clouds in the sky looked like a picture of an object or animal?

What kinds of cloud pictures have you seen? Answers will vary.

Do all clouds look alike? no

What are some different shapes or colors of clouds that you have seen? Answers will vary.

Clouds appear differently in different weather conditions. Clouds are classified based on these differences. Each type of cloud has a different name.

Preparation for Reading

- Allow time for the students to mentally make questions using the headings, bold and italicized words, and captions.
- Direct them to read silently pages 126–31 to find the answers to these questions.

  What does water vapor become as it cools?

  How do clouds form?

Teach for Understanding

- Discuss a metaphor about precipitation.
- Ask a volunteer to read aloud Isaiah 55:10–11.

  What do rain and snow do to the plants on the earth? water them, make them grow

  Do the rain and snow continue to fulfill this purpose? yes

  In verse 11, what is God's Word compared to? rain and snow

  Summarize how verse 11 compares the Bible to rain and snow. Just as the rain and snow fulfill their jobs, the message of the Bible fulfills what God has planned. The Bible's message is never void (empty) because God will prosper (do or fulfill) all of its promises.

Moisture in the Air

In God's plan, most living things on the earth depend on water. All the water on the earth follows a cycle, or repeated path. Water falls to the earth as *precipitation*. Plants, animals, and humans need this water to survive. Some water returns to the atmosphere through *evaporation*. Most of the water in the air is evaporated ocean water. When conditions are right, the water in the air changes to droplets and forms clouds through *condensation*. The clouds then produce more precipitation, continuing the cycle.

Precipitation

Water vapor, or water in the air, cools at high altitudes. As water vapor cools, it forms water droplets or ice crystals around dust particles in the air. Gravity pulls the larger water drops and ice crystals toward the earth and causes precipitation. **Precipitation** is any type of solid or liquid water that falls from the atmosphere and reaches the ground. Rain, sleet, snow, and hail are all types of precipitation.

Rain forms when water droplets in the clouds join together to form larger drops. Raindrops can be big or small. Their sizes are determined by air temperatures, winds, and the types of clouds.

Sometimes the air is warm enough for the rain to fall as a liquid, but the

Rain

surface of the earth is below freezing. When this happens, the raindrops freeze as they hit the earth. This *freezing rain* forms an icy glaze that coats objects near the ground, such as tree branches and fences.

Sleet occurs when raindrops fall through air that is below freezing. Unlike freezing rain, the drops freeze on their way to the earth and land as tiny ice pellets. If the air temperature borders on freezing, a winter storm may have a mixture of freezing rain and sleet.

TRY IT YOURSELF

Get a spray bottle of water. Adjust the nozzle to spray a fine mist. Find a smooth outside wall or window. Aim the bottle at the wall or window and spray once or twice. Look at the tiny droplets. Spray more water. What happens to the size of the droplets? Compare the movement of your drops to that of falling rain.

126

Background

Water Cycle

Precipitation, evaporation, and condensation are part of the water cycle. The water cycle is discussed more completely in Chapter 8, *Changes in an Ecosystem*.

Supercooled Water

Sometimes water can exist in a liquid form at temperatures below freezing. This is called supercooling. For example, freezing rain forms when water droplets are supercooled in a cool layer of air. These water droplets will immediately freeze when they hit any surface.

Snow forms when ice crystals in the clouds join together and fall to the earth. The air between the clouds and the earth must be at or below freezing for snow to fall. The size of snowflakes varies depending on the temperature and the amount of moisture in the air. Large, fluffy-looking snowflakes form at temperatures near 0 °C (32 °F). As temperatures drop, the flakes become smaller because there is usually less water vapor available in cold air to form snowflakes.

Because of the need for cold air temperatures, most frozen precipitation occurs during cold weather. But one type, *hail*, usually falls when weather is warm on the ground. In summer thunderstorms, the upward-moving warmer air pushes raindrops higher into the tall clouds. There the raindrops freeze into tiny pellets that fall back to the lower clouds. The upward-moving air pushes the pellets back up, starting the cycle again. Moisture in the clouds continues to add layers of ice to the pellets each time they are blown upward. Finally, the ice becomes too heavy and falls to the earth as hail. Hailstones vary in size from as small as a pea to even larger than a grapefruit.

Ice from freezing rain or sleet

Snow

Beth Hamel

Layers in a large hailstone

127

Background

Water Vapor Machines

Scientists have invented machines that can add or remove water vapor from the air. A *vaporizer* is a device that produces a mist of water vapor and adds humidity to the air. A *dehumidifier* cools the air so that the moisture (humidity) is removed through condensation.

Vapor and *Evaporate*

Both words are from a Latin word that means "steam." The prefix *e-* has the same meaning as *ex-*, so it gives the idea of "outside" or "away." As water evaporates, it forms vapor and moves away from the water.

Hail

Hailstones can damage crops and other objects that they hit. In Bible times, God sometimes used hail as a judgment against sin. In the time of Moses, God used hail as a plague to judge Egypt (Exodus 9:12–35). The hail destroyed many of Egypt's crops.

God used the plagues to demonstrate His power to the Israelites and build their trust in Him. He often uses His creation to show people His power, love, and salvation.

- Continue the discussion by directing the students to create a chart of Bible promises for display. Make a section for promises that God has already kept (e.g., Jesus' first coming as promised in Genesis 3:15 and Isaiah 7:14; 9:1–2; fulfilled in Luke 2) and ones that have not yet been fulfilled (e.g., Jesus' Second Coming in Isaiah 9:3–7; John 14:1–7; Acts 1:11).

  What are the steps of the water cycle? precipitation, evaporation, and condensation

  In which step of the water cycle are clouds formed? condensation

  What is precipitation? any type of solid or liquid water that falls from the atmosphere and reaches the ground

  What are some types of precipitation? rain, hail, sleet, snow

  How does rain form? Water droplets in the clouds join together to form larger drops.

  What determines the sizes of raindrops? air temperatures, winds, and types of clouds

- Explain that raindrops do not form in the teardrop shape that people usually draw. Most raindrops are round and slightly flat on the bottom.

  What condition can cause freezing rain? when the air is above freezing but the temperature of the earth's surface is below freezing

  Freezing rain can be dangerous because it forms a glaze on objects. Sometimes this glaze is not visible. This "black ice" can cause accidents.

  How is sleet different from freezing rain? The air temperature is below freezing for sleet and above freezing for freezing rain. Sleet lands as ice pellets, but freezing rain forms a glaze over objects.

  What type of precipitation is formed when ice crystals in the clouds join together and fall toward the earth? snow

  What kind of air temperature is necessary in order for snow to fall? The air temperature must be at or below freezing.

  As the air temperature decreases, what happens to the size of the snowflakes? They become smaller.

  What is unusual about hail? It usually falls during warm weather.

  How does hail form? Warm air pushes raindrops higher into tall clouds and freezes the drops into pellets. The pellets begin to fall but are blown up into the clouds again. Additional layers of ice are added to the pellets. The pellets continue to rise and fall until they are heavy enough to drop to the earth.

  What are some things that can be damaged by hailstones? Possible answers: crops, vehicles, homes

69
70

Why are dew and frost not classified as precipitation? They do not fall from the sky.

How are dew and frost formed? from water vapor close to the earth's surface that has collected in a liquid or solid state

How is frost different from dew? The water vapor goes directly to the solid state, never becoming a liquid.

- Discuss *Science & the Bible.*

  People have observed God's creation of dew and humidity. What useful invention did their observations of humidity lead them to create? the hygrometer

  What does the hygrometer measure? humidity

  What is humidity? water vapor in the atmosphere

  What is relative humidity? the amount of water vapor in the air compared to the amount that the air could hold at that temperature

  Which can hold more moisture, warm air or cool air? warm air

Dew and Frost

Dew and frost are not types of precipitation. Precipitation falls from the sky, but the moisture that forms dew and frost does not fall. Instead, dew and frost form from water vapor in the air. Dew appears when water vapor condenses on the surface of objects. Early on a cool morning, you may see tiny dew drops on the blades of grass. During colder weather frost may form from the vapor. Frost is ice that forms without passing through the liquid state.

Humidity

As the sun warms the earth, some of the water evaporates into the atmosphere. Water vapor in the atmosphere is called **humidity**. The temperature and air pressure influence the amount of humidity that the air can hold at a given time.

You may have heard a weather forecaster speak of relative humidity. *Relative humidity* is the actual amount of water vapor in the air compared with the amount that the air could hold at that temperature. Warmer air can hold much more water vapor than cold air. For a given amount of water vapor in the air, cold air has a higher relative humidity than warm air does. Relative humidity is identified as a percentage.

SCIENCE & THE BIBLE

A person's curiosity and desire to solve a problem can lead to useful ideas and inventions. One such invention is the hygrometer. It can be used to measure the amount of humidity in the air. The hygrometer was invented in the early 1400s by Nicholas of Cusa, a German philosopher and clergyman. His hygrometer was designed to measure the mass of moisture in the air that collected on a piece of wool. Leonardo da Vinci made one around 1480. His was similar to Nicholas's design but used a sponge instead of wool. Today there are a variety of hygrometers that use many different materials to measure humidity.

128

Background

Dew Point

The term *dew point* is used to refer to the temperature at which water vapor in the air reaches a saturation point. This temperature is always lower than the actual temperature of the air. If the air cools to the dew point temperature, some of the water vapor is removed from the air through the process of condensation. This condensation is known as dew. The temperature of the dew point varies in relationship to the amount of moisture in the air. If the humidity is high, the dew point will be at a higher temperature. The condensation point is discussed in Chapter 3, *Matter*.

Types of Clouds

There are over a dozen classifications of clouds based on the appearance, the altitude, and the type of precipitation that is produced. Only the basic types are discussed in the Student Text.

Deposition and Sublimation

In certain situations, the liquid phase or state is skipped in a phase change. When a gas goes directly to a solid, as in frost formation, deposition has occurred. When a solid goes directly to a gas, sublimation has occurred.

Clouds

Clouds form from condensed water vapor or ice crystals. One way clouds are classified is by shape. The three basic shapes of clouds are stratus, cumulus, and cirrus. Clouds can also be classified according to altitude. The prefix *alto-* is added to the names of some clouds. This prefix shows that the clouds are part of the middle altitude. They are neither very high nor very low clouds. A form of the word *nimbus* is added to the names of clouds that produce precipitation.

Stratus clouds look like flat blankets and are usually the lowest clouds in the sky. Sometimes they actually lie on the surface of the earth. In this case, the cloud is called fog. Nimbostratus clouds often produce light rain.

Probably the most familiar clouds are cumulus clouds. **Cumulus clouds** are large, fluffy-looking clouds with flat bottoms. These clouds mostly form in the middle of the troposphere. The tall, dark clouds that produce thunderstorms are called cumulonimbus clouds.

The thin, curly-looking clouds seen high in the sky are called **cirrus clouds**. Because they are high in the troposphere, cirrus clouds are made of ice crystals rather than water droplets. Cirrus clouds often indicate that a warm front is moving into the area.

• MEET THE SCIENTIST • Luke Howard

A scientist does not have to work in a science profession. For example, Luke Howard was a British businessman. But he enjoyed studying nature, especially clouds. He kept accurate notes on weather observations for more than thirty years.

In the early 1800s, he wrote a paper based on his observations. He classified the clouds and gave them Latin names. The paper was published and used by other scientists. As a result, Mr. Howard's groups and names are the basis for the cloud classifications used today. He also wrote and gave lectures on many weather topics. His lectures were printed and became the first textbook on meteorology.

129

- Discuss *Meet the Scientist.*
  Do all scientists work in laboratories and science professions? no
  Who was Luke Howard? a British businessman who enjoyed studying nature
  How do we know about what Luke Howard observed? He kept accurate notes about the weather for more than thirty years.
  What did he do with his observations about clouds? He wrote a paper based on his observations. He classified the clouds and gave them Latin names.
  Are his classifications and names still used today? yes
  What is meteorology? the study of weather
- Refer to the *Types of Clouds* diagram on pages 130–31 as clouds are discussed.
  How do clouds form? from condensed water vapor or ice crystals
- You will want to do the *How Clouds Form* demonstration at this time.
  What two ways are clouds classified? by shape and altitude
  What are the three basic shapes of clouds? stratus, cumulus, and cirrus
- Explain that Luke Howard named the basic types of clouds based on Latin words that described their appearances.
  What do you know about a cloud whose name begins with the prefix *alto-*? It is part of the middle altitude of clouds. It is not very high or very low.
  Why is *nimbus* added to the names of some clouds? to identify them as clouds that produce precipitation
- Explain that in Latin *nimbus* means "rain."
  Describe stratus clouds. They look like flat blankets and are low in the sky.
- Explain that in Latin *stratus* means "layer."
  What is fog? a stratus cloud that lies on the surface of the earth
  Describe cumulus clouds. large, fluffy clouds with flat bottoms
- Explain that in Latin *cumulus* means "heap."
  What are tall, dark clouds that produce thunderstorms called? cumulonimbus clouds
  Describe cirrus clouds. thin, curly clouds seen high in the sky that are made of ice crystals
- Explain that in Latin *cirrus* means "curl."
  What kind of front is usually moving into the area when cirrus clouds are seen? a warm front

Demonstration

How Clouds Form

Materials: large glass jar with lid, hot tap water, ice, aerosol air freshener or hair spray, black paper, flashlight (optional)

Put about 3 cm of hot water into the jar and cover. Let it sit for several minutes. Rotate the jar, swirling the water to remove condensation on the sides.

Remove the lid. Turn the lid over. Fill it with ice and set it on the jar. Let it sit for several minutes.

Remove the lid and immediately spray a small amount of the aerosol into the jar. Set the lid of ice back on the jar. Within a few minutes, the curling wisps of a cloud should form.

Hold the black paper behind the jar to see the cloud more easily. You may need to use the flashlight to help you see the cloud. Remove the lid and watch the cloud float out of the jar.

The hot water produces humidity. The ice causes a layer of cool air to form. The aerosol supplies the particles needed for the humidity to condense and form visible droplets.

69
70

- Direct attention to the *Types of Clouds* diagram.
  Which type of cloud is lowest in the sky? stratus
  Which type of cloud is highest in the sky? cirrus
  What nickname is given to cirrus clouds? mares' tails
  Why do you think cirrus clouds are called mares' tails? because they are thin and wispy like horses' tails

Background

Cloud Classifications

Clouds have more classifications than those discussed in the Student Text. This chart shows the twelve basic cloud classifications and their approximate altitudes.

Legend			
1	fair-weather cumulus	7	nimbostratus
2	stratocumulus	8	cumulus congestus
3	stratus	9	cirrocumulus
4	swelling cumulus	10	cirrostratus
5	altocumulus	11	cirrus
6	altostratus	12	cumulonimbus

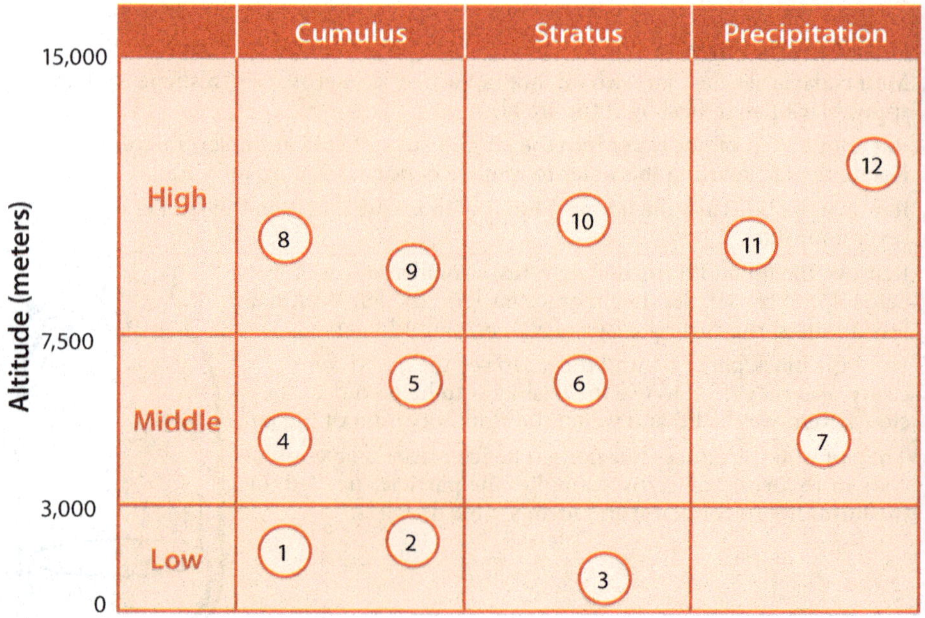

QUICK CHECK

1. What is precipitation?
2. What influences humidity?
3. Describe a cumulus cloud.

Cirrus
Thin, wispy clouds sometimes called "mares' tails"

Stratus
Wide, thin layers of clouds that form at lower altitudes

Nimbostratus
Thick mass of rain clouds that seems to cover the sky

131

What kind of weather will you probably have when you see cumulonimbus clouds? heavy rain or hail

Which type of cloud is found in the middle of the sky? cumulus

What kind of weather will you probably have when you see cumulus clouds in the sky? fair weather

What does *fair weather* mean? Possible answers: sunny, nice, pleasant

You may have heard someone called a "fair-weather" friend. What do you think a fair-weather friend is like? Possible answers: someone who is friends with you only when it is convenient for him; someone who is friends with you only during the good times

- Ask a volunteer to read aloud Proverbs 17:17. [BAT: 5e Friendliness]

Should a friend be a fair-weather friend? No, a friend loves at all times.

ANSWERS

1. any type of moisture that falls from the atmosphere and reaches the ground
2. water vapor in the air
3. tall and fluffy with a flat bottom

70

- The *How Clouds Form* demonstration and *Cloud Crayon Resist Artwork* activity found at the bottom of the pages in this lesson may also be incorporated into the lesson.
- Suggestions for additional activities for this two-day lesson are on the Teacher's Toolkit CD.

Activity Manual

Review, page 88

This page reviews Lessons 69–70.

Assessment

Quiz 5-C

The quiz may be given any time after completing this lesson.

Activity

Cloud Crayon Resist Artwork

Materials: white paper, crayons, watercolors

Direct each student to choose a type of cloud to illustrate. He then uses crayons to color clouds and other features of the sky such as birds or lightning. He is not to color the main part of the sky. The horizon and ground may be added if desired.

After the coloring is complete, direct each student to paint over the crayon and blank places of the sky with diluted watercolors. The colors of the paint chosen by the student should correspond to the colors that would be seen in a sky with that type of cloud. For example, if cumulonimbus clouds are drawn, the sky should have a grayish, stormy look. A bright, sunny day may have areas of yellow and pink around the clouds.

Objectives

- Describe characteristics of thunderstorms, tornadoes, and hurricanes
- Differentiate between a weather watch and a weather warning

Materials

- weather radio or other battery-powered radio
- flashlight
- extra batteries
- blanket
- canned food with easy-open lids
- bottled water

Vocabulary

- thunderstorm
- lightning
- thunder
- tornado
- hurricane

Introduction

- Display the items.
  Why might someone collect these items? Accept reasonable answers.
- You may choose to turn on the weather radio. Allow the student to listen to the current weather report.
  A weather radio can alert you if severe weather is coming to your area. Some cellular phones and other electronic devices also alert you about severe weather. The radio and these other items may be useful when preparing for some types of storms.

Preparation for Reading

- Direct the students to mentally make questions using the headings, bold and italicized words, and captions.
- Instruct them to read silently pages 132–34 to find out the answers to these questions.
  What is another name for clouds that produce thunderstorms?
  What is the difference between a weather watch and a weather warning?

Teach for Understanding

What are some characteristics of severe weather? high winds, large amounts of precipitation, or extreme temperatures

In what direction do the prevailing westerlies cause most of the weather in the United States to move? from west to east

When do thunderstorms occur? when a large, warm air mass quickly rises above a cold air mass

What are some characteristics of thunderstorms? thunder, lightning, strong winds, heavy rains, sometimes hail

What type of cloud is called a thunderhead? a tall, dark cumulonimbus cloud

Severe Weather

Thunderstorm

Weather includes temperature, winds, air pressure, and humidity. Severe weather conditions occur when various aspects of weather combine to form high winds, large amounts of precipitation, or extreme temperatures. Some types of severe weather can be very harmful. But knowing how and why this weather occurs can reduce the damage and loss of life.

Global winds influence weather patterns around the world. For example, the prevailing westerlies cause most of the weather in the United States to move from west to east. Sometimes cold air masses move southward from Canada. Warm, wet air masses can blow toward the East Coast from the Gulf of Mexico. When large air masses like these meet, severe weather may result.

Thunderstorms

A **thunderstorm** often occurs when a large, warm air mass quickly rises above a cold air mass. All thunderstorms have lightning and thunder. Most also have strong winds, heavy rain, and sometimes hail. The tall, dark cumulonimbus clouds that produce thunderstorms are called thunderheads.

During a storm, air temperatures change rapidly. This causes strong air movements among the clouds. The resulting friction creates **lightning**, a type of static electricity. Lightning moves either between clouds or between the clouds and the earth. The temperature of lightning can reach up to 30,000 °C (54,000 °F). Imagine how quickly the lightning bolt heats the air around it! The hot air instantly expands and then the cooler air rushes in with a crash. The resulting rumble we hear is **thunder**.

FANTASTIC FACTS

You can easily tell if a thunderstorm is moving toward you or away from you. Since light travels faster than sound does, you see lightning before you hear the thunder. Count the seconds between each flash of lightning and the sound of its thunder. As the thunderstorm gets closer, there are fewer seconds between the lightning and the thunder.

132

Background

Seasonal Storms

Severe storms are often seasonal. Thunderstorms usually occur in spring and summer. Tornadoes can happen at any time but are more frequent from late winter to midsummer. Hurricane season in the Atlantic Ocean is from June 1 to November 30.

Whirling Winds

The whirling winds of a storm are called a vortex. Tornadoes and hurricanes are not the only winds with whirling air. Other whirling winds include dust devils and waterspouts.

Radio Weather Reports

Severe-weather radio reports may not be available for your area. Consult an Internet weather site that has national severe-weather reports. Choose a report for an area that you know is currently experiencing severe weather. (Coastal areas usually have some type of report.) Reports can be printed and read to the students.

Tornado Alley

Most tornadoes in the United States occur in the central plains and the Southeast. This area is called "Tornado Alley" because of the frequent tornadoes. Texas, Oklahoma, Kansas, and Nebraska have the most severe tornadoes.

Tornadoes and Hurricanes

Depending on the speed and strength with which the fronts hit, a thunderstorm may also produce tornadoes. A **tornado** is a funnel-shaped cloud of swirling winds that reaches down to the ground. Tornado winds are powerful and can reach speeds of 400 km/h (250 mi/h). Their force may destroy anything in the tornado's path, including houses and trees. Most tornadoes, though, are small and last only a short time.

Tornado

In the Northern Hemisphere, hurricane winds spiral in a counter-clockwise direction. The *eye* of a hurricane is a calm, clear area in the center of the storm.

Another type of severe storm may form over a low-pressure area of an ocean. Winds gather warm, wet air from near the ocean surface. As the warm air rises, towering cumulonimbus clouds form. Heavy rains pour from the clouds, and strong winds swirl around the low-pressure area. In the Atlantic Ocean or the eastern Pacific Ocean, a storm with spiraling winds that forms over the ocean is called a **hurricane**. A hurricane can cover hundreds of miles. In the western Pacific Ocean, these storms are called *typhoons*. When a hurricane comes on land, it is very dangerous. The strong winds often blow down trees and damage buildings. The ocean rises and covers the shore in flood water. This is called storm surge. The Galvenston storm that happened in 1900 was actually a large hurricane. It was the strong winds and storm surge that were responsible for most of the people who died. Eventually, the storm moved northeast and disappeared over the Atlantic Ocean.

What is lightning? a type of static electricity that moves between clouds or between clouds and the earth

How is lightning formed? Rapid temperature changes cause air to move among the clouds, producing the friction that causes lightning.

How does lightning produce thunder? Hot air instantly expands and then cooler air rushes in with a crash.

What are some safety tips to observe during thunderstorms? Possible answers: Go inside. Get out of water. Do not stand under trees.

What is a tornado? a funnel-shaped cloud of swirling winds that reaches down to the ground

- Explain that not all tornadoes are funnel-shaped. Some tornadoes are wide with straight sides.

  What makes a tornado so dangerous? Its force may destroy anything in its path.

- You may want to do the *Tornado Vortex* demonstration at this time.

  How do hurricanes form? Winds gather warm, wet air over a low-pressure area of the ocean. The warm air rises and forms cumulonimbus clouds. Heavy rains pour from clouds and strong winds swirl around the low-pressure area.

  What determines whether a storm is a hurricane or a typhoon? Its location. A storm in the Atlantic Ocean or the eastern Pacific Ocean is called a hurricane. One in the western Pacific Ocean is called a typhoon.

- Direct attention to the hurricane photo.

  How is the eye of the hurricane different from other parts of the storm? The eye is a calm, clear area in the center of strong, swirling winds and rain.

Demonstration

Tornado Vortex

Materials: two 2 L clear plastic soda bottles, water, glitter, seed beads, waterproof tape

Pour 1 L of water into one bottle. Add a capful of glitter and a capful of beads to the water. Turn the empty bottle upside down so that the mouths of the bottles are together. Tape securely. Invert the bottles quickly.

What does the water look like as it flows into the bottom bottle? Elicit that it looks like a tornado.

A tornado moves in much the same way as the water does. What do you observe about the glitter and beads? The glitter remains in the main swirl, and the beads are farther out and settle to the bottom faster.

How might the glitter and beads compare to items that are carried by a tornado? Accept reasonable answers.

71

What is a blizzard? a snowstorm with strong, freezing winds and blowing snow

Why are blizzards so dangerous? Possible answers: Blowing snow reduces visibility. Temperatures are often very cold.

What is a heat wave? A period of time with higher-than-average temperatures. The humidity is also usually higher than normal.

Why is a heat wave dangerous? People outdoors for long periods of time or elderly people without air conditioning or fans may overheat or dehydrate.

What organization tracks the weather in the United States? the National Weather Service

What is the difference between a weather watch and a weather warning? A weather watch means conditions are right for severe weather, but that the weather is not currently happening. A weather warning means that a storm has begun.

What are some safety procedures for severe weather? Possible answers: Plan ahead. Stay inside. Go to a basement or another sturdy place during very strong winds.

- Review any weather safety procedures recommended for your school, home, or town. These procedures may include a reminder to remain calm, the safety routes and locations for safe shelter within the building, the items to take or not take to a location of safety, emergency telephone numbers, and the location of emergency shelters.
- Discuss *Science & the Bible.*

  Who controls the weather? God

  What can storms show us about God? Storms can show us God's power and control.
- Ask a volunteer to read aloud Genesis 1:28.

  How can people have dominion over the weather? with timely, accurate predictions of the weather
- Ask a volunteer to read aloud Matthew 5:16.

  As a Christian, how can someone use accurate predictions for God? to give glory to God and allow other people to see Christ in him

ANSWERS

1. lightning, thunder, wind, rain, hail, rapid temperature changes
2. A tornado is swirling winds that reach down to the ground. A hurricane is spiraling winds that form over water and bring heavy rain.
3. an announcement that conditions are right for a type of severe weather that is not currently happening

Other Severe Weather

Some areas of the earth may experience other types of severe weather. One type of severe weather is a blizzard. A *blizzard* is a snowstorm with strong, freezing winds and blowing snow. The blowing snow is dangerous because it reduces a person's ability to see very far and can build up in large drifts. Another type of severe weather is a heat wave. A *heat wave* is a period of time with higher-than-average temperatures. The humidity is also usually higher than normal. During a heat wave, people should avoid being outdoors for long periods of time.

SCIENCE & THE BIBLE

Throughout the Bible, God uses storms to show His power. God has given man dominion over the earth to use and care for it. But man cannot control the weather. Usually, the best dominion we can have over the weather is timely, accurate prediction. The person who does a good job of predicting the weather is doing a good job of living out Genesis 1:28. The Christian who does this sort of thing is engaged in a good work that God may use to glorify Himself (Matthew 5:16).

Storm Safety

Severe weather can be very dangerous. An organization that tracks the weather in the United States is the National Weather Service. Sometimes the National Weather Service issues a *weather watch*. This means that the conditions are right for a type of severe weather but that the weather is not currently happening. A watch alerts people to get ready for the possibility of a storm. Once the storm begins and people report having seen it, a *weather warning* is announced. You should always look for safe shelter if you hear a weather warning.

Listen to the radio for watches and warnings if you think a severe storm is on its way. Always try to stay inside a building during a storm. If the storm has very strong winds, go to a basement, bathroom, or other sturdy place inside the building. Plan ahead. Find out about what to do and where to go if severe weather comes.

QUICK CHECK

1. What are some characteristics of a thunderstorm?
2. Compare a tornado with a hurricane.
3. What is a weather watch?

134

Background

National Weather Service

In addition to announcements on special weather radios, the National Weather Service's severe-weather bulletins are also announced on radio and television stations, cellular phones, and other technical devices.

Blue Hill Observatory

The Blue Hill Observatory in Milton, Massachusetts, opened for weather observations on February 1, 1885. Since that day, it has remained an active weather observatory. It is the oldest weather observatory in continuous use in the United States.

Activity

Weather Sayings

Direct each student to ask his parents and other adults to recall weather-related sayings, old wives' tales, poems, etc. The student should then compile the information into a booklet entitled *Weather or Not: Fact or Fiction?* Each student should research to find whether the information is scientifically accurate.

Dangerous Extremes

EXPLORATION

Extreme weather can bring many dangers. Severe weather conditions such as tornadoes and hurricanes have strong winds and heavy rains. Strong winds may destroy buildings. Heavy rains may result in flooding or mudslides. Extreme conditions such as blizzards and heat waves can be dangerous to people, livestock, and plants.

People were made in the image of God, and will live somewhere forever—either with or without God. Because of this, Christians should be concerned for all human life. In this exploration, you will make and present an awareness poster or pamphlet for a specific type of severe, or extreme, weather. Helping others learn about extreme weather and how to be prepared for it is a way to care for others.

What to Do

1. Choose a type of severe weather.
2. Make an awareness poster or pamphlet about your chosen type of weather. Provide a description of the weather conditions that people should watch for. Include the dangers caused by that type of weather. List actions people should take to avoid these dangers. Include telephone numbers or other information that people living in your area would need in case of this type of emergency.
3. Present your poster or pamphlet. Emphasize the weather conditions to watch for and the precautions that should be taken to avoid injury.

135

Objectives

- Research the safety precautions for a type of severe weather
- Make and present a poster or pamphlet

Materials

- sample weather-safety posters or pamphlets

Teacher's Toolkit CD

- Rubrics: *Dangerous Extremes*, for each student

Sources of safety information include the American Red Cross, the National Weather Service, local disaster services agencies, the library, and the Internet.

Introduction

What types of severe weather do we have in our area? Answers will vary.

Each type of severe weather can be dangerous. There are specific safety guidelines that should be followed when severe weather is coming.

Preparation for Reading

- Direct the students to read silently page 135 before beginning the exploration.

Teach for Understanding

- Display the posters or pamphlets.

  You will choose a type of severe weather and research to find safety precautions for that type of weather. Then you will design and make a poster or pamphlet that can alert others to those precautions.

  You are not limited to the types of weather that we have in our area.

Other weather-related disasters not mentioned in this chapter include droughts, floods, and mudslides. Droughts and floods are discussed in Chapter 8, *Changes in an Ecosystem*.

What to Do

- Guide the students in their research. Provide resources as needed.

Assessment

Rubrics

Select the prepared rubric, or design a rubric to include your chosen criteria.

73

Student Text pages 136–37
Activity Manual pages 89–90

Objectives

- Describe the job of a meteorologist
- Read and interpret basic symbols on a weather map

Materials

- various weather maps from newspapers, almanacs, or the Internet

Teacher's Toolkit CD

- Quiz 5-D

Vocabulary

- weather forecast
- meteorologist

Introduction

- Display the maps.

  Have you seen maps like these before? Where have you seen them? Accept reasonable answers.

  You may have seen a person giving a weather report on television. The maps used in a weather report may look like one of these.

  Today we will learn how to read some of the information that is found on a weather map and apply the information to what we have learned in this chapter.

Preparation for Reading

- Display the word *meteorologist*. Pronounce the word together.
- Direct the students to mentally make questions using the headings, bold words, and captions.
- Instruct them to read silently pages 136–37 to find the answers to these questions.

  What is a meteorologist?

  What are some types of equipment that a meteorologist uses?

Teach for Understanding

Who designed and maintains the weather patterns? God

How does understanding and predicting weather patterns help glorify God? It helps people better protect life and allows them to use the knowledge of weather to benefit others.

What are some predictable weather patterns? Possible answers: global winds; the way that air collects moisture; the movement of cool air as it replaces warm, rising air

What is a weather forecast? a prediction of future weather conditions

What is a meteorologist? a scientist who studies the atmosphere and weather

Does a meteorologist control the weather? No, he only predicts what it might be. Who controls the weather? God

Forecasting the Weather

God values human life. Understanding and predicting the patterns that weather moves in helps us better protect life and allows us to use our knowledge of the weather in ways that benefit others.

Air moving over water collects moisture. Cooler air moves to replace warmer, rising air. Global winds blow in set paths around the earth. These, and other predictable conditions, provide some of the information needed to forecast the weather. A **weather forecast** is a prediction of future weather conditions.

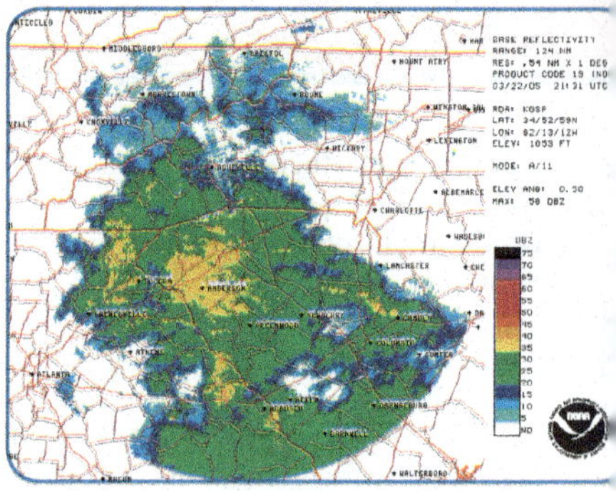

Radar image of northwest South Carolina

Making Predictions

The scientists who study the atmosphere and weather are called **meteorologists**. Much of their work involves gathering data, or information, about the atmosphere and environment. They use thermometers, barometers, anemometers, and other basic equipment to provide some of this data. More advanced equipment such as radar and satellites can also provide useful data.

Throughout history people have watched the skies for weather patterns. By studying these patterns, people hoped to predict the coming weather. Farmers and sailors, who depended on the weather, found many predictable patterns. One of these patterns is even mentioned in the Bible. Matthew 16:2–3 tells us that ancient people had observed that the color of the sky could indicate what type of weather was coming.

Modern meteorologists work in a similar way. They daily gather weather data to find and study weather patterns. In this way meteorologists can use present conditions to make fairly accurate predictions about future weather.

Satellite map of North America

136

Background

Meteorologist

Meteorologists study the atmosphere and its characteristics, movements, and other processes. Only about one-third of the meteorologists forecast the weather. Others do research to help with pollution control or to plan air and sea transportation. Some study the climate. Their research can help engineers plan appropriate heating and cooling for buildings.

Early Weather Maps

The first weather maps were developed in the 1870s by the US Army Signal Service. This department eventually became the National Weather Service that we have today. The early maps were made from data that was sent to Washington, DC, from stations around the United States. Many of these maps were incomplete since the less populated areas of the country did not have stations.

Reading Weather Maps

Meteorologists put some of their data onto weather maps. These maps allow us to understand the data more easily. Some weather maps are colored to show differences in temperature. Other maps show differences in air pressure with lines that connect points of equal pressure. Lines with triangle and half-circle symbols are used to show fronts. Many weather maps also have symbols to indicate different types of precipitation.

QUICK CHECK

1. Why should a Christian be concerned with predicting the weather?
2. What does much of the work of a meteorologist involve?

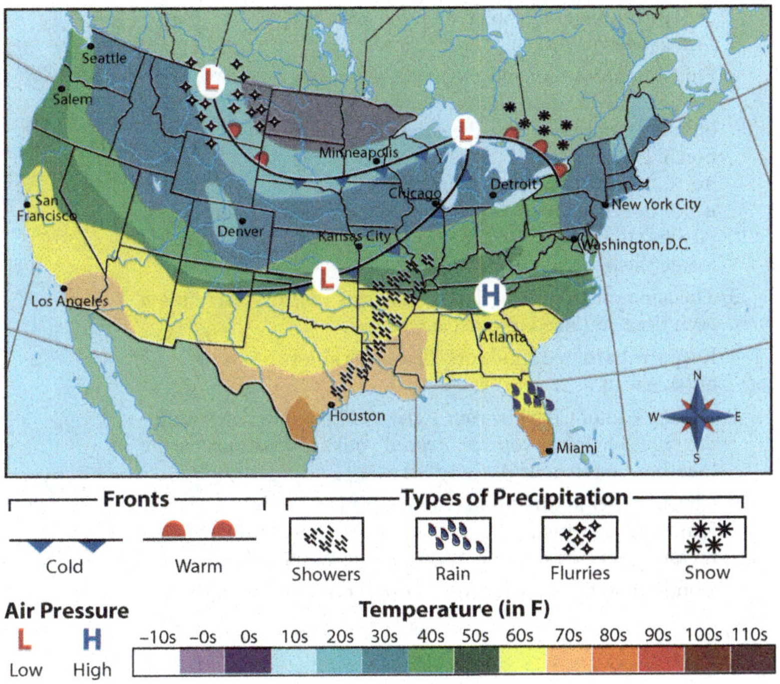

137

What types of equipment do meteorologists use? Possible answers: thermometers, barometers, anemometers, radar, satellites, computers

Before the science of meteorology, why did farmers and sailors study the weather? Their jobs depended on weather conditions.

- Ask a volunteer to read aloud Matthew 16:2–3.

  What is one weather pattern mentioned in the Bible? That the weather will be fair if the evening sky is red but stormy if the morning sky is red.

  How do meteorologists use records of daily weather conditions? to study weather patterns and make predictions

  What symbols do meteorologists use on weather maps? Possible answers: colored areas, circular lines, lines with shape symbols

- Explain that the symbols on weather maps may vary. Understanding and using the key is important in reading a weather map.
- Discuss the weather map.

  What symbols show precipitation? slanted lines, rain drops, diamond shapes, and snowflakes

  How is a cold front indicated? by a line with triangles underneath it

  What do the *H* and *L* symbols mean? high pressure and low pressure

  According to this map, which area of the country is experiencing temperatures below 40 °F? the North

  According to this map, what temperatures is most of the state of New York experiencing? 30s

  Do you think this map is from a winter day or a summer day? winter How do you know? Possible answers: the snowflakes on the map, the lower temperatures

ANSWERS

1. to help better protect human life and use the knowledge learned in ways that benefit others
2. the daily gathering of weather data to find and study weather patterns

Activity

Weather Map Symbols

Instruct the students to design their own set of weather map symbols. Provide a list of the weather words, such as the different types of precipitation and fronts, to be illustrated. The students should present the finished symbols in a key.

Activity Manual

Reinforcement, page 89

This page gives the students practice in reading a weather map.

Review, page 90

This page reviews Lessons 71 and 73.

Assessment

Quiz 5-D

The quiz may be given any time after completion of this lesson.

Objectives

- Make working weather instruments
- Correctly use the instruments to gather information about the weather
- Record data
- Use data to make weather predictions

Materials

- See *Weather Instruments* pages for individual instrument materials

Teacher's Toolkit CD

- Instructional Aids 5.2–5.4: *Weather Instruments*, for each student
- Rubrics: *Weather Observatory*, for each student

Two lesson days are allotted for this activity. On the first day, introduce and set up the activity. After the observation period is complete, use the second lesson day to discuss student observations and conclusions. See the Background section on this page.

Introduction

- Display the barometer.

  What is the name of this weather instrument? a barometer

  What does it measure? air pressure

  A high-pressure reading on the barometer usually indicates good weather. A low-pressure reading usually indicates stormy or rainy weather.

  Today we will be making other weather instruments. Using those instruments and this barometer, we will set up our own weather station.

Preparation for Reading

- Direct the students to read silently Student Text pages 138–39 and Activity Manual pages 91–92 before beginning the activity.

Teach for Understanding

Procedure

- Guide the students as the instruments are constructed.
- Find an out-of-the-way place to set up the station so that the instruments are not disturbed but are exposed to wind and rain.
- Review the types of data to record on the *Weather Log.*

  Air Pressure

  Use the barometer. The reading is probably in inches. After moving a barometer, you may need to reset it. Contact the local weather service for the current pressure and adjust it accordingly.

ACTIVITY

Weather Observatory

Meteorologists use a variety of instruments to collect data related to weather conditions. In this activity you are the meteorologist. You even make some of the equipment!

Process Skills
- Measuring and using numbers
- Making and using models
- Observing
- Collecting, recording, and interpreting data

Purpose

Make and use weather instruments.

Materials
barometer
Weather Instruments pages
Activity Manual

Procedure

1. Construct the wind vane, rain gauge, thermometer holder, and anemometer as directed on the *Weather Instruments* pages.
2. Find an outdoor location to use as a weather station. The location should be away from buildings and trees. Use the compass to find north. Place the wind vane with the *N* facing north. Place the rain gauge in a level place so it will not easily tip over. Set up the thermometer holder by pushing the dowel into the ground. Place the thermometer inside the holder.
3. The anemometer is not designed to stay outside. Carry it out each time you need to make a reading.
4. Keep the barometer indoors. Place it near a window or outside wall.
5. Use the Cloud Chart in your Activity Manual to determine the types of clouds you see. Record your observations on the Weather Log in your Activity Manual.
6. Check the wind direction, amount of precipitation, temperature, wind speed, and air pressure. Record your measurements and observations on the Weather Log.
7. Continue to record your weather observations for 10 days.

138

Background

Activity Helps

Directions for making the instruments are available from numerous sources. You may choose to use other sources or adjust the directions on the *Weather Instructions* pages as needed. This activity uses a manufactured barometer. Directions for making a barometer appear in the *Make a Barometer* activity on page 135.

You may choose to have each group make a different instrument. Depending on class size, this would provide instruments for one or more weather stations. Or you may assign each student to make and set up all the instruments at home.

Here are some suggestions for gathering and recording daily weather observations.

- Provide a chart on a clipboard to be carried out to the station when one or more students are gathering data.
- Make a large classroom chart to record daily observations. Students may complete their individual charts from information on the large one.
- Keep a classroom chart of instrument readings and have the students keep personal charts of data gathered from official weather sources, such as a radio, a newspaper, or the Internet.

Conclusions

- What patterns or relationships can you identify between your observations and the types of weather that occurred each day?
- Can you predict the weather for tomorrow based on your observations from today?
- What benefits can your observation and predictions be to others?

Follow-up

- Compare your observations each day with the daily weather maps and reports from newspapers, the Internet, or other sources.
- Make your own weather maps based on your observations.

139

Temperature
Use the thermometer. The bulb part of the thermometer should be kept inside the thermometer holder so that it stays out of direct sunlight. Use the clay to hold the thermometer in place.

Wind Direction
Use the wind vane. Remember that the arrow points toward the direction that the wind blows from.

Wind Speed
Use the anemometer. Read the degree along the curved edge. Use the Wind Speed Chart on *Weather Instruments* (5.3) to convert the degree measurement to km/h or mi/h.

Rainfall
Use the rain gauge. Remember to keep the water level at the line. After a rainfall, pour some water out. If the weather is very dry, add additional water.

Cloud Description
Use the Cloud Chart on Activity Manual page 92 to identify and describe the basic types of clouds observed. The descriptions should include the types and amounts of clouds seen.

Other Observations
Include wording such as *clear*, *sunny*, *fair*, or *cold*, as well as any types of precipitation. Include the color of the sky.

Check the weather equipment at the same time(s) each day for the best comparisons. Readings from midmorning to afternoon are best. (Early morning is still a nighttime reading.)

Process Skill

The process skill emphasized in this activity is collecting, recording, and interpreting data.

- Explain that recording data allows others to view and evaluate your results.

  What kinds of data are you collecting in this activity? wind direction, wind speed, amount of precipitation, temperature, air pressure, cloud descriptions

  Which of these kinds of data are based on your opinion? cloud descriptions, the color of the sky

  Which of these kinds of data are based on a measurement? air pressure, temperature, wind direction, wind speed, rainfall

  How can we use this data? to compare the information to actual weather reports, to look for patterns in the data, to try to make weather predictions

Conclusions

- Discuss the conclusions from the students' observation.

Activity Manual

Activity, pages 91–92

Assessment

Rubrics

Select the prepared rubric, or design a rubric to include your chosen criteria.

76

Activity Manual pages 93–94

Objectives

- Explain how clouds form
- Defend a biblical view of evidence for one ice age against a secular view of evidence for multiple ice ages

Materials

- 2-liter bottle with a cap
- wooden match
- 1 cup warm water

Teacher's Toolkit CD

- Visual 5.3: *Timelines*

Introduction

- Display the *Timelines* visual. Remind the students about the earlier discussion from Chapter 2 regarding the biblical view of the age of the earth compared to the non-Christian view of the age of the earth.

Do you know what an ice age is? Accept any reasonable answer. An ice age is a time when a large portion of the land is covered by ice in the form of glaciers.

- Explain that the non-Christian viewpoint sees many ice ages over billions of years, with the most recent ice age ending about 10,000 years ago. Those that start from the Bible conclude that there was a single ice age that was a result of the Flood.

Teach for Understanding

- Read Job 38:19–30 together. Explain that God is speaking to Job.
- Invite the students to list all the weather-related terms in the verses on Activity Manual page 93.
- Discuss the terms that were identified.

Cloud in a Bottle Demonstration

- Add the warm water to the bottle and close the cap tightly. Ask several students to squeeze and release the bottle rapidly several times. No cloud should form. Unscrew the cap and direct a student to squeeze the bottle, with another student standing by holding the cap. Light a wooden match, and let it burn for a few seconds. Bring it to the mouth of the bottle. As you blow out the match, direct the student to release the sides of the bottle, which will draw in the smoke. The other student should cap the bottle quickly.
- Allow the students to squeeze the bottle and release it rapidly, observing how clouds form only when there are particles in the air.
- Allow time for the students to answer questions 2 and 3 on the Activity Manual page.
- Instruct the students to read silently the paragraphs on Activity Manual page 93 to find out why clouds formed when the smoke was added.

Why did the clouds form? The smoke provided particles for the water drops to stick to and form clouds.

A lesson from
Answers in Genesis Name ______________ 1:1

Ice Age

The Bible has much to teach us about weather.

1. Read Job 38:19–30. List all the weather related terms that God uses.
   snow, hail, light parted, east wind, lightning, thunder, rain, drops of dew, ice, frost, frozen water

Watch as your teacher does the Cloud in the Bottle demonstration.

2. Before the smoke was added, did a cloud form when the bottle was squeezed? *no*
3. What happened while the bottle was squeezed after the smoke was added?
   Clouds formed.

The cloud only forms when there are tiny particles in the air for the water to stick to. When you squeeze the bottle it gets hotter. When you release the bottle, the air cools. The water drops form on the smoke particles. There would have been similar conditions to these in the atmosphere at the beginning of Noah's Flood.

If you have ever seen a geyser or a hot spring, you have a hint at what happened when Noah's Flood began. In Genesis 7:11 we read that God caused "the fountains of the great deep" to burst open. Hot water along with magma was released and combined with ocean waters, producing many clouds. The clouds then brought the rain that fell for forty days and nights. Just as you saw the cloud form in the bottle, massive clouds would have formed to create such a severe storm. All the hot water in the oceans evaporated and cooled as it rose and collected on particles high in the atmosphere. The ocean waters would have remained hot for hundreds of years. Because hot water evaporates faster than cold water, there was lots of water in the atmosphere.

Where did all the water go? It fell as snow and rain over the land. The ash, debris, and cloud cover reflected much of the sun's energy. The climate was colder after the Flood. Over many years the snowfall over the cold continents accumulated. In the North, massive ice sheets miles thick built up and caused the Ice Age. For an Ice Age to occur, hot oceans and reduced sunlight for an extended period of time are needed.

The earth may have looked this way during the Ice Age.

Science 5 Chapter 5 | Lesson 76 Answers in Genesis 93

Background

Activity Helps

To avoid lighting a match, the demonstration on page 147 may be substituted.

Cold Weather in Job

Though the Bible does not clearly refer to an ice age following the Flood, the book of Job, one of the oldest books in the Bible, has multiple references to ice, snow, and frost. Here are a few: Job 6:16; 9:30; 24:7, 19; 37:9, 10; 38:22, 29, 30.

In Job 31:19, Job notes that one could perish from a lack of clothing. This would be a comment made more likely in a cold climate. Also, there are twenty-one references to clouds, especially dark or thick clouds, which would be consistent with a post-Flood ice age.

We can see the scars that these ice sheets and glaciers left on the surface of the earth. This picture shows striations or scratches left in solid rock as a glacier moved across it.

5. Why do you think all the striations are pointing in the same direction?
The striations tell us which direction the glacier was moving—the same direction as the lines.

The Ice Age must have happened after the Flood, or we would not be able to see moraines, kettle lakes, and other features. These would have been washed away. The Ice Age probably happened about 4,000 years ago. The glaciers and ice sheets that we see today are mostly the remnants of that amazing event.

Scientists who do not trust the Bible or who do not believe Noah's Flood covered the whole Earth think that there have been many ice ages. They believe that the last ice age ended about 10,000 years ago. Creationary scientists believe that date cannot be right since God created the earth around 6,000 years ago. Examining history through Bible glasses helps us see things clearly.

When we begin with what the Bible tells us about the history of the earth, we can give a reasonable explanation for the Ice Age and when it occurred.

When would conditions have been similar to these during Bible times? Noah's Flood

What did God cause to burst open during the Flood? "the fountains of the great deep"

What did the fountains of the great deep release? hot water and magma

What did the hot water do to the oceans? It heated them.

- Explain that when the warm water heated the oceans, lots of water evaporated.

  What happened as the water evaporated? Clouds formed, and it rained for forty days and nights.

  What happened after the Flood? The hot water began to evaporate. It began to cool and to collect on particles.

  How long would the ocean waters have remained hot? for hundreds of years

  Where did the water go? It fell as snow and rain over the land.

- Explain that the ash, debris, and cloud cover reflected much of the sun's energy, cooling the atmosphere.

  Because the climate was colder after the Flood, what happened over the cold continents? The snowfall accumulated.

  What did these ice sheets cause? the Ice Age

- Discuss the picture of the earth on Activity Manual page 93.
- Explain that scientists cannot use the conditions of the present climate to determine how the ice sheets built up. There is not enough snowfall in these areas today to accumulate to the thickness of the ice sheets that existed during the Ice Age.
- Choose a volunteer to read aloud the first paragraph on Activity Manual page 94.
- Discuss the picture of striations.

  Why do you think all the striations are pointing in the same direction? The striations show which direction the glacier was moving—the same direction as the lines.

- Direct the students to read silently the last two paragraphs to find out what scientists who do not believe the Bible think about the Ice Age.

  What do scientists who do not trust the Bible believe about the Ice Age? that there were many ice ages

- Direct attention to Timeline B on the *Timelines* visual.

  When was the Flood? about 4350 BC

  From a biblical worldview, when is a reasonable estimate for when the Ice Age occurred? about 4,000 years ago

Activity Manual

Answers in Genesis, pages 93–94

77

Student Text page 140
Activity Manual page 95

Objectives
- Recall concepts and terms from Chapter 5
- Apply knowledge to everyday situations

Review
- Materials for the Chapter 5 Test will be taken from Student Text page 140 and Activity Manual pages 83, 85–86, 88, 90, and 95. You may review any or all of the material during the lesson. Questions similar to the ones in Thinking It Through, Activity Manual page 95, may appear on the test.
- You may choose to review Chapter 5 by playing the review game or a game from the Game Bank on the Teacher's Toolkit CD.

Activity Manual
Review, page 95
This page requires written responses to application questions. Page 96 has been left blank intentionally.

78

Objective
- Demonstrate knowledge of concepts taught in Chapter 1

Assessment
Chapter 5 Test

CHAPTER SUMMARY

Words to Know

atmosphere	front	thunder
air	global winds	tornado
air pressure	local winds	hurricane
altitude	precipitation	weather forecast
weather	humidity	meteorologist
wind	thunderstorm	
air mass	lightning	

Key Ideas
Describe the relationship of altitude and air pressure
Identify the layers of the atmosphere
Describe characteristics of the troposphere
Describe what ozone does
Identify characteristics of air masses
Identify cold, warm, and stationary fronts
Identify examples of global winds
Explain how winds are identified
Describe air movements of different local winds
Identify the use of weather instruments
Identify types of precipitation
Identify what clouds form from
Identify and describe different kinds of clouds
Compare a weather watch and weather warning
Identify why learning about weather is important

140

Review Game
Blow, Wind, Blow
Materials: masking tape, Ping-Pong ball

Make two lines with masking tape on the floor about 5 m apart. One line represents Europe, and the other line represents North America. The space between the lines represents the Atlantic Ocean. Place a Ping-Pong ball (the "ship") in the center of the ocean.

Divide the class in half. One team is the prevailing easterlies and the other is the prevailing westerlies. Alternate questions between the two teams. Each correct answer earns the student a chance to provide the ship with one gust of wind (one blow) to move it toward the opposite shore. The first team to get the ship to the other continent wins.

After one team has won, direct the teams to switch continents and represent the other type of wind. Continue asking questions as time allows.

Variation: Draw vertical lines about 2 m apart on the chalkboard. Label one line *Europe* and the other line *North America*. Draw additional lines to divide the space between Europe and North America into 10 equal spaces. Each correct answer represents a gust of wind that allows the team to move their marker (or ship) one space closer to the opposite shore.

Biomes 6

WORLDVIEW THINKING

Janaiya did a project about land use in the Mississippi River delta. She explained how the delta formed from sediment being deposited at the mouth of the river. There are many marshes and wetlands with a wide variety of plant and animal life on the delta. She told how the environment that the wildlife depends on has changed because of the construction of levees and the projects of some oil companies. Unfortunately, these changes have stopped new sediment from being deposited. Parts of the river delta are slowly disappearing.

Janaiya wants to protect the river delta and make sure that the animals, birds, and fish have the food they need. To do this she wants to get rid of the oil companies and levees. Some students agree with Janaiya. Other students disagree because they live in areas protected by levees or have parents who work for the oil company.

Consider

- What problems is Janaiya trying to address?
- What kinds of solutions might be able to meet the needs of the people as well as the wildlife?

141

Background

Weblinks

The BJU Press website offers additional information and links you may find helpful throughout the chapter.
bjupress.com/resources/science

Chapter Preview

Preview and prereading activities may include using a K-W-L chart, a probe, or an anticipation guide.

Chapter Photo

The photo on page 141 shows a misty forest. The miniature scientist is looking at a hologram, or three-dimensional image, of a section of the earth.

Objectives

- Appreciate the effect of human intervention on a wetland biome
- Apply the Bible's teaching of stewardship of creation to biomes
- Generate possible solutions to the concerns about destroying or using biomes

Introduction

- Discuss characteristics of the area you live in. Include things such as land features, weather, and the kinds of plants and animals native to the area.
- Choose several volunteers to describe places that they have traveled to.
- Discuss similarities and differences between the places.

Preparation for Reading

- Direct the students to read the introduction to find the purpose of Janaiya's (juh NA yuh) project.

Teach for Understanding

- Guide the students to apply God's truth to manage biomes in God's world. Identify the problems Janaiya is trying to address. Display a chart for cause and effect. Direct the students to identify the causes and effects that are presented. Suggestion provided.

Cause	Effect
levee construction	environment changes
oil company projects	environment changes
environment changes	stop new sediment
less sediment	delta disappearing

- Identify Janiaya's solution and discuss the pros and cons of doing it.

  Is it a good thing to take care of the property and animals? yes Explain your answer. God has told us to be good stewards of things He has put under our authority (Genesis 1:28).

  Is it good to build levees to control flooding and drill for oil? yes

  How do you know? Levees control flooding to protect homes, businesses, and lives. Oil is a resource that has many uses and meets many needs.

  Do you think only one good thing has to be chosen? no Explain your answer. Guide the students in concluding that people should work together to come up with solutions that have some benefit for everyone involved.

Activity Manual

Preview, page 97

The Looking Ahead page is intended to assess the student's prior knowledge before beginning the chapter.

80

Student Text pages 142–45
Activity Manual page 98

Objectives

- Differentiate between a biome and the biosphere
- Identify climate as a major influence on land biomes
- Describe basic characteristics of the tundra
- Name some ways that animals and plants survive on the tundra

Materials

- globe or world map showing latitude

Teacher's Toolkit CD

- Visuals 6.1–6.2: *Land Biomes of the World; Biome Characteristics*

Vocabulary

- biosphere
- biome
- climate
- permafrost

Introduction

In how many places have you lived? Did each place have the same kind of weather? Answers will vary.

- Explain that the weather can be very different at different places on the earth. Some places are hot; other places are cold.
  Where on the earth do you think that the weather is very cold? Possible answers: near the poles, high on mountains

Today we will identify and describe the coldest section of the earth.

Preparation for Reading

- Display the word *biome*. Pronounce the word together.
- Direct the students to mentally make questions using the headings, bold words, and captions.
- Instruct them to read silently pages 142–45 to find the answers to these questions.
  Why is climate important to a biome?
  Why do some plants on the tundra have red leaves?

Teach for Understanding

What is the scientific name for all the different areas of the earth where life can exist and the living organisms that live there? the biosphere

Why does every living creature have exactly the right living conditions that it needs? God planned it that way.

What is a biome? a large area of the earth in which plants and animals share a similar environment

What are some examples of biomes? deserts, forests, prairies, oceans

How is a biome related to the biosphere? A biome is a section of the biosphere.

What is one reason that creationary scientists work with biomes? to exercise dominion and wise stewardship over the earth as God commanded

When God created the earth, He made many kinds of plants and animals. Just as there are many types of these plants and animals, there are many places where they live. Scientists refer to all of the areas where life can exist and the living organisms that live there as the **biosphere**.

The biosphere is very large, though. A person could not study all of it. So scientists divide it into sections called biomes. A **biome** is a large area of the earth in which plants and animals share a similar environment. A biome might be a desert, a forest, a prairie, or even an ocean. These smaller sections are easier for a scientist to study as he makes his observations and tries to solve problems. One reason a Creationary scientist works with biomes is to exercise dominion and wise stewardship over the earth as God commanded.

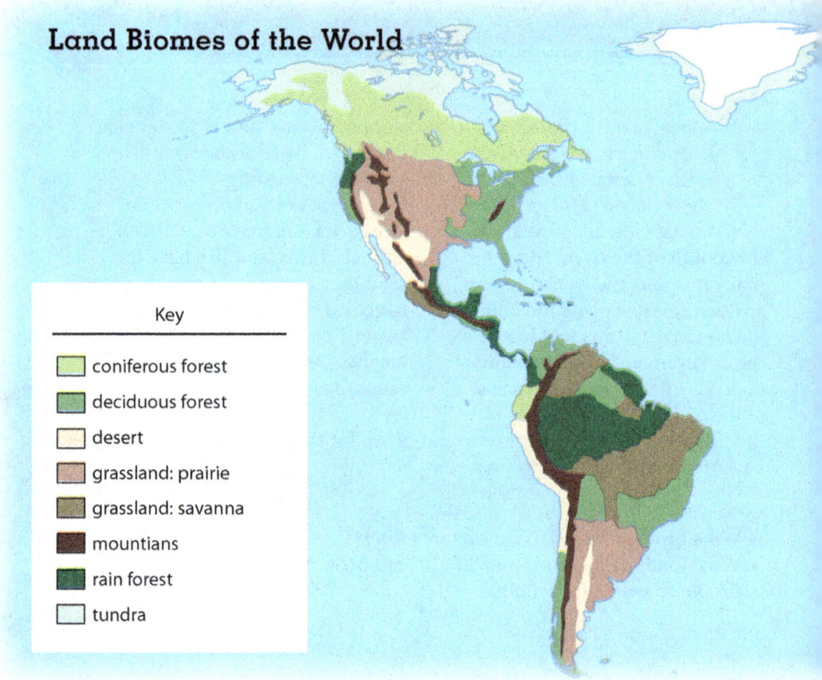

142

Background

Biome Classification

Biomes are only a general way to classify areas of the earth. The types of divisions and their criteria vary greatly. Some classifications include the water biomes and others include additional land biomes such as the chaparral. Different names are often used to describe the same biome.

Some biomes are classified by climate. Others are classified by region, zone, or latitude. Classifications by latitude include names such as temperate forests, temperate grasslands, tropical forests, tropical grasslands, and polar deserts.

Organisms in Biomes

A biome is usually known for specific animal and plant life, but many organisms are found in more than one biome.

Biome Boundaries

Biomes often gradually merge into each other. These transition areas are called *ecotones*. The boundaries of a biome can change or be affected by human activity or changes in weather patterns and air currents.

Land Biomes

Each land biome is influenced by its climate. The **climate** is the usual weather of a region over a long period of time. It includes the temperature and the amount of precipitation in an area. The climate determines what types of plants and animals can live in a biome.

Scientists have divided the earth's land into several major biomes. These land biomes can be very different. The tundra around the Arctic Circle is cold and frozen. But the tropical rain forests near the equator are very hot and humid.

The Northern Hemisphere has more biomes than the Southern Hemisphere does. The main reason for this is that there is more land in the Northern Hemisphere. It stretches from the equator all the way to the Arctic Circle. Most of the land in the Southern Hemisphere is in or near the tropics. Because the conditions do not vary much, there are only a few types of biomes.

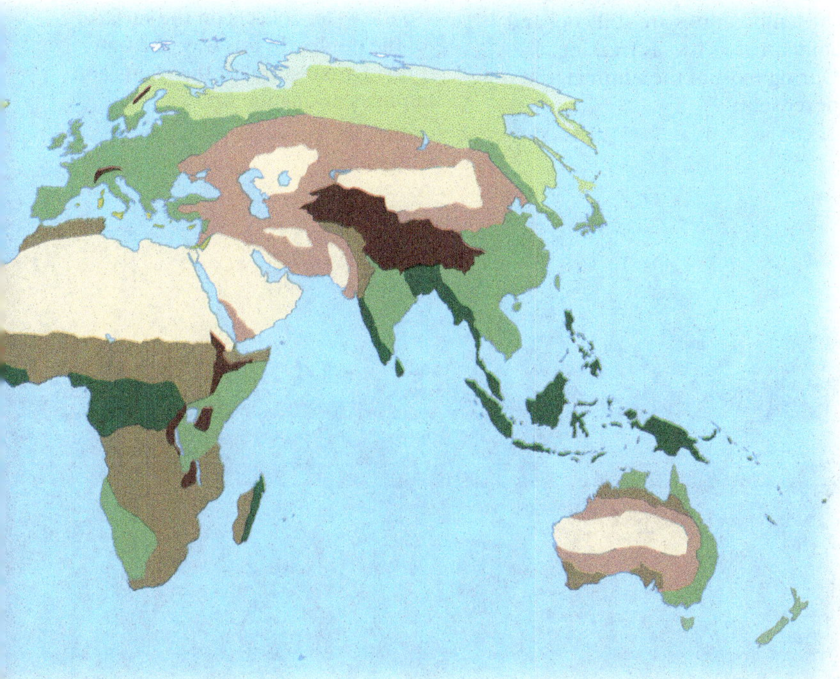

143

What is the climate? the usual weather of a region over a long period of time

Why is climate important to a biome? The climate determines the types of plants and animals that can live in a biome.

What might happen to a plant or animal that is moved from its natural biome? It might have difficulty surviving in a new place.

In which hemisphere are most biomes found? the Northern Hemisphere

Why are there more biomes in the Northern Hemisphere than in the Southern Hemisphere? The Northern Hemisphere has more land.

- Review the names and locations of places such as the Arctic Circle, equator, Northern Hemisphere, Southern Hemisphere, and tropics using the *Land of Biomes of the World* visual and the globe.
- Allow the students to study the map key on the *Land Biomes of the World* map in their books.

Maps that name biomes and show their locations can vary. For *SCIENCE 5*, the emphasis is on the general locations rather than on the specific locations of the biomes.

Is the tundra located near the North Pole or the equator? the North Pole

What type of forest is usually found near the equator? a rain forest

What are the other two types of forest? the coniferous forest and the deciduous forest

Background

Scientist Wladimir Köppen

Wladimir Köppen (1846–1940) was born in Russia to German parents. He studied botany in St. Petersburg and received a PhD at Heidelberg for his studies on the effect of temperature on plant growth. Köppen was a meteorologist, climatologist, and botanist. In the early 1900s, he developed a system to classify climates. This classification was based on temperature and precipitation. He used letters to identify major climate zones. With a few modifications, some of his classifications are still used to classify biomes.

- A chart organizer is on Activity Manual page 98. You may choose to display the copy of the organizer from the Teacher's Toolkit CD.
- The organizer provides a place for the students to write the important characteristics of each biome. You may guide the students in making notes in the chart as you teach or allow them to complete the page at the end of the lesson. The organizer will be used again in Lessons 81–82.
- Point out the topics listed across the top of the chart: *Climate*, *Location*, *Interesting Facts*, and *Common Animals and Plants*.

Which biome is the coldest? the tundra

Where is the tundra located? in the northern latitudes around the Arctic Circle

What are latitudes? Latitudes are the imaginary lines that run east and west around the earth.

- Show the students that the northern latitudes are near the Arctic Circle on the *Land of Biomes of the World* visual.
- Ask volunteers to point out the tundra areas on the visual.

What are the two main seasons in the tundra? winter and summer

According to the calendar, the seasons in the Northern Hemisphere begin in March, June, October, and December. Do you think that the term *seasons* as it is used with the tundra refers to these "calendar seasons"? no Explain your answer. Calendar seasons last for only three months. The tundra winter is described as lasting for nine months, and the tundra summer as lasting for less than two months.

For about how many months of the year are temperatures in the tundra below freezing? at least nine months

What do we mean when we say that the temperature is below freezing? The temperature is below 0 °C (32 °F).

What is unusual about the sunshine on the tundra? For most of the summer, the sun never sets. For part of the winter, the sun never rises.

Why do you think that the tundra has times when the sun never sets or never rises? because of the tilt of the earth and the angle at which sunlight hits the earth's surface

- Use the globe to demonstrate how the tilt of the earth affects the amount of sunlight in the tundra. Refer to the *How Rays of Sunlight Hit the Earth* demonstration in the next lesson if needed.

What is permafrost? soil that is permanently frozen

Can trees grow on the tundra? no Why? Their roots cannot get through the permafrost.

Tundra

The coldest biome is the tundra. This biome is found in the northern latitudes around the Arctic Circle. Winter and summer are the two main seasons in this biome.

The tundra has long, cold winters, with temperatures that are far below freezing. In fact, the tundra stays below freezing for at least nine months of the year. For part of the winter, the sun never rises.

The summer is the growing season. But in the tundra it is very short—sometimes less than two months. And, even during the summer, the temperatures are still cool and sometimes as low as freezing. During most of the summer, the sun never sets.

Because of the cold, most of the ground stays frozen all year long. This permanently frozen soil is called **permafrost**. A thin layer of soil on the surface thaws a little during the summer. However, the permafrost below is still frozen. Trees cannot grow on the tundra because their roots cannot get through the frozen soil.

The tundra is not just cold, though. It is also very dry. Only a small amount of precipitation falls during the year. Most of this precipitation is in the form of snow. When the snow melts in the summer, the permafrost is too frozen to soak up the moisture. So the water collects on the surface of the tundra. In this way, the water provides moisture for the plants and animals during the summer.

Tundra in the summer in Denali National Park, Alaska

144

Background

Tundra

The term *tundra* is derived from a Finnish word, *tunturia*, meaning "treeless plain" or "barren land." Very few trees grow on the tundra. Most trees need deep root systems, which cannot develop in Arctic tundra areas because of the permafrost. The tundra covers a large area of land in the regions around the North Pole. The South Pole, Antarctica, has some small tundra-like areas. However, Antarctica is much colder than the Arctic Circle. The snow and ice covering the land in Antarctica do not melt. For this reason, some scientists do not consider Antarctica to be a true tundra.

Permafrost

When the temperature of the ground remains below freezing for two or more years, the ground is considered to be permafrost. Cold air temperatures do not necessarily indicate permafrost because ground and air temperatures may not be the same. Permafrost can be found in the tundra and farther south into parts of the coniferous forest.

Tundra Plants

Some plants are able to live in the tundra because they can grow beneath the snow. They can carry on photosynthesis even in cold temperatures. Once summer begins, the plants can grow and quickly produce flowers and seeds.

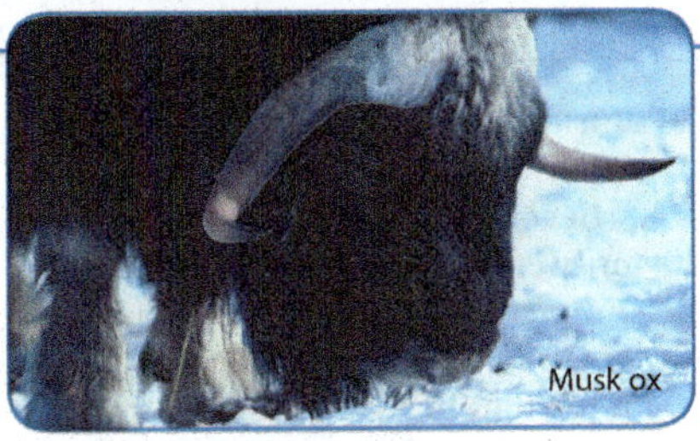
Musk ox

God designed some animals and plants to be able to keep warm even in cold weather. One tundra animal, the musk ox, has a double coat of hair. This animal has a thick undercoat of soft fur and a thick overcoat of coarse, straight hair. The undercoat provides warm insulation during the winter months. During the warm months, the musk ox sheds its undercoat, but it grows back again in time for winter. The outer hair is so long that it reaches almost to the ground. The hair resists water and keeps the musk ox from getting wet and cold.

The woolly lousewort has "hair" that helps trap heat in the cold tundra.

Plants also need to keep warm in order to grow. Some tundra plants are "hairy." The woolly lousewort, for instance, has fuzzy coverings on its buds and stems. This "hair" traps heat and keeps the surface of the plant warmer. Other plants have dark red leaves instead of green leaves. The red color helps the plants absorb more sunlight and keeps them warmer. Most plants on the tundra are also small and grow in groups. Being close to the ground protects them from the wind and allows the plants to absorb heat from the ground.

QUICK CHECK

1. How is a biome related to a biosphere?
2. How does the climate affect a biome?
3. What is permafrost?

Tundra in the winter in Alberta, Canada

145

Describe the climate of the tundra. cold and very dry; small amount of precipitation; most precipitation in the form of snow

When snow melts, why does the water remain on the surface of the tundra? It cannot drain through the permafrost.

How is this surface water used? to provide moisture for plants and animals

- Direct attention to the picture of the musk ox on page 145.

What is one special need that animals and plants on the tundra have? They must be able to keep warm in cold temperatures.

How has God designed the musk ox to survive life on the tundra? It has a double coat of fur. The undercoat provides insulation, and the outer coat is long and water resistant.

Why do you think that the water-resistant fur helps keep a musk ox warm? Possible answers: The musk ox would feel colder if its skin got wet. Moisture freezes in cold temperatures.

How does having "hairy" stems benefit a plant? Hairs trap heat and keep the plant warmer.

What is the name of a "hairy" tundra plant? the woolly lousewort

- Explain that "hair" on a plant refers to a fuzzy coating. It is not the same as hair on a person or animal.

How do red leaves help a plant stay warmer? Red leaves can absorb more sunlight than green leaves can.

Why are most plants on the tundra small? This size helps protect them from the wind and helps them absorb heat from the ground to stay warm.

- Direct attention to the photos of the tundra in the summer and in the winter at the bottom of both Student Text pages. Find the locations of the places on the globe. Denali National Park in Alaska is between Anchorage and Fairbanks.

Background

Tundra Animals

The animals that live in the tundra have many ways to keep warm during cold weather. Many tundra animals have small ears. This helps them lose less body heat. Thick fur provides insulation. Extra food is often stored as body fat to provide warmth and energy during the winter. Some animals, such as lemmings, line their underground tunnels with grass and moss. This lining helps keep the burrow at an even temperature and makes the burrow a little warmer than the outside air.

Tundra Insects

Many parts of the tundra are covered with insects during the warm season. But during the cold season, many of the insects keep from freezing by making a chemical in their bodies that works like antifreeze.

Alpine Tundra

Areas of alpine tundra found high up on mountains are similar, but not identical, to the Arctic tundra biome discussed on Student Text pages 144–45.

ANSWERS

1. A biome is a section of the biosphere.
2. The climate determines the types of plants and animals that can live in a biome.
3. permanently frozen soil

Activity Manual

Study Skill and Reinforcement, page 98

This page is used again with Lessons 81–82.

Objectives
- Describe basic characteristics of the coniferous forest
- Describe basic characteristics of the deciduous forest
- Differentiate between conifers and deciduous trees
- Name two ways that animals in the deciduous forest survive the changing seasons

Materials
- small branch of evergreen needles (leaves)
- leaves from a deciduous tree
- globe or world map

Teacher's Toolkit CD
- Visuals 6.1–6.2: *Land Biomes of the World; Biome Characteristics*
- Quiz 6-A

Vocabulary
- conifer
- deciduous

Two days have been allotted for this lesson. You may want to teach Lesson 81 the first day and use ideas from Lesson 82 on the second day. You may also incorporate the ideas from Lesson 82 into the text materials, making the lesson two days.

Introduction
- Display the two types of leaves.
  What are some differences that you notice between these leaves? Answers may include differences in size, shape, texture, and color.
  Which of these leaves belongs to a tree that you think could live in a very cold climate? the evergreen needles Why? Possible answers: The other leaves fall off trees in cold weather. Evergreen trees are often in pictures with snow.
- Explain that these leaves belong to trees that are often found in different forest biomes.

Preparation for Reading
- Display the words *coniferous*, *taiga* (TIE guh), and *deciduous*. Pronounce the words together.
- Direct the students to read silently pages 146–49 to find the answers to these questions.
  What kind of trees are most of the trees in the coniferous forest?
  What does *deciduous* mean?

Teach for Understanding
- Refer to the globe and the *Land Biomes of the World* visual as you discuss the forest biomes. Refer the students to the *Land Biomes of the World* map on Student Text pages 142–43. Continue use of the *Biome Characteristics* organizer on Activity Manual page 98.

Coniferous Forest
The coniferous forest is the largest land biome. This biome, also called the *taiga*, is located just south of the tundra. The biome is so large that it stretches across most of Canada and into some of the northern United States. It also covers large parts of Europe and Asia.

The coniferous forest has two main seasons: summer and winter. Spring and autumn do occur. They are so short, however, that they are barely noticed.

The winters are long and cold. The temperatures usually stay below freezing for at least six months of the year. Winter days are also short. During part of the winter the sun is up for only four to six hours!

In contrast, the sun may shine for twenty hours a day during the summer. The growing season lasts longer in the coniferous forest than in the tundra. The summer is still short and cool, however. Most of the year's precipitation occurs during the summer.

Circumpolar map of the tundra and the taiga

Coniferous forest in Ontario, Canada

Science Misconceptions
Identifying a Coniferous Forest
Conifers grow in places other than the coniferous forest. A group of conifers growing together does not necessarily mean that the biome is a coniferous forest. The coniferous forest biome is also characterized by its latitudinal location and specific climate conditions.

Background
Coniferous Forest Names
The northern coniferous forest has many different names. It is often called the *taiga* or the *boreal forest*. The term *taiga* is a Russian term for a swampy, marshy forest. The term *boreal forest* is most often used in North America. *Boreal* means "northern." This biome is also nicknamed the "spruce-moose" biome. Spruce is a common tree found in the forest, and the moose is one of the most common animals.

FANTASTIC FACTS

The taiga has many ponds, bogs, and lakes. One type of bog, a *muskeg*, can be found in some areas of the taiga where water does not drain through the soil. This swamplike ground may appear firm because of the moss, grass, or trees growing on it. But, underneath, the land is soggy and wet. In some areas in Manitoba, Canada, stepping on certain parts of the ground causes the trees to move.

Snowshoe hare

Most of the trees in the coniferous forest are **conifers**, or cone-bearing trees. Some of the most common conifers are spruce, fir, and pine trees. Many conifers are evergreens. These trees do not lose their leaves during the winter. This enables evergreens to make food even during the long winter season.

Winter temperatures in the coniferous forest are usually so cold that the snow does not melt between storms. So the trees must be able to support a lot of heavy snow. Otherwise, the weight of the snow would break the tree branches.

Many conifers are shaped like cones. This shape helps some of the snow to slide off the tree without causing damage. Other trees are able to bend with the weight of the snow. Some trees can bend so much that they become almost flat to the ground. But when the snow melts, the trees stand straight again.

God provided some animals with the ability to adapt to extreme environments. For example, the snowshoe hare's fur changes to brown in the summer and then to white in the winter. This color change helps camouflage the hare. The white fur is not only for camouflage, however. Inside the white hairs are air-filled spaces. These spaces help insulate the hare and keep it warmer during the winter. The hare also has large back feet that act like snowshoes, helping it stay on top of the snow. Stiff hairs on the bottom of its feet form fur pads that keep the hare's feet warm and insulated.

147

Demonstration

How Rays of Sunlight Hit the Earth

Materials: globe, flashlight

Direct a student to hold the globe so that it is tilted at a slight angle. If the globe you are using is already tilted, you could just place the globe on a desk or table. Darken the room and stand about 1 m away from the globe. Holding the flashlight parallel to the floor, shine the flashlight toward the equator.

If the flashlight represents the sun, which part of the earth receives the most sun? the equator

Is the light as bright at the poles? no

Rotate the earth on its axis while continuing to shine the light toward the equator. Locate the areas that have tundra and coniferous forest biomes.

As the earth turns, look at the light near the North Pole. Does the light reach every area during each rotation? no

How do you think the slant of the light and its brightness affect the growing seasons for the tundra and the coniferous forest? Possible answers: The growing season would be shorter. Less light from the sun means that the climate is cooler.

Where is the coniferous forest located? just south of the tundra

What is another name for the coniferous forest? the taiga

- Guide the students in making comparisons between characteristics of the climate of a biome, its location on the earth, and the climate region that it is in.

  How does the size of the coniferous forest compare with the size of other biomes? It is the largest.

  What are the two main seasons in the coniferous forest? winter and summer

  Does the coniferous forest have spring and autumn seasons? Yes, but they are very short.

  Is the coniferous forest a cold biome or a warm biome? a cold biome

- Refer to the diagram on Student Text page 118. Discuss how the sun's rays warm the tropical and temperate latitudes more than the poles. The slant at which the sun's rays strike the earth's surface near the poles spreads the heat over a greater area, resulting in a cooler climate.

  Does the coniferous forest receive most of its precipitation as rain or snow? rain How do you know? Most of its precipitation falls during the summer months.

  What type of tree is most common in the coniferous forest? conifers

  What are conifers? trees that have cones

  What term describes trees that do not lose their leaves during the winter? evergreen

  How does the cold climate affect snowfall in the coniferous forest? Winter temperatures are so cold that snow does not melt between storms. This causes the snow to pile up.

  How are most conifers shaped? like cones

  How does the shape of a conifer help it during the winter? The snow can slide off without breaking the branches.

  How did God design some of the other trees to handle the weight of snow? They bend with the weight of the snow and then stand straight again when the snow is gone.

  What special characteristics does the snowshoe hare have that help it survive during the winter? Possible answers: Its white fur provides camouflage, and the air-filled spaces in its fur provide insulation for warmth. Its large back feet act like snowshoes, and the stiff hairs on the bottom of its feet help keep its feet warm.

  How do the changing colors of the snowshoe hare's fur help camouflage it? The fur helps the hare to blend in with its surroundings in both summer and winter for protection from enemies.

- Discuss *Fantastic Facts*.

  What is a muskeg? a type of bog that is found in the taiga

  Why does the surface of a muskeg appear firm? Moss, grass, or trees can grow on it.

  What happens in some muskegs when a person walks on them? The trees move.

Where are deciduous forests found? along the eastern coast of North America and in other parts of the earth such as Europe and Asia

Would deciduous forests most likely be found in a polar region, a temperate region, or a tropical region? a temperate region

In each hemisphere, where is the temperate zone in relation to the polar region and the tropical region? It is between them.

How did the deciduous forest get its name? Most of the trees there are deciduous.

What are deciduous trees? trees that lose their leaves in the winter

Are deciduous forests classified as a wet biome or a dry biome? a wet biome Why? A deciduous forest receives more rain than any biome except a rain forest.

How many seasons do deciduous forests have? four

Which biome has a longer growing season, the coniferous forest or the deciduous forest? the deciduous forest

Why does the deciduous forest have a longer growing season than the coniferous forest? The deciduous forest receives more direct sunlight, causing it to have a warmer climate, more rainfall, and a longer growing season than the coniferous forest.

How long does the growing season in the deciduous forest last? about half of the year

Other than trees, what kinds of plants are commonly found in a deciduous forest? shrubs, berry bushes, wildflowers

On the globe or world map, locate the countries and continents where deciduous forests are found. Deciduous forests in the Northern Hemisphere are found in the eastern United States, Great Britain, France, central Europe, and in parts of Russia, China, Korea, and Japan. The Southern Hemisphere has deciduous forests in South America, Australia, and Africa. The deciduous forests of different countries have different species of trees.

Deciduous Forest

Deciduous forests line the eastern coast of North America. They are also found in other parts of the earth, such as in Europe and in Asia. This biome gets its name from the trees that make up the forest. Most of these trees are **deciduous**, meaning that they lose their leaves in the winter. Deciduous trees usually have large, wide leaves.

The weather in the deciduous forest is not constant but changes with the seasons. Spring is usually warm and rainy. Summer is hot and humid. Autumn is cool and crisp. Winters are often cold and snowy.

The deciduous forest is a wet biome. It receives precipitation throughout the whole year. In fact, a deciduous forest gets more rain than any other type of biome except a rain forest.

The warm growing season lasts about half of the year. This longer time allows the biome to get warmer than those biomes with shorter summers. This longer growing season provides time for plants and trees to grow. Many shrubs, berry bushes, flowers, and other plants cannot survive in colder biomes. But these plants flourish in the warmer climate of the deciduous forest.

The plants and trees in a deciduous forest must change with the seasons. Deciduous trees lose their leaves in cold weather. If deciduous trees kept their broad leaves throughout the winter, the water in the leaves would freeze. This could damage the tree.

Deciduous trees in Quebec, Canada

148

Background

Autumn Leaves

Plants need sunlight to produce the green-colored chlorophyll. The seasonal changes, especially decreased daylight, signal deciduous trees to begin growing a layer of cells at the base of each leaf. This layer limits nutrients and the leaf's production of chlorophyll. As the green decreases, other colors within the leaf are seen. The layer of cells at the base of the leaves gradually dries, and the leaves separate and fall from the trees.

Hibernation

Hibernation is discussed in Chapter 7, *Interactions in an Ecosystem*.

A cardinal is a common bird in the deciduous forest.

During the summer, deciduous tree leaves use sunlight to make food. Some of this food is used for growing. But some is stored in the trees' roots. In autumn, the days get shorter, and the temperatures become cooler. These changes cause deciduous trees to block nutrients and water from the leaves. The leaves can no longer make chlorophyll, so their green color begins to fade. The other colors that had been hidden by the green begin to show. Leaves fall to the ground, decay, and add nutrients back into the soil. In the spring the trees use their stored food energy to form new leaves.

Some birds, butterflies, and large mammals move to warmer areas for the winter. The animals then return to the forest in the spring when it is warm again.

Other animals remain in the forest year-round. Some, such as the deer, grow thick winter coats of fur. They shed these winter coats when the temperature gets warmer. Other animals, such as the woodchuck, hibernate.

A deer grows a thick winter coat to stay warm in the deciduous forest during cold weather.

QUICK CHECK

1. What is the largest land biome?
2. How are the trees in deciduous forests different from the trees in coniferous forests?

149

81
82

What would happen if broad-leaved trees kept their leaves through the winter? The water in the leaves would freeze. This could damage the tree.

What conditions in autumn cause deciduous trees to block nutrients and water from their leaves? the shorter days and cooler temperatures

What causes the leaves of deciduous trees to change colors? The leaves can no longer make chlorophyll. This causes the colors hidden by the green chlorophyll to appear.

How do deciduous trees improve soil in the forest? Fallen leaves decay and add nutrients back into the soil, helping new plants to grow.

Name some animals that may move to another biome in the winter. Possible answers: some birds, butterflies, large mammals

How do some animals that remain in the forest year-round survive the changing seasons? Possible answers: grow thick fur, hibernate

ANSWERS

1. the coniferous forest, or taiga
2. Most of the trees in a deciduous forest lose their leaves every winter. Most of the trees in a coniferous forest are evergreens.

82

- The *How Rays of Sunlight Hit the Earth* demonstration, the *Reveal the Colors in Leaves Masked by Chlorophyll* activity, and the *Make Leaf Rubbings* activity found at the bottom of the pages in Lessons 81–82 may be incorporated into the lesson.
- Suggestions for additional activities for this two-day lesson are on the Teacher's Toolkit CD.

Activity

Reveal the Colors in Leaves Masked by Chlorophyll

Materials: paper coffee filter, assorted green leaves, baby food jars, rubbing alcohol or acetone nail polish remover, small paper clips, toothpicks

Cut the coffee filter into 2.5 × 9 cm strips. Rub a green leaf against the middle of one filter strip until the leaf makes a green smudge. Pour a small amount of rubbing alcohol or nail polish remover into a jar. Hook a paper clip through one end of the filter, and hang it from a toothpick balanced across the top of the jar so that the lower end of the filter strip touches the liquid. Repeat with the other leaves.

Set the jars aside and check on them every 15 minutes. As the liquid is absorbed, other colors in the green smudges will be revealed. These different bands of colors are pigments other than chlorophyll that also function in photosynthesis.

Make Leaf Rubbings

Materials: deciduous leaves, newspaper, white paper, pencil or crayon

Collect leaves from several different deciduous trees. Place one leaf on newspaper with the underside of the leaf up. Cover the leaf with a thin sheet of white paper. Rub a pencil or a crayon lightly over the area of the paper that is over the leaf. Label the leaf rubbing with the type of tree the leaf came from. You may choose to cut out the leaf rubbings and use them to design note cards, stationery, or bookmarks.

Activity Manual

Study Skill and Reinforcement, page 98

Review, pages 99–100

These pages review Lessons 80–82.

Assessment

Quiz 6-A

The quiz may be given any time after completion of this lesson.

83
84

Student Text pages 150–53
Activity Manual page 101–2

Objectives

- Describe basic characteristics of grasslands
- Compare and contrast prairies and savannas
- Name ways some savanna grasses and trees survive the dry season
- Describe characteristics that all deserts have in common
- Name some ways that desert animals and plants survive the extreme temperatures and dryness

Materials

- globe or world map

Teacher's Toolkit CD

- Visuals 6.1, 6.3: *Land Biomes of the World*; *Biome Characteristics*
- Quiz 6-B

Vocabulary

- nocturnal
- succulent

Two days have been allotted for this lesson. You may want to teach Lesson 83 the first day and use ideas from Lesson 84 on the second day. You may also incorporate the ideas from Lesson 84 into the text materials, making the lesson two days.

Introduction

Where would you usually find animals such as prairie dogs and bison? Answers will vary but should include the prairies of North America.

Other than zoos and nature parks, where would you usually find animals such as lions and zebras? Answers will vary.

One type of biome can be a home for both prairie dogs and lions.

Preparation for Reading

- Display the words *nocturnal* and *succulent*. Pronounce the words together. Direct the students to read silently pages 150–53 to find the answers to these questions.

  What are some differences between prairies and savannas?

  How are some animals able to live in desert places?

Teach for Understanding

- A second organizer for biome characteristics is on Activity Manual page 102. You may choose to display the copy of the organizer from the Teacher's Toolkit CD.
- You may guide the students in working through the chart as you teach or allow them to complete the page at the end of the lesson. The organizer will be used again in Lessons 86–87.
- Refer to the globe and the *Land Biomes of the World* visual as you discuss grasslands and deserts. Direct the students to compare the climate regions with the *Land Biomes of the World* map on Student Text pages 142–43.

Grasslands

Grasslands are usually found between deserts and forests. Some grasslands are flat. Others have gently rolling hills. The main *vegetation*, or plant life, is grass. Grasslands also have wildflowers and small bushes but very few trees.

Prairies

The prairie is one kind of temperate grassland. Temperate areas are located between the polar and tropical regions. Much of the central area of North America is covered by wide, grassy prairies. Prairies have four regular seasons. The winters are usually cold, and the summers are warm.

Most precipitation on the prairie falls in late spring and early summer. The amount of rainfall is usually low to moderate. A prairie receives about the same amount of precipitation as a coniferous forest does. However, water evaporates faster on the prairie than in the forest. This means that prairie plants do not have as much water available as the trees in a coniferous forest do.

Prairies have a wide variety of grasses. Most prairie grasses, though, do not look like the grass growing on people's lawns. Instead, prairie grasses vary in height. Some, such as buffalo grass, are short. Others, such as prairie cord grass, can grow over 3 m (10 ft) tall. The wheat, corn, rice, and rye that we use as food are also types of grasses.

The roots of prairie grasses usually are matted together and spread deep underground. This system prevents seeds from trees and other plants from taking root. It also enables the grasses to grow again even after a fire or after grazing animals have eaten them.

Many kinds of animals live on the prairies. However, prairies are very open spaces, so animals can easily be seen by other animals that could eat

SCIENCE & THE BIBLE

In Psalm 103:15–16, the Bible compares man's life to the grass. Both grow quickly and then are gone. But God's mercy lasts forever. The next verse says, "But the mercy of the Lord is from everlasting to everlasting upon them that fear him."

Prairie in Manitoba, Canada

150

Background

Names of Grasslands

Grasslands are known by a variety of names. Most grasslands are classified as either temperate or tropical. Temperate grasslands are classified as tall grasses (prairies) or short grasses (steppes). Other names for temperate grasslands include veldts and pampas. The grasslands region in North America is referred to as the Great Plains. Tropical grasslands are called savannas. Some types of savannas include woodland savannas, shrub savannas, pine savannas, and acacia savannas.

Grassland Plants and Trees

There are at least 9,000 different kinds of grasses found on prairies. Common prairie grasses include a variety of bluestem grasses, blue grama, switch grass, Indian grass, and buffalo grass.

Not all grassland plants are grasses. Some are soft-stemmed, broad-leaved, flowering plants. These plants include thistles and wildflowers such as clovers, sunflowers, and black-eyed Susans. A prairie may have a few trees, usually located near streams and rivers. Some common prairie trees include the cottonwood, elm, ash, and box elder. Trees such as the baobab and the acacia, are scattered across savannas, growing alone or in small groups.

them. To protect themselves, some animals, such as prairie dogs, hide in underground holes. Snakes and birds often hide in the grass. Other animals simply run fast to escape. The pronghorn antelope, for instance, can run as fast as 96 km/h (60 mi/h).

Savannas

Tropical grasslands are called savannas. They are located in tropical areas of Africa, Australia, South America, and India. The vegetation is similar to that of a prairie. Savannas have a much warmer climate, though. Because savannas are close to the equator, they are hot all year.

Savannas have only two seasons—dry and wet. Savannas receive anywhere from 64 to 152 cm (25 to 60 in.) of rain. Most of that rain falls during the wet season. During the dry season, these tropical grasslands usually receive less than 10 cm (4 in.) of rain.

This long dry season makes it hard for plants. The grasses and trees that grow on a savanna have to be able to survive through the long dry season. Grasses grow quickly during the wet season. During this time, they store food and water in their roots. Later, during the dry season, the grasses turn brown and appear to die. But the roots are still alive. When the wet season returns, the grasses will grow again.

One of the biggest differences between savannas and prairies is the number of trees each can support. Savannas have more trees than prairies do. These trees have special features that allow them to survive the dry season. For instance, the baobab tree, found on the African savanna, stores water in its trunk during the wet season. During the dry season, the tree depends on the stored water.

The baobab tree loses its leaves during the dry season. Some people think that it looks as if it has been planted upside down.

Serengeti Plain, Tanzania, Africa

151

Where are grasslands usually found? between deserts and forests

Are all grasslands alike? no

- Share the *Names of Grasslands* background information.

What is the main vegetation of grasslands? grass

Where are prairies found? in the central area of North America

Why are prairies called temperate grasslands? They are located in the temperate zone, or the area between the polar and tropical regions.

How many seasons do prairies have? four

What other biome receives about as much rain as the prairie does? the coniferous forest

Are the temperatures on the prairie warmer or colder than in the coniferous forest? warmer Why? The prairie is closer to the equator.

If the prairie and coniferous forest get about the same amount of rain every year, why do you think there are so few trees on the prairie? Possible answer: Water evaporates faster when it is warm, so grasslands have less water available for plants. Trees need more water than grasses do.

What are some common prairie grasses? Possible answers: buffalo grass, prairie cordgrass, wheat, corn, rice, rye

How do some grasses keep other plants and trees from growing on grasslands? They have deep roots that are matted together.

What are three ways that some prairie animals escape from animals that want to eat them? burrowing, hiding, and running

- Discuss *Science & the Bible*. Read Psalm 103:15–16. Discuss the frailty of mankind and the eternality of God's mercy.

Are savannas temperate or tropical grasslands? tropical

How are savannas different from prairies? Savannas have a warmer climate, have only two main seasons instead of four, and have more trees.

During which season does the savanna receive the most precipitation? the wet season

How do savanna grasses survive during the dry season? They store food and water in their roots.

How does the baobab tree get enough water during the dry season? It loses its leaves and uses the water stored in its hollow trunk during the wet season.

- Discuss the picture of the baobab tree.

Why do you think that losing its leaves helps the baobab tree survive the dry season? Possible answer: Leaves use some of the tree's energy and water. Not having leaves provides more water for other parts of the tree.

- The hollow trunks of baobab trees can be quite large. These trees have been used as shops, prisons, houses, barns, bus shelters, places of refuge, and tombs. At times, as many as forty people have fit inside a baobab tree.

Background

Sod Houses

Prairie grasses form a thick layer of matted roots and soil called sod. American pioneers settling on the prairie often built homes out of sod because trees were scarce. Sod was cut from the ground in chunks in the shape of bricks and set out to dry. The "bricks" were then stacked to form walls. Sod homes were warm in the winter and cool in the summer.

Waves of Grain

Katharine Lee Bates (1859–1929) wrote the poem "America the Beautiful" after her train trip to Colorado Springs. During the long trip, the train had stopped in various cities and states, and Katherine had eagerly observed many new and beautiful sights. When the train passed through Kansas, she noticed that the fields of amber-colored grain moved like waves in the ocean when the wind blew. This observation of "amber waves of grain" was included in her poem. The poem was later set to music and has become one of America's most loved patriotic songs.

Growing Seasons

A savanna's growing season is determined by the length of the rainy season. However, for temperate grasslands, temperature determines the length of the growing season.

83
84

Are all deserts alike? no Describe some different kinds of deserts. Possible answers: cold, snowy, hot, dry, sandy, rocky, gravelly

What do all deserts have in common? They receive very little precipitation.

About how much precipitation does a desert receive in one year? less than 25 cm of rain

What causes the temperatures in a desert to be so extreme? Deserts are dry.

How would more moisture help keep the temperatures more even? Water warms and cools more slowly than land does.

Where are hot deserts usually located? in the tropical regions, near the equator

Where are cold deserts usually located? between the tropics and the polar regions

- Compare hot and cold deserts.

How is the weather different for a hot desert than for a cold desert? A hot desert is warm year-round and just gets hotter in the summer. A cold desert has cool winters and warm summers.

What do you think causes the difference in climate for hot deserts and cold deserts? where they are located, how near they are to the equator

- Use the globe or world map to locate the deserts as they are discussed.

Which desert is the largest desert in Africa? the Sahara Desert

Is the Sahara Desert a hot desert or a cold desert? a hot desert Why? It is near the equator.

The Atacama Desert is one of the driest deserts in the world. In which country is this desert located? Chile

Where are most cold deserts located? between the tropics and the polar regions

What cold desert is found in China and Mongolia? the Gobi Desert

- Share with the students that deserts can be classified in different ways. Some classifications identify deserts by dominant climate conditions such as arid deserts, semiarid deserts, or cold deserts. Other classifications identify deserts by geographic locations such as subtropical deserts, coastal deserts, cool coastal deserts, cold winter deserts, polar deserts, or rain shadow deserts.

Deserts

What are deserts like? Do you think of hot, dry places when you think of deserts? Although some deserts are hot and dry, others are very different. Some are cold and covered with snow! Others have rocks and gravel instead of sand.

All deserts, though, have one thing in common. They receive very little precipitation. Most get less than 25 cm (10 in.) of rain each year. Because deserts are so dry, the temperatures are usually extreme. During the day the temperature might reach 38 °C (100 °F) or higher. But at night the temperature drops quickly. Sometimes it can even get below freezing.

Deserts can be found all over the world. Many deserts are located in tropical areas on either side of the equator. These hot deserts are warm year-round but get even hotter in the summer. The Sahara Desert, the largest desert in Africa, is an example of a hot desert.

Cold deserts have cool winters and warm summers. These deserts are usually located between the tropics and the polar regions. One cold, dry desert is the Atacama Desert in Chile. The Gobi Desert, another cold desert, is located in China and Mongolia. Though these are warmer regions, deserts can at times be found in very cold places. For example, most of Antarctica is considered a desert.

Great Basin Desert, Nevada

152

Background

Coastal Deserts

Coastal deserts, such as the Atacama Desert, are found on the western edges of continents near the tropics. Coastal deserts occur where cold ocean water meets warm air from the land. Coastal deserts are often foggy. The cold ocean water chills the air, causing the air to hold less moisture. This dry air moves inland, resulting in coastal deserts.

Polar Deserts

Polar deserts, such as those in Antarctica and Greenland, have long winters and short summers. These areas receive lots of snow during the winter. However, most of the water in polar deserts is frozen and is not available for plants and animals to use.

Rain Shadow Deserts

Some deserts, such as the Great Basin Desert, form next to high mountains. As warm, moist air moves up the mountains, the moisture cools and falls as precipitation. This side of a mountain, called the windward side, receives almost all of the moisture that is in the air. The dry air then moves over the mountain and warms as it moves down the other side, causing a desert to form from lack of precipitation. This side of the mountain is called the leeward side, or shadow of the mountain.

Sonoran Desert, Arizona

Animals demonstrate God's amazing design in living with extreme desert temperatures. The jackrabbit, for example, has large ears that help its body get rid of excess heat. Many desert animals are **nocturnal**, or active at night. During the day, they sleep. Some desert animals, such as kangaroo rats, burrow under the ground to stay cool. They come out to hunt and eat at night. Other animals, such as rattlesnakes, are most active at dusk and at dawn. Resting during the hot part of the day helps them conserve energy and water.

Desert plants can also survive in extreme temperatures. Some have the ability to store water. The saguaro cactus is able to store as much as 757 L (200 gal) of water in its stem. This water can provide enough moisture for the cactus for an entire year! A plant that can store water in its stem or leaves is said to be **succulent**.

Desert plants usually have either long, deep roots or a system of wide, shallow roots. Some plants have both. Long, deep roots allow plants to reach water located deep under the ground. Shallow roots spread out in many directions and can get water from a larger area of soil.

Plants lose water through the small pores on their leaves and stems. Because of this, many desert plants, such as cactus and aloe, have waxy surfaces that help keep water inside the plant. Some plants have small leaves or only a few leaves. Other plants remain as seeds until the rain comes. Then the plants quickly grow, flower, and produce new seeds while water is available.

QUICK CHECK

1. How are savannas different from prairies?
2. What is true about all deserts?

153

83–84

What are three ways that some desert animals handle the extreme desert temperatures? They may have large ears, be active at night, or be most active during dusk and dawn.

How do large ears help animals such as jackrabbits remain cooler? Large ears help the animal's body get rid of excess heat.

Would a nocturnal animal sleep during the day or during the night? during the day How do you know? Nocturnal means "active at night."

What does the term *succulent* tell you about a plant? that it is able to store water in its stem or leaves

About how much water can a saguaro cactus store in its stem? 757 L (200 gal)

What are two types of roots that desert plants have? long, deep roots or wide, shallow roots

How do these roots help plants get water? Long, deep roots can reach underground water sources. Wide, shallow roots spread out and allow the plant to get water from more soil.

What are some other ways that desert plants can decrease the amount of water they lose through their leaves? Possible answers: have waxy leaves; have small leaves or only a few leaves; only grow, flower, and produce seeds during times of rain

ANSWERS

1. Savannas have a warmer, drier climate than prairies do. Savannas have only two main seasons and prairies have four. More trees are found on savannas than on prairies.
2. All deserts receive very little precipitation.

Background

Scientist Friedrich Welwitsch

Friedrich Welwitsch (1806–72) was both an explorer and a botanist. He was born in Austria but moved to Portugal. The Portuguese government sent him to southern Africa to collect plants. While exploring the Namib Desert in 1859, he discovered an unusual desert plant that had only two leathery, straplike leaves, a stem, and roots. Later other people named the plant the *Welwitschia mirabilis* after Friedrich Welwitsch in recognition of his botanical work and because he found the plant first. The species part of the name, *mirabilis*, means "marvelous" or "wonderful." When Welwitsch left Africa, he moved to London and worked for the Natural History Museum until his death in 1872.

Activity

How the Saguaro Cactus Expands to Store Water

Materials: strip of paper about 1 × 20 cm, tape, paper plate, sand

Fold the strip of paper accordion style, and then tape the ends together. This represents your cactus. Place the paper loop on a paper plate. Pour sand into the paper loop, packing the sand down as it is poured in. The sand represents the water that the cactus would collect. Observe what happens to the folds in the "cactus" as you fill it with sand.

84

- The *How the Saguaro Cactus Expands to Store Water* activity found on this page may be incorporated into the lesson.
- Suggestions for additional activities for this two-day lesson are on the Teacher's Toolkit CD.

Activity Manual

Study Skill and Reinforcement, page 101

This page is used again with Lessons 86–87.

Review, page 102

This page reviews Lessons 83–84.

Assessment

Quiz 6-B

The quiz may be given any time after completion of this lesson.

Objectives
- Identify some characteristics of water-efficient plants
- Predict how waxy surfaces on plants affect water loss
- Relate the effectiveness of a petroleum-jelly coating on a sponge to the waxy surfaces on some leaves and stems

Materials
- container of lotion

Teacher's Toolkit CD
- Rubrics: *Help Prevent Water Loss!,* for each student

This activity will take several hours or overnight to complete.

Introduction
- Display the lotion.
  Why do people use lotion on their skin? Possible answers: to keep their skin from getting too dry, to trap moisture

  Losing too much moisture can cause skin to become dry and sore. Lotion often helps retain moisture or restore some of the lost moisture.

  Plants also lose moisture. But a plant does not put on lotion to prevent losing too much water. In this activity you will model one way that God designed some plants to lose less moisture through their leaves.

Preparation for Reading
- Direct the students to read Student Text pages 154–55 and Activity Manual pages 103–4 before beginning.

Teach for Understanding
- Discuss what stomata are and how plants use them.

Procedure
- Guide the students in forming their hypotheses.
- As needed, help the students measure the two sponges. Be sure that the bottoms of their jelly-coated sponges have not been coated. Provide a "drying place" in the classroom or in an outside area where the sponges will not be disturbed.
- Decide on the additional lengths of time needed for step 10. Choose times that fit your classroom schedule. You may want to have the sponges checked every 15 minutes, every 30 minutes, or every hour. You may choose to let the sponges sit overnight and have the students check them the next morning.
- As needed, help the students complete the line graph.
- Although the students should be able to see on the graph the difference between the two sponges, you may choose to have them subtract to find the difference in the mass at each time.

ACTIVITY

Help Prevent Water Loss!

Process Skills
- Predicting
- Measuring
- Making and using models
- Inferring
- Recording data

Part of studying a biome is finding out how the needs of living things are met in that environment. Understanding these needs and how resources are used can help people apply what they have learned.

All plants lose water through tiny pores in their leaves called *stomata.* But if a particular biome does not receive much precipitation, the plants in that biome could lose more water than they receive. So God made some plants to be very water efficient. These plants have characteristics that help them lose very little water.

Some plants, such as cactuses, have fewer stomata than other plants do. Other plants open their stomata only at night when temperatures are cooler. A waxy surface also helps some desert plants keep water inside.

In this activity you will use a model to discover how much an outer coating affects the amount of water that evaporates.

Materials
- 2 flexible, thin sponges
- rubber bands
- petroleum jelly
- 100 mL water
- metric measuring cups
- container
- waxed paper
- balance
- pie plate
- Activity Manual

Problem
How will coating the surface of a sponge affect the amount of water that evaporates from the sponge?

Procedure: Part 1
1. Complete the hypothesis in your Activity Manual.
2. Roll each sponge into a tube shape and fasten it with a rubber band. Arrange each rubber band so that the two sponges look the same.
3. Coat the sides and top of the first sponge with petroleum jelly. Make sure one end is not coated. Do not coat the second sponge.
4. Pour 100 mL of water into the container.

154

Background

Activity Helps
The petroleum jelly can be messy to work with. Provide paper towels or other materials to clean hands and surfaces that may get petroleum jelly on them.

The waxed paper is to help keep the balance clean. The waxed paper could also be used to hold the coated sponge as it is moved around.

Tongs may be used to move the coated sponge.

The container should be large enough that the rolled sponges can stand without touching each other.

All of the spaces on the chart do not have to be filled. However, to gain the benefit of the graphing exercise, at least three measurements are needed.

Cuticle
The name for the waxy covering on many leaves is the *cuticle.* The cuticle is most prominent in the thickest, shiniest leaves. Southern magnolias, camellias, and hollies are good examples. The cuticle causes water to bead up and roll off the surface, cleaning the leaf of debris and removing possible pathogens and chemicals that could harm the plant. It also prevents the evaporation of internal water. The cuticle is usually more prominent on the upper surface of a leaf since that part of the leaf is more exposed to the drying effects of the sun.

5. Stand the sponges in the water. Be sure that the coated sponge is standing on its uncoated end. Do not allow the sponges to touch each other. Allow the sponges to absorb water for three minutes.
6. Remove the sponges from the water. Place waxed paper on the balance. Measure and record the mass of each sponge.
7. Stand each sponge on its end in the pie plate. Place the plate in a sunny or well-lighted area.

Procedure: Part 2

8. Let the water evaporate from the sponges for one hour.
9. Measure and record the mass of each sponge.
10. Continue to measure and record the mass of each sponge at the times given by your teacher.
11. Graph the mass of each sponge for each given time.

Conclusions

- Did your results support your hypothesis?
- How did this activity model the way some plants survive?

Follow-up

- Place the wet sponges in a windy area.
- Use flat sponges instead of rolled sponges.

155

Process Skill

- Predicting is forming an idea about a future event, based on previous knowledge.

  On what did you base your predictions? Answers will vary, but elicit that the predictions were based on previous knowledge about plants.

  Suppose you were asked to predict something for which you have no background knowledge. Would you be making a good or reasonable prediction? no Why? It would be a guess since it is not based on an observation or knowledge.

  What could you do to make your prediction more valid? Possible answers: conduct research, observe similar situations

  Why do you think scientists extensively research their topics and the research results of other scientists before planning and making their own predictions about their own experiments? Since predicting implies background knowledge, scientists must gain this knowledge before they can predict a result.

Conclusions

- Discuss the evaporation of water.

  Which sponge lost more water? the plain sponge

  Why do you think more water evaporated from one sponge than from the other sponge? Guide the student to conclude that water could evaporate from all of the surfaces of the plain sponge. Water could not evaporate as easily through the petroleum jelly on the other sponge.

- Compare the model with real plants.

  How is this activity similar to the way that some plants survive? God designed some plants with waxy coatings that help keep water inside the plant rather than letting the water evaporate.

Activity Manual

Activity, pages 103–4

Assessment

Rubrics

Select the prepared rubric, or design a rubric to include your chosen criteria.

86
87

Student Text pages 156–58
Activity Manual pages 101, 105–7

Tropical Rain Forests

Rain forests receive abundant rainfall throughout the year. More rain falls in this biome than in any other type. A few rain forests are located in temperate areas. But most are located near the equator and are very humid and rainy. They are known as tropical rain forests.

Rain forests are known for their great biodiversity. **Biodiversity** refers to how many species of plants and animals are found in a specific area. Scientists think that about half of all the plant and animal species on the earth live in the rain forest.

Scientists have divided the rain forest into four main layers. The crowns, or tops, of tall trees form a green canopy for the rain forest. A few taller trees stick out above this canopy. These trees form the *emergent layer*. They receive the most sunlight, but they also must face the strongest winds and the hottest temperatures.

In the *canopy layer*, the trees grow closely together. The upper branches of these trees are often covered with vines. Plants such as ferns, orchids, and mosses often grow on the branches in the trees of the canopy. Very little sunlight gets through the canopy to the plants below. The canopy also keeps moisture from evaporating. The little sunlight and the extra moisture make the areas under the canopy very warm, dark, and damp.

Underneath the canopy is a third layer made up of younger trees and leafy plant The plants in this *understory layer* do not need much sunlight. Most have large leav to help gather sunlight. Many leaf-eating animals live in this section of the rain forest and eat these large leaves.

Even less sunlight reaches the fourth layer, the *forest floor*. A few plants grow there, but most cannot. The forest floor is home to organisms such as termites, fungi, and bacteri

156

Objectives

- Describe basic characteristics of a tropical rain forest
- Identify the layers of the rain forest
- Name ways that roots benefit the rain forest trees
- Recognize that biomes are only a general way to classify sections of the biosphere
- Explain how a mountain can have several biomes

Materials

- timer or clock
- globe or world map

Teacher's Toolkit CD

- Visuals 6.1, 6.3: *Land Biomes of the World*; *Biome Characteristics*
- Quiz 6-C

Vocabulary

- biodiversity

Two days have been allotted for this lesson. You may want to teach Lesson 86 the first day and use ideas from Lesson 87 on the second day. You may also incorporate the ideas from Lesson 87 into the text materials, making the lesson two days.

Introduction

- Direct the students to fold the paper into two columns. They should label one column *plants* and the other column *animals*. Set the timer for two minutes.
  List as many species of animals and plants as you can think of in two minutes. The animals and plants can be from any biome.
- Start the timer. After two minutes, allow students to see whose list is longest. Read and discuss some of the lists.
  You have mentioned many different species of plants and animals from several biomes. The biome we will study today is home to a great variety of plant and animal species.

Preparation for Reading

- Display the word *biodiversity* and pronounce it together. Direct the students to read silently pages 156–58 to find the answers to these questions.
  What does *biodiversity* mean?
  Why is the rain forest floor so dark?
  How does temperature change with mountain altitudes?

Teach for Understanding

- Refer to the globe and the *Land Biomes of the World* visual as you discuss the forest biomes. Refer the students to the *Land Biomes of the World* map on Student Text pages 142–43. Continue use of the *Biome Characteristics* organizer on Activity Manual page 101.

Background

Temperate Rain Forests

Temperate rain forests are located along the coasts of some continents in the mid-latitudes. They can be found in Norway, New Zealand, Australia, and along the Pacific coast of North America. Unlike tropical rain forests, temperate rain forests have cooler temperatures. The Sitka spruce, the bigleaf maple, the western hemlock, and the Douglas fir are some common trees found in temperate rain forests. Club moss is often found growing on the trees. The most common epiphytes in temperate rain forests are mosses and ferns. Most species of temperate rain forest animals are ground dwellers.

Activity

Discuss Differing Opinions About Rain Forest Preservation

Materials: books and other resources

Some people want to cut down rain forests to aid development and economy. Others want to preserve the rain forests as valuable natural resources. Divide the students into two groups. Let one group represent government officials and developers wanting to cut down part of a certain rain forest. The other group wants to keep it from being cut down. Provide resource materials for each group to research and find facts to support their position. Then allow each side to present its information in a simplified debate.

Some trees in the rain forest have stilt roots to help support them and to help them gather nutrients from the shallow soil.

The trees in a rain forest need sunlight, but they also need nutrients to grow and stay healthy. Only the top layer of the rain forest soil has the nutrients that the trees need. So the roots spread out horizontally under the ground and intertwine with other roots. Because the roots are so near the surface of the soil, the trees need extra support to keep from falling over. Some tall trees have roots that grow out of the sides of their trunks. These aboveground roots give the trees extra support.

Many rain forest animals live in the treetops and rarely touch the ground. Some of these animals use their tails to swing through the trees. Other animals glide or jump from branch to branch. Some tree frogs have large, webbed feet that they use like parachutes as they glide. Their feet also have soft, flat toes that act like suction cups to help the frogs stick to the branches. One kind of tree frog, Wallace's flying frog, has flaps of skin along the sides of its body. These flaps of skin and the frog's webbed feet enable this flying frog to glide for distances of up to 15 m (50 ft).

Wallace's flying frog

FANTASTIC FACTS

Have you ever heard of a pond on top of a tree? Bromeliads, a plant family that includes pineapples, often grow on the branches of trees in rain forests. Their long, stiff leaves collect water. Some bromeliads can hold as much as 7 L (2 gal) of water! The small ponds that form in the leaves are homes to many kinds of insects and amphibians.

157

Where are tropical rain forests located? near the equator

Are most rain forests tropical or temperate? tropical

What kind of climate is found in a tropical rain forest? hot, humid, and rainy

Rain forests are known for their great biodiversity. What does the term *biodiversity* refer to? the variety of plant and animal species found in an area

- Discuss the different layers of the rain forest.

What is the uppermost layer of the rain forest called? the emergent layer

What name is given to the second layer of the rain forest? the canopy layer

Why do you think this layer is called the canopy? A canopy is a type of roof. The trees in the rain forest canopy grow so close together that the tops form a covering like a roof or canopy.

Why do some plants grow on branches of trees in the canopy? Since little sunlight gets through the canopy, some plants grow on the branches so that they can get enough sunlight.

The canopy prevents sunlight from getting through. What else does the canopy do? prevents moisture from evaporating

How do the limited sunlight and the extra moisture affect the layers below the canopy? They make the area under the canopy very warm, dark, and damp.

What name is given to the smaller trees and plants that grow underneath the canopy? the understory layer

Why do many of the trees and plants of the understory have large leaves? Little sunlight gets through, so bigger leaves help gather more sunlight for the plant to use.

What is the fourth, or bottom, layer of the rain forest called? the forest floor

How do the roots of some rain forest trees help support the trees? The underground roots spread out and intertwine with the roots of other trees. Some roots grow out of the sides of trees and spread out above the ground.

Where do most rain forest animals live? in the treetops

- Discuss effects of destruction to different layers of a rain forest.

What might happen to the lower layers of a rain forest if the canopy were destroyed or damaged? Possible answer: Many species of plants and animals might not be able to survive the change of conditions.

- Discuss *Fantastic Facts*.

What is unusual about some bromeliads in the rain forest? Plants such as bromeliads grow on tree branches and collect water, forming small ponds that are homes to many different insects and animals.

Activities

Make a Bromeliad

Materials: small paper tube; green and blue construction paper; glue; tape; scissors; small plastic or gummy insects, amphibians, or other animals

Cover the paper tube with green construction paper. Cut long, jagged leaf shapes out of green construction paper. Curl each leaf around a pencil to give a realistic shape. Glue the ends of the leaves around the tube, starting at the top and overlapping the leaf ends in some places. Use blue construction paper to make a small pond at the top of the bromeliad. Attach small plastic insects, amphibians, or other appropriate gummy animals to the pond.

Write About a Rain Forest Animal

Direct the student to research a rain forest animal and write a short report describing that animal and how it is able to live in this biome. Listed below are some interesting animals to research.

Reptiles: paradise tree snake, green tree python, Jackson's chameleon, green iguana
Amphibians: red-eyed tree frog, poison dart frog
Birds: scarlet macaw, hoatzin
Insects: blue morpho butterfly, leaf-cutter ant
Mammals: kinkajou, capybara, Malayan tapir, okapi

86–87

How do scientists classify sections of the biosphere? into biomes, based on climate

Why do mountains have more than one biome? Changes in altitude affect the climate and the plants and animals that can live there.

What changes in the air affect the climate at higher altitudes? the air becomes thinner and colder

How does the temperature change as you go up a mountain? It decreases as the altitude increases.

List some biomes that may be on or near mountains in dry regions. Possible answers: deserts, grasslands, deciduous forests, coniferous forests, alpine tundras

Describe how the biomes of a mountain in a wet region near the equator might differ from the one pictured. Possible answer: The mountain might have a tropical rainforest instead of a deciduous forest.

What is another name for mountain forests? montane forests

Are the biomes found on mountains just like the other land biomes? No, they are similar, but not identical, to the major land biomes.

ANSWERS

1. hot, humid, and rainy
2. the emergent layer, the canopy layer, the understory layer, and the forest floor
3. Changes in altitude affect the climate and what lives there.

87

- The *Discuss Differing Opinions about Rain Forest Preservation, Make a Bromeliad,* and *Write About a Rain Forest Animal* activities at the bottom of the pages in this lesson may also be incorporated into the lesson.
- Suggestions for additional activities for this two-day lesson are on the Teacher's Toolkit CD.

Activity Manual

Study Skill and Reinforcement, page 101

Review, page 105

This page reviews Lessons 86–87.

Expansion, pages 106–7

These pages use climatographs to compare the climates of two biomes.

Assessment

Quiz 6-C

The quiz may be given any time after this lesson.

Mountains

Biomes are a general way to classify sections of the biosphere based on climate. But sometimes a large area such as a mountain may have more than one biome in it. As you travel up a mountain, the air becomes thinner and colder. These changes create different climates. The climates, in turn, affect what types of plants and animals can survive.

In dry regions, desert or grassland biomes are usually at the base of a mountain. As you go up the mountain, desert or grassland vegetation becomes scarce, and more trees grow. These mountain forests are usually deciduous forests.

Farther up the mountainside, the climate may change again. The deciduous forest now changes to a coniferous forest. Mountain forests are often called *montane forests.*

Above the coniferous forest is a biome that is similar to the tundra. This cold biome is called the alpine tundra. Above the alpine tundra is the snowy top of the mountain. The mountain biomes are similar, but not identical, to the major land biomes.

QUICK CHECK

1. What kind of climate do tropical rain forests have?
2. What are the four layers of the rain forest?
3. Why do biomes change as you go up a mountain?

158

Background

Mountains

Many scientists consider mountains to be biomes because some of the animals that live in them are found only on mountains and do not live in other areas. For example, mountain goats and chinchillas are found only in mountain environments. Biomes encompass both the plants (flora) and animals (fauna) of an environment.

Cloud Forests

A cloud forest is a specific type of rain forest found only on some steep mountain slopes in tropical and subtropical areas. The cooler temperatures on the mountain cause low clouds to form, and the entire forest is immersed in the clouds most of the time. The moisture from the clouds is deposited directly on the leaves of trees and plants. The water then drips off the leaves. The photo at the top of Student Text page 141 is a cloud forest.

Alpine Tundra

The alpine tundra receives more precipitation than the Arctic tundra does. Permafrost is found in the Arctic tundra but is not always found in the alpine tundra. However, the alpine tundra land is usually covered with snow in winter.

Build a Biome

EXPLORATION

What do you think of when someone mentions a tropical rain forest? Maybe you think of parrots, monkeys, and thick vines. In contrast, you probably think of cactuses and sand as belonging in a desert. Each biome has characteristics that make it unique and different.

For this exploration you will build a model of one of the biomes.

What to Do

1. Choose one of the biomes mentioned in this chapter.
2. Decide on the location and characteristics of the biome that you chose. For example, if the chosen biome is a deciduous forest, will your model represent the forests of North America or those of Europe? Which species of broad-leaved trees will you include?
3. Display the dominant vegetation, or most common plants, for that biome. Include a variety of plants and animals in the model.
4. Research and find some interesting facts about at least two of the plants and two of the animals included in your model. Record those facts on note cards. Include this information in your model.
5. On your model, identify the type of biome and its location. Include information about the climate and other characteristics of the biome. Present your model.

159

Objectives

- Research a biome
- Create a model of that biome

Materials

- resource books about biomes

Teacher's Toolkit CD

- Rubrics: *Build a Biome,* for each student

Introduction

Which biome interests you the most? Why? Answers will vary.

Which do you like learning about more, land animals and plants or water animals and plants? Answers will vary.

For this exploration you will make a model of one of the biomes in this chapter.

Preparation for Reading

- The students should read the page before beginning the exploration.

Teach for Understanding

What to Do

- Establish guidelines for the size, content, and format of the display and written information. Set a date for the completion and presentation of the biome.
- Guide the students in planning and building their biomes. The students may choose to model a land biome or an aquatic biome. However, if an aquatic biome is chosen and water is used in the model, you may wish to give guidelines about containing the water. An aquatic biome display can be made without water. Fish and other floating or swimming objects can be hung from the top of the box with thread or fishing line.
- Students may work together in groups to complete one biome per group. You may choose to assign specific tasks to each student so that all contribute to the project.

Assessment

Rubrics

Select the prepared rubric, or design a rubric to include your chosen criteria.

Student Text pages 160–65
Activity Manual pages 108–10

90 91

Objectives
- Name the two categories of aquatic biomes
- Explain why coral reefs are called "the rain forests of the sea"
- Identify the force that keeps river water moving
- Describe kinds of wetlands
- Recognize that people have the God-given responsibility to be good stewards of the earth

Teacher's Toolkit CD
- Visuals 6.4–6.5: *Zones of the Ocean*; *Classifying Freshwater*
- Quiz 6-D

Vocabulary
- salinity
- marine biome
- wetland
- peat

Two days have been allotted for this lesson. You may want to teach Lesson 90 the first day and use ideas from Lesson 91 on the second day. You may also incorporate the ideas from Lesson 91 into the text materials, making the lesson two days.

Introduction

What do you usually think of when someone mentions a forest? Answers will vary.

Have you ever heard of a rain forest in the sea? Answers will vary.

In this lesson we will find out about "rain forests of the sea" and also learn about some other aquatic biomes.

Preparation for Reading

- Display the word *salinity* and pronounce it together. Direct students to mentally make questions using the headings, bold words, and captions. Instruct them to read silently pages 160–65 to find the answers to these questions.

What are "the rain forests of the sea"?

How are ponds different from lakes?

What are three types of wetlands?

Teach for Understanding

What influences land biomes? the climate

How is this different for aquatic biomes? A water biome does not have one specific climate.

What are the two main classifications of aquatic biomes? freshwater biomes and saltwater biomes

What does the term *salinity* mean? the amount of dissolved salt in the water

Which type of aquatic biome has a higher salinity? saltwater

Aquatic Biomes

Land biomes are influenced directly by the climate. However, aquatic, or water, biomes do not have one specific climate. One example is the Pacific Ocean. It stretches from cold Alaska to the warm tropics and then to the cold shores of Antarctica.

Aquatic biomes are classified as either saltwater or freshwater. The **salinity**, or amount of dissolved salt in the water, varies with each biome. Salt water has a higher concentration of salt than fresh water does. But even fresh water contains a little salt.

Most fish can live in only one type of water. However, some aquatic animals and plants are able to adjust to both. These animals and plants can adjust to the salinity, temperature, and speed of the water more easily than others can. Salmon and eels, for example, are able to live in both kinds of water.

Saltwater Biomes

Saltwater biomes are also called **marine biomes**. The largest one is the ocean. Because it is so large, scientists divide it into different zones, or sections. One of these zones is the place where the ocean and shoreline meet.

As the ocean tide comes in and out, it sometimes forms tide pools. Tide pools are low places along the shore that are exposed by low tides. The living things in the tide pool change their behavior as the water goes in and out. When the water is low, clams snap their shells shut to keep water inside. Sea anemones pull in their tentacles and keep moist by covering themselves with slimy mucus. As the waves rush back into the tide pool, sea urchins wedge themselves between rocks, and sand dollars burrow into the sand to keep from being pulled back into the ocean.

Tide pool

160

Background

Coral
Coral is a living organism. It is a tiny, soft animal that looks like it has a jellyfish attached to its head. Thousands of these animals live together. They absorb and excrete the calcium carbonate that builds up to form a coral reef.

Tides
Ocean tides are influenced by the gravitational forces between the earth, moon, and sun. Since the earth's rotation and the moon's orbit are predictable patterns, tides are also predictable.

Activity

Write About an Ocean Animal
Research an ocean animal. Determine which zone of the ocean the animal lives in. Write a short paragraph describing the animal, where it lives, what it eats, and how it gets its food.

Coral reefs are another section of the ocean and are found in or near the tropics. These reefs are formed by living animals called corals. Corals prefer the clear, warm, shallow salt water of the tropics. Certain types of corals attach themselves to underwater objects. When the coral dies, its limestone skeleton remains. Other corals attach themselves to the skeletons and the cycle continues. The result is a beautiful coral reef.

Many kinds of organisms live in coral reefs. In fact, coral reefs have such great biodiversity that they are sometimes called "the rain forests of the sea." Many brightly colored fish, such as angelfish and parrotfish, live in coral reefs.

Another zone is the open ocean. It is often divided into layers based on the amount of sunlight each layer receives. The top layer receives the most sunlight. Some tiny organisms can live in only the top layer of the ocean. They need sunlight to grow. Many fish and sea mammals also live in this layer because they depend on these organisms for food.

The deeper layers of the ocean are dark and cold. They also have a lot of water pressure. God designed some animals that can thrive even in these conditions. Some fish that live deep in the ocean have large eyes to help them see in the dark waters. Other fish, such as the firefly squid, have light-producing organs that help them attract their prey.

Firefly squid

FANTASTIC FACTS

Have you ever seen a fish fly? Some surface fish, such as the California flying fish, can "fly" to escape being a larger fish's meal. These fish have tails that allow them to swim very rapidly in the water. When in danger they can gather speed, head to the surface, and launch themselves out of the water. Their large side fins act like wings and allow the fish to soar above the water. The fish do not actually fly but glide. Some can stay aloft for distances of several hundred meters. That is longer than two football fields! Sometimes the fish unexpectedly land on the decks of small boats.

161

What is another name for saltwater biomes? marine biomes

What is the largest saltwater, or marine, biome? the ocean

- Tide pools are found in the zone where the ocean and the shoreline meet. This zone is known as the intertidal zone.

What are tide pools? low spots that are exposed by low tides

Describe a "low tide." Possible answer: A low tide is the time when the level of the ocean water is lower along the shore. More of the land or beach is seen.

What are some ways that animals in tide pools adjust to the lower water when the tide goes out? Possible answers: Clams close their shells. Sea anemones pull in their tentacles and cover themselves with mucus. Sea urchins wedge between rocks. Sand dollars burrow into the sand.

- Coral reefs form in the neritic (near shore) zone located between the shoreline and the open ocean.

What are sometimes called "the rain forests of the sea"? coral reefs

Why do you think that coral reefs are compared to rain forests? Both exhibit high biodiversity.

Where are most coral reefs found? in or near the tropics

What kind of water do corals prefer? clear, warm, shallow salt water of the tropics

How are coral reefs formed? Some corals take salt from the sea and build limestone skeletons on underwater objects.

What determines the different layers of the open ocean? the amount of sunlight each layer receives

In which layer of the ocean are you most likely to find tiny floating organisms? the top layer Why? The top layer receives the most sunlight, and many organisms need sunlight to grow.

What are some characteristics of the deep ocean layers? dark, cold, and under a lot of water pressure

Why are the deeper layers of the ocean cold and dark? Less sunlight penetrates these layers, so there is less light and heat.

- Display the *Zones of the Ocean* visual and share the *Zones of the Ocean* background information.
- Discuss *Fantastic Facts*.

What helps the California flying fish "fly"? its large side fins

How far can some California flying fish glide through the air? several hundred meters

Background

Zones of the Ocean

The ocean can be divided into three main zones: the intertidal zone, the neritic (near shore) zone, and the oceanic (open ocean) zone. The intertidal zone is the area of land that is covered and uncovered by the tides. The neritic zone consists of the waters over the continental shelf. Beyond the continental shelf is the oceanic zone.

The oceanic zone is also divided into layers. These layers are often referred to as the sunlit zone (epipelagic layer), twilight zone (mesopelagic layer), and the midnight zone (bathypelagic layer).

90
91

- An organizer is on Activity Manual page 108. You may choose to display the copy of the organizer from the Teacher's Toolkit CD.
- Guide the students in making notes in the chart as you teach or allow them to complete the page at the end of the lesson.
- Review the organization of the page. Instruct a volunteer to read aloud the two subheadings that are provided. Direct the students to skim the first paragraph to find what to write as the other two subheadings (lakes and rivers). Guide the students as they make notes for each subheading.

  How does the amount of fresh water on the earth compare with the amount of salt water? There is less fresh water.

  What are two ways to classify fresh water? as standing water or moving water

  Which kind of water do ponds and lakes have? standing water

  What does *standing water* mean? water that is not constantly moving

  How do ponds and lakes get their water? from rivers, springs, rain, and melting snow

  Which are deeper, ponds or lakes? lakes

  Where do most lake plants grow? along the edges of the lake

  How are ponds different from lakes? Ponds are not as deep as lakes. Plants can grow in the middle of a pond as well as at the edges. Lake plants grow only at the edges.

  If you see plants growing in a body of water as well as around the water's edge, is this body of water most likely a pond or a lake? a pond

  Some plants grow on the surface of ponds and lakes. What is one reason these plants often have rounded leaves? to protect them from the wind
- Discuss *Creation Corner.*

  What helps water-lily leaves float? air pockets

  Why are water-lily flowers closed on rainy days? Too much water on top of the plant could cause the plant to sink.

  Why do water lilies need strong, flexible stems? to adjust to the depth and movement of the water

Great Lakes, North America

Freshwater Biomes

About three-fourths of the earth's surface is covered by water. But only a small amount of that water is fresh water. Freshwater biomes include ponds, lakes, streams, and rivers. Most of the plants and animals that live in fresh water could not survive in marine biomes.

Freshwater biomes are classified as either standing water or moving water. The term *standing water* is used for any water that is not constantly moving. Ponds and lakes have standing water. This water can come from rivers, springs, rain, and melting snow. These bodies of water can be found in both cold and hot climates.

Ponds and lakes are similar, but ponds are not as deep. This allows plants such as cattails and reeds to grow in the pond as well as at its edge. Because lakes are deeper, many lake plants grow only at the edge of the lake. Some plants, such as the water lily, live on top of ponds and lakes. The rounded leaves of these plants help protect them from the wind.

Many animals also live in ponds and lakes. Dragonflies dart among the plants. Bullfrogs and snapping turtles live along the water's edge. Freshwater fish, such as perch and largemouth bass, swim in the water. Beavers may use nearby trees to build their dams in the water.

CREATION CORNER

If you put a leaf on the surface of a pond or stream, the leaf will eventually sink. But water-lily leaves have air pockets that help them float. The leaves are covered with a waxy surface that helps water slide off the leaf. The lily's flower also helps it stay on top of the water by not opening on rainy days. If it did, water would collect on top of the plant and cause it to sink.

Underneath the water, the water lily's stem is anchored in the mud. This stem is very strong but is also flexible. The flexibility helps the plant adjust to the height and movement of the water.

162

Background

Amount of Fresh Water

Less than 3% of the water on the earth is fresh water. About 90% of that fresh water is frozen at the poles or in Greenland.

Salinity of Fresh Water

Even fresh water has salt. Fresh water has a low salt content, usually less than 1%. Most of the plants and animals that live in fresh water could not survive in water with a high salt content.

Lakes

Lakes can be fresh water or salt water. Large lakes are sometimes called seas. The Caspian Sea is the largest lake, but Lake Baikal in Russia is the deepest with a depth of 1,620 m (5,314 ft). When measured by surface area, Lake Superior (visible in the upper left of the satellite photo on Student Text page 162) is the world's largest freshwater lake. When measured by volume, Lake Baikal is the largest freshwater lake.

Limnologist

A *limnologist* is a scientist who studies lakes and ponds.

The water in other freshwater biomes, such as rivers, is constantly moving. River water can come from snow, rain, glaciers, and ground water. Gravity keeps this water flowing downhill toward the ocean.

The beginning of a river, or the *headwater*, is usually a little stream. This water is usually clear and moves quickly. As the river winds along, other smaller rivers, called tributaries, join it. The river gets wider, slower, and warmer as it flows toward the ocean. The river may also get muddy as dirt and debris get mixed in. The *mouth*, or end, of the river empties into the ocean.

The temperature and speed of the water affect what kinds of plants and animals can live in freshwater biomes. Plants and animals that live in fast-flowing water must be able to either move with the water or anchor themselves against the current. Plants such as mosses attach themselves to rocks in a stream. Many fish, including some kinds of trout, prefer the cool, fast-moving water.

The plants and animals that live in the slower water can be very different from those in the fast-flowing water. Plants along the edges of slower water may be similar to those around ponds and lakes. Some fish, such as certain kinds of catfish, feed on the bottom of rivers. They prefer the slow-moving water. Some large river mammals, such as the manatee, also live in slow-moving rivers. They eat the water plants that can grow in slower water.

163

What force keeps river water moving? gravity

Where does river water come from? snow, rain, glaciers, and ground water

What is the beginning of a river called? the headwater

What smaller freshwater biome is usually the headwater of a river? a stream

Does the headwater of a river move slowly or quickly? quickly

What happens to the temperature of the water as the speed of the river slows? The temperature rises.

What are tributaries? smaller rivers that join a main river as it flows toward the ocean

What is the end of a river called? the mouth

Where do rivers end? in the ocean

What two factors help determine the types of animals and plants that can live in freshwater biomes? the temperature and speed of the water

What kind of fish are sometimes found in fast-moving water? trout

What are some animals that live in slow-moving water? Possible answers: catfish, manatees

Activity

Viewing Freshwater Organisms

Materials: container of water from a pond, lake, or river; microscope slide; microscope; paper; pencil

Observe the water, looking to see if it contains any small plants or animals, such as tadpoles. Put a sample of the water on the microscope slide. Use a microscope to view the slide. Direct the student to draw a picture of any organisms or other things that are seen on the slide. You may choose to have the student research to identify the organisms that were drawn. Return the water to the pond or other body of water when the student has finished observing it.

Demonstration

How a Water-Lily Flower Floats

Materials: 2 coffee filters, string, bowl of water, dropper

Loosely tie the top of a coffee filter together to resemble a flower blossom. Gently place the "blossom" on the surface of the water. Place the other (open) filter on the water. Drop water on both and observe which sinks more easily, the closed "blossom" or the open "flower."

- Discuss reasons scientists may not agree on the number of biomes that exist.
  Do biomes have set boundaries? no What usually happens where two biomes meet? The biomes merge, or gradually change.
- Some scientists classify wetlands as a biome. Other scientists do not because they consider wetlands to be part of other biomes, such as the bogs found in the coniferous forest. Some wetlands, such as marshes and swamps, are sometimes considered to be a transition zone, or ecotone, between two biomes.
  What are wetlands? land that is almost always wet
  Are all wetlands alike? no
  What are the three main types of wetlands? marshes, swamps, and bogs
- Describe and compare marshes and swamps.
  What type of wetland is the most common? marshes
  Are marshes and swamps salt water or fresh water? They can be either.
  What is one main difference between marshes and swamps? Marshes usually do not have any trees, but swamps have many trees.
  What types of vegetation are usually found in marshes? grasses such as reeds and cattails
  What type of tree is often found in saltwater swamps? the mangrove
  What types of trees are usually found in freshwater swamps? the maple, willow, or bald cypress
- Discuss characteristics of bogs.
  What type of wetland always has fresh water? a bog
  In what climate areas are bogs often found? in cooler northern climates
  Where do bogs usually form? near lakes and ponds and in places where the soil does not allow water to drain easily
  What is peat? the partially decayed plant matter found in bogs
  What is deceptive about the appearance of a bog's surface? Mosses, grasses, and some trees grow in the layers of peat and cause the surface to appear solid even though it is mostly water.
  What kind of bog occurs in the taiga? a muskeg
  Why do people often disagree about the use of wetland areas? Some people want them preserved as natural habitats, and others want the land for buildings or farms.

Florida Everglades

Wetlands

Scientists often disagree about how many land and water biomes exist. This is because biomes do not have set boundaries. They often merge into each other. One biome gradually turns into a different one. This transition zone can happen in places such as where grasslands meet a forest or where water meets the land. It can even happen where a river flows into the ocean.

Janaiya's science project was about protecting wetlands. **Wetlands** are areas of land that are almost always wet. Some wetlands are covered by water all year round. Other wetlands do not have water for part of the year. The water in some wetlands comes and goes with the ocean tides. Wetlands are usually classified as marshes, swamps, or bogs.

Marshes and Swamps

The most common wetlands are marshes. They are found in many places and can have either fresh water or salt water. Most marsh plants are grasses such as reeds and cattails. Trees usually do not grow in marshes. Animals such as muskrats and alligators sometimes live in freshwater marshes. Saltwater marshes are home to oysters, crabs, and water birds such as pelicans and cranes.

Swamps, on the other hand, have many trees. The types of trees vary depending on where the swamp is located. The type of water also determines what kinds of trees can grow there. Saltwater swamps usually have mangrove trees. Freshwater swamps may have maple, willow, or bald cypress trees. Wading birds and snakes are common in swamps. Some also have alligators.

Much of the Mississippi River delta is covered with marshes and swamps. Because there is so much plant and

Bond Swamp National Wildlife Refuge, Georgia

164

Background

Estuaries

Estuaries occur in places where fresh water from rivers and streams mixes with salt water from an ocean. These areas get salty ocean water during high tides and fresh river water during low tides. Animals and plants that live in estuaries have a constantly changing environment. Certain types of crabs and prawns can swim out of the estuary when there is less salt in the water. Other animals close their shells or burrow into the mud or sand when the salinity changes.

Estuaries often serve as "nurseries" for many species of fish. Some species of fish lay their eggs in the sheltered waters of an estuary. The eggs hatch, and the young fish mature in the protection of the estuary before swimming out to sea.

Bogs and Preservation

Some peat bogs are found in northwestern Europe. Many things have been found preserved in these bogs. The acidic water and lack of oxygen in bogs can preserve the soft body tissues, clothing, and other artifacts of a person or animal even after the bones decay. In Ireland and Scotland, butter was sometimes buried in bogs. They apparently protected the butter from the heat and kept mold from forming. Buried butter has been found in containers made from a variety of materials including wood, cloth, wicker, and animal skins.

Bog

animal life, even a small change can have big consequences. People have to be careful when making changes to swamps and marshes.

Bogs

Swamps and marshes can have fresh water or salt water, but bogs can have only fresh water. The water is usually cold as well. This is mainly because bogs are often found in cooler northern climates. Bogs sometimes form near lakes, ponds, or other places where water cannot drain easily through the soil. Most of the water in bogs comes from precipitation.

When leaves fall or plants die, the dead plant material usually decays, or rots. Sometimes, though, it does not completely decay. When this happens, the partially decayed plant material called **peat** piles up into dense layers. The swampy soil in bogs forms from this peat. Moss, grass, and even some trees can grow in it. The surface of the bog may look firm and solid, but it is really mostly water.

Man's Stewardship

People face difficult decisions about how to use wetlands and biomes. They may have strong opinions and disagree. The land, plants, and animals are affected by any natural or manmade changes. God values people, so it is important that their needs be met as they use and manage resources. It is their responsibility to glorify God and to act wisely and not misuse or waste what God has given them.

QUICK CHECK

1. What are the two ways water biomes are classified?
2. What keeps river water flowing toward the ocean?
3. What are the three main types of wetlands?

90–91

What responsibility has God given people? to glorify God and to act wisely and not misuse or waste what God has given them

- Guide the students in compiling a list of ways that they can demonstrate wise stewardship as they interact with nature. The list may include items such as correctly disposing of trash, planting seed-bearing plants for birds and animals to feed on, providing nesting boxes for birds and small animals, and not breaking branches or damaging plants in parks or forests.
- Continue the activity by compiling a second list of ways they can demonstrate wise stewardship in other areas of their lives. The list may include items such as taking only the amount of food they will eat for meals and snacks, giving unused clothing or toys to others, and donating time to serve and help someone complete a project or need.

ANSWERS

1. as saltwater biomes or freshwater biomes
2. gravity
3. marshes, swamps, and bogs

91

- The *Viewing Freshwater Organisms* activity and the *How a Water-Lily Flower Floats* demonstration found at the bottom of the pages in Lessons 90–91 may be incorporated into the lesson.
- Suggestions for additional activities for this two-day lesson are on the Teacher's Toolkit CD.

Background

Mangrove Swamps

Saltwater swamps are usually called *mangrove swamps* because of all the mangrove trees that grow there. Mangrove swamps are found along the coastlines in warm climates. Water from the ocean tides overflows into these swamps. Only trees and plants that can tolerate salt water are able to grow in these swamps.

Scientist G. Henry Mühlenberg

G. Henry Mühlenberg (1753–1815) was a Lutheran preacher and an American patriot. During the Revolutionary War, Mühlenberg hid in forests in Pennsylvania in order to evade the British army. It was during this time that he became seriously interested in botany. He studied grasses, sedges, and many wetland plants. He named over one hundred species of plants. In 1813 Mühlenberg listed all the known native and naturalized plants of America. His collections and notes were very helpful to others studying plant taxonomy. Many scientists used and published his information without giving him the proper credit. While hiding from the British, he had discovered a wetland turtle that had not been known before. He gave this turtle to a friend, and the species of turtle was later named for Mühlenberg. In addition to the turtle, many species of plants and lichens as well as a fungus have been named for him.

Activity Manual

Study Skill and Reinforcement, page 108

Reinforcement, page 109

Students practice vocabulary by completing a crossword puzzle.

Review, page 110

This page reviews Lessons 90–91.

Assessment

Quiz 6-D

The quiz may be given any time after completion of this lesson.

92

Activity Manual pages 111–12

Objectives

- Compare the description of the Garden of Eden to a map of modern-day Iraq
- Explain why the climate and biomes changed after the Flood

Teacher's Toolkit CD

- Visual 6.6: *Modern-Day Iraq*

Introduction

Have you ever thought of the Garden of Eden as a biome?

What kind of biome would you classify Eden as?

- Review the different kinds of biomes and discuss which seems to be the probable classification.

Preparation for Reading

- Direct the students to read the paragraphs on Activity Manual page 111 silently to find a source for locating the Garden of Eden.

Teach for Understanding

- Explain that we have very little information on the pre-Flood Earth in the Bible. From the description of the Garden of Eden, we can know that it was a very lush biome. Since the original couple was living there without clothing, we can infer that the climate was stable and mild. Studying the fossil record, the massive amount of vegetation that we find buried by the Flood gives another indication that there were fewer extremes in climates and biomes than we see today.

> The name *Tigris* is also translated *Hiddekel* in some Bibles. The students need to use "Tigris" for the purpose of this lesson.

- Read together Genesis 2:8–14 and record for display a list of the characteristics of the Garden of Eden that are given.
- Guide the students as they draw their interpretations of the descriptions of the rivers.
- Allow students to share their drawings and explain their reasoning.

  Do you recognize the names of any of the rivers? Possible answers: Tigris and Euphrates

A lesson from

Answers in Genesis Name ______________________ **1:1**

Where Was the Garden of Eden?

Have you ever been reading the Bible and found yourself wishing you could be standing where the events are taking place? The Garden of Eden was one of the most beautiful places on Earth. Imagine standing in a grove of lush trees next to a beautiful river where everything was absolutely perfect. It was its own unique biome. If you wanted to go on an expedition to find the Garden of Eden, where would you begin looking? Hopefully you would use the Bible as your starting point. What does the Bible have to say about where the garden was?

Open your Bible to Genesis 2:8–14 and answer the questions.

1. What do these verses say grew in the garden? *trees*
2. How are the trees described?
   Some were nice to look at, some were for food, one was the tree of life, and one was the tree of the knowledge of good and evil.
3. Draw a picture of how verse 10 describes the rivers flowing into and through the garden. Label the rivers with the names found in verses 11–14.

Resources

The BJU Press website offers additional links from Answers in Genesis that you may find helpful with this lesson.
bjupress.com/resources/science

4. Some people believe that the Garden of Eden was somewhere in what we call Iraq. Look at a map of Iraq. Why might some people think the Garden of Eden was in Iraq?

The names of the two rivers are Tigris and Euphrates.

5. Does this map match the description from Genesis 2?

No, it only has two rivers, not four.

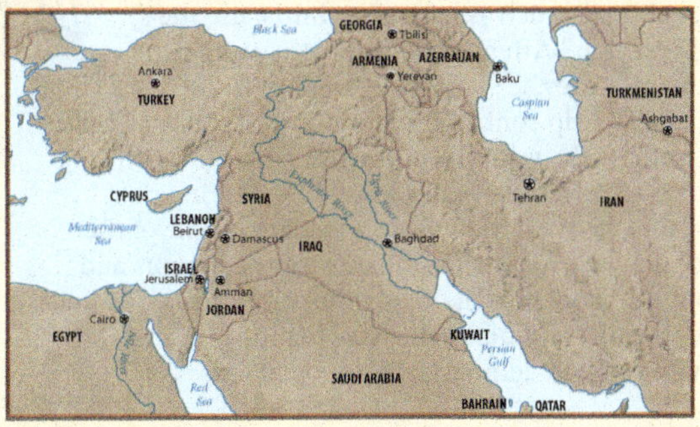

Genesis 2 describes one river that splits into four. The map of Iraq shows two rivers joining to become one—they are not describing the same places. So why do the rivers have the same names? Again, we can look to the Bible to help us understand the answer.

The Tigris and Euphrates were names of rivers before the Flood described in Genesis 6–9. During the Flood, the entire earth was covered with water, and many new sedimentary rock layers were formed all over the earth. During the last part of the Flood, the water flowed off the land and into the oceans (Genesis 8:3). New mountains were formed and much erosion occurred. Water moving toward the oceans formed valleys. These valleys became channels for rivers after the Flood. As Noah's descendants moved to the east, they settled in Shinar (Genesis 11:1–2). Although the land had been totally changed by the Flood, there were still rivers just like before.

So what do you name a river you have never seen before? People who came to America from Europe often named the towns and rivers after places in their homeland. For example, New Jersey was named after Jersey Island in the English Channel. The people who settled in Shinar after the Flood named the two rivers there Tigris and Euphrates—names of rivers Noah and his family would have known from before the Flood.

6. So, why can we not really know where the Garden of Eden was?

Possible answers: The Flood would have completely destroyed and buried its original location. The climate of modern-day Iraq, a desert, is very different from the pleasant climate of the Garden of Eden.

Even if we cannot know the location of the Garden, what we do know is that God has promised to create a new heaven and Earth where we can worship and enjoy Him forever. To enjoy that perfect world, we must become sons and daughters of God (Ephesians 1:3–14).

- Direct attention to the *Modern-Day Iraq* map and the map on Activity Manual page 112.
  Some people believe that the Garden of Eden was somewhere in Iraq. Looking at this map, why might some people think the Garden was in Iraq? The Tigris and Euphrates Rivers are located there.
  Just like in Genesis 2, the Tigris and Euphrates Rivers are mentioned. Many people use this as support for the Garden being in Iraq.
  Compare your drawing to the map. What is the main difference? The Pishon and Gihon Rivers are not on the modern map.
- Direct the students to read the paragraphs on Activity Manual page 112 silently to find why the rivers are named Tigris and Euphrates.
  Why do you think the rivers in Iraq are named the Tigris River and the Euphrates River? Noah and his family reused names that they would have known from before the Flood.
- Explain that another reason why the Garden of Eden could never have been located in Iraq is the fact that the rock layers below the rivers are filled with fossils that were buried during the Flood. Since we know that those fossils are only possible after the Fall, and that the Flood happened after that, those rivers cannot be the ones described in Genesis 2 even though they share the same name.

Activity Manual

Answers in Genesis, pages 111–12

Activity

New Place, Old Name

Direct the students to research the history of the names of places in your area or other places of interest to find ones that were named after other places. The students should write a brief paragraph about each and compile them in a book for display.

Student Text pages 166–67
Activity Manual page 113

Objectives

- Demonstrate how wetlands purify water
- Infer how the activity models the purifying process of a real wetland

Materials

- pictures of wading birds, such as a blue heron, sandhill crane, and snowy egret

> Any type of dried moss, such as sphagnum moss or orchid moss, can be used. Dried mosses can be found with floral-arranging supplies in department or craft stores and at plant nurseries or garden centers.

Teachers Toolkit CD

- Rubrics: *From Dirty to Clean*, for each student

Introduction

- Display the pictures of the birds. Identify the name of each species of bird.

  What are some physical characteristics that these birds share? Possible answers: body shape, leg length, beak structure

  What do you think these birds eat? Answers will vary but should include fish and small crustaceans.

  Where do you think birds like these live? Answers will vary, but elicit that these birds live in wetlands or coastal areas.

- Many species of wading birds depend on wetlands for their survival. In today's activity we will discover another reason why wetlands are so important.

Preparation for Reading

- The students should read all the pages before beginning the activity.

Teach for Understanding

Procedure

- Guide the students in preparing their "wetlands." Be sure that some of the cheesecloth is hanging over the sides of the cup.
- The hole at the bottom of the cup should be about 2.5 cm in diameter.
- Remind the students to shake the jar of water to mix the soil each time before pouring it. They may also need to shake the jar of dirty water when comparing the filtered water and the dirty water.

ACTIVITY

From Dirty to Clean

Process Skills
- Making and using models
- Observing
- Inferring

Marshes, bogs, and swamps are important to both people and animals. Many animals lay their eggs or raise their young in wetlands. Some animals, such as migrating birds, use the wetlands as a "rest stop." After a little while, the animals continue on their journey.

Wetlands can also help control flooding. They act like sponges by absorbing some of the water from the tides or an overflowing river. Wetlands often clean the water as well by filtering out dirt and pollutants. As water flows into a wetland, some of the sediment carried by the water sinks. Plants trap sediments and pollutants from the water as well. When the water flows back out of the wetland, the dirt and pollutants stay behind.

In this activity you will demonstrate how wetlands filter water.

Materials
- foam cup, 16 oz
- scissors
- cheesecloth
- dried moss
- 4 jars with lids
- 500 mL water
- metric measuring cups
- potting soil
- spoon
- Activity Manual

Purpose

Demonstrate how wetlands filter water.

Procedure

1. Cut a hole in the bottom of the foam cup. Line the cup with cheesecloth. Some of the cheesecloth should hang over the sides of the cup.
2. Pack moss into the bottom one-third of the cup. This is your "wetland." Balance the cup over one of the empty jars.
3. Select two other jars and pour 250 mL of water into each. Add a spoonful of soil to each jar of water. Close the jars and shake well.
4. Set one jar of dirty water aside. You will be comparing this jar of water to the water that you filter through your wetland.
5. Shake the other jar of dirty water again. Pour the water from that jar into the cup so that the water will drip down through the moss and into the empty jar.

166

6. Compare this filtered water with the jar of dirty water. Record your observations in your Activity Manual.
7. Once all the water has dripped through, move the cup to the clean jar that has not been used yet. Put the lid on the jar of water that has been filtered. Shake the jar and pour the water through the wetland again.
8. Compare the jars of water and record your observations.
9. Carefully remove the moss and cheesecloth from the cup. Examine the moss and cheesecloth and record your observations.

Conclusions

- How does the activity model the way wetlands filter water?

Follow-up

- Use a different amount of dirt in the water.
- Add a few drops of food coloring to the water to represent pollutants.
- Use what you have learned to defend an opinion about the use of wetlands and people's dominion over them.

167

Process Skill

Inferring involves interpreting observations and drawing conclusions based on previous knowledge and experience.

How might the results affect how you set up other related experiments? Answers will vary, but elicit that knowledge gained from one experiment can influence later experiments.

Conclusions

- Compare this model with the way wetlands clean water.

What part of this model would represent the plant material in an actual wetland? the moss and cheesecloth inside the cup

How do the plants and vegetation in a wetland filter water? They trap sediments and pollutants that did not sink to the bottom.

Did the cheesecloth catch all of the dirt? no

Do you think that all of the sediments and pollutants can be filtered out by wetlands? no

Some organisms in wetlands are able to take many harmful substances from the water and change them to less harmful substances. However, pollutants added by mankind often prevent natural wetlands from working properly.

Activity Manual

Activity, page 113

Assessment

Rubrics

Select the prepared rubric, or design a rubric to include your chosen criteria.

94

Student Text page 168
Activity Manual page 114

Objectives

- Recall concepts and terms from Chapter 6
- Apply knowledge to everyday situations

Review

- Materials for the Chapter 6 Test will be taken from Student Text page 168 and Activity Manual pages 99–100, 102, 105, 110, and 114. You may review any or all of the material during the lesson. Questions similar to the ones in Thinking It Through, Activity Manual page 114, may appear on the test.
- You may choose to review Chapter 6 by playing the review game or a game from the Game Bank on the Teacher's Toolkit CD.

Activity Manual

Review, page 114

This page requires written responses to application questions.

95

Objective

- Demonstrate knowledge of concepts taught in Chapter 6

Assessment

Chapter 6 Test

CHAPTER SUMMARY

Words to Know

biosphere
biome
climate
permafrost
conifer
deciduous
nocturnal
succulent
biodiversity
salinity
marine biome
wetland
peat

Key Ideas

Recognize that the biosphere is divided into sections called biomes
Recognize that climate determines what types of plants and animals live in a land biome
Describe identifying characteristics for each land biome
Compare and contrast coniferous and deciduous forests
Compare and contrast a prairie and a savanna
Recognize that deserts can be hot or cold
Identify and describe the layers of a tropical rain forest
Identify a mountain as an area that may have more than one biome
Identify saltwater and freshwater as two main categories of aquatic biomes
Identify the ocean as the largest marine biome, divided into several zones and layers
Identify ponds, lakes, streams, and rivers as freshwater biomes
Identify gravity as the force that keeps fresh water flowing downhill toward the ocean
Explain why scientists may disagree about the number of land and water biomes
Identify marshes, swamps, and bogs as transition zones called wetlands
Compare and contrast marshes and swamps
Explain how bogs are different from marshes and swamps
Recognize that people have the God-given responsibility to be good stewards of the earth

168

Review Game

Biome Puzzlers

Find magazine pictures of different biomes or print some from the Internet. Draw lines for nine or more puzzle-shaped pieces on the back of each picture. Cut out along the lines and keep the pieces for each puzzle in a separate envelope or bag.

Divide the class into two teams. Place an unsolved puzzle in front of each team. For each correct answer, allow a team member to choose a puzzle piece. When all the pieces have been chosen, allow the team to assemble the puzzle and identify the biome. Give bonus points if the biome is correctly identified. Continue with other puzzles as time allows.

Teacher Notes

Objectives

- Recognize the interrelationship of science concepts
- Explain the relationship between the study of ecosystems and Genesis 1:28
- Apply the Bible's teaching of stewardship to creatures in a habitat

Teacher's Toolkit CD

- Visuals 7.1–7.2: *Horseshoe Crab*; *Red Knot*

Unit Introduction

- Chapter 7 defines an ecosystem and explains its parts. This chapter also explores many interactions in an ecosystem. These interactions can be between living organisms or between the living organisms and their environments. Chapter 8 continues discussing ecosystems, emphasizing the cycles and stresses that occur.
  Look through Unit 4. What kinds of topics do you think you will be studying in this unit? Possible answers: interactions in an ecosystem, changes in an ecosystem

  Why do you think these chapters are organized into the same unit? Answers will vary. Elicit that the chapters are about things that happen in ecosystems.
- God designed all the parts and organisms of an ecosystem to work together and to be dependent on each other.
- Ask a volunteer to read aloud Genesis 1:28.
  What does it mean to have dominion? Elicit that it means to be a good steward and to use things wisely and constructively.

  Does God want us to have dominion over ecosystems? Yes, Genesis 1:28 says that mankind is to subdue the earth and have dominion over fish, birds, and all animals.
- Explain that studying ecosystems is important to Christians because of God's command to exercise dominion over the whole earth. You cannot be as good a steward without learning how all the parts and organisms of God's amazing ecosystems work together and depend on each other. As these chapters are studied, emphasize God's command to have dominion (Genesis 1:28).

Background

Unit Photo

The unit photo on page 169 is a grizzly bear with a salmon in Katmai National Park in Alaska.

Weblinks

The BJU Press website offers additional information and links you may find helpful throughout the unit.
bjupress.com/resources/science

- Ask volunteers to read the teasers. Discuss the information.

Project Idea

- The project idea presented at the beginning of each unit is designed to incorporate elements of both chapters as well as information gathered from other resources. You may choose to have the students do the project as a culminating activity at the end of the unit or as an ongoing activity while the chapters are being taught.

Unit 4: Terrarium Ecosystems

- Set up one or more terrariums to observe interactions and cycles. Each terrarium should model an ecosystem. Research to find the plants and soil types to include.
- Plans for setting up terrariums are available from many sources. Each terrarium should have a bottom layer of gravel that is covered by a thin layer of aquarium charcoal to provide drainage for moisture. Add soil and plant grass seed or other plants appropriate for the ecosystem. Add enough water to moisten the soil. No water should pool at the bottom of the container. Organisms such as earthworms, snails, or other small creatures may be added.
- Keep the terrarium in an area that is well lighted but out of direct sunlight. If organisms other than plants are added, provide a food source and a small opening covered with screen or gauze for air to circulate. After several weeks, more water may be added if the soil has become dry.
- Observe and record interactions and cycles that occur within each miniature ecosystem. The water cycle is the easiest to observe.

This lesson continues with the Chapter 7 introduction on page 171.

Unit 4 Overview

Lesson	Teacher's Edition	Student Text	Activity Manual	Assessment	Content
Chapter 7: Interactions in an Ecosystem					
96	194–97	169–71	115		• Worldview thinking • Unit and chapter preview
97–98	198–201	172–75	116–18	Quiz 7-A	• Ecosystem parts: individual, population, and community • Ecosystem roles: producer, consumer, and decomposer
99	202–3	176–77	119–20	Rubric	**Activity: Habitat Investigation** • Process skill: observing
100	204–7	178–81	121–22	Quiz 7-B	• Food chains and food webs • Energy pyramid
101	208	182			**Activity: Food-Web Connections** • Process skill: defining operationally
102–3	209	183		Rubric	**Exploration: A Tangled Web** • Make a visual representation of a food web
104	210–11		123–24		**Answers in Genesis** • A biblical viewpoint of the history of carnivores
105–6	212–15	184–87	125		• Plant adaptations • Animal adaptations
107	216–19	188–91	126–29	Quiz 7-C	• Parasitism, mutualism, and commensalism • Instincts and learned behaviors
108	220	192	130		**Chapter Review** • Apply knowledge to everyday situations
109	220			Chapter 7 Test	**Chapter 7 Test**
Chapter 8: Changes in an Ecosystem					
110	221	193	131		• Worldview thinking • Chapter preview
111–12	222–25	194–97	132–34	Quiz 8-A	• Seasons • Carbon cycle
113–14	226–29	198–201	135–36	Quiz 8-B	• Nitrogen cycle • Water cycle
115–16	230–31	202–3	137–38	Rubric	**Activity: Decomposers at Work** • Process skill: identifying and controlling variables
117–18	232–35	204–7	139	Quiz 8-C	• Stresses on an Ecosystem: fire, flood, and drought • Succession
119–20	236	208	140	Rubric	**Exploration: Stress Alert** • Present a mock news report from research on a historical stress event
121–22	237	209		Rubric	**Activity: Current Events** • Process skill: communicating
123	238–39		141–42		**Answers in Genesis** • A biblical worldview relating cycles in nature to God's care of creation
124	240–43	210–13	143	Quiz 8-D	• Manmade stresses • Native species, invasive species, and extinction
125	244	214	144		**Chapter Review** • Apply knowledge to everyday situations
126	244			Chapter 2 Test	**Chapter 2 Test**

Interactions in an Ecosystem 7

Horseshoe crabs are very useful to people. Millions are used every year as bait to catch other types of seafood.

Scientists also gather them and collect some of the blood from each crab. Their unique blood is used to test thousands of medicines. This testing helps provide safe medicines for many people. The crabs are then released back into the wild.

Horseshoe crabs are also important to animals such as red knots. Red knots are birds that migrate each spring. They fly from South America to the Arctic Circle, stopping at places such as Delaware Bay in North America. When they reach the bay, they are dangerously thin. But year after year they arrive at about the same time that thousands of horseshoe crabs lay their eggs on the beach. These eggs provide the nutrition that the red knots need to complete their trip to the Arctic Circle.

Consider

- What might happen to the red knots if more horseshoe crabs are used for bait?
- What truths might guide a Christian fisherman or scientist in his use of horseshoe crabs?

171

Background

Chapter Preview

Preview and prereading activities may include using a K-W-L chart, a probe, or an anticipation guide.

Chapter Photo

The photo on page 171 shows a terrapin sitting on a fallen tree on a river. The miniature scientist is examining baby horseshoe crabs and horseshoe crab eggs.

Activity

Creating a Management Plan

Divide the students into groups. Direct each group to create a solution for the question.

What can we do so that the red knots, the scientists, and the fishermen each have what they need? How would this plan be implemented?

Emphasize that the goal is to create a solution that helps all groups. Allow the students to be creative. It is not important that their ideas be financially or politically feasible. Accept any reasonable plan. Encourage them to draw pictures or maps that would help explain their plan.

Chapter Introduction

Have you seen birds migrating?

What birds have you seen?

Preparation for Reading

- Direct the students to read the page silently to discover what birds migrate from South America.

Teach for Understanding

- Display the *Horseshoe Crab* visual.

  Have you ever seen a horseshoe crab?

  How can horseshoe crabs help us? provide blood that scientists use to test medicine; used as bait to catch other types of seafood

  Why is it important to help people? People are made in God's image and are important to Him.

  Think about Genesis 1:28. Do people have authority to use horseshoe crabs for bait? Yes, God gave people authority over all creation including the fish of the sea.

- Display the *Red Knot* visual.

  Have you ever seen a red knot?

  What are the red knots like when they arrive in Delaware Bay? They are dangerously thin.

  What is happening with the horseshoe crabs at the same time the red knots arrive at Delaware Bay? The horseshoe crabs are laying their eggs on the beach.

  What might happen if the horseshoe crabs laid their eggs at a different time of year? The red knots would not have enough nutrition to complete the trip to the Arctic Circle.

  Does it matter if all the red knots die? Yes, people are given the responsibility to exercise dominion over the earth. If the red knots die because people are selfish or irresponsible then people are not being good stewards.

  What might happen to the medicines if horseshoe crabs were killed and used for bait? Scientists would have a more difficult task of testing medicines. It might be more difficult to get safe medicines.

  Why do useful and safe medicines matter? People are made in God's image and are important to God.

  In what three ways can people show good stewardship in this story? Keep the horseshoe crab eggs available for the red knots, have the horseshoe crab blood available for scientists, and have the bait available for fishermen.

Activity Manual

Preview, page 115

The Looking Ahead page is intended to assess the student's prior knowledge before beginning the chapter.

Student Text pages 172–75
Activity Manual pages 116–18

Objectives

- Identify the two parts of an ecosystem
- Explain the relationships between individuals, communities, and populations
- Identify the functions of producers, consumers, and decomposers
- Explain why scavengers and decomposers are important to an ecosystem

Teacher's Toolkit CD

- Visual 7.3: *Ecosystem Organizer*
- Quiz 7-A

Vocabulary

- ecosystem
- individual
- population
- community
- habitat
- niche
- producer
- consumer
- decomposer

Two days have been allotted for this lesson. You may want to teach Lesson 97 the first day and use ideas from Lesson 98 on the second day. You may also incorporate the ideas from Lesson 98 into the text material, making the lesson two days.

Introduction

What are the different parts of a birthday cake? the cake, the frosting, decorations, candles

All of these things together make up a birthday cake. In a similar way, an ecosystem is made up of different parts.

Preparation for Reading

- Display the word *niche*. Pronounce the word together.
- Direct the students to mentally make questions using the bold words.
- Instruct them to read silently pages 172–75 to find the answers to these questions.
  What is an ecosystem?
  What are the three main roles in an ecosystem?

Teach for Understanding

- The organizer found on Activity Manual page 116 will be used in the teaching of this lesson. You may choose to display the copy of the organizer found on the Teacher's Toolkit CD.
- Ask a volunteer to read the section titles on the organizer.
  What is an ecosystem? all the living organisms and their environment in a certain section of the earth

At first, the forest seems to be quiet. The leaves of tall trees rustle softly in the evening breeze. Then the silence is broken by the soft hoot of an owl. As you look for the owl, you become aware of other animals. A small mouse scurries across the forest floor. A red fox watches the mouse. These are just some of the living organisms in the forest. Many organisms are so small that you cannot see them.

Each organism, whether big or small, is part of an ecosystem. An **ecosystem** consists of all the living organisms and their environment in a certain section of the earth. Ecosystems vary in size. Some, such as a tropical rain forest, are as large as a biome. Others, such as a rotting tree stump on the forest floor, are small. Even your skin is an ecosystem for many tiny organisms. Part of being a good steward in God's world is learning about ecosystems.

Parts of an Ecosystem

One part of an ecosystem is its environment. The *environment* is the nonliving part of an ecosystem. This part includes the soil, water, sunlight, temperature, and air. The environment determines the kinds of organisms that can live in that ecosystem. For example, plants that need a large amount of water could not live in a desert. An animal with thick, long fur would not survive in the hot, humid climate of a tropical rain forest. Nor could a reptile that depends on warmth from the sun for its body heat live in the cold Arctic tundra.

The other part of an ecosystem is made up of *living things*. These living things include plants and animals, as well as organisms such as bacteria that are too small to be seen. Some of the living members in a forest ecosystem may include trees,

172

Background

Ecosystems

Ecosystems do not have set boundaries and distinct classifications. A group of ecosystems that have similar vegetation and climates are sometimes considered to be a biome.

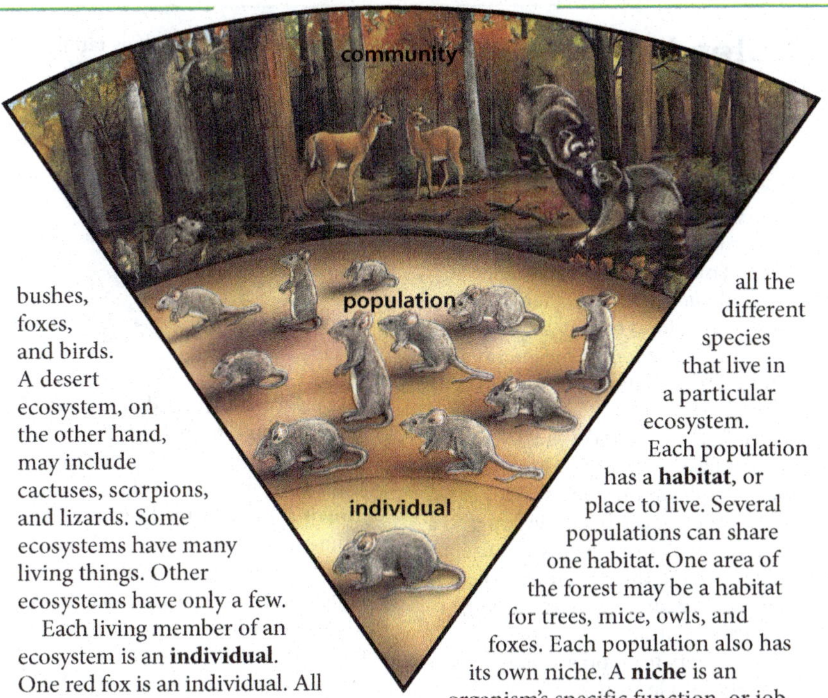

bushes, foxes, and birds. A desert ecosystem, on the other hand, may include cactuses, scorpions, and lizards. Some ecosystems have many living things. Other ecosystems have only a few.

Each living member of an ecosystem is an **individual**. One red fox is an individual. All the red foxes in the forest, though, make up the forest's red fox population. A **population** consists of all the organisms of the same species that live in an ecosystem.

Populations can be different sizes. One forest might have a population of only a few foxes. But that same forest could have a red oak tree population of hundreds and a white-footed mouse population of thousands. Together, the populations of the foxes, mice, trees, and other living things form a community. A **community** includes all the different species that live in a particular ecosystem.

Each population has a **habitat**, or place to live. Several populations can share one habitat. One area of the forest may be a habitat for trees, mice, owls, and foxes. Each population also has its own niche. A **niche** is an organism's specific function, or job, in the ecosystem. A niche includes all the different ways that an organism uses its *resources*, or available supplies. An animal's niche includes what the animal eats, when it eats, how it protects itself, and how it raises its young.

Populations can share a habitat and may even eat similar foods. But two populations cannot share exactly the same niche. For example, both hawks and owls eat mice. However, hawks hunt by day and owls hunt at night.

- Display the prefix *eco-* and the suffix *-logy*. Explain that *eco-* comes from a Greek word meaning "house" and *-logy* means "a study of." Ecology is the study of the house, or home. A system can be defined as "parts that work together as a whole." An ecosystem is a home with parts that work together.
  How big are ecosystems? Possible answers: Their sizes vary. They can be as big as a biome or as small as a rotting tree stump.
- Allow time for the students to complete the "Ecosystem" section on the organizer.
  What are the two parts of an ecosystem? living things and nonliving things
  What term is used to describe all the nonliving parts of an ecosystem? the environment
  What are some parts of the environment? Possible answers: soil, water, sunlight, temperature, air
  How does the environment affect the ecosystem? The environment determines the kinds of organisms that can live in the ecosystem.
  Besides animals, what other organisms are living things? Possible answers: plants, organisms too small to be seen, humans

> As you teach about the parts of an ecosystem, help the students understand the relationship between the levels by comparing the levels to the types of relationships that a person has. You may extend the relationships by reviewing a biome and biosphere from Chapter 6.
>
> | | |
> |---|---|
> | individual—person | ecosystem—state |
> | population—family | biome—country |
> | community—town | biosphere—world |

What is an individual? each living member of an ecosystem

- Guide the students as they complete the "Individual" section of the organizer.
  What is the difference between a population and a community? A population is all the organisms of the same species that live in an ecosystem, but a community includes all the species that live in a particular ecosystem.

What is a species? a specific kind of organism

- Explain that an example of different species is red squirrels and gray squirrels. They are similar, but each forms its own population.
- Allow time for the students to complete the "Population" and "Community" sections on the organizer.
  Are all populations the same size? no
  What term describes where an organism lives? habitat
  Can populations share a habitat? yes
  What can populations not share? the exact same niche
  What is a niche? an organism's function, or job, in the ecosystem
  What kinds of things are included in an animal's niche? Possible answers: what the animal eats, when it eats, how it protects itself, how it raises its young

Background

Niches

An organism's niche is actually a broad description of its characteristics and function within its environment. An animal's niche includes where it lives, where it hunts, when it hunts, what it eats, how it gets shelter, and how it raises its young. For example, an owl is a nocturnal bird usually found in forests. It eats small mammals, insects, amphibians, and other animals. Many owls nest in trees and raise their young during the spring.

Sharing Niches

Organisms cannot share the exact same niche, but their niches often overlap. For instance, a forest may have many species of insect-eating birds competing for food and nesting sites. Foxes and coyotes also have very similar niches and may compete. Competition is discussed in Lesson 100.

97
98

What is the source of an ecosystem's energy? the sun

- Remind the students that energy cannot be created or destroyed.

  How do plants use the sun's energy? to make food from carbon dioxide and water

  What is this process called? photosynthesis

  How do plants use most of their food energy? to grow and reproduce

  Where do plants store their unused food energy? in their roots, stems, or leaves

  Why are plants called producers? They make, or produce, their own food.

  What are consumers? living things that depend on producers for food

  Why must consumers eat food in order to get energy? They cannot get energy directly from the sun.

  What name is given to consumers that eat only plants? herbivores

  Do all herbivores eat only the leaves of the plants? no What else might they eat? Some herbivores eat the seeds or roots.

  How are omnivores different from herbivores? Omnivores eat both plants and animals. Herbivores eat only plants.

  What are consumers that eat other consumers called? carnivores or meat eaters

- Direct attention to the picture on page 172.

  What are the producers? the grass and the tree

  What are the consumers? the owl, fox, and mouse

  What kind of consumer is the fox? an omnivore

  Do carnivores eat plants? not usually

  How do carnivores get their energy? by eating other consumers

- Share with the students that most carnivores hunt and kill the animals that they eat. Some carnivores may steal freshly killed meat from other carnivores. This is usually meat that they normally would have hunted.

  Do scavengers usually eat plants or other animals? other animals How do you know? because most scavengers are carnivores

Roles in an Ecosystem

Animals eat food to get energy. But where does this energy come from? The source of this energy is the sun. Plants use energy from sunlight to make food from carbon dioxide and water. This process of making food is called *photosynthesis*. Plants use most of this food energy to grow and reproduce. The rest of the energy is stored in the roots, stems, or leaves of the plant.

Because plants make their own food, they are called **producers**. All life depends on producers. They change the energy from the sun into a form that other organisms can use.

Many living things cannot get their energy directly from the sun. These living things that depend on producers for food are called **consumers**. Consumers, such as animals and humans, must eat food in order to get energy. In an ecosystem there are usually more producers than consumers.

174

Some consumers, such as rabbits and grasshoppers, eat only plants.

Herbivore

These animals are called *herbivores*. They get their energy from the plants. However, not all herbivores eat the same part of the plant. Some herbivores eat only the leaves. Others eat the seeds or roots.

Some consumers eat both plants and other animals. These consumers are called *omnivores*. For example, the red fox usually eats mice, but it also eats berries and grass. Many omnivores change their eating habits when the seasons change. The red knot eats grass shoots and seeds in the cold Arctic until warm weather brings insects for it to eat.

Consumers that eat other consumers are called *carnivores*, or meat eaters. These animals usually do not eat plants. Instead

Omnivore

Background

Producers

Producers can be divided into two groups, and not all carry on photosynthesis. Some producers, such as plants and algae, are *photoautotrophs*. They get energy from light. Other producers, such as some bacteria, are *chemoautotrophs*. They use inorganic compounds for energy.

Consumers

Some consumers change roles as they become adults. For example, young green sea turtles are omnivores. Adult green sea turtles, though, are carnivores.

Activity

Rain Forest Ecosystem

Materials: Instructional Aids 7.1, 7.3: *Toucans in the Rain Forest*; *Tropical Rain Forest Ecosystem*, for each student; Instructional Aids 7.2, 7.4: *Toucans in the Rain Forest Answer Key*; *Tropical Rain Forest Ecosystem Answer Key*

The students may work these pages independently to reinforce the concept of the parts of an ecosystem.

Carnivore

these animals get their energy by eating other consumers. Many carnivores, such as cougars and weasels, eat several different species of animals.

Some consumers help keep the ecosystem clean. These animals, called *scavengers*, eat things that have already died. Cockroaches and vultures are two common kinds of scavengers. Most scavengers are carnivores.

Another group of organisms are called decomposers. **Decomposers** help break down dead things and wastes. Decomposing adds minerals and nutrients back into the environment. Plants then use the nutrients to grow and produce more food. There are many kinds of decomposers. Most, like bacteria, cannot be seen without a microscope. Other decomposers include molds, mushrooms, and earthworms.

Scavengers and decomposers are an important part of the ecosystem. Without them, the ecosystem would be full of wastes and dead things.

QUICK CHECK

1. What are the two parts of an ecosystem?
2. Which is larger—a population or a community?
3. What is different about the way producers and consumers get energy?

175

97
98

- Explain that scavengers generally eat the remains of meat from an animal that has been killed and left by carnivores. Scavengers may also eat animals that have been killed on a road or that have died from illness.

Which group of consumers helps break down dead things and wastes? decomposers

What are some examples of decomposers? bacteria, molds, mushrooms, earthworms

Are decomposers large organisms? Not usually. Most decomposers cannot be seen without a microscope.

Why are scavengers and decomposers important to an ecosystem? They help keep it clean by getting rid of dead things and wastes.

ANSWERS

1. living things and nonliving things
2. a community
3. Producers get energy directly from the sun. Consumers get energy by eating producers or other consumers.

98

- The *Rain Forest Ecosytem* activity found at the bottom of the pages in Lessons 97–98 may also be incorporated into the lesson.
- Suggestions for additional activities for this two-day lesson are on the Teacher's Toolkit CD.

Background

Scavengers

Scavengers usually do not kill their prey. Instead, they eat animals that have recently died.

Decomposers

Decomposers and scavengers could also be considered consumers since they do not make their own food.

Science Misconceptions

Plants

Some plants are both producers and carnivores. There are at least 500 species of carnivorous plants. Some of these plants are discussed in Lessons 105–6.

Activity Manual

Study Skill, page 116

Guide the students as they complete the organizer about the parts of the ecosystem.

Review, pages 117–18

These pages review Lessons 97–98.

Assessment

Quiz 7-A

The quiz may be given any time after completing this lesson.

99

Student Text pages 176–77
Activity Manual pages 119–20

Objectives

- Investigate a habitat
- Distinguish between living things and nonliving things
- Identify producers and consumers
- Record interactions

Materials

Materials needed for this activity are listed on the Student Text page.

Teacher's Toolkit CD

- Rubrics: *Habitat Investigation*, for each student

This activity will work best if the designated habitat area is grassy or is at the edge of a wooded area. Be aware of any stinging or biting insects or small animals that may be in the habitat areas.

Introduction

Have you ever watched fish swim in a fish tank or pond?

- Explain that if you watch closely for a while, you start to notice different things. You might notice how one fish looks different from another, or you might notice that one fish is hiding under a rock while another swims in the open.

The longer you watch the fish, the more you will notice. You might notice things about their behavior, the water that they are in, or other living things in the water, such as plants or insects.

In this activity you are not watching fish. But as you look at a different kind of ecosystem, you will need to do the same kind of careful observation.

Preparation for Reading

- Direct the students to read Student Text pages 176–77 and Activity Manual pages 119–20 before beginning the activity.

Teach for Understanding

Process Skill

The process skill emphasized in this activity is observing, which is using the senses to gather information.

What kinds of information can you gain by observing your habitat in this activity? Possible answers: the types of organisms living in the habitat; some interactions between living things; the behavior of some organisms

Why are observations important to scientific research and thinking? Possible answers: to increase knowledge, to make comparisons, to see changes over time

ACTIVITY

Habitat Investigation

Process Skills
- Observing
- Classifying
- Collecting and recording data
- Defining operationally

What do you think of when someone mentions habitats? Perhaps you think about large animals and their homes. Or you might think of a plant that is unique to a certain environment.

Both plants and animals have habitats. Some are big. For example, a pond, forest, or savanna can be a habitat for many different populations. But some habitats are so small that you might not even notice them at first.

In this activity you will be investigating a small habitat.

Purpose

Investigate a habitat.

Procedure

Materials
- meter stick
- 4 stakes
- string
- scissors
- toothpicks
- garden trowel
- magnifying glass
- Activity Manual

1. Go outside to an area that your teacher has selected. Choose a spot of ground to mark as your habitat.
2. Measure a square that is 1 m on each side. Push a stake into the ground to mark each corner of the square. Outline your habitat by connecting the stakes with string.
3. Carefully observe your habitat without disturbing it. Describe your habitat and its surrounding area in your Activity Manual. Use terms such as *grassy*, *wooded*, or *sandy*.
4. Notice whether any living things are interacting. Record your observations.
5. List the nonliving things that are part of the habitat.
6. Use toothpicks to carefully move pieces of plant matter or other surface material. Look for any living things that might be hiding underneath.
7. Use toothpicks or the trowel to gently move some of the soil. Look for living things that might be in the soil. Use the magnifying glass as needed.

176

8. List all the living things that you observe in your habitat. Remember to include plants as well as animals. If you cannot identify a plant or animal by name, describe it or draw a small picture of it.
9. Identify the living things as either producers, consumers, or decomposers. Count several populations and record the totals.

Conclusions

- How many different populations did you observe in your habitat?
- How does the environment affect the organisms that make up your habitat?

Follow-up

- Use a field guide or other resource to identify the consumers as herbivores, omnivores, or carnivores.
- Observe a different habitat. Compare this habitat to your earlier one.

177

99

Why is recording data from observations important in scientific research? Possible answers: to ensure accuracy, to make comparisons, for others to use the information that you observed

Procedure

- As needed, assist the students in measuring and staking a square. Remind them to draw and describe any living thing that they cannot identify.

> To save time, you may want to construct a frame of meter sticks taped together to help the students mark their square meter on the ground.

- When the students are identifying consumers as herbivores, omnivores, or carnivores, it might be helpful to remind them that some carnivores, such as spiders, are very small.

Conclusions

- Review that each population is a different species.
- Help the students find connections between the living things and the nonliving things found in their habitats so they can complete the questions.

Activity Manual

Activity, pages 119–20

Assessment

Rubrics

Select the prepared rubric, or design a rubric to include your chosen criteria.

100

Student Text pages 178–81
Activity Manual pages 121–22

Objectives

- Identify the predators and prey in a food chain
- Differentiate between a food chain and a food web
- Describe the transfer of energy from one organism to another
- Explain how competition affects population size

Materials

- 5 dominoes

Teacher's Toolkit CD

- Visual 7.4: *Forest Food Web*
- Quiz 7-B

Vocabulary

- food chain
- predator
- prey
- food web
- competition

Introduction

To show the introduction to an entire class, stand the dominoes beneath a document camera.

- Stand five dominoes in a straight line. Place the dominoes about 3 cm apart.
  What will happen if the first domino is knocked over?
  The other dominoes will fall.
- Push the first domino, asking the students to observe the dominoes as they fall.
  Did our results support our prediction? yes
- Stand the dominoes again and remove the third and fourth dominoes.
  I have removed two dominoes. Do you think the last domino will fall when the first one is knocked over?
  Answers will vary.
- Push the first domino. Allow the students to determine if the results supported their prediction.
  One falling domino can knock down the whole line of dominoes. But removing dominoes from the line prevents some from falling. In a similar way, all living things in an ecosystem interact with each other. A change to one part of an ecosystem affects the other organisms.

Preparation for Reading

- Direct the students to mentally make questions using the headings, bold words, and captions.
- Instruct them to read silently pages 178–81 to find the answers to these questions.
  How are food webs different from food chains?
  How much energy is transferred at each level of an energy pyramid?

Energy in an Ecosystem

Chains and Webs

A grasshopper nibbling on a leaf might become dinner for a mouse. Later, a fox might catch and eat the mouse. The energy that the leaf got from the sun has been passed from one organism to another. This transfer of energy and nutrients through a community is called a **food chain**.

Food Chain

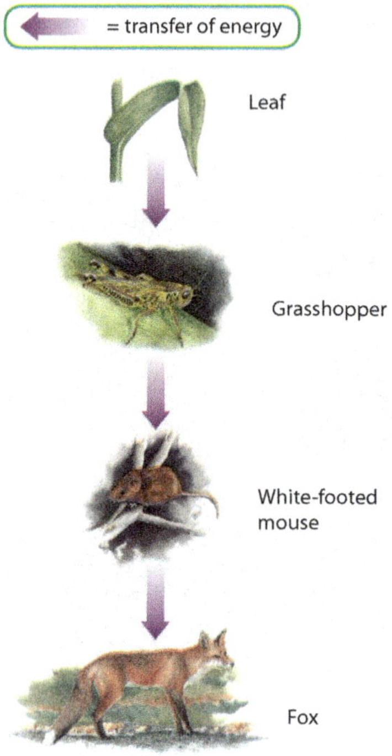

A food chain always begins with a producer. The second link of a food chain is usually a herbivore. The animals that make up any additional links are predators. A **predator** is any animal that hunts and eats other animals. Predators are either carnivores or omnivores. The animals that a predator hunts are called its **prey**.

A food chain shows only one source of food for each animal. However, most animals are part of more than one food chain. This is because animals usually eat more than one kind of food. And most animals are eaten by more than one kind of predator. A **food web** consists of several food chains linked together. For example, a grasshopper may be eaten by a mouse, but a grasshopper can also be food for a spider, bird, or salamander.

Changing one part of a food web affects all the other organisms in that web. For example, weasels are often eaten by hawks. If you removed all the hawks, the weasel population would probably increase. But then the ecosystem might not have enough resources for all the weasels. Weasels eat mice and small birds. If there were more weasels, there would be fewer mice and small birds. And other animals that eat mice and birds would have less food. Food webs show how organisms depend on each other.

178

Background

Trophic Levels

The term *trophic levels* is sometimes used when referring to the different links in a food chain. The major trophic levels are producers, consumers, and decomposers.

Levels of Consumers

Herbivores are generally considered primary consumers in any food chain. The consumer that eats the herbivore is the secondary consumer. The tertiary consumer eats the secondary consumer, and so on. Food chains do not usually have more than five links.

Science Misconceptions

Animal Stereotypes

People sometimes think that all herbivores are small and gentle and that carnivores are mean and aggressive. However, both herbivores and carnivores can be large or small. Also, some herbivores—such as geese, rhinoceros, and coral reef fish—can be aggressive. Some omnivores and carnivores—such as bearded dragons, spectacled bears, and dolphins—can be gentle.

Forest Food Web

179

Background

Technology and Food Webs

Scientists use computers to help manage the food webs of wildlife populations. One example is in the Chesapeake Bay area of Maryland and Virginia. Chesapeake Bay has many striped bass that prey on menhaden. If too many of these fish are caught, the populations of other organisms will change. Too many striped bass in the bay could cause the menhaden population to dwindle drastically. If too many menhaden are caught, the striped bass might not have enough food. The populations of other organisms, such as blue crabs and rockfish, might also be affected.

Scientists working with the National Oceanic and Atmospheric Administration use computer models to help predict these kinds of results before they happen. Studying the results of the computer models helps scientists estimate the maximum number of each species of fish that fishermen should catch.

100

Teach for Understanding

What do scientists call the transfer of energy and nutrients through a community? a food chain

- Direct attention to the *Food Chain* diagram on page 178.

  What do the arrows in the diagram tell you? The arrows show the direction that energy is moving.

  What is always the first link of a food chain? a producer

  What is the producer in this food chain? a leaf

  Which kind of consumer is often the second link in a food chain? a herbivore

- Ask a volunteer to name the herbivore in the *Food Chain* diagram. grasshopper

  Could an omnivore be the second link in a food chain? yes Why? They eat both plants and animals.

  What is a predator? an animal that hunts and eats other animals

  What are the predators in the *Food Chain* diagram? white-footed mouse and fox

  What are the animals that a predator hunts called? prey

  Trace the transfer of energy back to see which animals are eaten by other animals. Which animals in the *Food Chain* diagram are prey? grasshopper and white-footed mouse

  How are food chains different from food webs? Food chains show only one source of food for each animal. Food webs consist of several food chains linked together.

- Display the *Forest Food Web* visual.

  What do the arrows in the diagram tell you? The arrows show the direction that energy is moving.

  Look at the different pictures and captions. What are the producers in this food web? the grass and blueberries

  Look at each animal. See which transfer of energy arrows point to it. Which animals only transfer energy from the grass and blueberries? the grasshopper and the rabbit

  Which animals are herbivores? the grasshopper and the rabbit How do you know? They eat only producers.

  Look at the transfer of energy arrows pointing to the striped skunk in this food web. What do striped skunks eat? grass and blueberries, mice, rabbits, and grasshoppers

  Is the striped skunk a herbivore, an omnivore, or a carnivore? omnivore How do you know? It eats producers and other animals.

  Is the hawk a carnivore or an omnivore? a carnivore How do you know? It eats only other animals.

  What is a predator? any animal that hunts and eats other animals. What animals are predators of the robin? the hawk and the snake

  Which animal is the prey of the white-footed mouse? the grasshopper

- Continue asking questions about the food web as time allows.

What happens to the amount of energy each time the energy is transferred in a food chain? The amount of energy decreases.

Why is less energy available? Each organism in a chain uses most of the energy for its own needs. Only a small amount is transferred.

How do scientists show the way energy moves through an ecosystem? with an energy pyramid

How many food chains are shown in an energy pyramid? one

What kinds of organisms are found at the base of the pyramid? producers

- Direct attention to the *Energy Pyramid* diagram.

What is the producer in the energy pyramid? grass

Why do producers receive the most energy? They get their energy directly from the sun.

Which consumer would probably receive more energy, a herbivore or a carnivore? a herbivore Why? Herbivores eat plants. A carnivore would have to eat a herbivore that had eaten plants.

Are the mice the predators or the prey of the grasshoppers? the predators

What animal gets its energy by eating the mice? the fox

Are there more foxes or producers in this energy pyramid? producers Why are there more producers? Producers get their energy directly from the sun. There are fewer foxes than producers because less energy is available.

What does the shape of the pyramid tell you about the amount of energy at each level? The widest part of the pyramid has the most energy available. The top is small to show where the least amount of energy is available.

- Explain that predators at the top of an energy pyramid need to eat more organisms in order to have enough energy to stay alive.

If each mouse provides 10% of the energy needed by a fox, how many mice would a fox need to eat to have 100% of its energy needs met? 100% × 10% = 10

If 100 foxes populate an ecosystem, how large would you expect the mouse population to be for that ecosystem? 100 × 10 = 1,000

- You may want to do the *Competing for Resources* activity at this time.

Energy Pyramid

Energy is transferred at each link of a food chain. However, the amount of energy transferred becomes smaller each time. A grasshopper gets energy from eating grass. It uses most of that energy for its own needs. Some of the energy, though, is stored in the grasshopper's body. That stored energy later transfers to the mouse that eats the grasshopper.

Scientists use an energy pyramid to show how energy moves through an ecosystem. An energy pyramid shows only one food chain. Producers, such as plants, form the base of the pyramid. The producers receive the most energy because they get it directly from the sun.

The other levels represent the different links of a food chain. The pyramid shows how the amount of energy changes at each level. There are fewer animals at the top of the pyramid because less energy is available. Each organism uses about 90% of its energy to live and grow. That means only about 10% of its energy is passed on to the next organism.

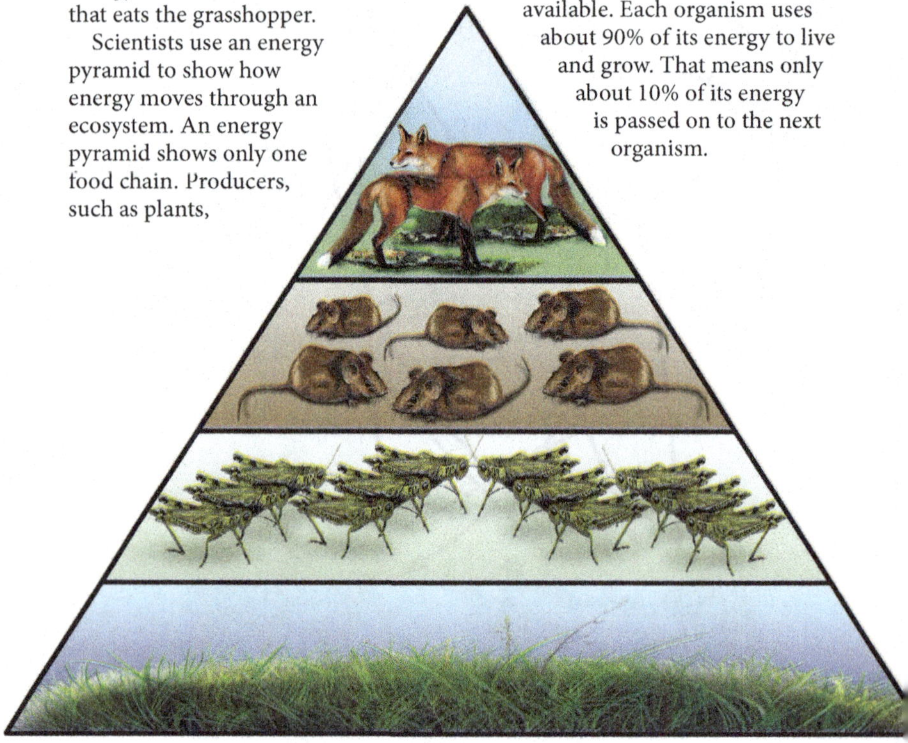

Energy Pyramid

180

Background

Ratio of Producers to Consumers

Most ecosystems have more producers than consumers. However, this is not always true. A giant oak tree is an ecosystem. The tree can be the single producer for thousands of insects, dozens of birds, several squirrels, and countless other organisms.

Uses of Energy

Even though only 10% of the energy is transferred at each level, the remaining energy is not lost. It is used for an organism's physical needs, such as growth and reproduction. At other times, the energy remains in the environment, but it may be stored in a form that is not usable to certain organisms.

On a prairie, grasses and wildflowers may have to compete for space, water, and sunlight.

Competition occurs when two or more organisms are trying to use the same resources. This happens when wildflowers compete with grasses for space, water, and sunlight. Competition also happens when two animals need to eat the same kind of food. Red knots and other kinds of sandpipers compete for horseshoe crab eggs. Competition can occur between members of the same species or between members of different species.

Competition helps keep the ecosystem balanced. For example, owls eat white-footed mice. If there are not enough mice, the owls will have to compete for the same limited prey. Some of them may not be able to find food in that ecosystem. They may move to another area, and some of them may die. The population of owls will adjust based on the populations of their food. If the population of mice increases, the population of owls may also increase again. Or other animals such as foxes might move into the area and help keep the mice population balanced.

Owl with prey

QUICK CHECK

1. How are food chains different from food webs?
2. Why are there fewer animals at the top of an energy pyramid?
3. When does competition occur?

181

100

When does competition occur? when two or more organisms are trying to use the same resources

What are some resources for which organisms might compete? Possible answers: space, sunlight, food, water, shelter

Does competition occur only between members of the same species? No, competition can also be between different species.

How can competition benefit the ecosystem? keeps it balanced

ANSWERS

1. Food chains show only one source of food for each animal, but food webs consist of several food chains linked together.
2. There is less energy available for them to survive.
3. when two or more organisms are trying to use the same resources

Activity Manual

Reinforcement, page 121

Review, page 122

This page reviews Lesson 100.

Assessment

Quiz 7-B

The quiz may be given any time after completion of this lesson.

Activity

Competing for Resources

Materials: resealable sandwich bags for three-fourths of the students, carnivore identification badges for one-fourth of the students, popcorn or packing peanuts, stopwatch or timer

Prepare each bag by making a line about 5 cm (2 in.) from the bottom edge. Choose three-fourths of the students to represent herbivores. Give each herbivore an empty bag. Give the remaining students a carnivore identification badge.

Scatter the popcorn "producers" around an area. Explain that, for 2–3 minutes, the herbivores will run around gathering producers. At the same time, they need to avoid the carnivores. The carnivores will chase the herbivores, trying to tag them. If a herbivore is tagged, he must give his bag to the carnivore and exit the game.

When time is called, the herbivores and carnivores should check their bags. For a herbivore to avoid starvation, his bag must be filled up to or above the line. Choose the number of bags that a carnivore must "capture" to avoid starvation. (Choose a number that some, but not all, of the carnivores can attain.)

After the game, compare the activity to the energy pyramid. Summarize that herbivores need a certain amount of the producer in order to have enough energy. Carnivores need an even greater amount. Because less energy is available for the higher levels of the pyramid, the population of carnivores is smaller.

101

Student Text page 182

Objectives

- Identify predators and prey within a food web
- Model a food web
- Recognize interrelationships among organisms in a food web
- Compare the model food web with an actual food web

Materials

- piece of string, 30 cm long, for each food-web card

Teacher's Toolkit CD

- Instructional Aids 7.5, 7.7: *Forest Food-Web Cards; Pond Food-Web Cards,* for each group of seven students
- Instructional Aids 7.6, 7.8: *Forest Food-Web Answer Key; Pond Food-Web Answer Key*

You may want to use either the forest food-web cards or the pond food-web cards according to what your students are familiar with. You may substitute a food web from an ecosystem near you. You will need at least 20 pieces of string for each group. Each student will use a different piece of string for each connection made to another organism. See the illustration on Activity Manual page 182 to help you with setup.

Introduction

What kinds of things do you think owls normally eat? Possible answers: mice, moles, insects, rabbits, squirrels, snakes, small birds

Owls eat more than one kind of food. The animals in the food web you will make also eat different kinds of food. See how many connections you can make in your food web.

Preparation for Reading

- Instruct the students to read silently page 182 before beginning the activity.

Teach for Understanding

Process Skill

The process skill emphasized in this activity is defining operationally, which is explaining an object or event properly but in terms of the student's observation and experience.

What is the purpose of the activity? modeling a food web

What are some of the proper terms to use as you explain the activity and your conclusions? Possible answers: producer, herbivore, omnivore, carnivore, transfer of energy, organisms

Procedure

- Divide the students into groups of seven. Give each student one food-web card to hang around his neck.

ACTIVITY

Food-Web Connections

Process Skills
- Making and using models
- Communicating
- Defining operationally

A food web consists of many food chains mixed together. Because animals eat more than one type of food, a food web can have many different connections.

In this activity you and your partners will make a model of a food web.

Purpose

Model a food web.

Procedure

Materials
food-web cards
pieces of string, each 2 m long

1. Get a food-web card and some pieces of string from your teacher. Look at your card and determine whether you represent a producer, an herbivore, an omnivore, or a carnivore.
2. Check the identities of the other members in your group. Decide which people represent your food sources. Use string to connect yourself to your "food sources." See how many different connections your group can make.

Conclusions

- How does your food web compare to an actual food web for an ecosystem?
- If something happened to one organism in your food web, how would that affect the other organisms?

Follow-up

- Identify the transfer of energy in your web.
- Research the consumers in your food web to find out which foods they prefer to eat.
- Add more consumers to your food web.

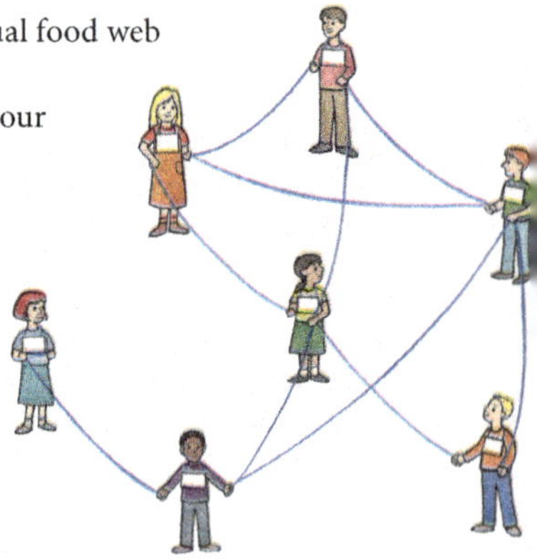

182

- Guide the students as they determine whether they represent a producer, a herbivore, an omnivore, or a carnivore. Use the answer key from the Instructional Aids to help as you guide the students to make connections in the food webs.

Conclusions

- Help each student explain his results in his own words. The key is for him to understand what is happening and be able to use the scientific terms accurately.
  What did you discover? Answers will vary but should include the idea that leaving out one food source affects all the other organisms in a food web.
  How does this food web compare to an actual food web for an ecosystem? An actual food web would have many more organisms and connections.
- Have one person who represents an organism drop the strings he is holding.
  If something happened to one organism in your food web, how would that affect the other organisms? Answers may vary but should include the fact that other populations would increase or decrease.

Assessment

An assessment, such as a rubric, is not recommended for this activity.

A Tangled Web

EXPLORATION

Ecologists are scientists who study how living things interact. This interaction may be between living things or between living things and their environment. Ecologists often use a food web to study predator and prey relationships.

In this exploration you are an ecologist who wants to exercise good dominion over God's world. You will make a visual to represent one of the food webs in an ecosystem.

What to Do

1. Choose an ecosystem that you would like to study. Make a list of several plants and animals that are common to that ecosystem. Be sure to include smaller living things as well as bigger living things. Separate your list into producers and consumers. Try to include at least eight different living things.
2. Research to find out which animals are herbivores. Identify the plants that each of those animals eats. Which animals on your list are omnivores? Identify the plants and animals that each of those animals eats. Next, look at the carnivores. Which consumers does each of them eat?
3. Using the information you gathered about the living things in your ecosystem, make a visual that shows the transfer of energy in the food web for your ecosystem.

183

Objectives

- Make a visual representation of a food web
- Identify producers, predators, and prey within a food web
- Identify animals as herbivores, omnivores, or carnivores

Materials

- books and other resource materials
- materials of each student's choice to make a visual

Teacher's Toolkit CD

- Rubrics: *A Tangled Web*, for each student

Two lesson days are allotted for this activity. You may choose for the students to do their research on the first day and prepare their visuals on the second day.

Introduction

Which kinds of animals interest you the most?

Do you like forest animals, such as deer and foxes?

Or are you interested in animals that live on the tropical savanna?

What about animals that live in the ocean or in a river?

Preparation for Reading

- Direct the students to read silently page 183 before beginning the exploration.

Teach for Understanding

What to Do

For this exploration, choose an ecosystem that you would enjoy learning more about. Then make a visual of a food web that includes animals and plants from that ecosystem.

- Guide the students in selecting an ecosystem and choosing the animals and plants. Provide books and other resources to help them identify predator-and-prey relationships.
- The students can find pictures for their food-web visuals from magazines, advertisements, or the Internet. They may also choose to draw or trace some of the pictures. The visual may use a mixture of pictures from different sources and media.
- Remind the students to use an arrow to link each prey to a predator. The arrow should point toward the predator. This shows the direction that the energy moves in the food chain. The students can use Student Text pages 179 and 183 as examples.

Assessment

Rubrics

Select the prepared rubric, or design a rubric to include your chosen criteria.

104

Activity Manual pages 123–24

Objectives

- Describe relationships among animals and plants in a simple ecosystem
- State the sources of food for both people and animals before the Fall
- Explain why the kinds of teeth in a skull may not determine the kinds of food an animal eats
- Compare and contrast the evolutionary and creationary views of the history of carnivores

Teacher's Toolkit CD

- Instructional Aids 7.9–7.10: *Prairie Food Web*, for each student; *Prairie Food-Web Answer Key*
- Visual 7.5: *Fruit Bat*

Introduction

Who created the world? God

Name a Bible verse that tells you that God created the world. Possible answer: Genesis 1:1

- Ask a volunteer to read aloud Genesis 1:31.

  What did God think about everything that He had created? It was very good.

Preparation for Reading

- Instruct the students to read silently the first two paragraphs on Activity Manual page 123 to discover some of the organisms that live on the South Dakota prairie.

Teach for Understanding

- Direct attention to the *Prairie Food Web* page. Ask a volunteer to read the names of the organisms in the prairie food web.

  We will use the food web to find out how the animals and plants are all related on the prairie.

  Which organisms on the food web are producers? grass, sunflowers

  Write the word *producers* on the line below grass and sunflowers.

  Look at prairie dogs. What do you think a prairie dog eats? grass, sunflowers, grasshoppers

  Draw an arrow to link each prey to a predator. In which direction should the arrow point? toward the predator

  Why does the arrow point toward the predator? to show the direction that the energy moves in the food chain

  Is the prairie dog a herbivore, an omnivore, or a carnivore? omnivore Why? It eats both plants and animals.

  Write *omnivore* on the line below prairie dogs.

- Follow the same procedure with mice, butterflies, grasshoppers, rattlesnakes, and coyotes.

A lesson from

Answers in Genesis Name ____________

1:1

Not What It Used to Be

When you look at nature, you can't help but see God's amazing work in creation. There are many complex interactions that make all these pieces work together so that everything survives and thrives.

Think about a prairie in South Dakota with organisms such as these living together: grasses, prairie dogs, rattlesnakes, mice, grasshoppers, coyotes, sunflowers, and butterflies.

1. Which prairie animals eat the plants?
   prairie dogs, mice, grasshoppers, and butterflies
2. Which prairie animals eat other animals?
   prairie dogs, rattlesnakes, mice, and coyotes
3. Which prairie animals eat both plants and other animals?
   prairie dogs, mice
4. What do you think would happen to the ecosystem if all the grass burned up in a certain area?
   Possible answer: Animals that depend on grass would migrate to another area that would provide food. Once the grass regrew, they would return. Other animals dependent on the grass eaters may also migrate or adapt to eat other foods.
5. What do you think would happen if all of the prairie dogs died or were killed?
   Possible answer: The populations that prairie dogs eat may expand uncontrolled, and the populations that depend on prairie dogs for food would need to find other foods or migrate.

Have things always been this way since the beginning of creation? Have coyotes always eaten other animals to survive? If we want to answer a question like this, the best place to start is with the Bible. Open to Genesis 1 and look for the answer there.

6. What did God tell Adam and Eve to eat in Genesis 1:29? *every herb (plant) and tree*
7. What did God tell all the animals to eat in Genesis 1:30? *every green herb*

Resources

The BJU Press website offers additional links from Answers in Genesis that you may find helpful with this lesson.
bjupress.com/resources/science

So everything God created was created to eat a vegetarian diet, and God called this "very good"! You might be thinking, "What about dinosaurs with their big, sharp teeth and rattlesnakes with their fangs? Doesn't every creature that has sharp teeth eat meat?"

8. What do you think this type of creature eats?
   Answers may vary.

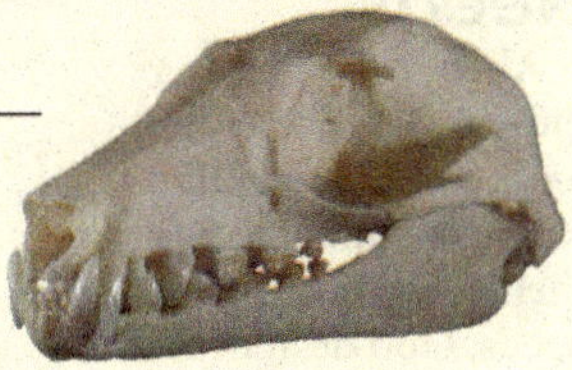

The shape of teeth a creature has may or may not tell you what it eats.

Read Genesis 3:14–19.

9. What caused the original creation to change?
   Adam's disobedience brought God's curse on all of creation, even the animals.

10. How was the original creation changed?
    Possible answers: Women would have sorrow when having children. Men would work hard to provide food.

11. Based on this passage, why do you think some animals eat one another today?

Evolutionary scientists believe that death and animals eating one another have always been part of life on Earth. There are also many Christians who believe that God used evolution in Creation. But this belief makes no sense when we look at the world through our Bible glasses. God would not call death and animals killing and eating each other "very good" before sin entered the world. The death and disease we see in ecosystems today is a result of the corruption that sin brought into the world. This is an important point to keep in mind as we study God's creation since it has been broken by sin.

One day those who are God's children will live with Him in a place where all this corruption is replaced by perfection—even better than what Adam and Eve knew.

124 Chapter 7 | Lesson 104 Answers in Genesis — Science 5

- The students are not expected to know what each animal eats. Encourage them to make logical guesses. Then help them with the correct answers.
- Allow time for the students to complete questions 1–3 on Activity Manual page 123.
- Discuss questions 4 and 5 together before giving the students time to complete the questions.
- Ask a volunteer to read aloud the next paragraph on the Activity Manual page.
  The ecosystems that we see today are quite different from the ones before the Fall. The original creation would have been very different because there would have been no death.
- Instruct a volunteer to read Genesis 1:29.
  What were Adam and Eve commanded to eat? herbs (plants) and fruits
- Direct the students to notice the words *every* and *everything* as a volunteer reads Genesis 1:30.
  All of the creatures were created to be vegetarians. All were commanded to eat the plants. This disagrees with the evolutionary scientists who believe that death and animals eating one another have always been part of the ecosystems on the earth.
- Allow time for the students to write the answers to questions 6–7 on the Activity Manual page.
- Ask a volunteer to read aloud the paragraph at the top of Activity Manual page 124.
- Direct attention to the picture of the skull. Discuss what the students think this creature might eat, allowing time for them to write their answers for question 8.
- Ask a volunteer to read aloud the next sentence on the Activity Manual page.
- Display the *Fruit Bat* visual. Reveal that the skull is from a fruit bat.
  What is the fruit bat eating? fruit
  The sharp teeth of the fruit bats are used to tear through the thick skin and flesh of fruits, not other animals.
- Read aloud Genesis 3:14–19.
- Discuss questions 9–11, allowing time for the students to write their answers.
- Instruct the students to read the remainder of the Activity Manual page. Discuss the paragraph.
  It is not until after the Fall and God's curse that animals began to eat one another. The carnivorous activity we see today is a part of the corruption that is the result of sin.
- Read Isaiah 11:6. Explain that God promises that one day animals will not eat each other.

Activity Manual

Answers in Genesis, pages 123–24

105
106

Student Text pages 184–87
Activity Manual page 125

Objectives

- Identify the basic needs of plants and animals
- Identify and describe adaptations that help plants survive
- Identify and describe adaptations that help animals survive

Vocabulary

- adaptation
- migration
- hibernation

Two days have been allotted for this lesson. You may want to teach Lesson 105 the first day and use ideas from Lesson 106 on the second day. You may also incorporate the ideas from Lesson 106 into the text material, making the lesson two days.

Introduction

What are some basic needs that all people have? Possible answers: air, water, food, shelter, protection

What would happen if a person did not have air, water, or food? He would not be alive.

Animals and plants also have these needs. God has given some plants and animals unique ways to meet these needs.

Preparation for Reading

- Display the words *acacia* (ah KAY shah) and *mimicry*. Pronounce the words together.
- Direct the students to make questions mentally using the headings, bold and italicized words, and captions.
- Instruct them to read silently pages 184–87 to find the answers to these questions.
  How are some acacia trees protected from giraffes and other animals?
  What are some reasons for migration?

Teach for Understanding

What are some of the basic needs of living things? Possible answers: food, water, protection from consumers, shelter

What are adaptations? any special characteristics or skills that help a living thing survive in its environment

How did living things get these adaptations? God designed them with differing abilities to adapt.

Why do you think that God equips living things with adaptations? so their needs can continue to be met in environments that are changed by sin and the Fall

What is photosynthesis? the process through which plants make their own food

What are three things that plants need for photosynthesis to take place? sunlight, carbon dioxide, and water

Meeting Needs

All living things have the same basic needs: food, water, protection from consumers, and shelter. To meet these needs, God designed living things with a variety of adaptations. **Adaptations** can include any special characteristics or skills that help a living thing survive in its environment.

Plant Adaptations

Most plants make food through photosynthesis. This process is not possible, though, without sunlight. Some plants need a lot of sunlight. Others need only a little.

Plants that grow in shady places sometimes have large leaves. These leaves have a large surface area for collecting sunlight. Vines often climb on objects or other plants. Climbing helps the vine get up higher so that it can get more sunlight.

However, plants need more than just sunlight for photosynthesis. They also need carbon dioxide and water. So plants have tiny pores in their leaves. These holes are called *stomata*. Stomata allow air and water to move in and out of the plant.

184

Background

Adaptations

Environments are constantly changing. The term *adaptation* is often used in an evolutionary context. Evolutionists believe that, due to mutations and genetic variability, some organisms are "selected" for survival. These organisms either change characteristics to adapt to environmental changes or become extinct. These changes supposedly formed new species. Creationary scientists recognize that plants and animals do adapt to changes in their environments. However, no new kinds of organisms are being produced. Creationary scientists believe that adaptations are limited to variation within the boundaries of the biblical "kind" described in Genesis 1:20–25.

Activity

Identifying Plant Adaptations

Materials: several different plants or pictures of plants

Display the plants. The students should identify a main adaptation of each plant and then classify the plants into groups based on their adaptations.

Venus flytrap

A few unusual plants, such as sundews and Venus flytraps, catch their food as well as make it. These carnivorous plants are able to make food through photosynthesis. However, the soil in their swampy habitats often lacks some of the nutrients these plants need. These plants attract insects and other small animals. When an insect or animal gets close enough, the plant traps and eats it. The trapped insect or animal provides extra nutrients for the plant.

Plants also need protection from predators. Some plants are protected by thorns, spines, or stinging hairs. Acacia trees have long thorns. These thorns protect the trees from many animals that would eat the leaves. Other plants, such as stinging nettles, are covered with tiny hairs. When touched, the tiny hairs "sting" the predator.

Sometimes plants use poisons to protect themselves. The poisonous plants often taste bad. For example, acacia trees are often eaten by giraffes. These trees release a chemical that makes the leaves taste bitter after a few mouthfuls. The giraffe soon moves to a different tree. A plant's poison can sometimes cause sickness or death. But poisons that harm one animal might not hurt another species. God carefully designed adaptations so that animals and plants can meet their basic needs.

Giraffe and acacia tree

185

Do all plants need the same amount of sunlight? No, some plants need more and some need less.

Why do plants in shady places sometimes have large leaves? The large leaves have a bigger surface area for collecting sunlight.

How does climbing on trees benefit vines? helps the plants get higher for more sunlight

What are stomata? tiny pores in leaves that allow water and air to move in and out of plants

- Discuss the picture of the stomata on page 184.
- Direct attention to the picture of the Venus flytrap on page 185.

What is unusual about plants such as Venus flytraps and sundews? Possible answers: They are carnivorous plants. The plants can catch food as well as make food.

Why do these plants need to catch some of their food? The soil in their swampy habitats may not have all the nutrients the plants need.

Why do plants need protection? They cannot run away when danger threatens.

What are some ways that plants protect themselves? thorns, spines, stinging hairs, poisons

Does a plant's "poison" always harm the animal eating the plant? no Explain. Sometimes the poison only makes the plant taste bad. A plant's poison might harm one species of animal but not hurt another species.

- Share the Background information about giraffes.

Background

Giraffes and Acacias

Many acacias are protected by thorns. While the thorns do slow the giraffe down, they do not stop it from browsing. A giraffe uses its tough, prehensile lips and fifteen-inch-long tongue to reach between the thorns and strip the leaves from the branches. Some giraffes even chew on the thorny twigs.

After a giraffe has eaten a few mouthfuls of leaves, the acacia tree releases an alkaloid into its leaves. This makes the leaves taste bitter. The tree also gives off a chemical that is carried by the wind. This chemical causes all the acacia trees downwind of the first tree to also pump the alkaloid into their leaves. However, the giraffe can still get his meal. Giraffes graze in the opposite direction from the wind. They move upwind to the next tree while the chemical alert is carried downwind.

Science Misconceptions

Carnivorous Plants

Although insects do provide minerals needed by the plant, the insects provide little energy for the plant.

105
106

- Emphasize that all adaptations are planned by God's design.

  What are some ways that animals protect themselves from predators? Possible answers: playing dead, camouflage, disguises, mimicry, living in groups

  How does playing dead give the opossum an advantage? Many animals that would eat an opossum will not eat a dead animal.

  How is camouflage useful to predators? Predators using camouflage can often sneak up on their prey.

  How is camouflage useful to prey? Camouflage helps prey avoid being seen.

  What is mimicry? the ability of a living thing to imitate another living thing

  How does mimicry help the scarlet king snake avoid predators? Animals that would eat it may avoid it because it looks like a poisonous coral snake.

- Explain that there are several old sayings that help a person remember which snake is the poisonous one. One saying is "Red and yellow kill a fellow. Red and black venom lack."
- Direct attention to the pictures of the king snake and the coral snake as you repeat the saying. Point out the color patterns on the snakes.

  How is looking like a tree branch an advantage for a tawny frogmouth? Predators do not realize that the bird is there.

  What are some names given to groups of different animals? pride, school, colony, herd

  What are the advantages of living in a group? It is easier for the animals to hunt for food, and it helps them protect and care for their young.

  Why might some animals move to different ecosystems? New ecosystems might have better living conditions or more food.

  What term describes the movement of animals from one ecosystem to another? migration

- Direct attention to each picture on the page, asking the students what each animal uses to protect itself.

Animal Adaptations

Animals use a variety of ways to find food or protect themselves from predators. Some animals, such as the opossum, lie still when threatened and pretend to be dead. Both predators and prey use camouflage to blend in with their environment. Predators use camouflage to sneak up on their prey. Prey use it to avoid being seen.

Other animals use *mimicry* and disguises. The scarlet king snake has colors and markings similar to those of the poisonous coral snake. This mimicry helps the scarlet king snake avoid hungry predators.

Scarlet king snake

Coral snake

The tawny frogmouth uses a disguise to hide from predators. The nocturnal bird's feathers are silver, gray, black, and white. During the day, the sleeping bird looks like a dead tree branch.

School of fish

Living in groups is another way that animals meet their needs. Many species of animals live in groups. These groups have different names. Lions live in *prides*. Groups of fish stay together in *schools*. Termites live in *colonies*. Elephants travel in *herds*. No matter how a group is named, though, its advantages are the same. Living in a group makes it easier for the animals to find food. Being in a group also helps the animals protect and care for their young.

Sometimes a whole group of animals leaves one ecosystem and moves to another ecosystem. This adaptation is called **migration**. Animals migrate for different reasons. Some animals, such as salmon and sea turtles, move to a new place in order to reproduce or to raise their young. This is the reason the red knots migrate each year from South

186

Background

Hibernation

Many animals experience periods of dormancy, times when they seem to be unaware of their surroundings. One type of dormancy is hibernation. Animals that hibernate experience a drop in body temperature, and they remain inactive in a sluggish, sleeplike state throughout the winter. Scientists often disagree about what happens during hibernation. The characteristics of hibernation seem to vary from species to species. Some animals hibernate but occasionally wake up to check the temperature, eat a little stored food, or care for other physical needs. Other animals, such as bears, maintain a fairly stable body temperature and can be easily aroused. Some animals remain in their dens throughout hibernation, but other animals occasionally come out for a short time and then return to their dens to continue hibernating.

Estivation

Estivation is another type of dormancy. Animals that estivate experience a period of dormancy in the summer. This protects the animal during long periods of hot, dry weather.

Groundhog Day

Groundhog Day originated in Europe, but it was originally called Candlemas. The legend about an animal and its shadow was associated with Candlemas, but the animal involved was a hedgehog, not a groundhog.

America to the Arctic Circle. Other animals migrate when their food supply is low. Animals on the African savanna often migrate in search of food and water. Some animals, such as birds and monarch butterflies, migrate when the seasons change. This migration helps them avoid weather that is too hot or too cold. Even huge gray whales move to warmer waters during the winter.

Some animals do not need to eat food during winter. They sleep through the winter and wake up in the spring. This deep sleep, called **hibernation**, is one way that animals adapt to their climate. Animals that hibernate eat large amounts of food during the summer and autumn. This food is stored as body fat and provides energy for the animal while it sleeps.

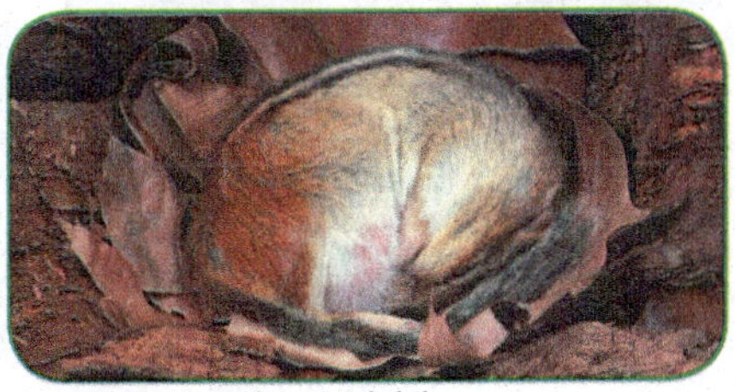
Most ground squirrels hibernate during winter.

When an animal hibernates, its body temperature drops. In fact, its body temperature might be almost as cold as the weather outside its den. The animal's breathing rate and heart rate also drop. These lowered rates allow the animal to use very little energy during the winter.

FANTASTIC FACTS

One hibernating animal has its own holiday! Americans celebrate Groundhog Day on February 2. Folklore and tradition state that a groundhog will come out of its hole on the second of February. If the groundhog sees its shadow, it will return to its hole. This supposedly means that there will be six more weeks of winter. If it does not see its shadow, then spring is coming and the worst of winter is over. Actually, groundhogs usually begin their hibernation in September or October and end in March.

QUICK CHECK

1. What are some of the basic needs of living things?
2. How are Venus flytraps adapted to their swampy habitats?
3. What are three reasons that animals migrate?

187

105 106

Name some animals that migrate regularly. Possible answers: birds, gray whales, salmon, sea turtles, monarch butterflies

What are three reasons that animals migrate? to reproduce and raise their young, to find food and water, and to avoid cold or hot temperatures (when the seasons change)

Some animals sleep all winter and wake up in the spring. What is this deep sleep called? hibernation

Why do animals that hibernate eat large amounts of food during the summer? Extra food is stored as body fat and provides energy for the animal throughout the winter.

How does the body temperature, breathing rate, and heart rate of an animal change during hibernation? The body temperature drops, and the breathing rate and heart rate slow down.

In which types of biomes would you probably find hibernating animals such as groundhogs? biomes with cold winter temperatures

- Discuss *Fantastic Facts*.

Which hibernating animal has its own holiday? the groundhog

When do groundhogs usually hibernate? September or October through March

Do you think that this folklore is based on scientific observation? Probably not. It may have been started with a single observation. Repeated observation is needed to confirm a fact.

ANSWERS

1. Possible answers: food, water, protection from consumers, shelter
2. They are carnivorous. They catch some of their food as well as produce it.
3. to reproduce and raise their young, to find food and water, and to avoid temperatures that are too hot or too cold

Background

Giraffes and Astronauts

God created giraffes with very tough skin and an amazing circulatory system. This design not only helps giraffes but also astronauts. Astronauts returning from space often have difficulty adjusting to the earth's gravity. Their cardiovascular system usually takes awhile to readjust to the earth.

NASA scientists studied giraffes as they worked on developing gravity suits for astronauts. A giraffe's leg has tightly stretched skin. This tight skin and the giraffe's strong leg muscles keep the blood from collecting in the giraffe's legs and feet. If a giraffe's leg is cut, it will not bleed profusely because the arteries and veins are located internally, away from the skin. Learning about the giraffe's circulatory system helped scientists develop technology to help people.

106

- The *Identifying Plant Adaptations* activity found at the bottom of the pages in Lessons 105–6 may also be incorporated into the lesson.
- Suggestions for additional activities for this two-day lesson are on the Teacher's Toolkit CD.

Activity Manual

Reinforcement, page 125

107

Student Text pages 188–91
Activity Manual pages 126–29

Objectives
- Identify different kinds of symbiosis
- Differentiate between instincts and learned behaviors
- Give examples of instincts and learned behaviors

Teacher's Toolkit CD
- Quiz 7-C

Vocabulary
- symbiosis
- parasite
- host
- instinct
- learned behavior

Introduction

Have you ever been part of a three-legged race? Answers will vary.

How many people are on a team for a three-legged race? two

What happens if the two team members do not work together? Possible answers: They fall. They cannot move forward.

Many sports and other athletic events require teamwork. Team members have to work together to reach the desired goal. Today we will look at how some living organisms work together.

Preparation for Reading
- Display the words *symbiosis*, *parasitism*, *mutualism*, and *commensalism*. Pronounce the words together.
- Direct the students to read silently pages 188–91 to find the answers to these questions.
  What is symbiosis?
  How are instincts different from learned behaviors?

Teach for Understanding

How can plants and animals meet their needs without competing? They can share their resources.

How do herbivores on a savanna share resources? Possible answers: by eating different plants, by eating at different times of the day, by eating different parts of the same plant

What name is given to a special partnership where two species interact with each other over a long period of time? symbiosis

Do these partnerships always benefit both partners? no Explain. Sometimes a partner is harmed. At other times, a partner is neither helped nor harmed.

Relationships

Plants and animals sometimes compete to meet their needs. But at other times, they share their resources. Several different forest animals might live in the same tree. Many different populations of herbivores live together on a savanna. Some of them prefer different plants. Giraffes feed on tree leaves while other animals eat grass. Some eat different parts of the same plant or eat at different times of the day. Zebras prefer to eat the top part of the grass. Wildebeests eat the middle stems and leaves, and gazelles prefer the part of the grass that is closest to the ground.

Some organisms share their resources by forming special partnerships. Scientists use the term *symbiosis* to describe this kind of relationship. **Symbiosis** occurs whenever two species interact with each other over a long period of time. There are several types of symbiosis.

One type of symbiosis is parasitism. *Parasitism* is an interaction that helps one partner and harms the other. A **parasite** is any organism that lives on or inside another organism and takes food from that organism. The plant or animal that a parasite lives on or in is called its **host**. The parasite usually does not kill its host, but it does cause harm to the other organism.

Parasitism

Fleas can really cause a lamb to itch! Fleas and lambs have a parasite/host relationship.

Lampreys are a type of fish that does not have jaws. A lamprey has a mouth that is like a suction cup and is filled with teeth. It uses its mouth to attach itself to a fish. The lamprey then uses its teeth to cut through the fish's scales and take blood and body fluids from the host fish.

188

Background

Symbiosis
The term *symbiosis* is not limited to relationships that benefit both partners. Symbiosis is any interaction between two species over a long period of time. There are many types of symbiosis, but the three discussed in this lesson are the most well known. There are many examples of parasitism and mutualism. However, true commensalism is extremely rare. Scientists sometimes disagree when identifying types of symbiotic relationships.

Other Mutualism Examples
Sometimes sea anemones will hitch a ride on a hermit crab's shell. This helps the sea anemone gather more food and travel to new places. The anemone, in turn, provides a type of camouflage for the crab.

The boxer crab often attaches some sea anemones to its claws. It uses these stinging claws to "box" with any would-be predators.

Wildebeest
Another name used for the wildebeest is the gnu.

In another type of symbiosis, the interactions benefit both partners. This type of symbiosis is called *mutualism*. A relationship that benefits one partner but does not help or harm the other partner is called *commensalism*.

Mutualism

Oxpeckers, a type of bird, have a special relationship with gazelles. The bird rides on a gazelle and eats ticks and harmful insects. The bird gets food and protection from predators, and the gazelle gets rid of its insect pests.

Bees visit flowers to collect pollen and nectar. When a honeybee visits a flower, some of the pollen sticks to the fuzzy hairs on the bee. This pollen usually falls off when the bee visits another flower. The flower gets pollinated, and the honeybee gets its food.

The scarlet cleaner shrimp runs an underwater cleaning service! It picks off parasites, bacteria, and dead skin from coral reef fish. Even fierce predator fish wait in line for cleaning. The fish get rid of parasites, and the scarlet cleaner shrimp gets a meal.

Commensalism

Very little sunlight comes through the canopy of a rain forest. Some plants, such as tropical orchids, get a boost from the tall trees. The flower is not rooted in the soil. Instead it attaches itself to the tree's broad branches and grows high on the tree. The tree supports the orchid, and the orchid is able to get sunlight without harming the tree.

189

107

What are the three types of symbiosis? parasitism, mutualism, and commensalism

Which type of symbiosis refers to a relationship in which both partners benefit? mutualism

How is parasitism different from commensalism? In parasitism, one partner is helped while the other is harmed. In commensalism, one partner is helped while the other is neither helped nor harmed.

What is a parasite? any organism that lives on or in another organism and takes nourishment from that organism

What name is given to the plant or animal that a parasite lives on? a host

- Refer to the pictures as you discuss each type of symbiosis.

Name two animals that are parasites. fleas and lampreys

Can a parasite have more than one species for a host? yes Give an example. Possible answer: Fleas live on various species of animals.

How many partners benefit in the relationship between an oxpecker and a gazelle? both partners

How does a flower benefit from the relationship with a bee? The flower gets pollinated.

How do scarlet cleaner shrimp interact with coral reef fish? The shrimp clean parasites off the fish.

Why do some tropical orchids need to grow high on a tree? in order to get enough sunlight

How does this relationship between the tree and the orchid affect the tree? The tree is neither benefited nor harmed.

Background

Scientist Heinrich Anton de Bary

The German scientist Heinrich Anton de Bary (1831–88) was a man of many scientific talents. He studied medicine and was a surgeon for a little while. However, his greater interest was in botany and mycology, the study of fungi. He became a botanist, a pioneer mycologist, a microbiologist, and a founding father of plant pathology. His work laid the foundation for the acceptance of Louis Pasteur's germ theory of disease. He also was the first to show that lichens are really a combination of a fungus and algae. In 1879, he coined the term *symbiosis* to describe the interactions or "living together" of different organisms.

Activity

Writing About Partnerships

Write a short story about a symbiotic relationship between two organisms. The characters can be real or fictional plants and/or animals. Draw at least one illustration for the story.

What are instincts? the God-given basic knowledge and skills that are needed for survival

If someone told you that animal instincts evolved, what response would you give? God designed His creatures with instincts. Some instincts did not become part of animals' behavior until after the Fall.

Are instincts taught or inherited? inherited

What are two common instincts? migration and hibernation

Are instincts unique to an individual, or are they shared by a population? They are shared by a population.

What is a learned behavior? a behavior that cannot be inherited

How are instincts different from learned behaviors? Animals are born with instincts. Learned behaviors are not inherited.

Can humans change an animal's instincts? No, the animal is born with the instincts.

What are some examples of behaviors that humans have taught animals? Possible answers: teaching animals to perform tricks, teaching birds to talk

Behaviors

How do animals know how to form partnerships? How are birds able to migrate thousands of miles without getting lost? What tells the squirrels and groundhogs when to hibernate? God designed His creatures with **instincts**, the basic knowledge and skills that are needed for survival.

Instincts are behaviors that are inherited. They are automatically passed down from generation to generation. Migration and hibernation are instincts. No one has to teach a groundhog how to hibernate. No one has to give hummingbirds travel directions from North America to Central America. No one has to lead the red knots to the horseshoe crabs' eggs at Delaware Bay. Instincts are not unique to one individual. Every member of a species is born with the instincts it needs to survive.

190

Background

Innate Behaviors

Instincts are sometimes referred to as innate behaviors.

Some behaviors, though, are learned. A **learned behavior** is a behavior that cannot be inherited. For example, a chimpanzee might learn how to use rocks to crack open nuts. This chimpanzee can teach its offspring how to crack nuts, but the baby chimpanzees are not born with this knowledge.

Sometimes it can be hard to decide whether a behavior is an instinct or is learned. For example, lions are born with the instinct to kill and eat meat for food. But to be good hunters, they need to learn from other lions. So lionesses teach their cubs how to hunt. Though lions seem to possess both, the instinct to hunt and the learned ability to hunt well are not the same.

SCIENCE & THE BIBLE

Before mankind's sin, animals did not hunt and kill each other for food. However, when man sinned, death became part of the natural order of the food chain. God provided animals with the necessary instincts to survive in a sin-cursed world. But one day God will change these instincts. Isaiah 11:6–9 tells us that predators and prey will live together peacefully on God's new earth. God tells us that this will be possible because in that day all "the earth shall be full of the knowledge of the Lord."

In God's creation, all living things interact with each other. Every plant and animal in an ecosystem has its specific niche. This does not mean that ecosystems never change. Ecosystems are constantly changing! Changing one part of an ecosystem affects all the other parts. But change is part of God's plan and purpose for the earth.

People are part of many ecosystems. Understanding the interactions within ecosystems allows people to exercise good dominion over the earth. This knowledge helps people find ways to use the living and nonliving things in ecosystems to serve others. As we use the parts of ecosystems, we also cause changes. Ecologists, other scientists, and fishermen work together to manage the changes to the horseshoe crab's ecosystem. They look for solutions that work best to meet the needs of both people and animals. When Christians do this, we bring glory to God.

QUICK CHECK

1. What name is given to special partnerships in which two different species interact with each other over a long period of time?
2. What are instincts?
3. How are instincts and learned behaviors different?

191

107

Is it always easy to decide whether a behavior is an instinct or a learned behavior? no

What is an instinct that lions are born with? to kill and eat meat

What is one reason that lion cubs need to observe other lions? to learn how to be effective hunters

Who is in control of all the interactions that happen in an ecosystem? God

- Discuss *Science & the Bible.*

  Why do animals hunt and kill each other? Death is one of the results of sin.

  Will these instincts ever change? yes Explain. God will change them. One day predators and prey will live peaceably together.

ANSWERS

1. symbiosis
2. the God-given basic knowledge and skills needed for survival
3. Instincts are inherited. Every member of a population has the same instincts. A learned behavior is not hereditary and does not affect an entire population.

Activity Manual

Expansion, pages 126–27

These pages use map skills to review migration.

Review, pages 128–29

These pages review Lessons 105–7.

Assessment

Quiz 7-C

The quiz may be given any time after completion of this lesson.

Activity

Ecosystems in the Bible

Materials: Instructional Aids 7.11–7.12: *Ecosystems in the Bible*, for each student; *Ecosystems in the Bible Answer Key*

The students will match some of the terms in the chapter to the correct definitions. Then they will look up Bible verses and match the verse to the correct science term.

108

Student Text page 192
Activity Manual page 130

Objectives

- Recall concepts and terms from Chapter 7
- Apply knowledge to everyday situations

Review

- Materials for the Chapter 7 Test will be taken from Student Text page 192 and Activity Manual pages 117–18, 122, 128–29, and 130. You may review any or all of the material during the lesson. Questions similar to the ones in Thinking It Through, Activity Manual page 130, may appear on the test.
- You may choose to review Chapter 7 by playing the review game or a game from the Game Bank on the Teacher's Toolkit CD.

Activity Manual

Review, page 192

This page requires written responses to application questions.

CHAPTER SUMMARY

Words to Know

ecosystem	producer	food web
individual	consumer	competition
population	decomposer	adaptations
community	food chain	symbiosis
habitat	predator	parasite
niche	prey	host

Key Ideas

Recognize that ecosystems vary in size and include both living and nonliving things
Recognize that the environment determines the kinds of living things in an ecosystem
Give an example of how a niche is unique to a population
Describe what is needed for photosynthesis to happen
Identify characteristics and examples of producers, consumers, and decomposers
Identify and describe different groups of consumers
Explain how producers and consumers get their energy
Explain why scavengers and decomposers are beneficial to ecosystems
Identify what types of living organisms are often the first two links of a food chain
Explain how changing one part of a food web affects other organisms
Explain what an energy pyramid shows
Discuss the benefits of competition
Identify four basic needs of all living things
Describe some ways that different plants meet their needs
Identify ways that some animals meet some of their needs
Explain reasons why some animals migrate
Describe what happens during hibernation
Describe three types of symbiosis and give an example of each
Contrast instincts and learned behaviors

192

Review Game

Make a Chain

Prepare pictures of one or more food chains the students would be familiar with. Include the sun and a variety of producers, herbivores, omnivores, and carnivores. Include the same number of pictures for each chain. Place each food chain in a separate envelope.

Divide the class into teams. Give each team an envelope. For each correct answer, award points and allow the team member to choose one picture. When all the pictures have been chosen, direct the team to assemble their food chain in the correct order. Give bonus points for each correctly assembled food chain. Continue playing as time allows.

Variation: If pictures are not available, write the names of the organisms on cards.

109

Objective

- Demonstrate knowledge of concepts taught in Chapter 7

Assessment

Chapter 7 Test

Changes in an Ecosystem

8

On July 14, 1991, a train crossed the Sacramento River. At a sharp bend in the tracks, several cars suddenly came off the tracks. About 19,000 gallons of a strong chemical spilled into the river. It polluted the water, and the reaction formed a cloud of gas above the river as the chemical slowly moved downstream. The pollutant immediately began killing plants and animals as it made its way toward Lake Shasta. People in a nearby town had to leave their homes. Signs were posted along the river and lake to prevent people from fishing or entering the water. It took two weeks to reduce the chemicals in the lake to acceptable levels. By that time, millions of fish and aquatic animals had died. On land, hundreds of thousands of trees were killed. Many land animals were forced to move in order to find food and shelter. Thankfully, over the course of several years, the ecosystem recovered and is once again open to recreation and fishing.

Consider

- Why should people try to prevent disasters like this from happening?
- What kinds of changes can people make to prevent similar accidents?

193

Background

Chapter Preview

Preview and prereading activities may include using a K-W-L chart, a probe, or an anticipation guide.

Chapter Photo

The photo on page 193 shows a forest fire. The miniature scientist is cleaning up chemical waste that has been disposed of improperly.

Current Events

You will want to preview the activity in Lessons 121–22. You may choose to have the students start collecting articles at any time during the chapter. Students will probably notice that people are involved with most of the changes that make the news.

Weblinks

The BJU Press website offers traditional information and links you may find helpful throughout the chapter.
bjupress.com/resources/science

Objectives

- Recognize ways that people can have dominion over the earth as God has commanded

Materials

- United States map or a map of California

Introduction

- Direct attention to the picture at the top of the page.
  What do you see? The picture shows a forest fire.
  How do you think a forest fire might change the forest food webs? Possible answers: Many of the producers (prey) may be reduced or eliminated. Consumers (predators) may go to other areas to find the producers (prey) they need. As certain webs recover, they will most likely have a greater variety of animals.

Preparation for Reading

- Direct the students to read page 193 to find out what happened at Lake Shasta.

Teach for Understanding

- Help the students locate the Sacramento River on a map. Explain that the track where the spill occurred is called Cantara Loop and looks similar to a horseshoe.
- Relay that the chemical tanker car had empty cars on either side of it. The train tried to proceed in a straight line rather than bend with the track.
  How did the chemical spill affect the Sacramento River? It killed plants and animals.
  What problems did the spill cause the people in the area? Many had to leave their homes because of the gas. It prevented people from fishing or entering the water.
  What changes did the land animals have to make? Many were forced to move to find food and shelter.
- Explain that an experimental technique was used at Lake Shasta. Air was pumped into the contaminated water. The water was also sprayed toward the sky. The exposure to air quickened the decay of the chemical.
  What changes do you think the railroad may have made to prevent future spills? Accept reasonable answers. Some of the changes concern the operation of trains, as well as the replacement of concrete tiles with wooden ones and the building of a massive guardrail.
- Discuss how God wants Christians to have dominion over the earth and that preventing spills similar to this one is part of that dominion.
- Direct attention to the miniature scientist.
  What do you think the scientist is doing? He is making sure the barrel of toxic waste is being disposed of properly.

Activity Manual

Preview, page 113

The Looking Ahead page is intended to assess the student's prior knowledge before beginning the chapter.

111 112

Student Text pages 194–97
Activity Manual pages 132–34

Objectives

- Recognize that the earth has many cycles
- Identify the seasonal changes that may occur in an ecosystem
- Explain the carbon cycle
- Differentiate between photosynthesis and respiration

Materials

- world map

Teachers Toolkit CD

- Visual 8.1: *Carbon Cycle*
- Instructional Aids 8.1–8.3: *Molecules*; *Photosynthesis*; *Respiration*, for each student or group of students
- Quiz 8-A

Vocabulary

- cycle
- seasons
- photosynthesis
- respiration
- decomposition

Two days have been allotted for this lesson. You may want to teach Lesson 111 the first day and use ideas from Lesson 112 on the second day. You may also incorporate the ideas from Lesson 112 into the text material, making the lesson two days.

Introduction

Which season is your favorite? Why? Answers will vary.

What are some things you like to do during that season? Answers will vary.

What are some things that usually happen or are done in a specific season of the year? Possible answers: planting flowers in spring, vacationing in summer, raking leaves in autumn, sledding in winter

Preparation for Reading

- Direct the students to mentally make questions from the headings, bold words, and captions.
- Instruct them to read silently pages 194–97 to find the answers to these questions.

What two weather conditions affect seasonal changes?

What gas is released through the process of photosynthesis?

Teach for Understanding

Who is in control of the changes in an ecosystem? God

What is a cycle? a regular pattern of change

What are the regular divisions of the year called? seasons

For most areas of the earth, what weather condition changes with the seasons? temperature

How many seasons do these areas of the earth usually have? four

Spring in Washington, DC

Summer in Seattle, Washington

Cycles of Change

Change is always happening in our world. Some changes occur quickly, like the train wreck at the Sacramento River. In just a few short hours chemicals killed wildlife for miles downstream. Fires can quickly destroy whole forests. Other changes, such as the growth of a plant, take place slowly. Sometimes changes happen so slowly that they are not even noticed for a long while. The weathering of rock is an example of this.

However, changes do not happen by chance. Each change, big or small, is planned and controlled by God. Some changes are part of a **cycle**, or regular pattern of change. God uses these cycles to maintain His earth. He promises that they will continue. In Genesis 8:22, God says, "While the earth remaineth, seedtime and harvest, and cold and heat, and summer and winter, and day and night shall not cease."

Seasonal Changes

One of the earth's cycles is the **seasons**, or the regular divisions of the year. Many parts of the earth change temperature as the seasons change. These areas usually have four seasons: spring, summer, autumn, and winter.

The season of spring is a time of new growth for living things. Fresh grass and new flowers appear. Trees put out new leaves. Migrating animals return in search of food. Other animals come out of hibernation. Many animals give birth to their young in the spring.

During the summer, the baby animals grow and learn. Plants grow and produce food both for themselves and for animals. Some animals begin to gather food and store it for the coming winter.

In autumn, the temperatures get cooler. The leaves on many trees change color. Animals continue

194

Background

Ecology

The term *ecology* has many definitions. Sometimes people use *ecology* as a synonym for *environmentalism*. As a science, *ecology* is the study of the interactions between living things and their environments. There are many branches of ecology, including landscape ecology, conservation ecology, plant ecology, and aquatic ecology.

Monsoons

The word *monsoon* comes from the Arabic word *mawsim*, meaning "season." The best-known and most severe monsoons are found in India and Southeast Asia, although monsoonal climates are also found in regions of Africa, North and Central America, and Australia. Monsoon winds are created by the changing patterns of atmospheric pressure and by the heating and cooling rates of land and oceans.

Scientist Sir Arthur Tansley

Sir Arthur Tansley (1871–1955), a British botanist, coined the term *ecosystem* in 1935. Tansley helped form many organizations to study ecology and protect wildlife. He founded and edited a scientific journal called the *New Phytologist* for thirty years. He was also a cofounder and president of the British Ecological Society and edited the *Journal of Ecology*. His knighthood in 1950 recognized him as a pioneer of modern ecology.

111
112

Autumn in West Virginia

Winter in the Midwest

preparing for the winter. Some animals migrate to warmer places. Other animals grow thick coats of fur.

Winter has the coldest temperatures of the year. Many tree branches are bare of leaves. Some animals stay in their holes or burrows and are inactive. Others hibernate in dens until the warm spring comes again.

In some areas of the earth, however, the temperature does not change very much, but the amount of precipitation changes a lot. The seasons in these areas are based mostly on rainfall. They usually have two seasons: the wet season and the dry season. During the wet season, most of the rain for the entire year falls. In the dry season, very little rain falls.

Monsoons are winds that change direction with the seasons. These winds control the wet and dry seasons in some areas. For example, the winds blow across the Indian Ocean for several months and bring heavy rains to India and Southeast Asia. Then the winds reverse their direction and blow across the land for the rest of the year. India and Southeast Asia have a very dry climate for these remaining months of the year.

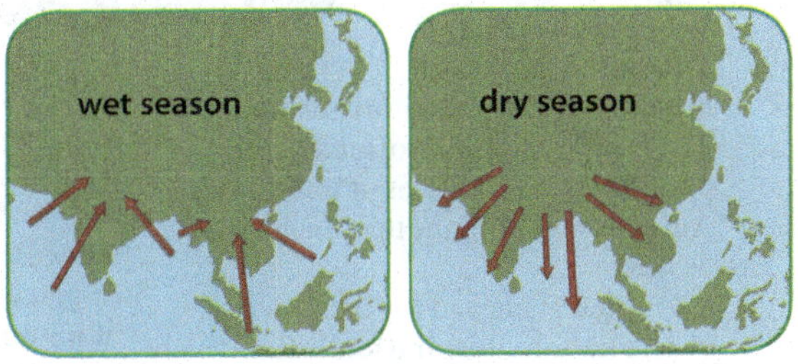

The arrows show the directions monsoons blow in India and Southeast Asia.

195

What are the four seasons? spring, summer, autumn, and winter

- Discuss how the seasons are reversed on the other side of the equator. When it is summer in the Northern Hemisphere, it is winter in the Southern Hemisphere. Other areas of the earth have only two main seasons. What weather condition changes with the seasons in these areas? the amount of rainfall

Describe how the temperatures in the two seasons compare. Possible answers: They do not change much. They are similar.

What are the two main seasons in these areas usually called? the wet season and the dry season

What controls some of these two-season areas? monsoons

What are monsoons? winds that change direction with the seasons

- Locate India and the southeastern part of Asia on a world map.
- Explain that both of the maps on page 195 show the same area of the southeastern part of Asia and India. Where are the monsoons blowing from during the wet season? from across the water or the Indian Ocean
- Explain that during the summer monsoon season in India and Southeast Asia, the amount of rain varies from 260 cm (100 in.) to 1,030 cm (400 in.).

Some years the rain comes late or not at all. How do you think the lateness of the rain affects the area's economy? Possible answers: crop failure, heat damage, heat stress on livestock

Some years there is too much rain. How do you think this affects the area's economy? Possible answers: severe flooding, crop destruction, need for families to move

Look at the maps. Where are the monsoons blowing from during the dry season? from across the land

Science Misconceptions

Winter Growth

Many plants continue to grow during the winter months. This growth is slower, however, and not always visible.

Activity

Identifying Seasonal Changes

Direct each student to divide a piece of paper into four sections and label the sections with the names of the seasons. In each section, the students are to list changes that occur in the local environment.

What substance is needed by every living thing? carbon

What gas is formed when carbon combines with oxygen? carbon dioxide

Which organisms use the carbon dioxide? plants

- Review the meanings of *atom*, *element*, and *molecule*. Refer to Student Text page 68 if needed.

  During photosynthesis, what three things do plants use to form sugar molecules and oxygen? energy from the sun, carbon dioxide, and water

  What do plants do with the sugar molecules? They use some for their own needs, and they store the rest.

  How do other organisms use the sugar molecules made by plants? as food

  What do plants do with the oxygen that forms during photosynthesis? Plants release oxygen back into the air.

  Which organisms need the oxygen that is released by plants? animals, humans

  What is respiration? the process in which organisms use sugar molecules and oxygen to produce energy

  In addition to energy, what else is produced during the process of respiration? carbon dioxide and water

- Distribute the *Molecules* page.

  Look at the carbon dioxide (CO_2) molecule. How many atoms of carbon are in each carbon dioxide molecule? 1

  How many atoms of oxygen are there? 2

- Follow the same procedure with the water (H_2O) molecules.

  Where does a plant get the carbon dioxide that it needs for photosynthesis? from the air

  Where does a plant get its water? from the ground

- Direct the students to cut apart the molecules into atoms of carbon, oxygen, and hydrogen.
- Distribute the *Photosynthesis* page.

  This leaf shows the atoms that a plant uses to form one sugar molecule.

- Discuss the amount of each atom that is needed to form the sugar molecule.
- Direct the students to use the atoms of hydrogen, oxygen, and carbon to match those on the leaf. Explain that this will give the plant what is needed to form a sugar molecule.
- Emphasize that the atoms were once molecules of carbon dioxide and water.

  After you make the sugar molecule, what do you have left? oxygen or 6 molecules of oxygen

  What does the plant do with the extra oxygen? releases it into the air

- Instruct the students to place the extra molecules of oxygen off the page to indicate that they were released into the air and to leave the sugar molecule as it is.
- Distribute the *Respiration* page.

  What do animals and humans need for respiration? oxygen and sugar

  What does respiration produce? carbon dioxide and water

Carbon Cycle

Another cycle is the carbon cycle. Every living thing needs carbon. Carbon combines with other elements to form the compounds that make up living things.

When carbon combines with oxygen, it forms a gas called carbon dioxide. Plants take in and use this gas. During the process of **photosynthesis**, plants use water, carbon dioxide from the air, and energy from the sun to form sugar molecules and oxygen. The sugar molecules contain carbon. The plants use some of the sugar molecules to provide for their own needs. The rest of the sugar molecules are stored in the plant. The stored sugar will provide food for other organisms. The oxygen that was formed during photosynthesis is released into the air. Animals and humans breathe in this oxygen.

Oxygen is needed for the process of respiration. **Respiration** is the process in which organisms use sugar molecules and oxygen to produce energy. This process also produces carbon dioxide and water. Respiration takes place in almost all living things. In plants, respiration is the opposite of photosynthesis.

196

Background

Carbon Sources

Carbon is found in every living thing, but it is also in seawater and the atmosphere. Carbon is released into the atmosphere in many different ways, such as by the burning of fossil fuels, plant and animal respiration, decomposition, volcanic eruptions, and the erosion of limestone, marble, and chalk.

Respiration

Internal respiration, also called cellular respiration, refers to the chemical reactions that take place in each individual cell in an organism. The respiration defined in Chapter 8, *Changes in an Ecosystem*, is cellular respiration. Plants, animals, and humans experience cellular respiration.

External respiration, or breathing, refers to the exchange of gases between the environment and the lungs or gills. Human external respiration is discussed in Chapter 11, *Respiratory System*.

Scientist Dr. Eugene Odum

Dr. Eugene Odum (1913–2002) was an American biologist and ecologist. He and his brother wrote the first ecology textbook, which for ten years was the only ecology textbook available. It was the first book to show how ecosystems and natural cycles were interrelated.

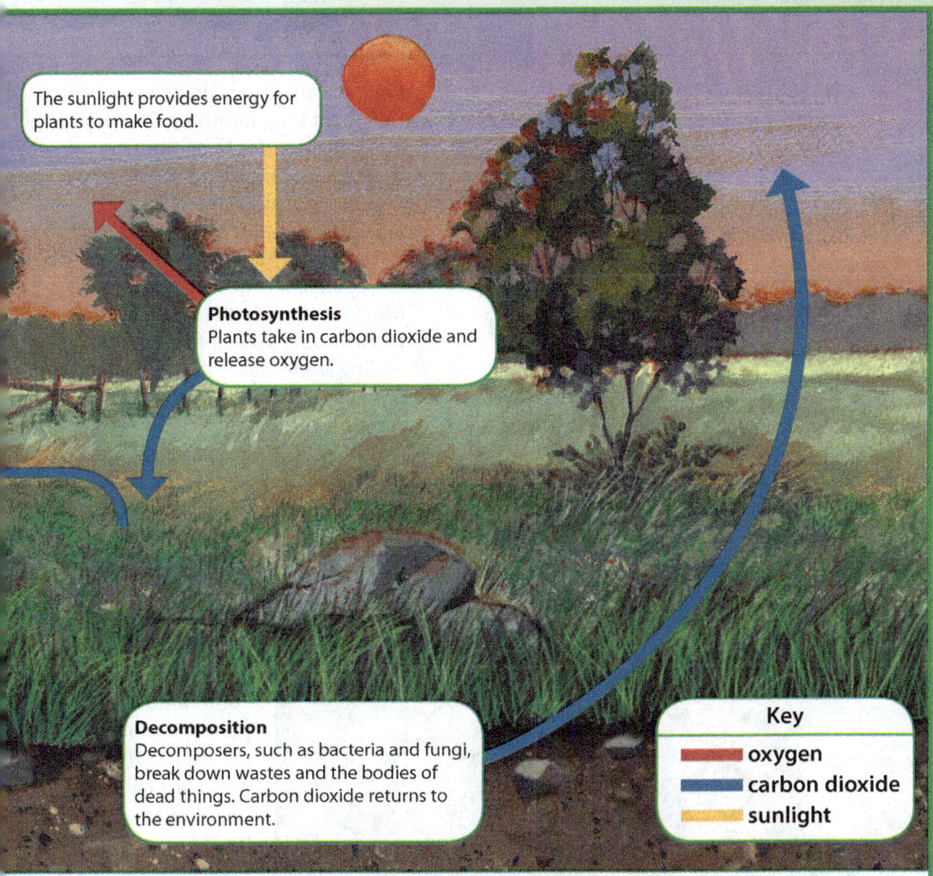

When living things die, carbon remains stored in their bodies for a while. Gradually, **decomposition** takes place as bacteria, fungi, and other decomposers break down the bodies of dead things. The carbon is then released into the air as carbon dioxide.

QUICK CHECK

1. What is a cycle?
2. What is the name of winds that change direction with seasons?
3. Which process in the carbon cycle uses sugar molecules and oxygen to produce carbon dioxide and water?

197

111 112

Notice that carbon dioxide and water molecules can be found on your *Respiration* page.

- Direct the students to take the atoms from the sugar molecule and the oxygen that was released into the air to form carbon dioxide and water molecules.
  What happens during photosynthesis? Plants use water, carbon dioxide, and energy from the sun to form sugar and oxygen.
  What happens during respiration? Animals use sugar and oxygen and produce energy, carbon dioxide, and water.
- Discuss the key on the *Carbon Cycle* visual. Refer to the arrows as you discuss the carbon cycle.
  What provides energy for plants? sunlight
  Explain what happens in photosynthesis. Plants use the sun's energy, water, and carbon dioxide to form sugar molecules. They release oxygen.
  Where do animals get carbon? from the plants they eat
  Look at the arrows. What else do the animals receive? oxygen from the air
  How did some of the oxygen get in the air? released by the plants during photosynthesis
  What do the animals release? carbon dioxide and waste products into the soil
  What happens to carbon when living things die? As decomposers break down the bodies of dead things, carbon is released into the air as carbon dioxide.

ANSWERS

1. a regular pattern of change
2. monsoons
3. respiration

Science Misconceptions

Exchanging Gases

Plants use carbon dioxide and release oxygen during photosynthesis. However, they also take in small amounts of oxygen and release small amounts of carbon dioxide during respiration. Humans and most animals use only oxygen and release only carbon dioxide.

112

- The *Identifying Seasonal Changes* activity found on page 223 may be incorporated into the lesson.
- Suggestions for additional activities for this two-day lesson are on the Teacher's Toolkit CD.

Activity Manual

Reinforcement, page 132

Before using this page to reinforce seasons and map skills, instruct the students to skim the *Seasonal Changes* section in the Student Text to contrast four seasons with two seasons.

Review, pages 133–34

These pages review Lessons 111–12.

Assessment

Quiz 8-A

The quiz may be given any time after completion of this lesson.

Student Text pages 198–201
Activity Manual pages 135–36

Objectives

- Name two ways that nitrogen is changed into usable compounds
- Describe the nitrogen cycle
- Identify the parts of the water cycle
- Identify and infer some ways that cycles work together in an ecosystem

Materials

- peanuts, unshelled

Teachers Toolkit CD

- Visuals 8.2–8.3: *Nitrogen Cycle*; *Water Cycle*
- Quiz 8-B

Vocabulary

- evaporation
- transpiration
- condensation
- precipitation
- ground water

Two days have been allotted for this lesson. You may want to teach Lesson 113 the first day and use ideas from Lesson 114 on the second day. You may also incorporate the ideas from Lesson 114 into the text material, making the lesson two days.

Introduction

- Display the peanuts.
  What are these? peanuts
  Peanuts grow under the ground. The roots of peanut plants are part of another important cycle in our environment.

Preparation for Reading

- Instruct the students to mentally make questions using the headings, bold words, and captions.
- Direct them to read silently pages 198–201 to find the answers to these questions.
  What are two ways that nitrogen is changed into usable compounds?
  What are the different parts of the water cycle?

Teach for Understanding

Like carbon and oxygen, what other gas is important for life? nitrogen

- Direct attention to the *Gases in the Atmosphere* pie graph on Student Text page 114.
  What gases are found in air? nitrogen, oxygen, and other gases including carbon dioxide
  Air is mostly made up of which gas? nitrogen
  Can plants, animals, and people use nitrogen when it is in the form of a gas? no

Nitrogen Cycle

The nitrogen cycle is also an important process. Like carbon, nitrogen is essential to life. Without it, living things could not grow and function normally.

Nitrogen gas is part of the air that we breathe. In fact, the air is mostly made up of nitrogen gas. However, plants, animals, and people cannot use nitrogen in this form. To be usable, nitrogen gas must combine with other elements.

One way nitrogen can change into a usable form is by lightning. Lightning causes nitrogen gas to combine with oxygen. Rain then carries these compounds of nitrogen and oxygen down to the earth. Lightning can change only a small amount of nitrogen at a time.

Most nitrogen is changed by bacteria. Many kinds of nitrogen-changing bacteria are found in the soil. They live in little nodules, or bumps, on the roots of plants such as peanuts, peas, and beans. The tiny bacteria absorb nitrogen gas from the soil and combine it with other gases to form nitrogen compounds. These compounds help plants to grow. The plants, in turn, provide a habitat for these helpful bacteria.

The nodules on the soybean plant contain bacteria that change nitrogen into useful compounds.

198

Background

Paths of the Nitrogen Cycle

The nitrogen cycle has two different paths, each with a different starting point. One path takes the nitrogen from the air to the nitrogen-fixing bacteria, and then to plants, animals, and finally decomposers. The other path starts with nitrogen from dead organisms and goes to decomposers, bacteria, and then plants.

Nitrogen Fixing

The process in which bacteria change nitrogen gas into usable compounds is called nitrogen fixation. Most plants that host the nitrogen-fixing bacteria are from the legume family, which includes peanuts, peas, clover, and soybeans. Some nitrogen-fixing bacteria are also found in the roots of alder and locust trees. Other nitrogen-fixing bacteria live free in the soil. In the ocean, blue-green algae fix nitrogen into usable forms.

Volcanic eruptions are another way that nitrogen is fixed into usable compounds. Volcanoes and lightning produce only a small amount of nitrates. Those nitrates are carried to the soil through precipitation.

Natural Fertilizer

Native Americans showed early settlers how to plant corn by burying a dead fish with the seeds. This improved the corn crops because dead organisms decompose and add nitrogen to the soil. The nitrogen fertilized the corn.

Plants use the nitrogen compounds both to grow and to make proteins. Animals get the proteins and nitrogen by eating the plants or other animals. In this way, animals get a form of nitrogen that they can use.

When plants and animals die, other kinds of bacteria act as decomposers. These bacteria form compounds from nitrogen and other gases. These compounds are used by some plants. Some of the compounds are broken down completely to form nitrogen gas. This returns nitrogen gas to the air so that the cycle can begin again.

199

What must happen to nitrogen for it to be useful? It has to be combined into compounds.

What are two ways that nitrogen is changed into a usable form? by lightning and by bacteria

What gas does lightning cause nitrogen gas to combine with? oxygen

How are compounds of nitrogen and oxygen carried to the earth? by rain

Why is lightning not the best source for usable nitrogen? Lightning changes only small amounts of nitrogen at a time.

What changes most nitrogen? bacteria

Where do these nitrogen-changing bacteria live? in the soil, on the roots of certain plants

- Display the *Nitrogen Cycle* visual. Be sure the students understand that there are two paths of the nitrogen cycle. Share the Background information with them as you discuss the nitrogen cycle.
  What do the arrows indicate on the diagram? the path of the nitrogen cycle
- Point out the inset picture of the roots and nodules.
  What are nodules? the little bumps on the roots
- Ask a volunteer to read aloud the bottom right box on the visual.
  Name some plants that have roots that help change nitrogen into usable compounds. Possible answers: peanuts, peas, beans

  The nitrogen compounds formed by bacteria help plants to grow, and the plants provide a habitat for the bacteria. This is an example of which type of symbiosis—parasitism, mutualism, or commensalism? mutualism
- Ask a volunteer to read aloud the bottom left box and the one above it.
  What happens after the bacteria breaks down the wastes and plant remains? Nitrogen compounds are returned to the soil and nitrogen is released back into the air.

  What do plants make with the nitrogen compounds? proteins

  How do animals get the proteins and nitrogen from plants? by eating plants or other animals
- Explain that proteins are used by living things for energy.
  What happens to nitrogen when plants and animals die? Bacteria act as decomposers and form more nitrogen compounds. Some of these compounds get broken down completely, forming nitrogen gas. This nitrogen gas is released into the air.
- Direct attention to the picture on Activity Manual page 135.
  What has been added to this nitrogen cycle diagram to make it more complete? lightning and rain

  What else could you add to the *Nitrogen Cycle* diagram to make it more complete? remains of a dead animal

Background

Scientist George Washington Carver

George Washington Carver (1864–1943) became concerned about the Southern farmers who were planting cotton year after year. Cotton fields were becoming worn out, and crops were failing. He encouraged farmers to plant different crops, such as peanuts and soybeans, so that the fields could be replenished with nitrogen and other nutrients. He taught them to plow under a crop instead of burning it after harvest. He also encouraged them to plant ground cover in the fields to keep topsoil from eroding.

While trying to create a market for peanuts the farmers had planted, he found over 300 uses for the peanut. Products that can be made from peanuts include linoleum, imitation marble, paper, milk, diesel fuel, cheese, meatloaf, coffee, Worcestershire sauce, shampoo, paper, washing powder, nitroglycerin, quinine, dyes, paints, and lotions.

- Display the *Water Cycle* visual. Use it to point out each step as you discuss the water cycle.
  What do the arrows show? the path of water between the atmosphere and the earth
- Point to the body of water on the visual.
  What causes evaporation from bodies of water? The sun heats the water.
  What is evaporation? the process of water changing to a vapor or gas
  What is the next step in the cycle? condensation
  What is condensation? the process of water vapor changing to a liquid
  What causes water vapor to condense? the cooling of the water vapor
  Why would water vapor get cooler as it rises? Possible answer: The atmosphere is cooler at higher altitudes.
  What term describes condensed water that falls back to the earth? precipitation
- Ask a student to point out the precipitation step on the cycle.
  What is precipitation that has soaked into the ground called? ground water
  How is ground water useful? Ground water provides water for plants, streams, rivers, and oceans.
  What is the process through which plants release water vapor into the air? transpiration
  What helps plants release water vapor? stomata
  What are stomata? tiny openings in plant stems and leaves
- Direct attention to the picture of stomata on Student Text page 184.
- The water cycle will be used in Lesson 123 to demonstrate God's faithfulness.

Water Cycle

Water is always moving between the atmosphere and the earth. The sun warms bodies of water, such as oceans, lakes, and rivers. This heats the water and causes some of it to evaporate. **Evaporation** is the process of liquid water changing into a vapor, or gas.

Plants, animals, and humans also release water vapor into the air. Plant roots absorb water from the soil. The water is then carried to the stems and leaves. Tiny openings, called stomata, allow water vapor to move out of the plant. This is called **transpiration**. Animals and humans also breathe out water vapor when they exhale.

As water vapor rises high above the earth, it cools. The cooling causes it to change back to a liquid. The process of water vapor changing to a liquid is

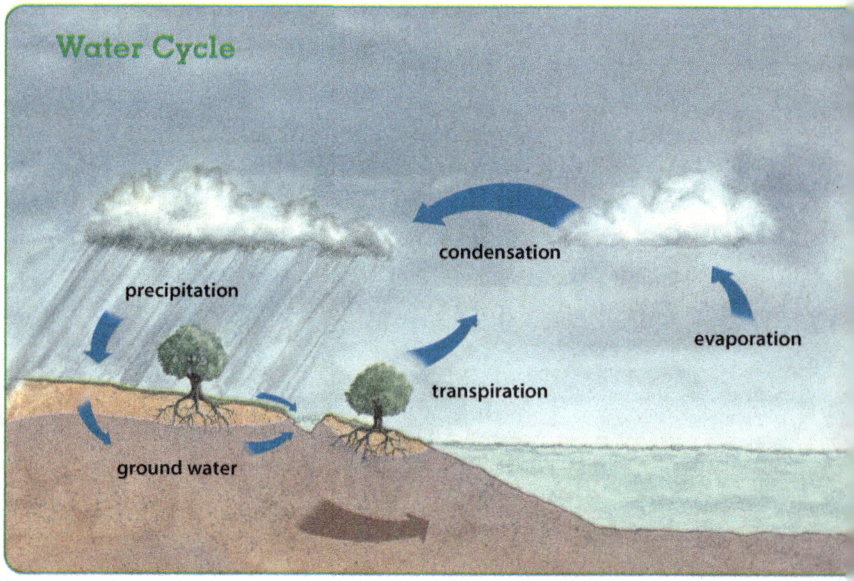

200

Background

Evaporation
As discussed in Chapter 3, *Matter*, vaporization is the process of a liquid changing to a gas at the liquid's boiling point. Evaporation is a form of vaporization that occurs when the liquid is not boiling.

Transpiration
Transpiration is the formation and release of water vapor from the pores of any living organism. Although transpiration is frequently associated with plants, animals and humans also experience transpiration when water is released through the pores in their skin and evaporates. Technically, the releasing of water vapor during exhaling is not transpiration although it does add water vapor back into the atmosphere.

Precipitation
Precipitation is discussed in Chapter 5, *Weather*.

Hydrological Cycle
Hydrology is the scientific study of water on and under the earth's surface and in the atmosphere. *Hydro-* means "water" and *-logy* means "the study of." Another name for the water cycle is the hydrological cycle.

called **condensation**. Rain droplets form from condensation. The droplets collect in the clouds and then fall to the earth as **precipitation**.

Some precipitation soaks into the ground and is stored there. This water is known as **ground water**. It provides water for plants, streams, rivers, and oceans. Water that does not soak into the ground collects on the earth's surface. The collected water may be as small as a puddle or as large as an ocean. Evaporation of some of this water begins the cycle again.

CREATION CORNER

You may be surprised to see all the benefits that God has provided for us in a single tree. A tree absorbs and uses carbon dioxide and then releases oxygen. However, through its leaves and trunk, the tree also absorbs some of the excess carbon dioxide and other gases that are harmful to life. Even a tree's roots are useful. They help filter and clean ground water.

A tree may also change the temperature around it. For example, a single average-sized oak tree not only provides shade from the sun to lower the temperature, it also transpires a lot of water. In fact, it can transpire as much as 375 L (99 gal) of water each day! Just like evaporation, transpiration helps cool the tree and the air around it. And, of course, the tree provides beauty for us and a habitat for animals. A tree shows how God's creation can be both beautiful and useful.

God's Perfect Plan

Our world did not begin by chance. Our great Creator planned and created every detail. He planned for each cycle to depend on the other cycles. Water is needed for the water cycle. Water is also used in the carbon cycle during photosynthesis. Nitrogen compounds are made available to plants through precipitation and by being dissolved in ground water.

Seasonal changes also affect other cycles. Autumn leaves fall to the ground. Bacteria break down the leaves to form nitrogen compounds, adding many nutrients to the soil. Plants need these nutrients so that they can grow. During the spring, plants grow and produce new leaves. The leaves are needed for photosynthesis to occur. Through these cycles, God is continually providing for what He has created.

QUICK CHECK

1. How is most nitrogen changed into usable compounds?
2. What is the process through which plants release water vapor into the air?

113–114

Who planned and created every detail of our universe? God, our Creator

Does each cycle happen separately, or do they depend on each other? They depend on each other.

What is part of both the carbon cycle and the nitrogen cycle? Possible answers: water, oxygen, bacteria

How is water used in the carbon cycle and the nitrogen cycle? In the carbon cycle, water is used during photosynthesis. In the nitrogen cycle, water provides a way for nitrogen compounds to be available to plants.

What gas is used in both the carbon cycle and the nitrogen cycle? oxygen

What do dead leaves add to the soil? nutrients that help plants to grow

- You may want to review the cycles that have been discussed and find other ways that these cycles are connected.

  What does God do through the cycles of the earth? He continually provides for His creation.

- Discuss *Creation Corner*.

  What are some benefits of a single tree? Possible answers: A tree absorbs and uses carbon dioxide and releases oxygen. It absorbs excess carbon dioxide and other harmful gases. Its roots filter and clean ground water. It can lower the temperature around itself. It transpires water. It provides beauty.

  How do the benefits of a tree help us appreciate God as the Creator? Answers will vary, but elicit that God's creation is a marvel of efficiency.

ANSWERS

1. through bacteria
2. transpiration

Demonstration

Plants Release Water Vapor

Materials: leafy potted plant; large, clear plastic bag; string

Tie the plastic bag over a large section of the plant. Allow some of the plant to remain uncovered. Set the plant in a sunny place. Leave the bag on the plant for at least one hour. The condensation inside the bag comes from the water vapor that the plant released through transpiration.

114

- The *Plants Release Water Vapor* demonstration found at the bottom of the pages in Lessons 113–14 may also be incorporated into the lesson.
- Suggestions for additional activities for this two-day lesson are on the Teacher's Toolkit CD.

Activity Manual

Review, pages 135–36

These pages review Lessons 113–14.

Assessment

Quiz 8-B

The quiz may be given any time after completion of this lesson.

Student Text pages 202–3
Activity Manual pages 137–38

Objectives

- Recognize that decomposers are a part of many cycles
- Identify water as a variable that affects decomposition
- Analyze the effects of water on the rate of decomposition

Materials

Materials needed for this activity are listed on the Student Text page. Also see the Activity Helps under Background on this page.

Teacher's Toolkit CD

- Rubrics: *Decomposers at Work*, for each student

Two lesson days are allotted for this activity. Use the first day to set up the activity. After the observation period, you may choose to use the second day to discuss the Conclusions.

Introduction

What do you do with things such as carrot tops, peelings from vegetables, and egg shells? Possible answers: throw them in the trash; give some of them to pets such as guinea pigs or rabbits; use them in the garden

Gardeners often use grass clippings and kitchen scraps such as egg shells, vegetable peelings, and cabbage leaves to make a compost pile. As organic substances decompose, they add nutrients to the soil. These nutrients can be used to help new plants grow.

Different factors in the environment affect how quickly organic substances decompose.

Name the cycles in which decomposers are a key part. the carbon cycle, the nitrogen cycle

Preparation for Reading

- Direct the students to read silently Student Text pages 202–3 and Activity Manual pages 137–38 before beginning the activity.

Teach for Understanding

Process Skills

The process skill emphasized in this lesson is identifying and controlling variables.

What is a variable? a part of an experiment that can be changed

What variables are you controlling in this activity? Answers should include the following: The containers are the same and have the same amounts of soil, carrots, and lettuce. They are kept covered, stirred the same number of times, and kept in the same place.

What variable is changed? the amount of water

- Explain that a scientist tries to identify all the possible variables before doing an experiment.

ACTIVITY

Decomposers at Work

Decomposers are an important part of many cycles. Without decomposers, living things would not be able to get the nutrients that they need. Some decomposers, such as worms and mold, are large enough to be seen. But others are too small to be seen without a microscope.

In this activity you may not be able to see the decomposers, but you will be able to observe their work.

Process Skills
- Hypothesizing
- Experimenting
- Observing
- Identifying and controlling variables
- Recording data

Problem

How does water affect how quickly carrots and lettuce decompose?

Procedure

1. Label one container *wet*, another one *moist*, and the last one *dry*.
2. Predict which condition will allow the carrots and lettuce to decompose the fastest. Complete the hypothesis in your Activity Manual.
3. Put 3 cm of soil in each container.
4. Add three carrot pieces and 20 mL of shredded lettuce to each container.
5. Pour 40 mL of water into the container labeled *wet*. Add 20 mL of water to the container labeled *moist*.
6. Stir the mixture in each container.
7. Cover each container with plastic wrap, and secure it with a rubber band. With a pencil, poke an equal number of small holes through the plastic wrap on each container. Set the three containers in the same location.

Materials
3 containers
potting soil
centimeter ruler
9 carrot pieces
60 mL shredded lettuce
60 mL water
metric measuring cups
spoon
plastic wrap
3 rubber bands
Activity Manual

202

Background

Activity Helps

Prepare the carrots and lettuce ahead of time. The carrot pieces should be diced to about 1 cm or smaller. To avoid introducing additional variables, make sure that the pieces of carrot used by each group of students are about the same size and shape. The same number of pieces should be placed into each cup.

You may choose to use pieces of other plants. Avoid using vegetables from the onion or cabbage families; these will smell as they decompose.

Some processed potting soils have been sterilized to remove bacteria. If purchased potting soil is used, you may want to add some soil from outdoors to increase the amount of decomposers in the soil.

If the students use materials that vary from the list printed in the Activity Manual, they should put a line through materials that were not used and then write the materials that were used. This reinforces accurate recording.

8. Observe the containers every two or three days for two weeks. Record your observations. Stir the mixtures each time you check them. Be sure to cover each container again after observing it.

Conclusions

- In which container did the mixture decompose the fastest? Did your results support your hypothesis?
- What are some other things that could have influenced how quickly the mixture decomposed?

Follow-up

- Test the mixtures in a cooler or warmer area.
- Add fertilizer and observe whether it affects how quickly the mixtures decompose.
- Use the same amount of water in each container, but use a different kind of soil in each.

203

What is the variable tested in this activity? the amount of water

- Explain that the scientist sets up a sample of his experiment without any variables and sets it aside. The sample that is set aside is called a *control*. As the scientist tests other variables, he compares the results to the control.
  Is one of your containers a control? yes Which one? the dry container

Procedure

- Guide the students as they prepare the containers. Make sure that each cup contains the same amounts of soil, carrot, and lettuce.
- Provide time for the students to check the containers every two or three days.
- Remind the students to stir the mixtures each time that they check the containers. You may choose to keep a spoon near the containers.
- Remind the students to cover the containers after stirring.

Conclusions

In which container did the mixture decompose the most? the wet container

What was different about that mixture that may have caused it to decay more? It had more water.

Did your results support your hypothesis? Answers will vary.

What are some other things (variables) that could have influenced the amount of decomposition in each container? Possible answers: temperature, forgetting to re-cover a cup, not having the same amount of soil in each cup, the type of soil

Do you think that it is difficult to get all but one variable to be the same in an experiment? Answers will vary.

What are some other variables that could change in this activity? Possible answers: the amount of the materials, the type of soil, the amount of fertilizer in the soil, the amount of light

- Discuss what information could be gained from changing one of these variables. For example, discuss what might happen if the type of soil was changed instead of the amount of water.
  If different types of soil are used as the variable to be tested, how would that affect the amounts of water that you use? The same amount of water would need to be used in each container, just as the same amounts of carrots and lettuce are used.

Activity Manual

Activity, pages 137–38

Assessment

Rubrics

Select the prepared rubric, or design a rubric to include your chosen criteria.

117
118

Student Text pages 204–7
Activity Manual page 139

Objectives

- Identify three natural stresses on an ecosystem
- Explain how fires and floods can be beneficial to an ecosystem
- Identify some effects of a drought
- Describe the process of succession
- Recognize that sometimes what seems to us like a disaster is actually God's way of maintaining the earth

Materials

- news article or other information about a recent natural disaster

Teacher's Toolkit CD

- Quiz 8-C

Vocabulary

- stress
- succession

Two days have been allotted for this lesson. You may want to teach Lesson 117 the first day and use ideas from Lesson 118 on the second day. You may also incorporate the ideas from Lesson 118 into the text material, making the lesson two days.

Introduction

- Display the news article and discuss it.

  What kinds of damage resulted because of this occurence? Answers will vary.

  Events such as this often seem like a tragedy. But the effects of some disasters can be helpful to ecosystems.

Preparation for Reading

- Direct the student to mentally make questions using the headings, bold and italicized words, and captions.
- Instruct them to read silently pages 204–7 to find the answers to these questions.

  How is a fire or a flood beneficial?

  How is a drought different from other natural disasters?

Teach for Understanding

What is a stress? any physical hazard to life caused by having too much or too little of things needed for life

Were stresses part of God's original creation? no
When did stresses come into the world? when sin entered the world

Even though sin brought stresses on our environment, who controls and maintains the world? God

What are two causes of forest fires? people's carelessness and lightning

Name some of the benefits of a forest fire. Possible answers: Some fires get rid of dead trees and things that could cause larger fires. A fire can recycle nutrients from dead trees and plants. A fire is needed for the seeds of some plants to sprout.

Stresses on an Ecosystem

An unexpected change in the environment can cause an ecosystem to not be able to function normally. A river might overflow its banks and flood nearby fields and towns. Fires might burn forests and force animals and people to flee to safety. Droughts can dry up the water supply, causing crops to fail and soil to erode.

A **stress** is any physical hazard to life caused by having too much or too little of things needed for life. These stresses were not a part of the world when God created it. They have come into the world because of sin. Usually they are tragic disasters. Animals may die or be forced to move. Plants may be uprooted or burned. However, despite all these stresses, God maintains His earth.

Lodgepole pine

Fire

Many forest *fires* are started through man's carelessness. Other forest fires are caused by lightning. Either way, fires can cause ecosystems to undergo quick and sudden changes. Fires are often thought of as destructive and harmful. Sometimes, though, a fire can be beneficial to an ecosystem. Forest fires get rid of dead trees, branches, pine needles, and other things that could cause larger fires. For this reason, some fires are allowed to burn themselves out. People carefully watch these fires, though, so that they do not get out of control. Any fire that threatens the safety of people is put out as soon as possible.

New trees grow in a burned area of Yosemite National Park in California.

204

Background

Causes of Floods

Floods can be caused by heavy rainfall, dam failures, hurricanes and tropical storms, or sudden thaws that melt large amounts of snow or ice in a short time. Undersea earthquakes or volcanic eruptions can also create huge waves, such as tsunamis, that flood coastal areas.

Warfare Floods

Floods have sometimes been used in warfare. In 1574 the people in the Dutch city of Leyden were under siege by the Spanish. Instead of surrendering, they broke the dikes and ended the siege by drowning about 20,000 Spanish soldiers.

In June of 1938, a Chinese general ordered his troops to smash the dikes of the Hwang He (Yellow River) in order to hinder a Japanese invasion. The Japanese were stopped. However, the flood killed over 500,000 Chinese and left millions of Chinese homeless.

Fires cause nutrients from dead trees and plants to be recycled into the soil. These nutrients make the forest soil rich and fertile. Some plants have seeds that will sprout only after a fire. For instance, the lodgepole pine's cone needs heat in order to open and release its seeds.

Over time, new forests grow to replace those that were burned. This may take many years. The new forest, though, is not exactly like the one that it replaces. It often has a greater variety of plants and animals than the old one did.

Flood

A *flood* occurs when an ecosystem receives more water than it can use or store. A flood can move large amounts of earth and rock. The force of water can uproot trees and move houses. Sometimes a flood can carry away rich topsoil. Without topsoil, crops may not be able to grow. Floodwaters may also destroy crops and other plants by covering them with water.

Sometimes flooding occurs when a river overflows its banks into surrounding land. It may take a while for floodwaters to recede into the river. When they finally do, the river may have a new water level or even a new route.

Like fires, floods are often very harmful and disastrous. However, floods can be beneficial. Floods can distribute fertile soil and seeds, allowing plants to grow in areas where they may not have grown before. Some trees depend on annual floods. Without them, the trees' seeds could not sprout.

Mississippi River flood in 1993

205

Why are some forest fires allowed to burn themselves out? The fires get rid of dead trees and other things that might start larger fires.

- Explain that brush fires and field fires can also be beneficial. These fires are sometimes intentionally started to remove old plant growth and to provide nutrients for new plants.

How does fire affect the soil in a forest? Nutrients from the dead trees and plants make the soil rich and fertile.

Name a nutrient we have discussed that would be recycled into the soil. carbon

Other than forests what other ecosystems may be affected by fires? Possible answers: tundra areas, grasslands, some deserts

What is unique about the seeds of a lodgepole pine? The cones will release their seeds only after the extreme heat of a fire.

When do floods happen? A flood occurs when an ecosystem receives more water than it can use or store.

What are things that could happen as a result of a flood? Possible answers: Large amounts of rock and earth may move. Topsoil may be carried away. Plants and crops may be destroyed. Rivers may have a new water level or a new route.

How can floods be beneficial? Floods sometimes distribute fertile soil and seeds. Some trees need annual floods in order for their seeds to sprout.

Background

Bangladesh and Rice

In Bangladesh, floods are often caused by heavy monsoon rains. These floods add eroded fertile soil to the rice fields. Some farmers prepare for the floods by planting different kinds of rice in the areas that they expect to flood. One kind of rice grows best when submerged underwater. Another kind of rice does best in a dry environment. Farmers often plant both kinds of rice together. In years that have normal flooding, both types of rice will survive and be harvested. If there is a major flood or a drought, at least one of the rice crops will survive.

Activity

Forest Fire Opinions

Materials: research materials

Assign each student one of these positions: a tourist who visits forests, a homeowner who lives near a forest, a logger, a zoologist working in the area, or a hunter.

Provide time for the students to research and write a paragraph about forest fires from their assigned position.

Allow students to present their positions and discuss pros and cons of each.

117
118

When does a drought happen? when an ecosystem receives less rainfall than it needs

What are some things that might happen as a result of a drought? Possible answers: There is less water available for plants, animals, and humans. The soil becomes dry and dusty. Wind can cause erosion. Food becomes scarce, resulting in famine. Fires can easily spread.

Look at the picture. What are the results of a drought? The river has much less water. The soil is dry and cracking.

What does *scarce* mean? that there is not enough to meet the need

Why do you think that drought would lead to famine in poorer countries but not in wealthier countries? Possible answer: In poorer countries, the people may raise their own food and not have the money to buy food from others. The governments and people in wealthier countries would probably be able to buy food that is raised in other countries.

How do droughts make an area more susceptible to fires? The vegetation becomes dry and could easily burn.

How is a drought different from other natural disasters? The beginning and ending of a drought are difficult to determine. They occur gradually.

What are some other stresses that might cause sudden changes in an ecosystem? Possible answers: a volcanic eruption, an ice storm or blizzard, a severe thunderstorm or tornado

- Discuss *Science & The Bible.*

What is another way God uses natural disasters? as judgment for sin

How long did the drought announced by Elijah last? three years

Can you think of any other instances in the Bible where God used natural disasters in His dealings with people? Possible answers: Noah's Flood; the famine in Israel mentioned in Ruth 1; the earth's swallowing Korah, Dathan, and Abiram when they rebelled against Moses; the famine that sent Joseph's brothers to Egypt to buy grain; several of the plagues mentioned in Exodus; the earthquakes at the time of Jesus' crucifixion

What is succession? gradual changes in the populations of organisms in an ecosystem

Does succession happen quickly or slowly? slowly

What usually causes succession? a change in the environment

Drought

A *drought* occurs when an ecosystem receives less rainfall than it needs. Sometimes a drought is so severe that there is not enough water to meet the needs of the plants and animals. Lower water levels in rivers and streams cause many problems for the fish and wildlife that depend on the water.

Dried-up riverbed in San Luis Obispo, California

The lack of rain causes topsoil to be dry and dusty. It can be easily eroded by the wind. Some plants die without a good water supply. These factors all cause food for animals and humans to become scarce. In poorer countries, drought often leads to famine.

Droughts also make an area more susceptible to fires. Dry vegetation burns easily. So, once a fire starts, it can spread rapidly. In extreme heat, some fires can even start by themselves.

A drought does not have a definite beginning or ending. In this way, a drought is different from other natural disasters. The beginning of a drought may not be noticed for weeks or even months. The end of a drought also occurs gradually. Some droughts last only a few months. Others can last for years and completely change an ecosystem.

Succession

Ecosystems are changing all the time. Many of these changes are due to stresses such as fires, floods, and droughts. After a stress

Science & The Bible

Sometimes God uses natural disasters as a judgment for sin. During the reign of King Ahab of Israel, the king and the people of Israel had been involved in idol worship and great wickedness. In 1 Kings 17–18, the Bible tells us that God sent Elijah to King Ahab to announce a drought. Elijah said God would punish Israel for its rebellion against God. So for three years God did not send dew or rain. A disaster is not always a judgment of God. But it can be the tool that God uses to bring His people back to Himself.

206

Science Misconceptions

Drought

Droughts do not mean that the earth is running out of water. There is just less water than normal in one area; meanwhile, another area might be getting more water than normal.

Activity

Effects of Drought

Direct the students to read 1 Kings 17–18. List three effects of the drought mentioned in those chapters.

occurs, the size of the populations and the types of organisms are often different. For example, new trees and plants grow to replace a destroyed forest. However, the species of trees and plants may be different from those in the original forest. Animal populations also change according to what foods are available. These gradual changes in the populations of organisms in an ecosystem are called **succession**.

Succession happens when part of the environment changes. The environment determines what kinds of living things can exist in an ecosystem. A change in the climate, amount of sunlight, or available nutrients causes the living things in an ecosystem to change.

QUICK CHECK

1. Describe what a stress to an ecosystem is.
2. How can floods be beneficial?
3. What is succession?

Beavers cut down trees and dam a forest stream. The stream floods and makes a pond.

Trees in the flooded area die. Pond plants, such as reeds and cattails, grow at the edges of the pond.

Over the years, sediment builds up in the pond, and the pond becomes shallower. The dam eventually breaks, and most of the pond water drains away.

As the pond dries out, plants at the edges of the pond gradually take over the center of the pond. The pond has become a meadow.

Shrubs replace the meadow grasses. Fast-growing trees, such as pines and poplars, replace the shrubs.

207

117–118

What kinds of things in the environment might change? the climate, the amount of sunlight, and/or the available nutrients

What are some of the causes for most of these changes? stresses such as fires, floods, and droughts

How might the new ecosystem be different after a stress? There may be different types of plant and animal species or different amounts of the populations.

- Discuss the pictures and captions of the succession of a beaver's dam.

How do beavers change a forest stream into a pond? by building a dam across it; by preventing the water from flowing and causing a flood that makes a pond

What happens to the trees that are in the flooded area? They die because of the flood.

Why does the pond eventually become shallower? Sediment builds up in the pond.

What happens after the dam breaks? Most of the pond water drains away, causing the pond to dry up.

What may a pond become after it dries out? a meadow

What may the meadow eventually become? a forest

ANSWERS

1. any physical hazard to life caused by having too much or too little of things needed for life
2. Floods sometimes distribute fertile soil and seeds. Some trees need annual floods in order for their seeds to sprout.
3. a series of gradual changes in the populations of organisms in an ecosystem

Background

Beavers and Succession

When a beaver builds his dam and floods an area of the forest, this creates many new habitats for other animals and plants. The dam influences wetland succession by creating bogs, marshes, or ponds. As the beaver's lodge is being built, the amount of timber in the water increases. This attracts invertebrates, minnows, and young fish. These animals in turn attract larger fish and birds. Otters, osprey, foxes, raccoons, and other predators then come to the pond. Waterfowl and mammals often come for a safe place to feed and drink. The pond also provides a wet and marshy environment for plants.

A beaver's dam helps to prevent erosion, raise the water table, and purify the water. The beaver abandons the pond when it becomes too shallow or when his preferred food is scarce. The abandoned dam breaks, and the pond drains. The rich layers of silt cause herbaceous plants to flourish, and a beaver meadow is formed. Eventually shrubs and trees will grow, setting the stage for a beaver's return.

Sahel Drought

Most continents have gone through severe droughts at times. Africa, though, has frequently experienced droughts that led to famine. The Sahel region of Africa suffered a twelve-year drought that began in the late 1960s. Cattle owners had to move from country to country throughout the region in order to find food and water for their cattle.

118

- The *Forest Fire Opinions* and the *Effects of Drought* activities found at the bottom of the pages in Lessons 117–18 may also be incorporated into the lesson.
- Suggestions for additional activities for this two-day lesson are on the Teacher's Toolkit CD.

Activity Manual

Review, page 139

This page reviews Lessons 117–18.

Assessment

Quiz 8-C

The quiz may be given any time after completion of this lesson.

Objectives

- Research a historical stress, such as a famous fire, flood, or other disaster
- Organize and present information about the stress

Materials

- reference materials and resources

Teacher's Toolkit CD

- Instructional Aid 8.4: *Stressful Events*, for each student
- Rubrics: *Stress Alert*, for each student

Two lesson days are allotted for this exploration. On the first day, introduce the exploration, set guidelines, and allow the students to begin planning. Use the second lesson day for the presentation of the reports. See the Background section for additional suggestions.

Introduction

Do you know of any historical natural disasters? If so, which ones? Answers will vary but may include the Hurricanes Katrina and Rita in 2005, California wildfires in 2007, tornado in Joplin, Missouri in 2011, and Hurricane Sandy in 2012.

- Explain that we often hear about disasters soon after they happen. Newspapers and television and radio stations usually send reporters as close to the disaster as possible so that they can tell others what is happening. In this exploration you will be the reporter responsible for covering an important disaster.

Preparation for Reading

- Instruct the students to read silently Student Text page 208 and Activity Manual page 140 before beginning the exploration.

Teach for Understanding

What to Do

- Distribute the *Stressful Events* pages. Allow each student to choose a disaster to research.
- Provide resource books and other information as needed. Remind each student to record the information about his event in his Activity Manual. Allow time for each student to prepare and present his news report.

Activity Manual

Exploration, page 140

Assessment

Rubrics

Select the prepared rubric, or design a rubric to include your chosen criteria.

EXPLORATION

Stress Alert

Have you ever heard a news report about an emergency situation? News reporters cover these events so that the public can be informed. Often a reporter is sent "on location" to be near the storm or disaster. The reporter can then give an eyewitness report about what is going on. He might also warn people to evacuate the area or tell them where to go for help.

In this exploration you are the news reporter. You will cover a flood, a fire, a drought, or another kind of stress just as if you were on location.

Materials
Stressful Events page
resource books about stresses
Activity Manual

What to Do

1. Choose one event from the *Stressful Events* page.
2. Research to find out facts about your chosen event. Remember that a reporter always looks for information to answer the questions *who*, *what*, *when*, *where*, and *why*. Record your facts in your Activity Manual.
3. Organize your information into a news report. Pretend that you are presenting your news report for the evening news on the day that the event occurred. Include pictures or drawings of the event. You may include other materials as well to make your news report seem more realistic.
4. Present your report.

208

Background

Exploration Helps

You may choose to have students work in groups or as individuals. You may choose to add or delete events from the *Stressful Events* page based on what resources are available. Not all of the stresses included were caused by natural events. You may choose to have each student record the sources of information that he uses.

Presentations

This project provides an opportunity for the students to incorporate various forms of technology for researching, preparing the reports, and making the presentation.

The student should present his report as if he were a television or radio news reporter. You may choose to use a video camera to record television broadcasts and play them back for the students.

Current Events

ACTIVITY

Ecosystems never stay the same. Sometimes they change in small ways that are barely noticed. At other times, they change in big ways because of events such as floods and fires. Sometimes the changes are mentioned in the news.

In this activity you will collect current news articles and information about ecosystem changes that are happening right now.

Process Skills
- Classifying
- Communicating
- Defining operationally

Purpose

Identify current events that involve changes in ecosystems.

Procedure

Materials
current magazines and newspapers
notebook
scissors

1. Gather news articles about ecosystems. The articles may come from online or print resources. You may also listen to news reports. If you hear news about an ecosystem, record the date and time of the report as well as the program and station that provided the report. Then write a summary of what was said.
2. Organize your articles and information for presentation. Arrange the materials in a logical way. You might group your news reports by topic, by type of event, by date, or by geographical location. Include a table of contents.
3. Share your information with your classmates.

209

Background

Activity Helps

You may choose to have a group of students work together on one notebook or have one notebook for the whole class.

You may want to have the students use a standard list of headings.

You may want to have each student record the sources of information that he uses.

Newspapers and newsmagazines are good sources of information. If a student does not have access to these, he could visit the library and photocopy articles and information. The students may choose to listen to television or radio news reports and write down information from those sources. News articles can also be printed from Internet sources.

Objectives

- Collect and record information about ecosystems
- Organize the information into a notebook for presentation

Materials

- front page of a newspaper
- cover of a news magazine

Teacher's Toolkit CD

- Rubrics: *Current Events*, for each student

Two lesson days are allotted for this activity. On the first day, introduce the activity, set guidelines, and allow the students to begin researching. Use the second lesson day for planning, organizing, and sharing the notebooks. See the Background section for additional suggestions.

Introduction

- Display the newspaper and the magazine.
  What catches your attention and helps you know what the articles are about? headlines, pictures
  As you look for current events about ecosystems, use the headlines and pictures to help you locate articles quickly.

Preparation for Reading

- Instruct the students to read silently page 209 before beginning the activity.

Teach for Understanding

Process Skill

The process skill emphasized in this lesson is communicating.

What is communicating? using a written, oral, or graphic means to give information to others

What types of communication will you include in the notebook? written and graphic

How do you think that organizing the notebook helps the reader get information? Possible answer: He can more easily find the type of information that he wants rather than having to look at each article.

Procedure

- Guide each student as he organizes his notebook.
- Remind the students of ways that they can find current information about ecosystems.

Assessment

Rubrics

Select the prepared rubric, or design a rubric to include your chosen criteria.

Activity Manual pages 141–42

Objectives

- Explain the water cycle using a model
- Relate the cycles of nature to God's care of His creation

Materials

- clear two-liter bottle with label removed
- 500 mL hot water
- ice
- scissors

Introduction

Are the ecosystems the same today as when God created them? no

Looking at the world through our Bible glasses, what is one reason that the ecosystems are different? the effects of the Fall (sin)

- Explain that God still cares for His creation.

What did Adam and Eve eat in the beginning? a vegetarian diet

How does God provide a different diet now? There are still vegetables, but there is also meat to eat.

Can you think of another stress on the ecosystems mentioned in the Bible that reshaped the world? the global Flood

Preparation for Reading

- Direct the students to read the first three paragraphs on Activity Manual page 141 to find out what it might have been like to step out of the ark.

Teach for Understanding

- Direct a discussion about how the students think they may have felt getting out of the ark.
- Ask a volunteer to read aloud Genesis 8:20–22.

What did Noah do when he got out of the ark? He built an altar to God.

What promise did God make? not to curse the ground again or send another worldwide flood

- Ask the volunteer to reread Genesis 8:22.

What cycle of change is this verse referring to? seasons

- Allow time for the students to answer the first two questions on the Activity Manual page.
- Instruct a student to read aloud the last paragraph on Activity Manual page 141.

What sign does God use in Genesis 9? a rainbow

- Direct the students to read silently Psalm 104:10–14 to find out ways God provides for His creatures.
- Allow time for them to answer questions 3 and 4. Discuss the answers together.

A lesson from
Answers in Genesis Name ______________ **1:1**

God Ordained Change

Think about the time after the Flood. What would it have been like to step out of the ark alongside Noah and his family? The Flood had been a major stress on the environment. The world they knew had been totally rearranged. Noah, his family, and the animals with them were responsible for repopulating the earth from scratch. As the water continued to move to the oceans, the bare land was exposed. Plants and seeds that had been floating during the Flood began to sprout in the barren landscape and colonize the land.

As Noah and his family got off the ark, they probably had fears. If you had been with Noah, how might you have felt?

The Bible tells us that Noah didn't have to wait long before God gave him comfort. Read about what happened in Genesis 8:20–22.

1. Will God ever curse the ground or send another worldwide flood? *no*
2. How long will there be seasons and other natural cycles on the earth?
   as long as the earth remains

In Genesis 9 God also uses the rainbow as a sign of His everlasting covenant with all people and animals that will ever live on the earth. God showed grace to Noah and all of his descendants. He gave them a promise to keep the natural cycles going on the earth to provide for their needs. God is pleased to give good gifts to all people, even if they don't deserve them. Psalm 104 describes many of the ways that God provides for His creatures.

3. Read Psalm 104:10–14. List ways that God is providing for His creatures.
   Possible answers given: springs provide drink for animals; birds live near the springs; rain waters the earth; grass grows for cattle; herbs are for people; food grows from the ground
4. What cycle of nature is partially described in Psalm 104:10–13? *the water cycle*

Resources

The BJU Press website offers additional links from Answers in Genesis that you may find helpful with this lesson.
bjupress.com/resources/science

Look at the model of the water cycle your teacher has. Each part of the model represents a part of the natural cycle.

5. Draw the model.

6. Label the parts of the water cycle next to the parts of the model they represent.

When we see the cycles in nature, we are actually seeing a part of God's character on display (Job 36:24–33). God is a God of order and is faithful to keep His promises. He has promised us in the Bible that He will keep the world turning, plants growing, and people farming until He brings the world to an end. Because God is holy and perfect, He cannot lie (Titus 1:2; Hebrews 6:18).

If God has been faithful to fulfill His promises this far, we have great hope that He will keep His promises to us in the future. One of those promises is to restore the earth to its original perfect state. People who believe that the cycles of nature and matter are all that exist have no hope of life after this one. But you can use the truth of God's Word to share that hope with them.

7. List some promises of God that remind you of His faithfulness.

____

____

____

____

____

____

God's Word not only helps us make sense of the world we live in but also of the world that we look forward to living and worshiping God forever in. Who can you share that hope with?

Model the Water Cycle

- Keep the cap on the bottle. Carefully cut off the top of the bottle about 2 cm below the upper edge of the straight part of the sides.
- Pour the hot water into the bottom part of the bottle.
- Place the top part of the bottle inside the bottom part so that the bottle cap is pointing downward. Put ice in the top part of the bottle.

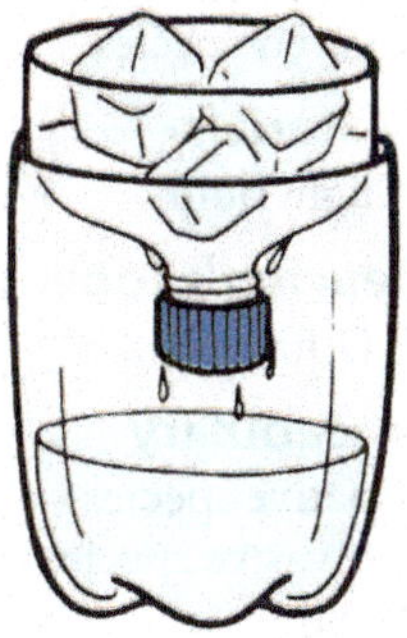

- Observe a small-scale water cycle as the water vapor from the hot water rises and condenses against the part that is holding the ice. Water droplets should form and fall from the top part.

  In this model, what does the steam represent? the water evaporating from lakes and oceans

  What does the lid represent? the condensation in clouds

  What does the dripping water represent? the rain (precipitation)

- Allow time for the students to draw and label the model as directed on Activity Manual page 142.
- Ask a volunteer to read aloud the two paragraphs on the Activity Manual page.

  What are some of the promises that God gives us in the Bible? Possible answers: will keep the world turning and the plants growing; will restore the earth to its perfect state

  Many people think that the cycles of nature are all that exist. They believe in living a good life, but they have no hope for eternity. What do Christians need to be telling these people? the hope of salvation in Christ and enjoying Him forever in heaven

Activity Manual

Answers in Genesis, pages 141–42

124

Student Text pages 210–13
Activity Manual page 143

Objectives
- Identify some manmade stresses
- List differing opinions about using natural resources
- Differentiate between an extinct species and an endangered species

Materials
- light bulb

Teacher's Toolkit CD
- Quiz 8-D

Vocabulary
- native species
- invasive species
- extinct
- endangered
- threatened

Introduction

- Direct attention to the light bulb.

  Suppose you knew that this light bulb was the last light bulb available in your area. When it burned out, you would not be able to have electric lights anymore.

  What would you do to make this bulb last longer? Possible answers: use it only when absolutely necessary; remember to turn it off when it is not being used; make a schedule of times when it can be used

  Do you think that anyone would start hunting for other light bulbs that might have been misplaced or forgotten? probably

  Maybe someone would not be concerned that it is the last light bulb. How might that person show that he is not concerned? Possible answers: would use the light bulb whenever he wanted to; would not try to remember to turn it off when he did not need it

  This light bulb is an example of some of our God-given resources. Just like with the light bulb, people always have different opinions about how to use our resources.

Preparation for Reading

- Display the words *nene* (NAY nay) and *hyacinth*. Pronounce the words together.
- Instruct the students to read silently pages 210–13 to find the answers to these questions.

  Which manmade stress is probably the largest stress?

  How is an endangered species different from an extinct species?

Teach for Understanding

What responsibility has God given people? to have dominion over His creation

What are three views, or opinions, about exercising this responsibility? People should be able to use and benefit freely from all of the earth's resources. Preserving an ecosystem is more important than any other benefit people might receive from the ecosystem. People should use God's resources wisely by having a balance between preserving and using.

Ecosystems and Man

Man's Responsibility

When God created the world, He placed Adam and Eve in the Garden of Eden. God gave humans dominion over His creation. But God also gave them responsibility. Adam's job was to be a steward, or keeper, of God's creation. Like Adam, we also must be good stewards of what God has given us.

However, people have very different views about how to use our resources. Some people believe that people should use and benefit freely from all of the earth's resources. This attitude can lead to a waste of natural resources. Other people believe that preserving an ecosystem is more important than any benefit we might receive from it. They do not believe that God gave people the earth to use. The Bible teaches that there should be a balance. Resources should be used but not wasted. Mankind honors God most by using His resources wisely (Genesis 1:28).

Mankind's Interactions

Ecosystems often undergo many kinds of stress. Some are natural stresses, such as droughts, floods, and fires. Many, though, are manmade stresses. A Christian who understands his responsibility will be careful about the stress that he puts on an ecosystem.

Probably the largest manmade stress is *habitat destruction*. People often destroy plant and animal habitats when expanding their own. Forests are cut down to provide lumber or farmland. Wetlands are drained to make room for shopping centers. Dams are built across some rivers to provide electricity.

Fish ladder at the Bonneville Dam in Washington

210

Background

Pollution Problems
Even useful chemicals such as fertilizer can cause problems. Fertilizer that is not used correctly can get into streams and lakes. This increases the growth of algae. Algae use up the oxygen that plants and animals in the water need.

Atlantic Puffins
Atlantic puffins, also called sea parrots, are small sea birds that spend most of their time in the northern Atlantic Ocean. The only times they usually come ashore are to breed and to raise their chicks. These birds, their eggs, and their chicks were often hunted for food. In the past, ladies also used puffin feathers to decorate hats. As a result, the number of puffins in North America declined rapidly. *Project Puffin*, an attempt to restore puffins to their old nesting places in Maine, was started in 1973 by the National Audubon Society.

Wildlife refuge in Crystal River, Florida

Construction and development are often good. However, they can be harmful if not done with care. There are many ways that man can lessen the harm done to an ecosystem during development. New trees can be planted to replace ones that are cut down. Areas of land can be set aside as wildlife refuges or national parks. Artificial wetlands can be built to replace wetlands that have been drained. Dams can have fish ladders added. These steplike waterways help fish travel safely around the dam. Remember the train wreck that damaged the Sacramento River system? To reduce the risk of future chemical spills, the railroad company built a large, strong fence along the bridge to prevent cars from falling into the river.

Pollution is another major manmade stress. As seen through the Sacramento River accident, living things need clean water, air, and land. Without them, the living things will die. Pollution is anything that makes the water, air, or land dirty. Most people think of oil spills and factory chemicals as pollution. But littering and car exhaust also pollute the environment. Other trash and chemicals can cause pollution as well. If not disposed of properly, they can harm plants, animals, and humans.

Excessive hunting and fishing can also be a problem. Laws today regulate, or control, fishing and hunting. These laws try to help keep the animal populations in an ecosystem balanced. For example, hundreds of years ago many Atlantic puffins lived in North America. By the early 1900s, though, the number of puffins had decreased greatly. Many puffins had died because of excessive hunting and habitat loss. Today, because of laws that protect them, the number of puffins in North America is increasing again.

Puffin

211

What term describes the stresses that usually occur in nature? natural stresses

What term describes the stresses that are caused by man? manmade stresses

Which manmade stress is probably the largest stress? habitat destruction

What does *habitat destruction* mean? when plant and animal habitats are destroyed

What is a habitat? the place where a plant or animal lives

What are some ways that people can lessen the harm done to an ecosystem during development? Possible answers: Plant new trees. Set land aside as wildlife refuges or national parks. Build artificial wetlands to replace wetlands that were drained. Add fish ladders to dams.

Why is a dam a problem for some fish? Possible answers: Some fish naturally swim upstream to breed and lay eggs. The fish would not be able to get over or through the dam.

- Direct attention to the picture on page 210.

How does a fish ladder help some kinds of fish? A fish ladder provides a way for fish to swim around a dam.

What is pollution? anything that makes the water, air, or land dirty

- Direct attention to the miniature scientist on page 193.

What are some causes of pollution? oil spills, factory chemicals, litter, car exhaust, trash and chemicals that are not disposed of properly

What is the miniature scientist doing? cleaning up a barrel that was not disposed of properly

Why are there laws to control, or regulate, fishing and hunting? to try to help keep the animal populations in an ecosystem balanced

What North American animal has been helped by these laws? the Atlantic puffin

- You may want to share some of the Background information about the puffin.

Science Misconceptions

Manmade Stresses

Much of the thinking in modern society says that the fate of the physical earth is determined by people. But mankind cannot destroy what God has created. God, not mankind, controls the fate of the earth.

Activities

Community Recycling

If possible, allow students to pick up litter within a certain area or participate in a community clean-up day. You may consider starting a recycling drive with your students. Contact a local recycling center for specific guidelines. Common items that are recycled include newspaper, glass, aluminum, and plastic.

Ecosystem Poster

Design a poster that promotes wise interaction with an ecosystem. Possible topics could be about stopping pollution, preserving land as a wildlife refuge, lessening the effects of construction, or not starting fires. Display and discuss the poster.

124

What are native species? species of plants and animals that originally live in an ecosystem

What are invasive species? species that are not native to a certain area and can cause harm

What was a problem that the nene, or Hawaiian goose, experienced? Ships coming to Hawaii accidentally brought rats, which ate the eggs of the nene.

What animal was brought to Hawaii to control the rat population? the mongoose

How did the mongoose make the problem worse? It preferred to eat the nene eggs.

What problems can occur when an invasive species is introduced in an area? Possible answers: The invasive species can bring diseases. The population size of the invasive species may grow rapidly, and there may be increased competition with native species. The invasive species may prey on native species.

- Discuss *Science & History*.

Why was the water hyacinth taken to China? as food for cattle

Why did water hyacinths become a problem? They grow rapidly and can double their population size in just twelve days. The plants block waterways, crowd out native plants, and can even kill fish by preventing sunlight and oxygen from reaching the water.

Unexpected Dangers

Plant and animal species that originally live in an ecosystem are called **native species**. Species that are not native to the area and can cause harm are called **invasive species**.

Introducing new species into an ecosystem often brings problems. For example, years ago ships accidentally brought rats to Hawaii. The rats were an invasive species. They preyed on the eggs of the nene, the Hawaiian goose. To control the rat population, the mongoose, a member of the weasel family, was brought to Hawaii. But the mongooses did not eat the rats. Instead, they preferred nene eggs. This introduction of a non-native species only made the problem worse.

Nene

Mongoose

Besides preying on native species, invasive species can also bring other problems. Sometimes native species die from diseases brought in by the new species. Many times the invasive species also has no natural predator in the new habitat. This allows the population of the invasive species to grow rapidly. With more species there is more competition for the same resources, such as water or food. Native populations may then decline.

Science & History

The water hyacinth is a plant native to South America. In 1930 the plant was taken to China to provide food for cattle. The plant is beautiful, and many people admired it. They used it to decorate ponds and other water ecosystems.

However, the water hyacinths quickly became a nuisance and danger to China's aquatic ecosystems. The plants grow very quickly. Sometimes the number of plants can double in just twelve days. The plants also block waterways and crowd out native plants. They can even kill fish by preventing sunlight and oxygen from reaching the water. Many methods are currently used to control the spread of these plants.

212

Background

Invasive Species

Many plants and animals that are considered to be native today were invasive species brought to America by the colonists. For example, European immigrants to New York in the 1890s brought and released starlings. These birds took over many bluebird nesting sites and food sources. Now in some places of the United States, starlings are abundant and bluebirds are rare.

Dandelions are another invasive species that can be found almost everywhere in the United States. They were introduced to North America by colonists, who valued the leaves for eating and for making tea. Today they are considered a weed by most homeowners.

Exotic Species

An exotic species is any species that is not native to an area. A new species is not always harmful. Honeybees, potatoes, and cotton are examples of exotic species that have been successfully introduced to North America.

Passenger pigeon

Extinction

At one time huge flocks of passenger pigeons used to fill the North American skies. But the last known passenger pigeon died in captivity in 1914. Most of the birds died from hunting or loss of habitat. The species is now considered extinct. Scientists consider a species to be **extinct** when its last known member dies.

A species becomes **endangered** when its population is so small that the species may become extinct. Many animal and plant species are considered to be endangered. Hunting, loss of habitat, and mankind's carelessness are most often the reasons a species becomes endangered.

A species is **threatened** if its population could become endangered in the near future. A species that can live in only one habitat is in more danger of becoming threatened, endangered, and eventually extinct than a species that can live in different areas is.

Mankind's Decisions

Ecosystems are always changing in one way or another. Some changes are part of a natural cycle. Other changes are sudden and unexpected. But God controls each change. He uses people to develop the earth He created. When God put people on the earth He told them to take care of it. But they disobeyed God and brought sin and death into the world.

There are a lot of challenges in a sinful world. A whole ecosystem can be destroyed by disaster or carelessness. Entire animal populations can be made extinct. Any change in an ecosystem affects other plants and animals.

Christians need to learn about the world and its ecosystems so that they can obey God's command. Learning how ecosystems work is the first step to managing them wisely. Managing them wisely is a way we can glorify God by imitating Him.

QUICK CHECK

1. What are some ways that people can lessen the harm done to an ecosystem during development?
2. Why would a species be considered invasive?
3. How is an extinct species different from an endangered species?

213

124

What does it mean for a species to be extinct? that its last known member has died

How is an endangered species different from an extinct species? An endangered species still has members that are living. However, its population is so small that it is in danger of becoming extinct. An extinct species has no living members.

When do scientists consider a species to be threatened? if the population of its species could become endangered in the near future

Why would a species found in only one type of habitat be in greater danger of becoming extinct than a species that could live in many different habitats? If that one habitat is destroyed, the species may die since it cannot live anywhere else.

Who plans and controls each change that happens on the earth? God

Why does it profit Christians to learn about ecosystems? to help obey God's command to take care of the earth; to manage the ecosystems wisely to glorify God

ANSWERS

1. Possible answers: Plant trees. Set aside land as wildlife refuges or national parks. Build artificial wetlands to replace wetlands that were drained. Add fish ladders to dams.
2. The species is not native to the area and causes harm.
3. A species is extinct when its last known member dies. A species is considered endangered when its population is so small that it may become extinct.

Activity Manual

Review, page 143

This page reviews Lesson 124.

Assessment

Quiz 8-D

The quiz may be given any time after completion of this lesson.

Activities

Extinct or Endangered Species

Direct the students to write and illustrate a paragraph about an animal or plant that is extinct or endangered. The paragraph should include a description of the species, where it lived or lives, and what was or is being done to try to protect it. You may choose to compile the paragraphs and pictures to make an illustrated encyclopedia of extinct and endangered animals.

Changes in the Bible

Materials: Instructional Aid 8.5: *Changes in the Ecosystem*, for each student

The students will match some of the terms in the chapter to the correct definitions. Then they will look up Bible verses and match the verse to the correct science term.

125

Student Text page 214
Activity Manual page 144

Objectives

- Recall concepts and terms from Chapter 8
- Apply knowledge to everyday situations

Review

- Materials for the Chapter 8 Test will be taken from Student Text page 214 and Activity Manual pages 133–36, 139, and 143–44. You may review any or all of the material during the lesson. Questions similar to the ones in Thinking It Through, Activity Manual page 144, may appear on the test.
- You may choose to review Chapter 8 by playing the review game or a game from the Game Bank on the Teacher's Toolkit CD.

Activity Manual

Review, page 144

This page requires written responses to application questions.

126

Objective

- Demonstrate knowledge of the concepts taught in Chapter 8

Assessment

Chapter 8 Test

CHAPTER SUMMARY

Words to Know

cycle	transpiration	native species
season	condensation	invasive species
photosynthesis	precipitation	extinct
respiration	ground water	endangered
decomposition	stress	threatened
evaporation	succession	

Key Ideas

Identify seasons, the carbon cycle, the nitrogen cycle, and the water cycle as regular patterns of change
Explain what monsoons are and what they affect
Describe the carbon cycle, nitrogen cycle, and water cycle
Explain why nitrogen is important and how it can be changed into a usable form
Give examples of how one cycle depends on another
Describe stresses that affect ecosystems
Identify possible benefits of fires and floods
Compare and contrast a flood with a drought
Describe manmade stresses on ecosystems
Explain how a species can be both a native species and an invasive species
Identify problems that an invasive species can cause
Describe the requirements for a species to be considered extinct, endangered, or threatened
Summarize the relationship Christians should have with ecosystems

214

Review Game

A Forest in Succession

Materials: two sets of cards with the names of living things (e.g., grass, bush, tree, bird, rabbit, fox, hawk, bear) that would return to a forest as it recovers from a stress

Divide the students into two teams. Designate an area to be the "forest" for each team. Scatter the cards on the floor in the "forest" area so that the words can be read.

For each correct answer, the student goes to his team's "forest" and stands on one of the cards. The first team to have a person on each card wins.

Teacher Notes

127

Student Text pages 215–17
Activity Manual page 145

Objectives

- Recognize the interrelationship of science concepts
- Recognize that technology can be designed to control sound because sound moves in predictable ways

Materials

- music recording

Unit Introduction

- Sound and light are two kinds of energy waves. Since both are waves, some of their characteristics are similar. However, light waves and sound waves are not the same. Each type of wave has specific characteristics that make it unique. Some of the similarities and differences of sound waves and light waves are discussed in Unit 5.
- Chapter 9 identifies sound waves as mechanical waves that are limited to moving through matter. The chapter also describes other characteristics of sound and the movement of sound waves.
- Chapter 10 discusses light energy and spectrums. Light waves are electromagnetic waves. They can travel through matter or through a vacuum. The interaction of light with other substances, resulting in reflection and refraction, is also discussed.

  Look through Unit 5. What kinds of topics do you think you will be studying in this unit? Possible answers: sound, light, waves

  Why do you think these chapters are organized into the same unit? Answers will vary. Elicit that the chapters are about waves and energy.

  What do you think the picture shows? Accept any reasonable answer. It shows fiber optics.

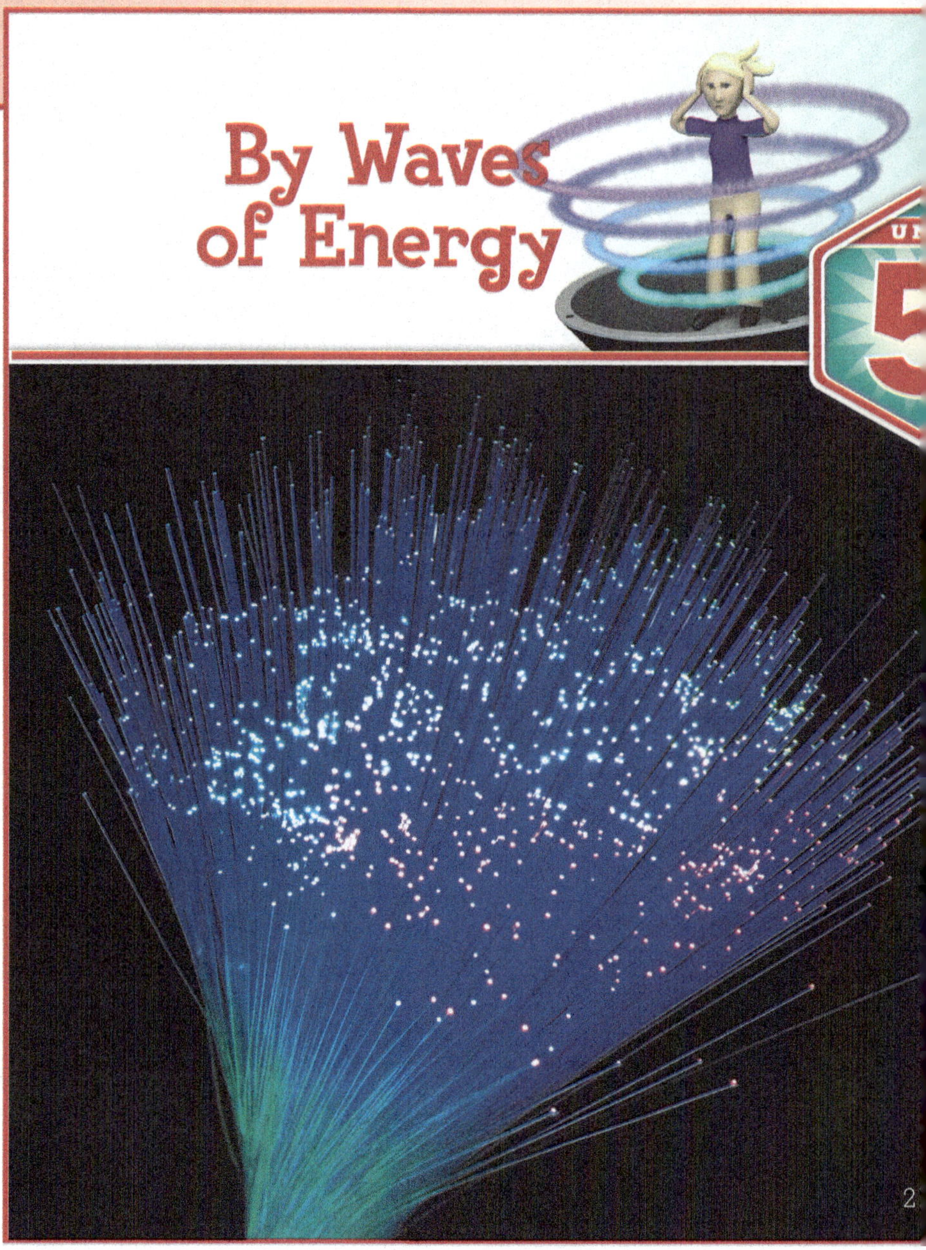

Background

Unit Photo

The photo on page 215 displays fiber optics showing data or an internet communication concept. The miniature scientist is reacting to the effects of sound waves while she stands on a speaker.

Weblinks

The BJU Press website offers additional information and links you may find helpful throughout the unit.
bjupress.com/resources/science

- Ask volunteers to read the teasers on page 216. Discuss the information.

Project Idea

- The project idea presented at the beginning of each unit is designed to incorporate elements of both chapters as well as information gathered from other resources. You may choose to have the student do the project as a culminating activity at the end of the unit or as an ongoing activity while the chapters are being taught.

Unit 5: Sounds and Lights

- Design a mural or pictorial timeline to show sounds, lights, and electromagnetic waves that were used in the past, are used in the present, and may possibly be used in the future. Include sounds from the past that are not often heard today. These sounds may include a fire crackling in a kitchen stove or the sound of cowboy boots on a wooden sidewalk. Lights from the past may include an oil lamp from Bible times or a candle on a miner's helmet. Present sounds include tapping on a computer keyboard and cell phones ringing. Lights and electromagnetic waves used today include lasers and x-rays. Future sounds and lights are left to the imagination of the student.

This lesson continues with the Chapter 9 introduction on page 249.

Unit 5 Overview

Lesson	Teacher's Edition	Student Text	Activity Manual	Assessment	Content
Chapter 9: Sound					
127	246–49	215–17	145		• Worldview thinking • Unit and chapter preview
128–29	250–53	218–21	146–48	Quiz 9-A	• Properties of sound • Vibrations, waves, frequency, and speed
130	254–55	222–23	149–50	Rubric	Activity: Sound Slide • Process skill: identifying and controlling variables
131–32	256–59	224–27	151–52	Quiz 9-B	• Characteristics of sound: pitch, volume, and quality
133	260–61	228–29	153–54	Rubric	Activity: Shhh, Quiet Please • Process skill: predicting
134	262–63		155–56		Answers in Genesis • A biblical viewpoint of problem solving
135	264–68	230–34	157	Quiz 9-C	• Seeing with sound: echolocation and ultrasounds • Communicating with sound: music, technology, acoustics
136	269	235		Rubric	Exploration: A "Medium" Exploration • Compare and contrast quality of sound in different mediums
137	270	236	158		**Chapter Review** • Apply knowledge to everyday situations
138	270			Chapter 9 Test	**Chapter 9 Test**

Lesson	Teacher's Edition	Student Text	Activity Manual	Assessment	Content
Chapter 10: Light					
139	271	237	159		• Worldview thinking • Chapter preview
140	272–75	238–41	160–61	Quiz 10-A	• Characteristics of light • Comparing waves
141–42	276–79	242–45	162–63	Quiz 10-B	• The visible spectrum: refraction, reflection, and colors
143	280–81	246–47	164	Rubric	Activity: Fog Vision • Process skill: hypothesizing
144–45	282–85	248–51	165		• Mirrors and lenses • Light and technology
146	286–87	252–53	166	Rubric	Activity: Angles of Reflection • Process skill: defining operationally
147–48	288–92	254–58	167–68	Quiz 10-C	• The electromagnetic spectrum • Electromagnetic-spectrum technology
149	293–94		169–70		Answers in Genesis • Recognize that the Bible calls Christians to defend their faith
150	295	259	171	Rubric	Exploration: Light at Work • Research and present uses of light in technology
151	296	260	172		**Chapter Review** • Apply knowledge to everyday situations
152	296			Chapter 10 Test	**Chapter 10 Test**

Sound 9

Lawrence J. Fogel was an engineer who worked for an aircraft company. In the mid-1900s he received patents for several instruments to be used with airplanes and helicopters. One design was especially useful. It reduced the noise that interfered with a pilot's ability to listen to people on the radio. Whenever pilots are taking off or landing they talk to a control tower for instructions. These instructions help keep the aircraft and everyone inside it safe. Mr. Fogel realized that aircraft equipment makes a lot of noise. He used his knowledge of the properties of sound to help him design a new type of technology. The company that manufactured his idea made a new kind of headset. The design included a microphone that picked up low unwanted sounds. The headset then produced opposite sounds for each. The opposite sounds canceled the original sounds. The pilot could then hear whatever was said on the radio without the additional noise.

Consider

- What problem did Mr. Fogel try to solve?
- How did his idea benefit others?

217

Background

Scientist Lawrence J. Fogel

Lawrence J. Fogel (1928–2007) developed flight-related technology in the 1950s. After that he began working with artificial intelligence. He devised evolutionary programming, which is an approach in artificial intelligence that actually evolves solutions to problems by replicating evolution. He founded Decision Science, Inc., the first company to solve real problems by evolutionary computation. He developed methods that created an experience-based technique of missile evasion in various types of warfare. Lawrence Fogel's research also led to the improvement of industrial production, the discovery of new medicines, and the enhancement of mission planning in defense software. Later he founded Natural Selection, Inc. The company addresses and solves problems in areas such as bioinformatics, medical diagnostics, data mining, perimeter security, autonomous vehicle capabilities, and risk management.

Chapter Preview

Preview and prereading activities may include using a K-W-L chart, a probe, or an anticipation guide.

Chapter Photo

The photo on page 217 shows tubas and baritones in a parade.

Chapter Introduction

- Play the music recording softly. Direct the students to talk with fellow students. Turn the volume up very loud for about 30 seconds, then turn it off.

  What difficulties did you have communicating with your friends while the music was loud? Answers will vary.

  What are ways that might fix the problem? Answers will vary.

  What kind of knowledge would be beneficial in solving the problem? how sound works

- Explain that God designed sound waves to move in predictable patterns. People can observe and study these patterns and then use them as they design helpful technology.

Preparation for Reading

- Instruct the students to read silently page 217 to find out what Lawrence Fogel designed.

Teach for Understanding

What problem was Lawrence Fogel trying to solve? a pilot being unable to hear the radio clearly over the noise of aircraft equipment

What did Fogel use to help him solve the problem? his knowledge of the properties of sound

- Explain that one characteristic of sound is that many sound waves can occupy the same space. By isolating and canceling harmful sounds, active noise reduction (ANR) technology not only protects a pilot's hearing but also helps the pilot perform his job better and more safely.

  What biblical principle does the invention of the headset for pilots portray? The invention shows how a person can exhibit dominion over the earth. The invention was done for the benefit of others and to glorify God.

- You may want to explain that Lawrence Fogel has recognition as an evolutionist. Point out that God's grace is so great that it even enables people that deny Him to do good things.

Activity Manual

Preview, page 217

The Looking Ahead page is intended to assess the student's prior knowledge before beginning the chapter.

128
129

Student Text pages 218–21
Activity Manual pages 146–48

Objectives
- Define *sound* and *wavelength*
- Identify a compression of a sound wave
- Differentiate between the frequency and speed of sound waves

Materials
- ruler
- paper, folded as a fan
- rubber band

Teacher's Toolkit CD
- Visuals 9.1–9.3: *Properties of Sound; Parts of a Sound Wave; Sound Waves*
- Instructional Aid 9.1–9.3: *Flip Book; Vibrations and Waves*, for each student; *Vibrations and Waves Answer Key*
- Quiz 9-A

Vocabulary
- vibration
- sound
- wave
- wavelength
- frequency
- speed of sound

Two days have been allotted for this lesson. You may want to teach Lesson 128 the first day and use ideas from Lesson 129 on the second day. You may also incorporate the ideas from Lesson 129 into the text material, making the lesson two days.

Introduction

Listen to the sound made by each item.

- Hit the ruler on the edge of a desk to make a sound.
- Wave the paper fan quickly to make a sound.
- Stretch the rubber band between your thumb and a finger and pluck it to make a sound.

Each item made a sound. What caused each item to make its sound? Each item vibrated, or moved.

Preparation for Reading

- Display the word *oscilloscope*. Pronounce the word together.
- Instruct the students to mentally make questions using the headings, bold and italicized words, and captions.
- Direct them to read silently pages 218–21 to find the answers to these questions.

What is a compression?

What unit of measurement is used to measure the frequency of sound waves?

Teach for Understanding

- Direct attention to the middle square on the web on Activity Manual page 146. The answers on this page may vary.

What topic is this web about? Properties of Sound

What is a vibration? a rapid back-and-forth movement

What is a sound? a vibration that is heard

Properties of Sound Waves

Engineers like Mr. Fogel must understand sound to design technology that can be used to help others. What do you know about sound? Sit quietly for a moment and listen. You may hear the hum of air moving through a vent or the ticking of a clock. Perhaps a noisy truck or train is passing nearby. A chair may squeak, or shoes may rustle on the floor. Although each sound is different, all sounds are made in a similar way. Each sound started because an object moved.

FANTASTIC FACTS

You may have heard the shrill trumpet sound that an elephant makes. But a deep growling or rumbling noise is a more common way that one elephant communicates with another. Infrasound is a sound below the level of human hearing. Many of an elephant's rumblings use infrasound. The vibrations of an elephant's infrasonic messages and warnings can be heard by family members miles away.

Both humans and elephants make sounds as air causes their vocal folds to vibrate. Vocal folds are strips of tissue that lie across their airways. An elephant's rumbling sounds are made as puffs of air cause its vocal folds to vibrate. The vocal folds of an elephant are about eight times larger than those of a human. Its very large folds vibrate much the way a flag flaps in the wind, resulting in the low infrasound.

Vibrations and Waves

The movement of an object causes vibrations. A **vibration** is a rapid, back-and-forth movement. **Sound** is a vibration that can be heard. The ticking that comes from a clock is caused by the movement of the parts inside the clock. The parts vibrate as they touch each other. They cause the air molecules around them to vibrate too. These vibrations blend together to make the ticking sound that you hear.

Sound moves through matter in waves. A **wave** is a disturbance that moves energy from place to place.

218

Background

Names of Waves

Waves can be classified and compared in different ways. For example, waves can be classified as either mechanical or electromagnetic. Waves also have names that describe the way in which they move. Some mechanical waves can be called compression waves or longitudinal waves. The terms *compression wave* and *longitudinal wave* are interchangeable. Electromagnetic waves are transverse waves. Electromagnetic waves are discussed in Chapter 10, *Light*.

Mechanical Waves

Any wave that can only travel through matter is a mechanical wave. Some examples of mechanical waves include water waves and the waves in a spring toy.

Rarefactions

In a mechanical wave, places where matter is being pushed together, called compressions, alternate with places where matter spreads apart, called rarefactions.

Parts of a Sound Wave

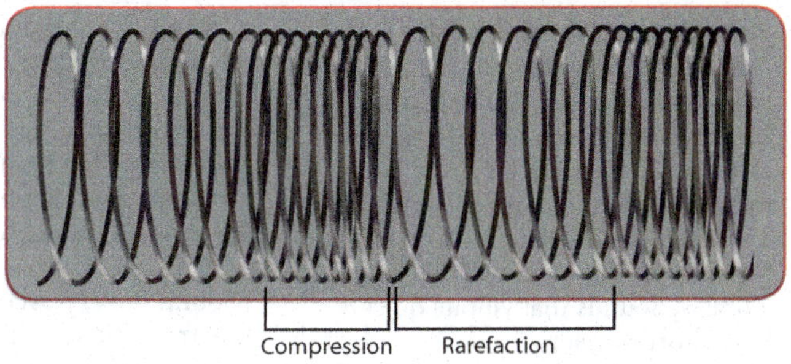

The matter through which a sound wave travels may be a solid, liquid, or gas. Most of the sounds that we hear travel through the air.

Sound energy moves from the vibrating object to the ear. You hear sounds because the vibrations of sound energy cause your eardrum to vibrate. As the vibrations are carried through the bones of your middle ear, the vibrations are changed into nerve messages. Your brain then interprets these messages and tells you what you hear.

Sound waves can be described as mechanical waves. Mechanical waves must have matter to travel through. Energy causes particles in the matter to move and vibrate. Sometimes you can feel sound waves. When you hum, you can feel the vibrations that your throat makes. Those vibrations cause the sound that you hear. Other sounds that you might feel are loud thunder or the beats of a bass drum.

In a sound wave the particles of matter vibrate back and forth in the same direction that the sound energy is moving. The motion of each vibration pushes together some of the particles of matter. This part of the wave is called a *compression*. As the motion of the vibration moves away from the compression, the particles spread apart again. This part of the wave is called a *rarefaction*. Waves continue to form as long as the object vibrates.

The way sound waves move is similar to the way a spring toy moves. If you stretch out a spring toy and push one end, the coils of the spring move back and forth as the wave passes to the end of the spring. The particles of matter are not carried along with the sound wave. Each particle vibrates but remains in its place. Imagine what would happen if air particles were carried with the sound waves. Sound waves from a choir's singing would cause a wind to blow!

219

Activities

How a Wave Moves Through Matter

Materials: chairs arranged in a straight line or bleachers

Seat volunteers in the chairs or along one bleacher bench. Direct the students to do a stadium-style wave by standing and sitting quickly, one right after the other along the line. The "wave" can be seen moving along the line. However, the students do not move to the end of the line. They stay at their seats and "vibrate" up and down.

How is the movement of the "wave" through the students like the movement of a sound wave through a medium? The students stayed at their seats as the wave moved, just as particles in a medium stay in their places as a sound wave moves through them.

Write a Script

Write a script or short story about what happens to a quiet "community" of particles of matter when a sound wave comes through and disturbs its peace.

- Encourage the students to write the definition for sound in the box with the topic. Then tell them to write the definition for vibration on the Activity Manual web.
  What causes vibrations? the movement of an object
  How does sound move? in waves
  What is a wave? a disturbance that moves energy from place to place
  What states of matter can a wave travel through? solids, liquids, and gases
  What do most of the sounds that we hear travel through? the air
  Which state of matter is air mostly made up of? gases
- Allow time for the students to complete the definition of a wave on the web.
  What happens in your ears as you hear sounds? The vibrations cause your eardrums to vibrate. The vibrations are carried through the middle ear bones and are changed into nerve impulses that the brain can understand.
  What kind of waves can move only through matter? mechanical
  Give an example of a time that you can feel a sound as well as hear it. Possible answers: humming, loud thunder, beats of a bass drum
  How do the particles of matter in a sound wave move? vibrate back and forth along the same path as the wave is moving
- Refer to the *Parts of a Sound Wave* visual as you discuss compression and rarefaction.
  What is the part of the sound wave called where the vibration pushes together some of the particles of the matter? a compression
  What is a rarefaction? the part of the wave where the motion of the vibration moves away from the compressions and particles spread apart again
  Do particles of matter move with a sound wave, or do they just vibrate in place? Each particle vibrates in its place.
- Allow time for the students to complete the definitions for a compression and a rarefaction on the web.
- You may want to do the *How a Wave Moves Through Matter* activity at this time.
- Discuss *Fantastic Facts* on page 218.
  What is the most common way that an elephant communicates? a deep growling or rumbling noise
  What is a sound below the level of human hearing called? infrasound
  How does an elephant make the rumbling sound? puffs of air cause its vocal folds to vibrate
  How much larger are an elephant's vocal folds than a human's vocal folds? eight times larger
  Who made the elephant's vocal folds so large? God
  Why do you think God designed an elephant's infrasonic sounds to be heard by other elephants miles away? Possible answers: to tell where food is located, to warn about enemies, to locate other family members

- Point out the wavelength on the *Parts of a Sound Wave* visual.
  What is a wavelength? the distance from one compression to the next
  What determines the length of a wavelength? how quickly the source of the sound vibrates
- Direct attention to the sound wave at the top of the *Sound Waves* visual.
- Ask volunteers to point out the compressions and the rarefactions on the sound wave. Show them how the wavelength is from one compression to the next one.
  What machine helps scientists study sound waves? an oscilloscope
  What does an oscilloscope do? It changes mechanical sound waves into electrical pictures that are seen on a graph.
- Direct attention to the oscilloscope screen on page 220.
  What are some types of equipment that might include an oscilloscope? Possible answers: stereos, computers, heart monitors and other medical equipment
- Point to the sound wave at the bottom of the *Sound Waves* visual.
  What do you think a wavelength is on the oscilloscope graph? Elicit that it is the distance from one high point to the next or from one low point to the next.
  Look at the wavelength that is marked. On the oscilloscope graph, the high points indicate the compressions of the compression wave.
- You may choose to use the *Flip Book* page to demonstrate the movement of waves at this time.
- Direct attention to the frequency and wavelength square on the Activity Manual page. Tell the students to complete the squares during the discussion.
  What is frequency? the number of sound waves that pass a point in one second
  What unit is used to measure frequency? the hertz
  In what direction do sound waves travel? in all directions away from the source
  What characteristic of sound changes as a sound moves through different types of matter? the speed of the wave
  What property of matter affects how fast and how far the sound wave can travel? the density of the matter
  What does the term *density* mean? the closeness of the molecules in matter
- Instruct the students to make notes in the box labeled *speed* on the Activity Manual page during the discussion about the speed of sound.
  In which state of matter does sound move most slowly? in a gas

Frequency and Speed

Sound waves cannot be seen, but they can be measured. The distance from one compression to the next is called a **wavelength**. Different sound waves have different wavelengths. These lengths are determined by how quickly the source of the sound vibrates. Sounds that vibrate quickly have short wavelengths.

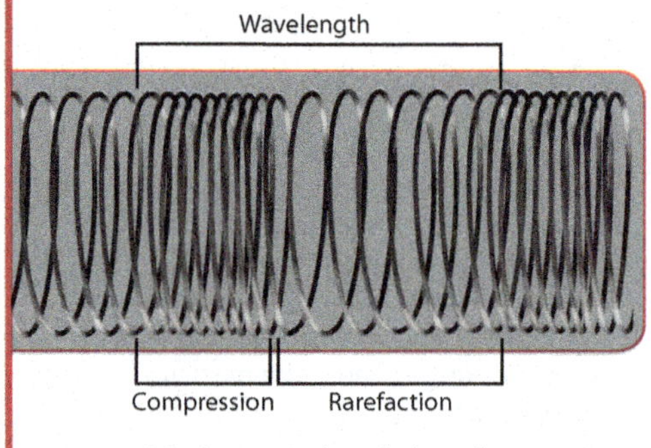

Machines and tools have been developed to help us "see" pictures of sound waves. An *oscilloscope* uses a type of microphone to convert mechanical sound waves into electrical impulses. The impulses are shown as a wavy line on a graph.

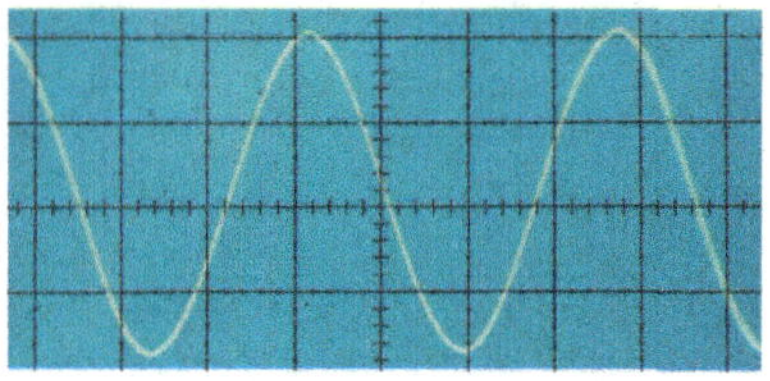
Oscilloscope screen

The number of sound waves per second produced by a vibration can vary. As the number of waves changes, so does the frequency of the sound. **Frequency** is the number of waves that pass a point in one second. Frequency is measured in a unit called hertz.

Sound waves travel in all directions away from the source of the sound. A sound wave can also travel through different types of matter. The siren of an ambulance can be heard whether you are in a house, in a tree, or underwater in a pool.

Sound waves move at different speeds in different types of matter, though. The density, or how close the matter molecules are, affects how quickly the vibrations move from one particle to the next. The density of the matter also affects how far the sound waves can travel.

Sound needs particles in order to move. Of the three states of matter, gas is the least dense, so sound moves the slowest through gases. The sound energy cannot easily move from particle to particle. Also, some of the energy gets used up as it travels through the spaces between the particles. This slows the movement of the waves.

Air is the most common type of matter through which we hear sounds. The **speed of sound** in air is about 335 m (1,100 ft) per second. This is how fast sound travels through

220

Background

Terms Referring to Waves

Wavelength and other terms relating to waves are not limited to sound waves. These terms are used with other mechanical waves as well as with electromagnetic waves. Electromagnetic waves are discussed in Chapter 10, *Light*.

Oscilloscope

Transducers are devices that convert one type of energy to another. In an oscilloscope, a transducer changes the energy to electrical impulses. An electrocardiograph is a type of oscilloscope that is used to monitor the heart. Electrocardiographs are discussed in Chapter 12, *Circulatory System*.

Physicist Heinrich Rudolf Hertz

Heinrich Rudolf Hertz (1857–94) was a German physicist. He experimented with the electromagnetic wave theory developed by the British scientist James Clerk Maxwell. As a result, Hertz was the first scientist to send and receive radio waves. He also studied similarities between different types of waves. Through his study, ways to measure the lengths and speeds of waves were developed. In 1960, a unit of measurement named in his honor, the hertz, was adopted as the standard unit for frequency.

Speed of Sound Through Matter

(Approximate distance traveled in one second)

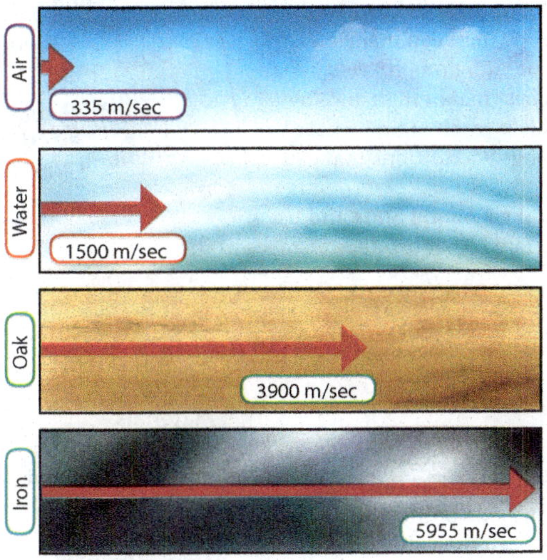

the air under specific atmospheric conditions. In different conditions, such as a hotter day, the speed of sound will be different.

The particles in liquids are denser than the particles in gases. So sound moves more quickly through liquids than it can through gases. The waves also travel farther than they can in a gas.

Solids are the densest state of matter. Their particles are very close together. The sound waves easily move from one particle to the next. So sound waves can also travel faster and farther in a solid than in a liquid or a gas.

The speed of sound is not the same as the frequency of sound vibrations. Speed tells you how fast something is moving. Think of watching cars traveling on a straight highway. Suppose all the cars on this highway are traveling at 50 km per hour (km/h). That is the speed of the cars. However, the frequency of the cars is the number of cars that pass you in a set period of time. Frequency is how often they pass. In the first minute, perhaps only one car passes. In the next minute, five cars may pass. The speed of the cars stayed the same, but the frequency changed. The speed and frequency of sound waves are related in much the same way as the speed and frequency of the cars on a highway are related.

QUICK CHECK

1. What is sound?
2. What is a wavelength?
3. Through which type of matter can sound travel faster—a gas or a solid?

221

128
129

Why does a sound wave travel most slowly through a gas? Gas particles are spread farthest apart, and some of the sound wave's energy is used up as it travels between particles.

What is the most common type of matter through which we hear sounds? air

What does the speed of sound measure? how fast sound travels through the air under specific atmospheric conditions

- Refer to the *Speed of Sound Through Matter* diagram as the speed of sound is discussed.

Do sound waves move faster in liquids or in solids? in solids

Why do sound waves move fastest through solids? Solids are denser. Their particles are closer together.

Look at the diagram. Does sound travel at the same speed through all solids? no

Which do you think is denser, oak or iron? iron Why? Sound moves faster through iron than through oak.

How is the speed of a sound wave different from its frequency? Speed is how fast the sound wave is moving. Frequency is the number of waves that pass a point in one second.

- You may want to do the *How Sound Moves at Different Speeds* activity at this time.

ANSWERS

1. a vibration that is heard
2. the distance from one compression to the next
3. solid

129

- The *How a Wave Moves Through Matter*, *Write a Script*, and *How Sound Moves at Different Speeds* activities found at the bottom of the pages in Lessons 128–29 may also be incorporated into the lesson.
- Suggestions for additional activities for this two-day lesson are on the Teacher's Toolkit CD.

Activity

How Sound Moves at Different Speeds

Materials: 3 beanbags, masking tape

Use masking tape to make three parallel lines three meters long on the floor.

Have four students stand on Line 1 to represent particles in a gas. These students should be spread about a meter apart from each other. Have four students stand on Line 2 to represent particles in a liquid. These students should stand close enough to almost, but not quite, touch each other. Line 3 should have four students standing so that their shoulders touch. They represent particles in a solid.

The beanbags represent sound energy (a wave). Race to see which state of matter can pass the sound energy the fastest. Beanbags must be passed from hand to hand without tossing.

Along which line did the beanbag move the fastest? the line with students standing close together (solid)

Compare the results of the race with how sound energy moves through different matter. The particles in a solid are closer together, so the energy moves through it faster. The farther apart the particles of the medium are, the slower the energy moves through that medium.

Activity Manual

Study Skill and Reinforcement, page 146

This page has a web organizer of the properties of sound.

Review, pages 147–48

These pages review Lessons 127–29.

Assessment

Quiz 9-A

The quiz may be given any time after completion of these lessons.

Objectives

- Observe how the size of a vibration affects its sound
- Change a variable and compare results
- Predict the highness or lowness of a sound

Materials

Materials needed for this activity are listed on the Student Text page.

Teacher's Toolkit CD

- Rubrics: *Sound Slide*, for each student

Introduction

Have you ever used things around your house to make musical sounds?

Maybe you hit a pot with a spoon, covered a comb with paper to make a kazoo, or made a whistle by blowing on a blade of grass or an acorn cap squeezed between your thumbs.

What are some other "instruments" that you have made?

Remember that sounds are made by any vibrating object. Today we will be using rubber bands and rulers to make sounds—maybe even music!

Preparation for Reading

- Instruct the students to read silently Student Text pages 222–23 and Activity Manual pages 149–50 before beginning the activity.

Teach for Understanding

Procedure

- Guide each student as he makes his Sound Slide. He may need someone to hold the ruler while he stretches the rubber band.
- A tight rubber band will vibrate quickly. It will not have very wide vibrations, so the vibrations may be hard to see. A looser rubber band will vibrate more slowly, so the movements of the rubber band may be a little easier to see.
- You may want to have a partner hold two index cards vertically on either side of the rubber band. The cards could be adjusted so that the rubber band touches them as it vibrates. The width could be measured. You may even hear tapping sounds on the card when the rubber band hits it.
- Reinforce the use of terms such as *wavelength*, *frequency*, and *speed* when describing the movements of the rubber band. Avoid using *pitch* and *volume*, since these terms are not introduced until Lessons 131–32. Use *high*, *low*, *loud*, and *soft* instead.

ACTIVITY

Sound Slide

Process Skills
- Predicting
- Experimenting
- Observing
- Identifying and controlling variables
- Communicating

Anything that moves makes sound. Sometimes one item can make a variety of sounds depending on different conditions. A guitarist makes high and low sounds. One way he does this is by changing where he places his fingers on the strings. Conditions that can change are called *variables*.

In this activity you will make a Sound Slide and use it to identify and control variables that produce changes in the sound.

Problem

How does changing the length of a rubber band affect the sound it makes when plucked?

Materials
- centimeter ruler
- assorted rubber bands
- plastic bottle cap
- marker
- Activity Manual

Procedure

1. Prepare your Sound Slide by stretching a rubber band lengthwise over the ruler. Use the marker to make a mark down the side of the bottle cap. Place the bottle cap between the rubber band and the side of the ruler that shows the measurements. Adjust the bottle cap so that the mark faces toward the lower numbers of the ruler. The bottle cap is called a slide.
2. Move the slide so that the mark is at 15 cm. Press the rubber band against the slide. Pluck the part of the rubber band that is over the lower numbers and listen to the sound. Observe the vibrations. This is the setting that you will be comparing the other sounds to. You may return the slide to 15 cm at any time to compare the sounds again.
3. Move the slide to 10 cm. Predict whether the sound will be higher or lower. Pluck the part of the rubber band that is over the lower numbers. Notice any changes in the sound and vibrations from when the slide was set at 15 cm. Record your observations.

222

Background

Activity Helps

This activity may be done individually or in groups.

The student should press down on the bottle cap while plucking the rubber band.

The "length of the rubber band" refers to the section of the rubber band that is plucked.

Be sure that the rubber band is long enough to fit the length of the ruler and does not sag or break.

4. Move the slide to 20 cm. Predict whether the sound will be higher or lower than the sound when the slide is at 15 cm. Pluck the part of the rubber band that is over the lower numbers. Observe the changes in sound and vibrations. Record your observations.
5. Choose three additional markings at which to set the slide. Predict the changes in sound and vibrations. Test your predictions and record your observations.

Conclusions

- Did your results support your predictions?
- How did the length of the rubber band affect the sound?

Follow-up

- Use other rulers, slides, or rubber bands. Test how changing a different variable affects the sound and vibrations.
- Try playing a simple song on your Sound Slide. Record the numbers that you use.

223

130

Process Skill

The process skill emphasized in this activity is identifying and controlling variables.

What are variables? the changeable parts of an experiment

An effective experiment begins with the variables set to specific criteria.

How many variables does the scientist change at a time? one

What variable did you change in this activity? the length of the rubber band

What are other variables that you could have changed? the type of ruler, the thickness of the rubber band, the type of bottle cap

Why is it important for scientists to identify and control as many variables as possible? Possible answer: Uncontrolled variables may give inaccurate results.

Conclusions

How did the rubber band change each time the slide moved? the length changed

How was the sound affected when the length of the rubber band got shorter? the sound got higher

How was the sound affected when the length of the rubber band got longer? the sound got lower

- Discuss differences between Sound Slides made with different rubber bands.

The Sound Slide may be used again in Lessons 131–32.

Activity Manual

Activity, pages 149–50

Assessment

Rubrics

Select the prepared rubric, or design a rubric to include your chosen criteria.

131 132

Student Text pages 224–27
Activity Manual pages 151–52

Objectives

- Define *pitch* and *volume*
- Explain how the pitch of a sound wave is related to its frequency
- Identify the frequency range of human hearing
- Explain how the volume of a sound is related to the intensity of its sound waves
- Define and describe *timbre*

Materials

- squeaky squeeze toy
- rattle or some nails in a can
- 2 wooden blocks
- folder
- Sound Slide from Lesson 130 (optional)

Other items that make distinct sounds may be used.

Teacher's Toolkit CD

- Visual 9.4: *Characteristics of Sound*
- Quiz 9-B

Vocabulary

- pitch
- volume
- decibel
- timbre

Two days have been allotted for this lesson. You may want to teach Lesson 131 the first day and use ideas from Lesson 132 on the second day. You may also incorporate the ideas from Lesson 132 into the text material, making the lesson two days.

Introduction

- Hide the squeeze toy, rattle, and blocks behind the open folder so they are out of the student's view.
  I am going to make some sounds. Try to guess what item is making each sound.
- Make sounds with the squeeze toy, rattle, and blocks. Accept guesses from the students. Reveal the items.
  Do all the items make the same sound? no
  Describe the sound made by each item and how that sound is different from those of the other items. Answers will vary.
  No two sounds are the same. Each has different characteristics.

Preparation for Reading

- Display the word *timbre* (TAM bur). Pronounce the word together.
- Direct the students to read silently pages 224–27 to find the answers to these questions.
  How is pitch different from volume?
  What is timbre?

Characteristics of Sound

Sound engineers understand and use the characteristics of sound. In a similar way the characteristics of sound help us compare and describe individual sounds. You experience many different sounds each day. A songbird trills a high, happy song. The horn of a tugboat makes long, low blasts. The siren of a police car may change between loud and soft sounds.

Pitch

Pitch is how high or low a sound is. The pitch of a sound is related to its frequency. Remember that the frequency is the number of waves that pass a point in one second. These waves result from vibrations. A slowly vibrating object has a low frequency. This produces a low pitch. As an object vibrates faster, its pitch gets higher.

You can hear changes in pitch, for example, by listening to the different sounds made by an electric mixer that has several settings. When the mixer is at the lowest setting, the motor vibrates slowly and you hear a low pitch. As the motor is turned to a higher setting, the motor vibrates faster. The pitch of the motor sounds higher.

The beating of a hummingbird's wings has a high pitch.

Sounds made by a grandfather clock often have a low pitch.

224

Background

Infrasonic and Ultrasonic

Sound waves below and above the range of human hearing have special names. Sound waves with frequencies below 20 Hz are called *infrasonic*, or subsonic. Sound waves with frequencies above 20,000 Hz are called *ultrasonic*.

Activities

Community Sounds

Write a paragraph that contrasts sounds heard around your community with sounds in another location. The other location can be in your country or in another country. If a student lives in the city, his sounds may include traffic, construction equipment, or foreign voices. A rural student may hear sounds such as those from a hay baler, chickens, or water in a stream.

Hearing Loss

After researching, write a paragraph about the sources and dangers of sound pollution, or write a paragraph about the potential hearing loss caused by consistently using earphones or headphones that are too loud.

God gave the human ear the ability to hear certain frequencies. Most people can hear sounds with frequencies between 20 and 20,000 hertz (Hz). The lowest note, or pitch, on a piano has a frequency of about 27 Hz. The frequency of the highest note on a piano is a little over 4,000 Hz. The pitch of most sounds is a blend of frequencies.

Some creatures, such as bats, dogs, and insects, can hear higher frequencies than humans can. If you blow a dog whistle, you cannot hear the sound, but a dog nearby can.

SCIENCE & HISTORY

When musical instruments play together, they must be tuned to the same pitch. If they are not, they will produce sounds that clash rather than blend. The standard pitch used today is not the same as the one musicians used in the past. The tuning fork of the eighteenth-century composer George Frideric Handel tuned the A above middle C to the frequency of 422.5 Hz. Today the same note is usually tuned to 440 Hz.

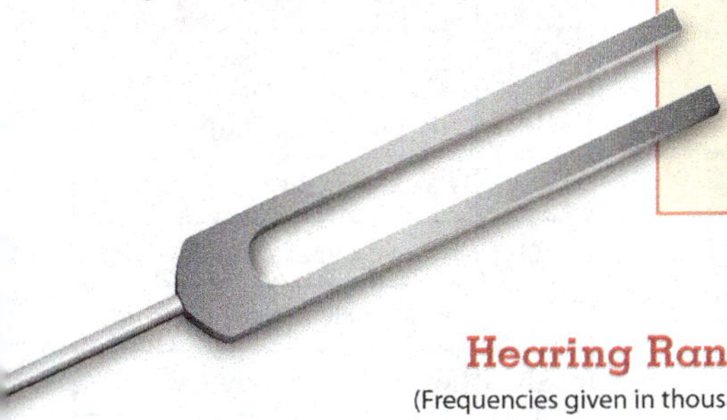

Hearing Ranges

(Frequencies given in thousands of Hz)

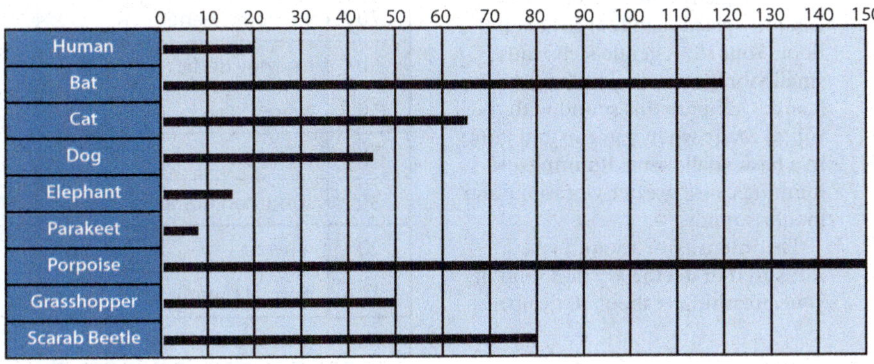

225

Teach for Understanding

- Direct the students to design their own organizers as you discuss the Activity Manual pages (see Background information). You may want to display the *Characteristics of Sound* visual.
  What is the topic of the section that you just read? Characteristics of Sound
- Allow the students time to write the topic in a geometric shape in the center of the page.
  What are three characteristics of sound? pitch, volume, and quality
- Give the students time to place each characteristic in a geometric shape attached to the topic shape. Remind them to make the shape large enough to write notes in.
- Tell the students to take notes as you continue the discussion.
  What is pitch? how high or low a sound is
  What is frequency? the number of waves in one second
  Will an object that is longer vibrate more quickly or more slowly than an object that is shorter? more slowly
- Display the Sound Slide. Place the slide at 15 cm and pluck the rubber band. Move the slide to 25 cm and pluck the rubber band.
  Does the rubber band have a higher or lower pitch and frequency as the slide moves farther down the ruler? lower Why? The rubber band is longer and vibrates more slowly.
  If I move the slide to 10 cm, will the pitch and frequency of the rubber band be higher or lower? higher Why? The rubber band is shorter and vibrates more quickly.
  What determines whether the sound wave has a high frequency or a low frequency? the speed at which an object vibrates
  How are pitch and frequency related? Low frequencies produce low-pitched sounds, and high frequencies produce high-pitched sounds.
  Why can a dog hear a dog whistle but a person cannot? Dogs and some other animals can hear higher frequencies than humans can.
- Discuss *Science & History*.
  What would happen if musical instruments in an orchestra were not tuned to the same pitch? The sounds would clash rather than blend.
  What has happened to the standard pitch? changed over time
- Discuss the chart of hearing ranges.
  Which animal has the highest hearing frequency? the porpoise
  Can an elephant hear a dog whistle? no Why not? The dog whistle has frequencies above what a human can hear, and a human can hear frequencies higher than an elephant can.

Background

Activity Manual Organizer

Activity Manual page 151 provides an opportunity for the students to make an organizer of their own. In this lesson the organizer will be used during the teaching and will follow the general pattern provided in the Activity Manual Answer Key. You may choose to use a different approach or tell the students to complete their own organizer at the end of the teaching.

Hearing Ranges

Hearing ranges for humans and animals do not actually start at 0. They begin at a small number that appears to be 0 on the chart since the increments are 10,000 Hz.

131
132

What is volume? how loud or soft a sound is

What changes the volume of a sound? the amount of energy used to make the sound waves

What does *intensity* mean? the amount of energy

Do small vibrations cause loud sounds or soft sounds? soft sounds

How is the intensity of sound measured? in decibels

How can loud volume damage your ears? Exposure to loud sounds over a period of time can damage your eardrums' ability to vibrate correctly.

- Discuss the *Decibel Levels* chart.

  How many decibels is a whisper? 20 decibels

  Which is louder, a vacuum cleaner or a circular saw? a circular saw

  Why do you think that a warning siren is 140 decibels? so that more people can hear it

  Why do you think that 120 decibels is labeled the "threshold of pain"? Sounds at this level or higher can cause pain in the human ear.

- Explain that this chart is for human hearing. The threshold of human hearing is 0 decibels. Some animals can hear below 0 decibels.

Volume

Volume is how loud or soft a sound is. The volume of a sound changes as the amount of energy used to make the sound waves changes. A greater energy, or *intensity*, will produce larger vibrations. Large vibrations make louder sounds than smaller vibrations do.

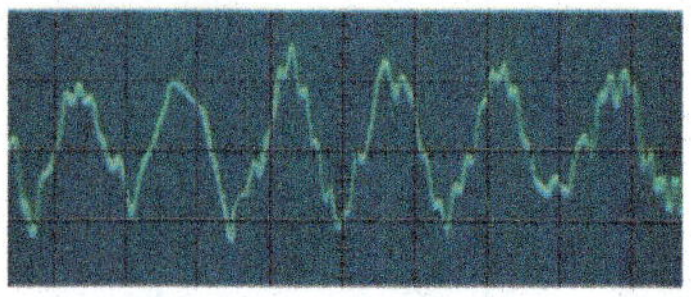

Loud sound

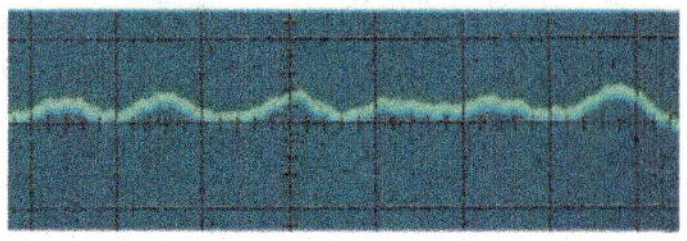

Soft sound

There are times that you may try to walk quietly through a room. You carefully place each foot softly on the floor. Your slow, gentle steps cause small vibrations. Very little sound is heard. Compare this sound with the sound made when you run and jump in a basketball game. Running and jumping cause greater vibrations and louder sounds.

The intensity of a sound is measured in **decibels**. The sound of your breathing is about 10 decibels. Talking in a normal voice to your friend is about 60 decibels. The sound of a jet taking off is about 160 decibels.

Decibel Levels

180	rocket engine
170	
160	jet plane taking off
150	
140	warning siren
130	rock music concert
120	(threshold of pain)
110	riveting machine
100	large circular saw
90	jet aircraft flyover
80	interior of a sports car
70	vacuum cleaner
60	freeway traffic
50	office noise
40	urban home
30	suburban home
20	whisper
10	normal breathing
0	(threshold of hearing)

226

Background

Mechanical Waves

Because sound waves are mechanical waves, they can be felt. The intensity, or force, of the vibrations is felt by the eardrum. This is why very loud sounds can damage the eardrum. The eardrum may rupture or lose its ability to vibrate.

Decibel

The term *decibel* means "one tenth of a bel." A bel is the minimum intensity of sound detectable at 1,000 Hz. The bel was named after Alexander Graham Bell. The lowest level that a human can hear is 0 decibels. This is known as the threshold for hearing. The maximum level for human hearing (120 decibels) is called the threshold of pain.

Timbre

Timbre is any characteristic of sound that does not have to do with loudness or softness, highness or lowness, or length. Timbre is sometimes referred to as the "color of sound."

Volume

The word *volume* has several different uses that are actually related. The first meaning refers to books. This meaning reflects the origin of the word *volume*, which meant "to roll or wrap." Early books were actually scrolls made of long pieces of animal skins that were rolled. Eventually, bound books became known as volumes.

The human ear can be damaged by too many large, loud vibrations. Because sounds cause your eardrum to vibrate, the force of strong vibrations can cause pain. Exposing your ears to intense sound for long periods of time may reduce the ability of delicate structures in the inner ear to sense sound. This may cause hearing loss. A sudden sound, such as a firecracker going off, can cause temporary hearing loss. You should wear hearing protection any time your ears are exposed to very loud sounds.

Quality

Timbre is the quality of a sound that distinguishes it from other sounds of the same pitch and volume. Before a band concert, the instruments tune to the same pitch. However, the trumpets sound different from the clarinets, and the clarinets sound different from the flutes. They all play the same pitch, but the timbre of each instrument is unique.

Most of the sounds that we hear are a blend of several waves. Each of these waves has different characteristics. As these waves blend, the sound has a certain timbre. If one sound wave is added or removed, the timbre will change.

A clarinet has a reed that vibrates. The reed is not the only part of the instrument that vibrates, though. When a musician blows into the clarinet, the long tube that forms the instrument vibrates as well. The air inside the clarinet also vibrates. A skilled musician must control all these vibrations so that they blend to produce a pleasant sound.

Clarinet

An oscilloscope image of a single, pure sound, such as that made by a tuning fork, looks smooth. Musical instruments, however, do not have a pure sound. Instead, musical instruments produce sounds that are a blend of different vibrations, or sounds. On an oscilloscope, these blended sound waves look rough or wiggly.

You recognize many sounds without thinking about it. You may be playing in a backyard with a group of friends when someone's mother calls from the house. You know immediately whether it is your mother or not. The timbre of her voice helps you know.

QUICK CHECK

1. What is another name for how high or low a sound is?
2. What causes the volume of sound to change?
3. What is timbre?

227

131–132

What is timbre? the quality of a sound that distinguishes it from other sounds of the same pitch and volume

What does *distinguish* mean? to make different or set apart

Why do some musical instruments sound different even when they are being played at the same volume and pitch? They have different timbres.

What determines timbre? the blend of different vibrations or sounds

What would an oscilloscope image of one pure sound look like? a smooth wave

What would an oscilloscope image of several musical instrumental sounds look like? a rough and wiggly wave

ANSWERS

1. pitch
2. the amount of energy used to make the sound waves
3. the quality of a sound that distinguishes it from other sounds of the same pitch and volume

132

- The *Community Sounds*, *Hearing Loss*, and *Coloring Sound* activities found at the bottom of the pages in Lessons 131–32 may also be incorporated into the lesson.
- Suggestions for additional activities for this two-day lesson are on the Teacher's Toolkit CD.

One book can be referred to as a volume, or the individual books in a set can be called volumes.

In the early 1500s, *volume* began being used to describe the size of a book. The term was later used to describe the size of any substance. (Volume as a measurement is discussed in Chapter 3, *Matter*.) By the early 1600s, *volume* was used to describe the amount, or loudness, of sound.

Activity

Coloring Sound

Materials: recordings of sound effects or assorted instrumental music

Sounds can cause pictures to form in the imagination. Play the musical recordings. As a specific sound is played, the students should draw and color pictures of what they think about. The chosen colors and events in the picture should reflect characteristics of the sound.

For example, a picture made during a lively band march may include bright, bold colors. The lines in the picture may appear straight and have sharp angles. A picture made during a sleepy violin concerto may include soft pastel colors and have rounded shapes and curved lines.

Activity Manual

Study Skill and Reinforcement, page 151

This page allows students to design their own organizer for the characteristics of sound.

Review, page 152

This page reviews Lessons 131–32.

Assessment

Quiz 9-B

The quiz may be given any time after completion of this lesson.

133

Student Text pages 228–29
Activity Manual pages 153–54

Objectives

- Compare the amount of sound absorbed by different materials
- Predict which material will absorb the most sound
- Rate the loudness of sounds
- Identify relationships between materials and their abilities to absorb sound

Materials

Materials needed for this activity are listed on the Student Text page.

Teacher's Toolkit CD

- Rubrics: *Shhh, Quiet Please*, for each student

Introduction

Have you ever noticed how sounds echo in a gymnasium?

Think about the materials that the walls, ceiling, and floor are made of. Are they different from materials used in a house?

Today we will test different materials that can be used in a room to see how easily sound waves move through them.

Preparation for Reading

- Explain that acoustics is the science of sound. An acoustical engineer is a person who applies the science of sound.
- Direct the students to read silently Student Text pages 228–29 and Activity Manual pages 153–54 before beginning the activity.

Teach for Understanding

Procedure

- Guide the students as they complete their hypotheses. Identify each of the materials and discuss characteristics of each as needed. The students should make any material changes in the Activity Manual chart.
- Decide how the materials will be held. You may want to have another student hold the material in place. Using only the fingertips is best. The more contact the material has with another item, such as the palm of a hand, the less the material will vibrate.
- Most carpet has a stiff backing. Direct the students as to which side should be held against the ear. You may choose to have the students try it both ways and then discuss possible reasons for any differences.
- The rating system is open-ended. Each student must decide the level of sound that is assigned to each number. He will probably have to test and compare the sounds of the materials several times before arriving at his decisions.

ACTIVITY

Shhh, Quiet Please

Process Skills
- Hypothesizing
- Predicting
- Observing
- Communicating

Acoustical engineers identify and control how sound waves move. The engineers know that some materials are able to reflect, slow, or stop the movement of sound. Some rooms would be very noisy if certain materials were not used to block or absorb sounds.

In this activity you are the acoustical engineer. Your task is to find which material absorbs the most sound. You will send sound waves through different materials that could be used in a room. Then you will determine which material absorbs sound the best.

Problem

Which building material absorbs sound the best?

Procedure

Materials
- tuning fork
- ceiling tile scrap
- carpet scrap
- metal, such as a baking pan
- wood, such as a piece of board
- Activity Manual

1. Predict which material will most effectively absorb sound. Complete the hypothesis in your Activity Manual.
2. Test the sound of the tuning fork. Place your ear against your desk. Have a partner hit a tine of the fork quickly on your knee and immediately place the tip of the handle of the fork on the desk. You should hear a high, vibrating sound. This is the sound that you will listen for as you do the activity.
3. Choose the first material to test. Hold the material against your ear. Have a partner hit a tine of the fork against his knee and place the tip of the handle on the material. Listen for the sound.
4. Rate the loudness of the sound that you hear on the chart in your Activity Manual. Use the numbers 0–4 to rate the sounds. A score of 0 means you heard no sound. A score of 4 means the sound was loud.
5. Repeat steps 3–4 for each material.

228

Background

Activity Materials

Inexpensive tuning forks are available from many science supply resources. See information on the BJU Press website for places to find tuning forks. bjupress.com/resources/science

If a tuning fork is unavailable, an all-metal dinner fork will work. Ones with plastic or wooden handles will not work as well.

Pieces of metal siding or other types of sheet metal may be used. If the metal is corrugated, it introduces another variable that would need to be discussed.

Materials should be cut into 10–15 cm pieces. Cover any sharp or ragged edges with tape.

You may choose to add other building materials to the activity, such as glass, brick, plastic, insulation, or ceramic tile.

6. If needed, repeat the test for any of the materials to determine their ratings.
7. As time permits, switch places with your partner and repeat the tests.

Conclusions

- Which material absorbed sound the best?
- What characteristic of that material made it most effective?
- Why might different people get different results?

Follow-up

- Use other materials or other sources of sound.
- Test the effect of layers of different materials between the sound and your ear.

229

Process Skill

The process skill emphasized in this activity is predictions.

What are predictions based on? Elicit that they are based on experience, study, and knowledge.

What is the key to good predictions? careful observations

What knowledge or experiences did you base your predictions on? Answers will vary, but elicit that the predictions were based on previous knowledge of rooms and buildings.

After you tested the first material, did you still think your predictions would be accurate? Why? Answers will vary.

Suppose you have no background knowledge about a material you are testing. Would you be making a good or reasonable prediction about the material? no Why? The prediction would be a guess since it would not be based on observation and knowledge.

What could you do to make your prediction more valid? Possible answers: conduct research, make observations of similar situations

Scientists spend a lot of time researching their topics and the results of other scientists. Why do you think they do this research before making predictions about their own experiments? Since predicting implies background knowledge, scientists must gain this knowledge before they can anticipate a result.

Conclusions

- Discuss which materials absorbed the most sound. Refer to the Background section if needed.
- If results between groups differ, discuss factors that may have caused the differences. Possible reasons include the different hearing abilities of the students, the way in which the materials were held, or background noises.

Background

Reflection and Absorption

Sound waves travel farthest and with the most strength in a straight line. Solid materials, such as metal, permit the waves to move easily through them with little reflection because the particles that make up the solid are tightly packed together. Some solids have spaces of air in them. You can see air spaces between the grains of wood and between the fibers of ceiling tiles and carpet. Sound energy reflects and slows as it moves from a solid part of the material and through an air space to the next solid part. The amount of reflection and slowing affects the material's ability to absorb sound.

Activity Manual

Activity, pages 153–54

Assessment

Rubrics

Select the prepared rubric, or design a rubric to include your chosen criteria.

134

Activity Manual pages 155–56

Objectives

- Summarize what the Bible says about hearing
- Explain why a creationary approach to science is a better approach to solving problems (like hearing loss) than an evolutionary approach

Materials

- video of a person's reaction to hearing for the first time after receiving a cochlear implant (optional)

Teacher's Toolkit CD

- Instructional Aids 9.4–9.5: *Ears and Hearing*, for each student; *Ears and Hearing Answer Key*

Introduction

God has formed every ear on every person in a unique way. The ability to hear sound is an amazing gift from God.

What is your favorite sound? Answers will vary.

What sounds could you do without? Answers will vary.

What sound would you miss the most if you could not hear? Answers will vary.

Preparation for Reading

- Instruct the students to read the two paragraphs on Activity Manual page 155.

Teach for Understanding

- You may choose to answer the first question on Activity Manual page 155 after question four as this lesson does.
- Choose a volunteer to read aloud Psalm 94:9.
  What does the verse tell you about the ear and hearing? God made the ear.
- Select another volunteer to read aloud Proverbs 20:12.
  What does the verse tell you about the ear and hearing? The Lord made ears to hear.
- Give the students time to complete question two.
  In what part of creation did God create the human ear? when He made people
  In what condition did God design the ear? Elicit that He designed the ear perfectly and to function perfectly.
- Select a volunteer to read aloud Genesis 3:17.
  What caused the world and everything in it, including some ears and hearing, to not function properly? sin and God's curse
- Read aloud Exodus 4:11.
  What does the verse tell us about God's sovereignty and the people who cannot hear? God made some people deaf.
- Explain that although deafness is a result of sin in the world, God's power is so great that He controls even the unpleasant things that happen in His world. By God's grace, He has allowed us to develop various cures and technologies to ease the effects of our sin.

A lesson from
Answers in Genesis Name ____________

1:1

The Hearing Ear

You probably take it for granted when you hear a sound. Your mom may have called you to breakfast this morning, or maybe you avoided stepping into the street because you could hear a car coming. Sound is an important part of our lives. But many people cannot hear a call for breakfast and are placed in harm's way when crossing the street.

1. If we want to help people who have hearing problems, would it be better to study hearing by thinking that it happened accidentally or by thinking that God had designed the ear in a special way? Why?
   Answers will vary. Elicit that when you study an accident you are 1) surprised to find things that look designed; 2) expect things to be broken/not work; 3) try to understand what you see so you can improve it. If you are studying something that is designed you 1) expect things to look designed; 2) are surprised when something is broken/does not work; 3) try to understand what you see so you can imitate it.

People who do not believe in God think that the ability to hear happened as the result of random chance over billions of years. Basically, they think the ability to hear is an accident. But God is a giver of good gifts, and the ability to hear sound is one of those gifts.

2. What does each of the verses below tell you about the ear and hearing?
   Psalm 94:9 *God made ears.*
   Proverbs 20:12 *The Lord made hearing ears.*
3. Do you think that physical conditions such as deafness existed when God created the world? What caused the world to change?
   No. Sin brought God's curse into the world (Genesis 3:17). All that is wrong in the world is a result of sin.
4. Do people who cannot hear bear less of the image of God? What does Exodus 4:11 tell us about God's sovereignty and people who cannot hear?
   God made some people deaf. Explain that although deafness is a result of sin in the world, God's power is so great that He controls even the unpleasant things that happen in His world.

Resources

The BJU Press website offers additional links from Answers in Genesis that you may find helpful with this lesson. Included on the resource page are links to videos showing the reactions of people hearing for the first time after receiving a cochlear implant.
bjupress.com/resources/science

Background

Cochlear Implant Video

Keywords to use to search the Internet for videos should include "cochlear first hearing."

Some people think that if you do not believe in evolution, you cannot be a real scientist or develop new cures. But there are many scientists who trust the Bible and reject evolution. Dr. Georgia Purdom has studied many areas of biology and taught many students to trust the Bible as they study science. But she was born with a hearing problem. This made it hard for her to communicate with the people she wanted to teach and share God's truth with. She was feeling the effects of the Fall and hoped for a cure.

Thankfully, she was able to find a solution. A simple hearing aid could not help her, but there was a special device placed deep in her ear that could! It is called a cochlear implant, and it is a thin wire that fits inside the cochlea deep within the ear. The cochlea is the part of the ear that turns vibrations into electrical signals that are sent to the brain.

Whether the designers of the device believe God exists or not, they imitated God's design of the ear. Because of the implant and therapy that followed, Dr. Purdom is now able to hear much better than before.

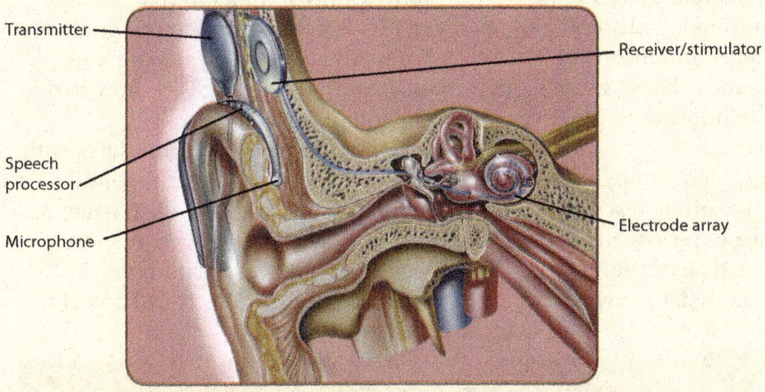

Someone once told Dr. Purdom that she would not be able to hear unless the person working on her implant believed in evolution. Dr. Purdom replied as follows:

> I am very thankful to all the scientists who have worked to develop the cochlear implant. It has profoundly changed my life for the better. It is likely that many of them did believe in evolution—however, believing that humans evolved from a single-celled ancestor over billions of years had no bearing on their research. I pray that they will "hear" the truth and acknowledge that God is the Creator of the human ear.

As a result of people studying God's design of the body, Dr. Purdom can now communicate better. Her hope is that all people will be able to hear the good news of salvation in Jesus Christ and fully put their trust in Him, obeying all that He has commanded in the Bible. How could God use you to share that good news?

- Allow time for the students to complete questions three and four on the Activity Manual page.
- Distribute the *Ear and Hearing* page. Direct the students to complete the chart during the discussion.
  When a creationary scientist looks at the ear, how does he view its design and appearance? The ear was originally created by God with a perfect design and function.

  What viewpoint would an evolutionary scientist have? Elicit that he sees the ear as something that has evolved or happened over a long period of time and works by chance.
- Explain that a creationary scientist would base his research and work on the truth that God's original design of the ear was perfect.
  How might he view an ear that was not shaped or functioning properly? Elicit that he would view that ear and its function as not working as designed.
- Discuss with the students that an evolutionary scientist thinks that the ear has not yet reached a design that works perfectly.
  How do you think he would view an ear that is not working correctly? He would expect it not to work correctly or be broken.
- Display the following statements and allow the students to determine the viewpoint for each.
  Use a correctly functioning ear as a model for correcting a damaged ear. creationary scientist's

  Improve the function of ears, especially those that are damaged. evolutionary scientist's

  Which viewpoint serves as a better starting point for studying hearing and trying to find cures? the creationary scientist's belief Why? Answers will vary.
- Explain that if a person believes the Bible, he knows that the ear has been designed by God, who made people in His image. He can study how healthy ears work and try to invent instruments that behave or work like the healthy ear. He can also pray to the God who made all ears and ask Him for help and guidance.
  For some people, hearing aids help by amplifying the sound that passes to the eardrum. But for many people, hearing aids are not effective.
- Direct the students to read Activity Manual page 156 to find out about another piece of technology used to help with hearing loss.
  What is a cochlear implant? a device comprised of a microphone, speech processor, and a small wire inserted into the cochlea that helps people hear
- Discuss the diagram and Dr. Purdom's reply to having a person who believed in evolution working on her implant.
- Show a video of the reaction of a person hearing for the first time. (optional)

Activity Manual

Answers in Genesis, pages 155–56

135

Student Text pages 230–34
Activity Manual page 157

Objectives

- Differentiate between *sound* and *noise*
- Recognize that a sound fades as its energy is used up
- List examples of how echoes are used in nature and technology
- Name examples of how an acoustical engineer uses his knowledge of sound

Materials

- piece of wrapped hard candy

Teacher's Toolkit CD

- Quiz 9-C

Vocabulary

- echo
- acoustics

Introduction

- Tell the students to be very quiet. Make as much noise as possible while opening the candy wrapper.

  Would you consider this sound pleasant or annoying? Answers will vary.

  Think about situations in which you may hear a wrapper crackling. As I read each situation, raise your hand if you think the sound would disturb you.

  during a church service
  at a birthday party
  at a ball game
  during a test

  Did the sound change for each situation? no What made the difference in your answers? Elicit that the difference often depends on the preferences of the individual hearers, the location, or the timing of the sound.

  What are some other sounds that annoy or disturb you when you hear them? Possible answers: a repetitive sound such as tapping fingers, the sound of scratching on a chalkboard, a high-pitched whine, the hum of a mosquito, the sound of someone slurping a drink, the vibration of a cell phone

Preparation for Reading

- Instruct the students to mentally make questions using the headings, bold and italicized words, and captions.
- Direct them to read silently pages 230–34 to find the answers to these questions.

  What is an echo?

  What name is given to the science of sound?

Teach for Understanding

What is the difference between *sound* and *noise*? Noise usually refers to unwanted or unpleasant sounds.

What happens to a sound wave as its energy is used up? It fades away.

Uses of Sound

Mr. Fogel used his knowledge of sound to develop a new type of technology. But sound influences technology in other areas. One example is how engineers who design bridges must understand the way sound moves through different materials to avoid a possible collapse.

Many of the sounds that you hear are pleasant. People usually enjoy talking with friends or listening to music. Voices and music are beneficial sounds. These sounds can communicate information or give us pleasure.

Some sounds are annoying and distracting. We often describe these sounds as noise. We usually think of a noise as a type of sound that is harsh, unwanted, or surprising. A jackhammer, children yelling, or the pop of a balloon are usually considered noises. But even these noises can be useful.

Seeing with Sound

Sounds convey meaning. The noise of a dog's barking lets you know that a dog is nearby. But sound waves themselves can also be useful. A sound wave uses up its energy as it moves away from its source. As the energy is used up, the sound gradually fades. You can try to estimate how close the barking dog is by how strong the sound waves are.

Sometimes an object interferes with the movement of a wave. A sound wave may bounce, or reflect, when it hits an object or a surface. A sound wave that bounces back toward its source and is heard is called an **echo**.

SCIENCE & HISTORY

Sonar was used during World War II to detect German U-boats. The word *sonar* comes from the first letters of SOund NAvigation and Ranging. The military attached the sonar transmitter-receivers to the bottom of submarines. The echoes from the sonar sent many false alarms from things such as whales, schools of fish, and sea currents. A good submarine operator learned the characteristics of the different sounds to stop the false alarms.

The Germans used different strategies to avoid the submarines' sonar. One strategy was to coat the U-boats with rubber. The rubber absorbed much of the sound waves and changed the way they echoed. In another strategy, the Germans released chemical pellets. The pellets made clouds of bubbles that reflected the sound waves so the U-boat could not be detected.

230

Background

Scientist Alexander Graham Bell

Alexander Graham Bell is best known for his invention of the telephone. But before the telephone, he tried to invent a musical telegraph, also called a harmonic telegraph. He knew that being able to send several messages over a single wire at one time would benefit the telegraph company. The company would be able to send more messages without having to add more poles and lines. He thought that if he could vary the pitch of the sound he would be able to send several telegraph messages at the same time.

His experiments attracted the attention of two prominent businessmen, both of whom provided funding for Bell's experiments on the telegraph. Bell worked on both the telegraph idea and his idea of a telephone for several years. However, Bell was more interested in the telephone and eventually stopped working on the idea of a musical telegraph.

Ultrasound

Because sound waves are mechanical waves, sound waves do not affect or harm the body tissues that they pass through. The molecules in the tissue only vibrate. For this reason, ultrasounds can be used to break up kidney stones and to kill some types of bacteria without harming surrounding body tissue.

Oilbirds use echolocation to help them "see" in caves.

Distance and the type of surface affect the clarity of an echo. To hear a clear repetition of the original sound, the distance between the source and the place of reflection must be at least 9 m (30 ft). A smooth surface produces a clearer echo than a rough surface does.

God designed several animals to use echoes. For example, a tropical bird called the oilbird lives in caves. It uses echoes to "see" inside its dark home. An oilbird makes clicking sounds at a frequency of about 7,000 Hz. These sounds hit surrounding objects and bounce back to the bird. The time it takes the echoes to return depends on how far the sound waves traveled before hitting an object. The oilbird hears and senses these echoes. In this way, the echoes tell the bird how far it is from objects.

Scientists call this ability to see with sound *echolocation*. Sonar is a type of echolocation developed by man. Sonar was first used to help locate items underwater. A ship or submarine sends beeping sounds through the water. The length of time the wave takes to travel back as an echo helps the listener tell how far away an object is. The echoes can also give other clues about the object, such as its size, the type of material it is made of, and how fast it is moving.

Ultrasound technology also uses echoes from high-frequency sound waves. Special instruments change the waves of the echoes to electrical images or pictures. Ultrasound images are often used by doctors to "see" inside a person's body without using surgery. For example, doctors can use ultrasound images to observe the growth of unborn babies. Many organs and even some large blood vessels inside the body can also be seen in this way.

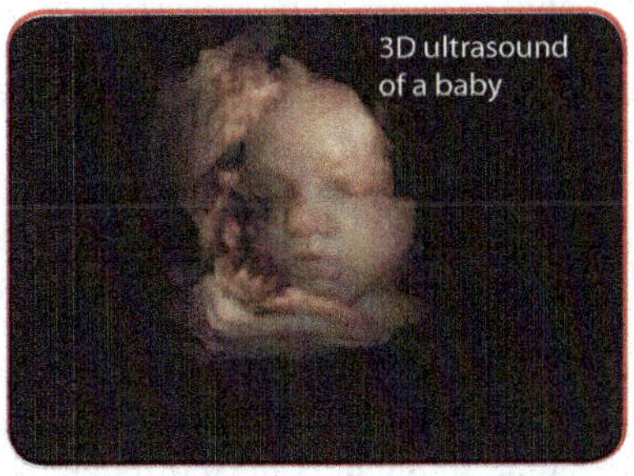
3D ultrasound of a baby

What sometimes happens when a sound wave hits an object or a surface? The sound reflects, or bounces.

What is an echo? a sound wave that bounces back toward its source and is heard

- Direct attention to the picture of the oilbird on page 231 as you continue to discuss echoes.

What two factors affect the clarity of an echo? the distance and the type of surface

Does a rough or a smooth surface produce a clearer echo? a smooth surface

How does an oilbird use echoes? to know how far away it is from objects, to help it "see" as it moves around its dark home

What do scientists call the ability to see with sound? echolocation

Name some other animals that use echolocation. Possible answers: dolphins, bats

What is sonar? a type of echolocation developed by people

What can scientists learn about an underwater object by using sonar? how far away it is, its size, the material that it is made of, and how fast it is moving

- Direct attention to *Science & History* on page 230.

Where does the word *sonar* come from? Sound Navigation and Ranging

Where did the military attach the sonar transmitter-receivers during World War II? to the bottom of submarines

What strategies did the Germans use to try to avoid the sonar? Possible answers: coated the U-boats with rubber; released chemical pellets

- Allow the students to focus on the bottom picture on page 231.

What is another type of technology that uses echoes of high-frequency sound waves? an ultrasound

How do doctors use ultrasounds? Possible answers: to "see" inside a person's body, to monitor the growth of unborn babies, to look at organs and large blood vessels

Background

Oilbird

The oilbird is related to the nighthawk and the whippoorwill. The oilbird lives in the South American tropics and subtropics. This bird is the only fruit-eating nocturnal bird. The oilbird is about 33 cm (13 in.) in length and has a 91 cm (36 in.) wingspan. It has short legs and does not perch. Instead, it hovers and uses its hooked beak to eat fruit. Young oilbirds have a lot of oily fat. In the past, the young birds were killed and their oily fat was used to burn in lamps. The oilbird probably got its name from this practice.

Activity

Drawing Synonyms

Use a thesaurus to find synonyms for *noise*. Direct the students to choose a synonym. Each student should draw an illustration or cartoon of a situation that his synonym describes. Possible synonyms: *din, racket, hubbub, clamor*

The students could also illustrate onomatopoeic words such as *buzz*, *hum*, or *kaboom*.

What are some ways that people can praise God with sound? singing and speaking

What body part enables people to sing or speak? the larynx, or voice box

How is sound produced in the larynx? The muscles of the larynx lengthen and shorten the vocal cords, producing different pitches. The force of air moving through the larynx controls volume.

Which came first, the study of music or ideas and theories about sound? the study of music

- You may choose to read Psalm 150 and discuss the ways that music is used to praise God.

What did David do that helped calm Saul? played on the harp

Give an example of a time when music that you had learned or heard comforted you in a stressful situation. Answers will vary.

- As time permits, allow the students to share some of the hymns or gospel songs that comfort them during difficult times.

What can music communicate to people? ideas and feelings

How do composers communicate ideas and feelings? by choosing specific notes and the instruments to play those notes

What do pleasant, calm sounds look like on an oscilloscope? smooth curves

What do harsh sounds look like on an oscilloscope? rough and jagged

- Discuss *Creation Corner*.

What do you think it would be like if you could hear only one sound at a time? What would a band concert sound like? Answers will vary.

Why can you hear multiple sounds at once? God designed sound waves to be able to occupy the same space without disturbing one another.

Communicating with Sound

God gave man the need to fellowship and communicate both with Him and with other people. One of the most common forms of communication is through sound. People can praise and glorify God through speaking and singing. With their voices, people can also tell others about God's love and salvation.

Singing and speaking are much alike. Both are a result of vibrations of the vocal cords inside the larynx, or voice box. The muscles of the larynx lengthen and shorten the vocal cords to produce different pitches. The force of the air moving through the larynx controls the volume.

Music

Music can communicate ideas and feelings. Psalm 150 says that music can be used to praise God. In 1 Samuel 16:14–23, the Bible tells us that David played music to calm the troubled spirit of King Saul. Today music is still used for these purposes. Hymns express our praise to God. A mother may sing a soft lullaby to help her baby calm down and go to sleep.

God created man with the desire to express his thoughts and feelings in song. People sang long before they understood pitch and frequency. They built instruments long before they knew how sound waves travel. In fact, many of the early ideas and theories about sound developed through the study of music, not the other way around.

Composers use the characteristics of sound as they write music. They choose specific notes as well as the voices or instruments to play those notes. The results of their choices combine to communicate ideas and feelings. The combinations of sound waves can blend, or they might clash. On an oscilloscope, the waves of calm sounds are usually seen as smooth curves. The waves of harsh sounds look rough and jagged.

CREATION CORNER

You know that you and your friend cannot both be in exactly the same place at the same time. However, God designed sound waves so that they can do just that. If you stand on a busy street corner, you might hear the sounds of traffic moving, people talking, and pigeons cooing. The sounds do not have to take turns entering your ear. They do it all at once. This is because many sound waves can occupy the same space and not disturb one another. Without this amazing design by God, we would not be able to enjoy combined sounds such as choirs, bands, or orchestras.

232

Background

Music Waves and Resonance

Music is a blend of many waves. For example, when a stringed instrument is played, the string, the air surrounding the string, and the other solid parts of the instrument all vibrate. Because the parts are made of different substances, each has a slightly different vibration. The basic vibration of the instrument is called its natural frequency. When the other parts begin to vibrate as a result of the natural frequency, the result is called resonance.

Natural Frequency

All objects, not just musical instruments, have a natural frequency and a resonant frequency. This is how a singer can cause something such as glass to break. When the singer's voice matches the natural frequency of the glass, the resulting resonant vibrations cause the glass to break.

Activity

Communicating with God Through Sound

Materials: Instructional Aids 9.6–9.7: *Singing to God*, for each student; *Singing to God Answer Key*

This page guides the students in interacting with several passages in the Bible about songs and singing.

Composers use many characteristics of sound to produce pleasing music.

Musical instruments and voices can imitate other sounds in nature and our surroundings. To do this, the instruments or voices must copy the pitch, volume, and timbre of the original sound. Some musical compositions can cause you to imagine specific scenes or events.

Electronic Sounds

God's design of the ear is remarkable. The ear receives sound vibrations. It then changes them into electrical messages that are interpreted by the brain. Microphones, recording equipment, and speakers are types of technology that use this same basic design.

A microphone receives sound vibrations. The vibrations are then changed into electrical messages. The messages can be copied by a recorder or sent out through speakers. Speakers produce new sounds that can travel to your ears.

Mr. Fogel used this technology to solve the problem of noise for pilots. He used microphones to receive sound and speakers to electronically produce an opposite sound. He recognized how important it is for pilots to hear their radios. Like Mr. Fogel, engineers today continue showing care for others by designing new technology.

A sound recording studio uses many types of sound technology.

233

In what ways do musical instruments imitate sounds in nature? by copying the pitch, volume, and timbre of the original sound

How does the human ear process sound? by changing the vibrations of sound into electrical messages that the brain interprets

What are some types of technology that work in the same manner as the human ear? Possible answers: microphones, recording equipment, speakers

How is the way that a microphone works similar to the way that an ear works? Both the microphone and the ear receive sound vibrations and change them into electrical messages.

How did Mr. Fogel use this technology to solve the problem of noise for pilots? He used microphones to receive sound and speakers to electronically produce an opposite sound.

Activities

Exploring Music and Matter

Materials: 4 or more identical glass bottles or jars, water, metric measuring cup, metal spoon

Arrange the bottles in a row. Tap on the first bottle with the spoon. Notice the sound produced. Discuss the types of matter that the sound waves are moving through—the glass and the air inside the bottle.

Pour 100 mL of water into the second bottle. Tap the bottle in the same way and in the same place that the first bottle was tapped. Notice changes in pitch and quality. Discuss the types of matter that the sound waves are moving through—the glass, the air, and the water.

Pour 200 mL of water into the third bottle and 300 mL into the fourth. Tap the bottles and evaluate as above. Determine if there is a predictable relationship between the pitch and the amount of water.

Continue with more bottles as time allows. Try to play a tune with the musical bottles.

Write About an Element of Music

Research the history of an element of sound, such as ultrasound, pitch, volume, or timbre. Write a short paragraph about that element. The paragraph should include any connections to music or musical instruments.

135

What is acoustics? the science of sound

What does an acoustical engineer do? studies sound and the ways in which it is carried in buildings

Are echoes usually a desired sound inside a building? no

What kind of surfaces usually cause echoes? smooth, hard surfaces

What kind of materials would absorb sound waves? rough, soft, fuzzy materials

Why are buildings built with materials that absorb sound? so that sounds will not echo

Why do acoustical engineers design some walls and floors to be slanted or curved? so sound waves can be directed toward particular areas of a room

What are ways to use sound for God's glory? Possible answers: singing, giving praise to God, sharing a comforting word with someone, telling others about God

- Discuss *Science & History.*

What are two ancient civilizations that used acoustics? the Greeks and Romans

How did the ancient Greeks and Romans use acoustics? Their theaters were designed so that a large audience could clearly hear the performers.

ANSWERS

1. a sound wave that bounces back toward its source and is heard
2. Both are produced by vibrations of the vocal cords in the larynx.
3. acoustics

Activity Manual

Review, page 157

This page reviews Lesson 135.

Assessment

Quiz 9-C

The quiz may be given any time after completion of this lesson.

Building Design and Sound

Acoustics is the science of sound. The term comes from a Greek word meaning "hearing." An acoustical engineer is a person who applies the science of sound.

Most sounds are intended to be heard only once. You would not want to hear a person's words repeated every few seconds. You would find it difficult to understand what he was saying.

The design of buildings affects how sound is heard. Sounds reflect more clearly from smooth, hard surfaces than from rough or soft surfaces. This is why echoes are often heard in rooms with smooth walls, such as gymnasiums. For most buildings, acoustical engineers use materials that absorb sound, or stop it from continuing, instead of reflecting it. Materials that absorb sound may be rough, soft, or fuzzy. The energy of the sound waves is absorbed into the material.

Acoustical engineers also try to control the direction and strength of sound waves within a room. They may design the walls, ceilings, and even the floors with a slant or curve. These slants and curves reflect sound waves to certain areas of the room and away from others.

God created sound and the ability for us to hear it. Without it, we would not be able to enjoy music or hear friendly voices. But God did not create sound just for our enjoyment. He also intended for us to use sound to help others and spread God's Word.

QUICK CHECK

1. What is an echo?
2. How are singing and speaking alike?
3. What is the name for the science of sound?

SCIENCE & HISTORY

People traveling to Greece and Rome are often amazed at the acoustics of the ancient outdoor theaters there. These amphitheaters use clever building designs that allow a large audience to hear the performers clearly. The seats are sloped, and the stage is placed in a precise location.

The theater at Epidaurus, Greece, was built in the fourth century BC. Its remarkable acoustic design can still be experienced today. Even visitors in the highest seating areas can hear sounds from the stage. Sounds as quiet as someone snapping his fingers can clearly be heard.

234

Background

Destructive Interference

The surfaces that reflect sound are not the only things that an acoustical engineer has to consider when designing a structure. The waves coming from the speakers or reflecting from the building's surfaces also interact with each other to cause other potential problems. One problem happens when the sound compressions of one wave and the rarefactions of other waves interact in destructive interference. This creates what is called a "dead spot." Although a person may be able to see that sound is being made, the destructive interference of the sound waves means that the compressions and rarefactions cancel each other and the volume level is greatly diminished in that spot. In rooms such as an auditorium or a concert hall, it is important that there are not dead spots where people cannot hear what is occurring on the stage.

A "Medium" Exploration

EXPLORATION

The matter through which a wave travels is called a *medium*. Sounds travel more quickly through solids than through liquids and gases. Does the type of matter also make a difference in how you hear the sounds?

To test this, try listening to the same sound through a variety of mediums. You may not be able to tell how quickly the sounds travel. However, try to hear other differences between the sounds.

What to Do

1. Get a stethoscope and two metal spoons.
2. Have a partner tap the two spoons together. Use the stethoscope to listen to the sound of the tapping as it travels through the air, through water, and through a table or other solid. For each test, make sure that your stethoscope is about the same distance from the tapping spoons.
3. Notice similarities and differences of the tapping sounds in the different mediums. Write a paragraph comparing and contrasting the characteristics of the sounds.
4. Present your findings and challenge your audience to conduct their own "medium" tests.

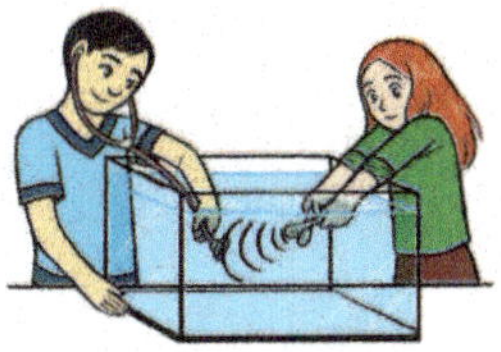

235

Background

Activity Materials

If a stethoscope is not available, you can make a listening tube with a piece of plastic tubing and two small funnels. Tape a funnel to each end of a 45 cm (18 in.) piece of plastic tubing. Hold one funnel against the ear and the other toward the sound.

A clicker or other noisemaker may be used instead of the spoons, as long as it can still make the sound while underwater.

Objectives

- Test the abilities of different mediums to carry sound
- Write a paragraph that compares and contrasts the results

Materials

- stethoscope (see Background)
- two metal spoons
- mediums that represent solids, liquids, and gases

Teacher's Toolkit CD

- Rubrics: *A "Medium" Exploration*, for each student

Introduction

Sounds travel differently through different types of matter. How do you think you could test this statement? Accept reasonable answers.

- Display the stethoscope.

How could we use a stethoscope to test the mediums? Accept reasonable answers.

Preparation for Reading

- Direct the students to read silently page 235 to find out what a medium is.

Teach for Understanding

What is a medium? the matter through which a wave travels

What to Do

- You may choose to set up stations in the classroom to test the mediums. The students could take turns at each station.

The variables need to be controlled as much as possible so that only one, the medium, is being tested.

What variables do you need to control? Answers should include making a consistent sound, having the same distance between the stethoscope and the sound source, having the same person make the sound, having the same person do all the listening, and eliminating background noises.

Assessment

Rubrics

Select the prepared rubric, or design a rubric to include your chosen criteria.

137

Student Text page 236
Activity Manual page 158

Objectives
- Recall concepts and terms from Chapter 9
- Apply knowledge to everyday situations

Review
- Materials for the Chapter 9 Test will be taken from Student Text page 236 and Activity Manual pages 147–48, 152, and 157–158. You may review any or all of the material during the lesson. Questions similar to the ones in Thinking It Through, Activity Manual page 158, may appear on the test.
- You may choose to review Chapter 9 by playing the review game or a game from the Game Bank on the Teacher's Toolkit CD.

Activity Manual
Review, page 158
The page requires written responses to application questions.

138

Objective
- Demonstrate knowledge of concepts taught in Chapter 9

Assessment
Chapter 9 Test

CHAPTER SUMMARY

Words to Know
vibration
sound
wave
wavelength
frequency
speed of sound
pitch
volume
decibel
timbre
echo
acoustics

Key Ideas
Summarize how sound moves
Explain how sound waves are measured
Describe how sound waves move through different substances
Identify characteristics and properties of sound
Describe how intensity affects volume
Identify how animals and people use echoes
Describe how sounds can be used to help others

236

Review Game
Moving Sound Along
Divide the students into teams. Each team will be asked five questions per round. The number of questions answered correctly represents the speed of sound through the different states of matter. The team represents a solid when they give four or five correct answers, a liquid when they give two or three correct answers, or a gas when they give one or no correct answers.

Ask the first team their five questions. Keep track of how many questions they answer correctly. Then ask the other team(s) their five questions. At the end of the round, allow the teams to determine whether they represented a solid, a liquid, or a gas.

At the end of the game, you may choose to total the scores for all the rounds and allow the teams to determine their state of matter for the entire game.

Light 10

During World War I there was an army dedicated to saving lives instead of taking them. This army of doctors and nurses cared for soldiers who had been wounded in battle. But how could they look inside a soldier to see what was wrong? Marie Curie, a Polish chemist and physicist, knew that the x-ray, a recent development in technology, would be helpful in solving this problem. She studied human anatomy and how to use the x-ray machines. She even researched how to work on cars. Then she gathered together twenty vans with the equipment for taking x-rays. She and her daughter drove to the front lines to help the doctors care for wounded soldiers. Her x-ray trucks were called "Little Curies." By the end of the war, she had set up 200 x-ray stations in military hospitals. She also directed the training of many people in the use of x-rays. Because of Marie Curie's work, many soldiers healed faster and many lives were saved.

Consider

- Why is it important to discover tools like x-rays?
- In what ways other than medicine are x-rays useful?

237

Background

Chapter Preview

Preview and prereading activities may include using a K-W-L chart, a probe, or an anticipation guide.

Chapter Photo

The photo on page 237 shows a lighthouse in the twilight on Sylt. It is the largest island in North Frisia, which is in northern Germany.

Weblinks

The BJU Press website offers additional information and links you may find helpful throughout the unit.
bjupress.com/resources/science

Chemist and Physicist Marie Sklodowska Curie

Marie Curie (1867–1934) was a French physicist and the first female professor of physics at Sorbonne. In 1898 Marie Curie and her husband discovered the element polonium. She was the first woman to win a Nobel Prize. In 1903 she shared the prize with two others (one was her husband) for her work on spontaneous radiation. In 1911 she won the Nobel Prize in Chemistry for her discovery of and investigations into the chemistry of radium and polonium. She died from a blood disease that is often caused by exposure to large amounts of radioactive materials.

Objectives

- Recognize that God provides for the needs of people

Materials

- child's cardboard or wooden puzzle with 10 or fewer pieces
- blindfold

Teacher's Toolkit CD

- Visual 10.1: *Little Curies*

Introduction

- Choose a student to put the puzzle together. Blindfold the student. Allow him to put the puzzle together again.
  How was it different putting the puzzle together this time? Possible answer: harder putting it together because you could not see the pieces

  The human body is more complex than the puzzle. Are we able to see inside the human body? no Why would it be important to see inside the human body? Possible answers: to be able to see what is wrong; to be able to repair damage to an organ

Preparation for Reading

- Instruct the students to read silently page 237 to find out a way to look inside the human body.

Teach for Understanding

What were Marie Curie's trucks called? Little Curies

- Display the *Little Curies* visual. Explain that this is one example of the trucks with x-ray equipment.
  What language is on the side of the truck? French

  What do you think the words say? radiology service

  Why do you think there is a red cross on the truck? to let everyone know that this vehicle was in the service of caring for the wounded and not part of the war itself

  What differences and similarities do you see compared to today's vehicles? Possible differences: open spokes on the tires; open sides on the van; elongated engine area Possible similarities: four tires; running board for ease in entering; drive on the same side of the road as Americans

  How were nurses and doctors able to see inside the human body? by using x-ray technology

  How did Marie Curie help save lives? She researched human anatomy, studied how to use the x-ray machines, and put together vans with needed equipment.

  What does God want people to use this technology for? to serve and care for others and glorify God

Activity Manual

Preview, page 159

The Looking Ahead page is intended to assess the student's prior knowledge before beginning the chapter.

140

Student Text pages 238–41
Activity Manual pages 160–61

Objectives

- Identify light as a form of energy
- Compare and contrast electromagnetic and mechanical waves
- Identify the four properties of waves: wavelength, amplitude, frequency, and speed
- Differentiate between the frequency of a wave and the speed of a wave

Materials

- picture of a cheetah
- picture of a peregrine falcon

Teacher's Toolkit CD

- Visuals 10.2–10.3: *Comparing Waves*; *Transverse Wave*
- Quiz 10-A

Vocabulary

- light
- transverse wave
- wavelength
- amplitude
- frequency
- speed

Introduction

Which land animal can run the fastest? Answers may vary. The cheetah is the fastest land animal.

- Display the pictures of the cheetah and the peregrine falcon.

The cheetah can run at a speed of 70 mi/h. However, it can run that fast for only about 300 yards before it must slow its pace.

The peregrine falcon is even faster than the cheetah. It can fly and dive for prey at speeds of 100–200 mi/h.

Both of these animals are extremely fast. Today, though, we will learn about something that travels much faster than either of these animals.

Preparation for Reading

- Tell the students to mentally make questions using the headings, bold and italicized words, and captions.
- Instruct them to read silently pages 238–41 to find the answers to these questions.

How are electromagnetic waves different from mechanical waves?

What is amplitude?

Teach for Understanding

- Read Genesis 1:3–5.

What was the first thing God made after He created the heaven and the earth? light

What is light? a form of wave energy

What do all waves carry? energy

- Explain that since that first day, energy has existed.

Light Energy

The people who first discovered x-rays many years before World War I were surprised and amazed by what these mysterious rays could do. It was as if they were a kind of invisible light that could shine right through solid objects! X-rays share many of the same characteristics with light that you can see.

Characteristics of Light

Light is a form of wave energy. It is constantly moving throughout the universe. A light ray travels in a straight line, but the matter around it moves up and down. Light waves spread out as they move away from their source like ripples of water move away from the place a pebble is dropped. However, light travels much faster than water. If you could travel at the speed of light, you would be able to go around the world about seven and a half times in just one second!

Though light waves carry energy just as other waves do, light waves are different from many other kinds of waves. The energy of water waves and sound waves must move through matter. Light waves, though, can move through both matter and empty space. Because light waves can travel through space, light from the sun and other stars can reach the earth.

Light waves also move faster than other types of waves, such as sound. Light can travel through space at about 300,000 km/sec (186,000 mi/sec). The sun is about 150 million km (93 million mi) away from the earth. Yet light from the sun reaches the earth in about eight minutes. Scientists do not know of anything that can travel faster than light does.

SCIENCE & HISTORY

One of the first scientists who tried to measure the speed of light was Galileo. In the 1600s, Galileo and his assistant stood on top of two hills that were about a mile apart. Each of them had a shuttered lantern. Galileo flashed his lantern. As soon as the assistant saw Galileo's light, he flashed his own lantern. Galileo tried to time how long it took before he saw the answering flash of light. Since light travels so fast, Galileo was not able to measure the speed.

More than 200 years later, another scientist, Albert Michelson, was able to make an accurate measurement of the speed of light. He used a special eight-sided revolving mirror on top of a mountain in California.

238

Background

Speed of Light

When scientists say that nothing travels faster than the speed of light, they are referring to the speed of light in a vacuum, or empty space. The speed of light in a vacuum is about 299,792,458 m/sec.

Measuring Light

On July 21, 1969, during the *Apollo 11* mission, astronauts Buzz Aldrin and Neil Armstrong placed mirrors on the moon. Other mirrors have since been placed on the moon. Scientists on the earth can aim lasers at these mirrors, and the mirrors reflect the light back to the earth. By doing this, scientists can measure the speed of light and monitor the path of the moon around the earth.

Waves or Particles

Scientists have discovered that light has a dual nature. Some scientists describe light as particles, or little individual packets of energy called photons. Other scientists describe light as a wave. Light exhibits characteristics of both particles and waves.

Physicist Albert Michelson

In 1907 Albert Michelson became the first American to receive the Nobel prize for physics. While an instructor at the US Naval Academy, he began conducting experiments with light. He conducted several experiments that measured the speed of

Comparing Waves

Sometimes scientists classify waves by whether or not they can move through empty space. Waves that can move through space are called *electromagnetic waves*. Light is an electromagnetic wave. Waves that can move only through matter, such as sound waves, are called *mechanical waves*.

Scientists also classify waves by the direction in which they move. Sound waves move in the same direction as the matter is vibrating. Waves that travel this way are called *longitudinal waves*.

Light waves move differently. If you make a wave with a rope, you can see the wave traveling from one end of the rope to the other. As the wave travels forward, the rope moves up and down. The direction that the wave travels is perpendicular to the up-and-down movement of the rope. This is an example of a transverse wave. Waves that move perpendicular to the way that the matter is moving are called **transverse waves**. Light waves are an example of transverse waves.

Longitudinal Wave (sound)

Transverse Wave (light)

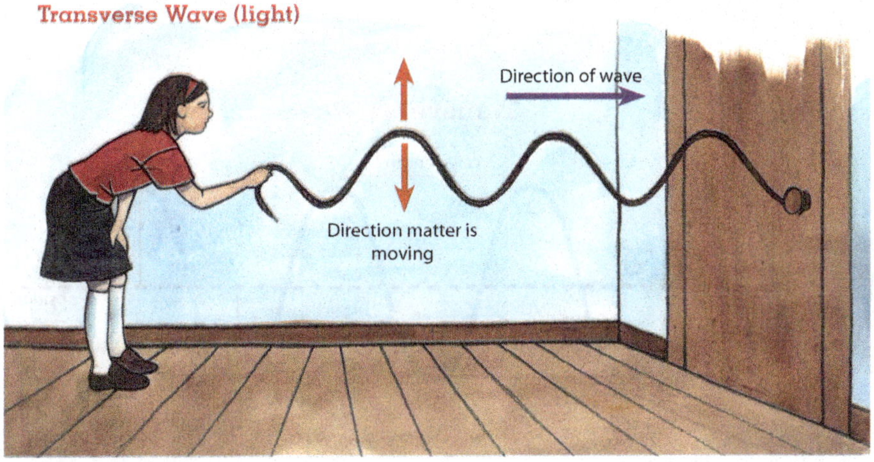

239

light. He also built an interferometer, an optical device that could split a beam of light into two pieces and then bring the two pieces back into one beam.

Demonstration

How Waves Move

Materials: metal spring toy, rope

Set the round side of the spring toy on top of a table or other hard surface. Hold one end of the toy and direct a student to hold the other end. Stretch it slightly so that there is space between the coils. Keeping the toy on the desk, gently push one end of the toy. Observe the coils moving as the wave travels to the other end of the toy. The coils and the wave are moving in the same direction. This is a longitudinal wave.

Tie one end of the rope to a fixed object, such as a doorknob. Direct a student to hold the other end of the rope. The rope should be stretched to its fullest length but should not be taut. Direct the student to move his hand up and down, starting a wave in the rope. Observe how the wave travels in a direction that is perpendicular to the way the student's hand is moving. This is a transverse wave.

What are two ways that light waves are different from other waves? They travel through matter and through empty space. They also travel much faster than other waves, such as sound, do.

About how long does it take for light from the sun to reach the earth? eight minutes

Why do we see lightning before we hear the sound of thunder? Light travels faster than sound.

- Discuss *Science & History* on page 238.

Who was one of the first scientists to try to measure the speed of light? Galileo

How did Galileo and his assistant attempt to measure the speed of light? by standing on top of different hills and timing how long it took to see a flash of light from the other person's lantern

Was this method successful? no Why not? Light travels too fast for Galileo's equipment to measure.

Who eventually made an accurate measurement of the speed of light? Albert Michelson

- Discuss classification of waves.

What are two different ways scientists classify waves? by whether or not the waves can move through empty space or by the direction in which the waves move

What can light travel through? matter and empty space

Are light waves mechanical or electromagnetic? electromagnetic

Since light is an electromagnetic wave, what can electromagnetic waves travel through? matter and empty space

What do scientists call the waves that can move only through matter? mechanical waves

- Refer to the *Comparing Waves* visual during this discussion.

What are longitudinal waves? waves that move in the same direction as the matter is vibrating

Which waves are longitudinal? sound waves

Which are transverse waves? waves that move perpendicular to the way that the matter is moving

Are light waves longitudinal waves or transverse waves? transverse waves

- You may want to do the *How Waves Move* demonstration at this time.

What do all waves have in common? the same basic properties, or characteristics

Name four basic properties of waves. wavelength, amplitude, frequency, and speed

- Sound waves also have these properties, although the crests and troughs can be seen only when the longitudinal waves are changed by an oscilloscope.
- Display the *Transverse Wave* visual. Allow students to point to the corresponding areas during the discussion.

  What does the dotted purple line represent? the rest position

  What is the highest point of the wave above the rest position called? the crest

  What is the lowest point of the wave below the rest position called? the trough

  What is the distance between two crests called? the wavelength

  What is the distance between two troughs called? the wavelength

  What is *amplitude*? the height of a wave from its rest position to its crest or trough
- You may want to use the *Parts of a Transverse Wave* activity to reinforce the terms.

Properties of Waves

There are many different kinds of waves. Some are tall. Others are short. Some move quickly, but others move slowly. All waves, though, have the same basic properties, or characteristics. These properties include wavelength, amplitude, frequency, and speed.

Transverse waves look like a series of hills and valleys. The top part of a "hill" is called the *crest*. This is the highest point of a wave. The *trough* is the lowest point, or "valley," of a wave. Scientists use the crests and troughs of a transverse wave to measure the length of that wave. Its **wavelength** is the distance between two crests or between two troughs. Waves with short wavelengths have more energy than waves with long wavelengths.

Another property of a wave is its amplitude. **Amplitude** is the height of a wave. If you stretch a rope tightly, it will be in a straight line. Think of this straight line as the *rest position* of a wave. Moving one end of the rope will make the rope vibrate up and down. Part of the rope will move above the rest position and make the crests of the waves. The rope will also move below the rest position, forming the troughs of the waves. The distance that the rope moves above or below the rest position is the amplitude, or height, of the wave.

Transverse Wave

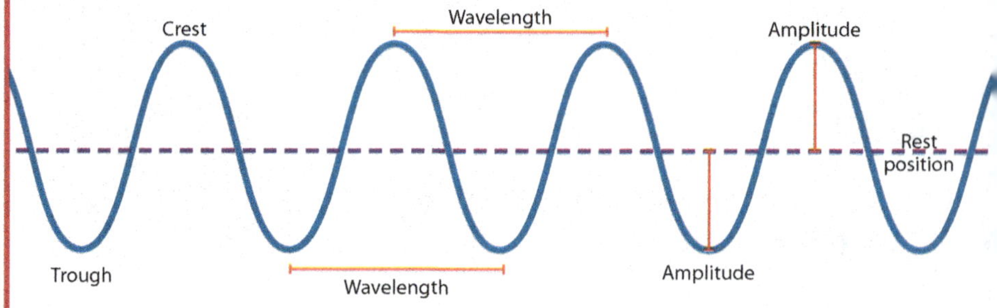

240

Background

Amplitude

Both transverse and longitudinal waves have amplitude, although it is more difficult to see amplitude in a longitudinal wave. Amplitude measures how much energy is in a wave. For a transverse wave, such as light, amplitude is measured by the distance that the wave moves above or below its rest position. For a longitudinal wave, amplitude refers to the change of pressure in the wave's medium. For a sound wave, amplitude is the intensity, or volume, of the sound.

Activity

Parts of a Transverse Wave

Materials: 30–50 cm string, for each student

Direct each student to shape his length of string into a transverse wave like the ones pictured on Student Text page 240. Identify the crests, troughs, wavelengths, amplitude, and frequency of the string wave. For amplitude the students may need a second piece of string to use as the rest position.

You may want to allow the students to glue the string to a sheet of paper, directing them to label the parts of a wave.

140

Wave Frequency

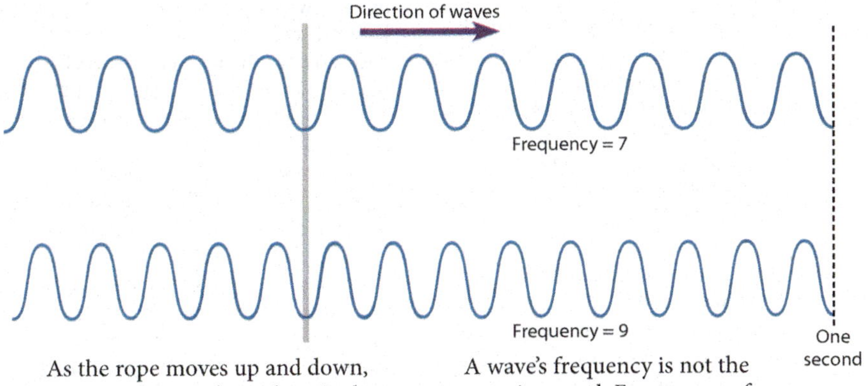

As the rope moves up and down, many waves move through it. Each wave has a crest and a trough. You can count the waves passing a point by counting either their crests or their troughs. The number of waves that pass a certain point in one second is the **frequency** of that wave.

A wave's frequency is not the same as its speed. Frequency refers to how many waves pass by a point in one second. **Speed** refers to the distance that one wave travels in one second. All waves have specific speeds, frequencies, amplitudes, and wavelengths.

SCIENCE & MATH

The speed, wavelength, and frequency of a wave are related mathematically. If you know the wavelength and frequency of a wave, you can calculate its speed. Suppose a wave had a wavelength of 6 cm and a frequency of 5 waves per second. The speed of the wave would be 30 cm/sec.

Wavelength × Frequency = Speed

QUICK CHECK

1. What are two ways that light waves are different from other waves?
2. What are transverse waves?
3. What is the distance between two crests of a wave called?

241

What is frequency? the number of waves that pass a certain point in one second

How is speed different from frequency? Speed is the distance one wave travels in one second. Frequency measures the number of waves that pass a certain point in one second.

- Discuss *Science & Math.*

What mathematical formula shows the relationship between speed, wavelength, and frequency? wavelength × frequency = speed

ANSWERS

1. Light waves can move through empty space as well as through matter. They move faster than other waves do.
2. waves that move perpendicular to the way that the matter is moving
3. a wavelength

Activity Manual

Reinforcement, page 160

Review, page 161

This page reviews Lesson 140.

Assessment

Quiz 10-A

The quiz may be given any time after completion of this lesson.

Science Misconceptions

Speed of a Wave

Speed depends on the characteristics of the matter, or medium, through which a wave is traveling. The only way to change the speed is to change the characteristics, or properties, of the medium. Speed can be calculated by multiplying the wavelength and the frequency of a wave. However, changing the wavelength does not change the speed. Changing the wavelength just increases or decreases the frequency.

Background

Frequency

The frequency of a wave is measured in hertz. The frequency of a sound wave is also its pitch. Sound waves are discussed in Chapter 9, *Sound*.

141
142

Student Text pages 242–45
Activity Manual pages 162–63

Objectives

- Differentiate between refraction and reflection
- Recognize that the color of an object depends on which colors of light are being reflected
- Identify the primary colors of light

Teacher's Toolkit CD

- Quiz 10-B

Vocabulary

- visible spectrum
- refraction
- reflection
- primary colors
- pigment

Two days have been allotted for this lesson. You may want to teach Lesson 141 the first day and use ideas from Lesson 142 on the second day. You may also incorporate the ideas from Lesson 142 into the text material, making the lesson two days.

Introduction

- Direct a student to pretend to be a robot that can move only in straight lines.
  Start walking. Whenever you bump into an object, turn and walk in a different direction, but you must keep walking in a straight line.
- Stop the student after he has changed direction at least twice.
  Just like our "robot," light generally travels in a straight line. It may change direction when it hits an object, but it will usually continue as a straight line.

Preparation for Reading

- Tell the students to mentally make questions using the headings, bold and italicized words, and captions.
- Instruct them to read silently pages 242–45 to find the answers to these questions.
  What determines the color of an object?
  What are the primary colors of light?

Teach for Understanding

What allows us to see objects around us and gives objects color? light

What do we call the colors of visible light arranged in order by their wavelengths? the visible spectrum

Who proved that light is made up of many different colors? Sir Isaac Newton

How could you describe the path of light? generally a straight line

What two things about light may change? its speed and direction

Visible Spectrum

If you try to walk through a dark room, you may stumble or bump into things. Without light, there is no vision. You see the objects around you because light bounces off them. Light makes things visible, or able to be seen. Light also gives objects color.

In 1672 Isaac Newton proved that light is made up of many different colors. Newton used a prism to separate white sunlight into a band of many colors. The different colors of light that we can see are called the **visible spectrum**.

Refraction and Reflection

Light generally travels in a straight line. The direction and speed may change, but its path will usually be straight. The matter through which a wave travels is called a *medium*. The speed of light changes as light enters different media. Light travels the fastest in empty space. It slows down a little when it enters air. Things such as a glass or water make light slow down even more. When it enters a new medium, light slows and can bend in a different direction. **Refraction** is the bending of light as it passes from one medium into another.

Prisms separate light into colors by the process of refraction. As sunlight passes through the air and enters a glass prism, the light slows down and bends. Each color of light has a different wavelength, so each color bends a little differently. This allows you to see the separate colors. Refraction also happens in a rainbow. Sunlight passes through raindrops and bends into the separate colors of the visible spectrum.

The colors in the visible spectrum are in order by their wavelengths. The reds have the longest wavelengths. The colors after the reds are oranges, yellows, greens, blues, indigos, and violets. The violets have the shortest wavelengths of the colors. You can remember the order of the color groups by thinking of the first letter of each. The first letters spell out the name *Roy G. Biv*.

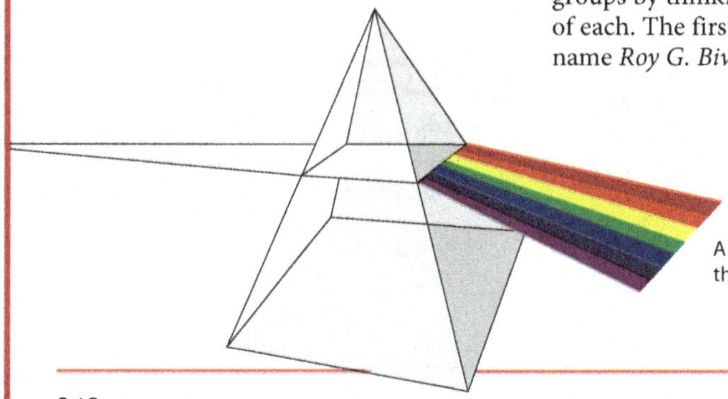

A prism bends light to show the color spectrum.

242

Background

Refraction

Scientists have measured how much light bends as it enters different mediums. This number is called the index of refraction.

Rainbows

Rainbows can be seen when there are water droplets in the air and the sun is behind you. Sunlight enters the raindrops and bends into the different colors of the spectrum. The most common type of rainbow is a primary rainbow. Its spectrum ranges from red to violet. Secondary rainbows occur when the sunlight reflects twice within the raindrop. The color spectrum is reversed, and the secondary rainbow appears next to the primary rainbow.

Rainbow

Refraction is one way that light changes its direction. When light hits *transparent* matter, the light passes through with little bending. Objects behind transparent matter are clearly seen. As light hits *translucent* matter, some light passes through and some is bent or blocked. Objects seen through translucent matter appear unclear. Light slows down and bends when it passes through transparent or translucent things. *Opaque* matter does not let any light pass through. All of the light either bounces off the object or is absorbed, or soaked in.

Light can also change direction by reflection. **Reflection** happens when light bounces off an object. The color that you see is determined by which colors are reflected and which are absorbed. You see the reflected wavelengths as the color of that object. A red book reflects the color red. All the other colors of light are absorbed. When all the colors of light are reflected, you see white. You see black when all the colors of light are absorbed.

Why are darker-colored clothes usually warmer? Darker colors absorb more of the light energy from the sun. The sun's energy makes the molecules in the clothing move faster. The temperature of the clothing rises and keeps the person warmer.

SCIENCE & THE BIBLE

God often uses light as a metaphor. In Psalm 119:105 the writer describes the Bible as a "light unto my path." Just like a light shows you where to walk at night, God's words show you which choices are right and safe. Sadly, many people do not like this "light" because they live in spiritual darkness. In John 8:12 Jesus said, "I am the light of the world: he that followeth me shall not walk in darkness, but shall have the light of life." Through Jesus, people can move from the world of darkness to the world of light (John 12:46).

243

141
142

Under what conditions does the speed of light change? as it enters a medium

What is a medium? the matter through which a wave travels

What happens when sunlight enters a prism? The light slows down and bends, showing the spectrum of colors.

What process is happening when prisms separate light into colors? refraction

What is refraction? the bending of light as it passes from one medium into another

How does refraction occur in a rainbow? Sunlight passes through raindrops and bends into the different colors.

What is the order of the colors in the visible spectrum? red, orange, yellow, green, blue, indigo, and violet

What name can help you remember the order of the colors in the visible spectrum? Roy G. Biv

What are the two ways that light can change its direction? through refraction and reflection

What kinds of objects refract light? transparent objects and translucent objects

Are objects behind transparent objects hard or easy to see? easy

- Ask the students to name some transparent objects. Possible answers: photo frame glass, eyeglasses, air, water, light bulb (if you can see the filament)

How much light can pass through translucent objects? only some of the light

- Ask the students to name some translucent objects. Possible answers: sunglasses, notebook paper, wax paper, stained glass, a lampshade, vegetable oil, lace

What kind of objects do not let any light pass through them? opaque objects

- Ask the students to name some opaque objects. Possible answers: book, table, walls, spoons, door, ceramic tile, desk, human body

What happens to the wavelengths of light when they strike an opaque object? Some are absorbed, and some are reflected.

What is reflection? the bouncing back of light off an object

What determines the color of an object? the colors that are reflected off the object

What is seen when all the colors of light are reflected? white

When do you see black? when all the colors of light are absorbed

- Discuss *Science & the Bible.*

What does the Bible often use as a metaphor? light

Who is the light of the world? Jesus

What light should Christians reflect? Jesus, the Light of the World

Background

Newton's Discovery

Even though many scientists before him had used prisms to experiment with color, Sir Isaac Newton was the first to realize that white light was actually many colors mixed together. He placed a prism in the sunlight and observed the spectrum that formed on the wall. Each color of light was bent a different amount by the prism. In order to prove that the prism did not color the light, Newton placed a screen between the prism and the spectrum. The screen had a narrow slit cut into it, so that only the green light could pass through the screen. He then placed a second prism in the path of the green light. The green light passed through the prism and did not change in color. Newton had proved that white light is made of many colors.

What are primary colors? the three colors that are used to make all the other colors

Can primary colors be made by combining any other colors? no

What are the primary colors of light? red, blue, and green

What color is seen when equal amounts of the three primary colors of light are combined together? white

How do televisions and computers show colored pictures and images? The picture is made up of groups of red, blue, and green lights.

- Discuss the *Colors of Light* diagram.

  What color is formed by mixing green light and blue light? cyan light

  What color is formed by mixing red light and green light? yellow light

  What color is formed by mixing red light and blue light? magenta light

  What are cyan, yellow, and magenta called? the complementary colors of light

  What is another name for complementary colors? secondary colors

  What color would you see if you combined magenta light and green light? white Why? Magenta light is a combination of red light and blue light, and combining red, blue, and green light results in white light.

- You may want to do the *Colors of Light* demonstration at this time.

  What are the colors that you see? those not absorbed by pigments in the objects

  What is a pigment? an opaque substance used to color other materials

  What are the primary colors of pigments? yellow, cyan, and magenta

  Why is a lime green? because its pigments absorb all the colors except green

- Discuss the *Colors of Pigment* diagram with similar questions as used above for the *Colors of Light* diagram.

Combining Colors

An amazing variety of color is all around you. Whether a beautiful blue sky or a bright yellow flower, the world is full of colors reflected from our sun and white lights. Scientists study color using the idea of a color model. There are different color models, but each has three primary colors. **Primary colors** cannot be made from other colors. God designed our eyes and brains to combine primary colors into all the other colors we see.

The primary colors of light are red, green, and blue. All the colors of light are formed by different combinations of these primary colors. If red light, green light, and blue light are mixed together in equal amounts, they reflect all colors and produce white, as shown in the center of the model.

Television screens and computer monitors take advantage of this color model. Tiny groups of red, green, and blue lights shining at different brightnesses can produce all the colors you see on a screen.

Each primary color has an opposite, or *complementary color*. The complementary color of red is cyan and the complementary color of green is magenta. What is the complementary color of blue?

Most objects in the world are not sources of light. Instead they reflect light. The colors you see are those that are not absorbed by pigments in the objects. **Pigments** are opaque substances used to color other materials. The primary colors of pigments are yellow, cyan, and magenta. A lemon is yellow because its pigments absorb all colors except yellow. Yellow is reflected. Notice that the primary colors of pigments are the complementary colors of light.

Colors of Light

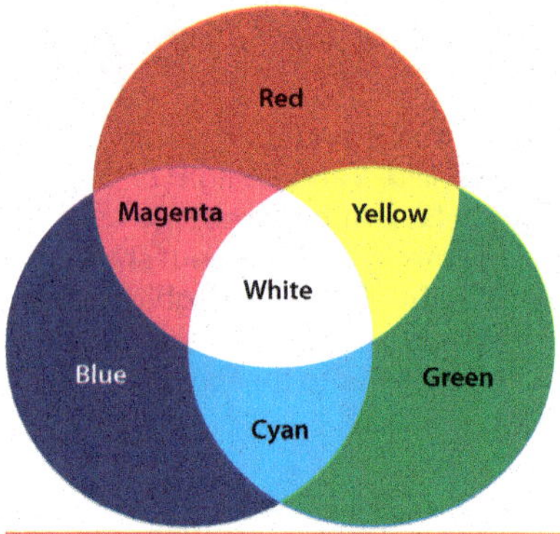

Colors of Pigment

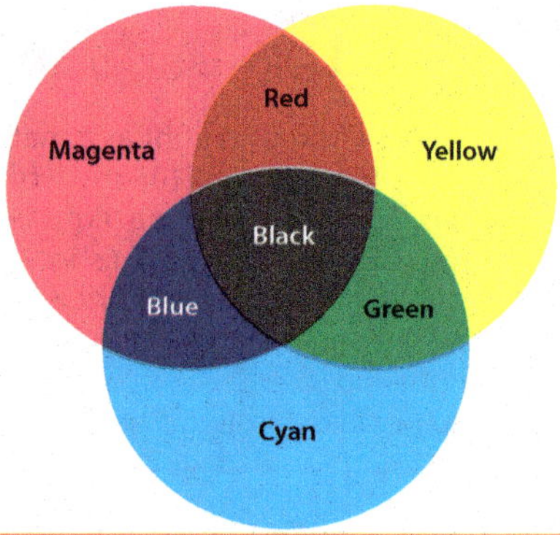

244

Background

Seeing Color

Special cells in the retina of the eye allow people to see color. A person actually sees only the primary colors of light. His eyes and brain mix the primary colors in certain amounts so that he "sees" all the other colors.

Blue Sky, Yellow Sun

When sunlight passes through the atmosphere, it bumps into gas particles and sometimes bounces off in different directions. The longer wavelengths of color can pass through the air, but the colors with shorter wavelengths, such as blue and violet, are often scattered by the atmosphere. This makes the color of the sky appear blue. The rest of the wavelengths mix together and cause the sun to appear yellow. However, out in space, the sun would appear white because space has no atmosphere to scatter the light.

Mixing Colors

Mixing colors of light is sometimes compared to addition because adding different amounts of the primary colors of light causes the eye to see different colors. Mixing colors of paint is a subtractive process. Each pigment reflects only the wavelength of light that gives it color. Adding more pigments results in more wavelengths being absorbed and fewer being reflected.

Paint pigments work the same way. You may have learned in art that the traditional primary colors are red, yellow, and blue. Pigments reflect these colors. Red paint looks red because it reflects red. A blue crayon reflects the color blue. When you mix pigments together you do not get white. You get a dark, almost black color. This is because they absorb nearly all of the colors of light shining on them.

Color is a wonderful gift from God. When He said in Genesis, "Let there be light," He had already planned for His earth to be beautiful with color. He created the pigments and light to work together. He knew that a wide variety of colors would bring pleasure to people.

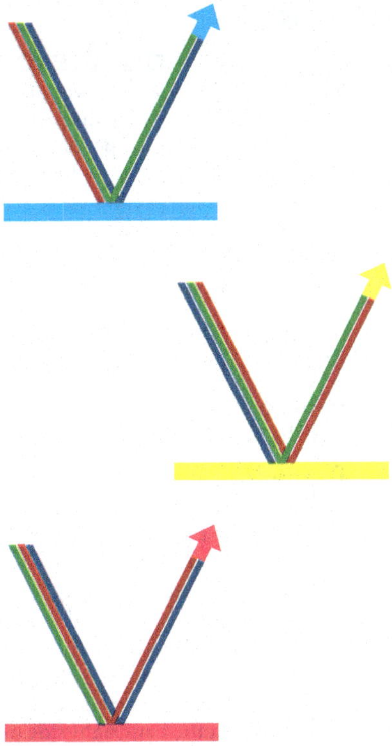

Light reflects from a surface and combines to form the colors we see.

TRY IT YOURSELF

Many books and papers are printed by a four-color printing process. The printers use combinations of cyan, yellow, magenta, and black ink to produce all the colors you see in a book. Use a magnifying glass to look at a colored picture in this textbook. Do you see all the individual dots of color? The dots are so close together that your eyes blend them together. This makes the color and shading that you see in the picture.

QUICK CHECK

1. How is reflection different from refraction?
2. What determines the color of an object?
3. What are the primary colors of light?

245

141–142

Are the primary colors of light the same as the primary colors used in art? no

Can the colors of paints and crayons give off light? no

What do the colors of paints and crayons do? reflect light

Why does blue paint look blue? It reflects the color blue.

What "color" is produced when the primary colors of paint are mixed? black

- Discuss *Try It Yourself.*

What four colors do many books and papers use? cyan, yellow, magenta, and black

How do your eyes see mixed colors? The color dots are printed close together. When you look at a picture, your eyes blend them together.

ANSWERS

1. Reflection is the bouncing of light off an object. Refraction is the bending of light as it passes from one medium into another.
2. the color that is reflected from that object
3. red, green, and blue

142

- The *Colors of Light* demonstration found at the bottom of this page may also be incorporated into the lesson.
- Suggestions for additional activities for this two-day lesson are on the Teacher's Toolkit CD.

Activity Manual

Reinforcement, page 162

Review, page 163

This page reviews Lessons 141–42.

Assessment

Quiz 10-B

The quiz may be given any time after completion of this lesson.

Science Misconceptions

Black and White

Black and white are not true colors. White is the presence of all light, and black is the absence of all light.

Demonstration

Colors of Light

Materials: red, blue, and green cellophane or color paddles; 3 flashlights with narrow beams; rubber bands (optional)

Note: The colored sandwich bags used in the *Fog Vision* activity on pages 280–81 will not work well for this demonstration. If color paddles are used, you will need only one flashlight. See information on the BJU Press website for places to find color paddles.
bjupress.com/resources/science

Cover one flashlight with red cellophane, one with blue cellophane, and one with green cellophane. Shine each flashlight onto a white surface so that each separate color is visible. Alternate overlapping the different colors so new colors are visible. Overlap all three colors to demonstrate how white light is formed.

Objectives

- Test the visibility of colors
- Infer which colors are most visible in fog

Materials

Materials listed on the student pages are needed for activity lessons. See Activity Helps under Background information.

Teacher's Toolkit CD

- Rubrics: *Fog Vision*, for each student

Introduction

A driver knows that lights of different colors have different meanings. What are some colored lights that have specific meanings to a driver? Possible answers: Red lights used for traffic signals, railroad crossings, or emergency vehicles mean to stop. Green traffic lights mean to go. Yellow or orange lights used for traffic signals and highway construction markers mean to slow down, move with caution, and be ready to stop. Blue lights are used to identify and signal the presence of law enforcement vehicles.

What are some road conditions that might make it hard for a driver to see? Possible answers: darkness, rain, snow, fog

These kinds of conditions can make it difficult not only for the driver to see but also for others to see the driver and his car.

Preparation for Reading

- Instruct the students to read silently Student Text pages 246–47 and Activity Manual page 164 before beginning the activity.

Teach for Understanding

Process Skill

The process skill emphasized in this lesson is hypothesizing.

What is a hypothesis? a testable statement that tells what a person thinks will happen

What is your hypothesis for this activity about? the color of light that is the easiest to see through fog

- Direct each student to complete the hypothesis in the Activity Manual.

If the hypothesis had not included the words "through the fog," would it be a good, testable hypothesis? no Why not? Possible answer: because you would not know what kind of situation you were using to test the colors

A good hypothesis should be as specific as possible.

ACTIVITY

Fog Vision

Process Skills
- Hypothesizing
- Predicting
- Experimenting
- Observing
- Inferring

Have you ever noticed how hard it can be to see through fog? Fog can make it hard for people to see the lights of the other cars around them. This can cause car accidents. To help prevent these accidents, vehicles have colored lights. Brake lights are usually red. Police cars often have red and blue lights. Construction vehicles may have yellow or orange lights. Colored lights like these help other drivers see the vehicles, even when visibility is not good.

In this activity you will test different colors of light to determine which color is best to use in foggy situations.

Problem

Which color of light is the easiest to see through fog?

Materials
- 5 plastic sandwich bags
- broad-tipped permanent markers (red, blue, green, yellow, and purple)
- small flashlight
- rubber band
- white wall (or a similar white surface)
- 15 translucent white plastic cups, 3 oz
- Activity Manual

Procedure

1. Predict which color of light you think will be the easiest to see through the fog. Complete the hypothesis in your Activity Manual.
2. Turn the plastic bags inside out. Use one of the overhead markers to color a large square on one side of one bag. On each of the other bags, color a similar square. Use a different color of marker for each bag. Wait a few minutes to let the ink dry. Turn the bags right side out.
3. Darken the room.
4. Choose one color to test first. Cover the end of the flashlight with the colored area of the bag. Use a rubber band to hold the bag on the flashlight.
5. Turn on the flashlight and aim it at a white wall or another white surface. Hold it about 45 cm (18 in.) away from the surface.

246

Background

Activity Helps

It is important to use broad-tipped permanent markers for this activity. Other markers will not produce the same results.

Be sure that you have a white surface to use for testing the colors. Some light walls have a hint of color. If the surface is not pure white, it will affect the results of the activity. You may choose to use a piece of white poster board as the white surface.

You may choose to mark the 45 cm distance from the surface so that the students will be able to keep the flashlight in the same position for each color test.

6. Place one translucent plastic cup over the flashlight. The plastic cups represent fog. Continue adding one cup at a time until you can no longer see the beam of light on the wall.
7. Count how many cups the colored light passes through before disappearing. Record that number in your Activity Manual.
8. Repeat steps 5–7 for each of the other colored bags. Remember to record the number of cups used for each color.

Conclusions

- Which light was the easiest to see through the "fog"?
- Did your results support your hypothesis?

Follow-up

- Put five cups on the flashlight. Use a light meter to measure how much light from each color can be seen through the "fog."

247

Procedure

- Guide the students as they color the bags. The ink wipes off easily, especially when it is wet. They need to color only an area large enough to cover the lens of the flashlight.
- The bags must be turned right-side out to avoid rubbing off the color. The square of color on the plastic bag should be a solid block of color. Any white space left in the color square will affect the results of the activity.
- Remind the students to keep the flashlight bulb the same distance away from the white surface for each color test.

Conclusions

Which color had you predicted would be the easiest to see? Answers will vary.

How many cups did it take before you could not see the colored light for your predicted color? Answers will vary.

Would any of the other colors be visible in a foggy situation? Which ones?

How does this activity relate to the colors of lights that are actually used at night and other times of low visibility? Yellow or orange lights are used as "fog" lights. Red is used for taillights.

Activity Manual

Activity, page 164

Assessment

Rubrics

Select the prepared rubric, or design a rubric to include your chosen criteria.

Student Text pages 248–51
Activity Manual page 165

Objectives

- Explain how light reflects off smooth and rough surfaces
- Identify and describe three kinds of mirrors
- Identify some technologies that use light
- Name some uses for lasers

Materials

- kickball

Vocabulary

- mirror
- concave
- convex
- lens

Two days have been allotted for this lesson. You may want to teach Lesson 144 the first day and use ideas from Lesson 145 on the second day. You may also incorporate the ideas from Lesson 144 into the text material, making the lesson two days.

Introduction

- Direct a student to stand about 3 meters away from you.
  Watch the path of the ball as I bounce it.
- Bounce the ball so that it reaches the student in one bounce.
  What did you notice about the path of the ball before and after it hit the floor? The height and angles were about the same.
  Where will the ball end up if I bounce it closer to my feet? Accept reasonable answers.
- Bounce the ball close to your feet and discuss how the path and angles changed.
  Light reflects off different surfaces in much the same way the ball bounced.

Preparation for Reading

- Tell the students to mentally make questions using the headings, bold and italicized words, and captions.
- Direct them to read silently pages 248–51 to find the answers to the questions.
  How is a convex mirror different from a concave mirror?
  What determines the color of a laser?

Teach for Understanding

How does light reflect off a smooth surface? It bounces off in the opposite direction at the same angle.

How does light reflect off a rough surface? It bounces off in many directions.

What is a mirror? any surface that can reflect light to form an image of an object

Mirrors

Have you ever bounced a ball against a wall? Light reflects off different surfaces in much the same way that a ball bounces. When light shines at an angle onto a smooth surface, the light bounces off at an equal angle in the opposite direction. Mirrors reflect light in this way.

When light shines on a rough or bumpy surface, the reflected light goes in many different directions. Some surfaces, such as paper, look smooth. However, if you use a microscope to look at a piece of paper, you will see that its surface is rough. So light that reflects off the paper will go in many directions.

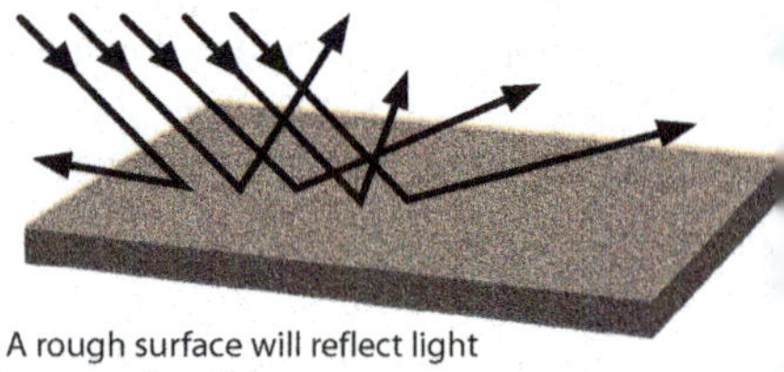

A rough surface will reflect light in many directions.

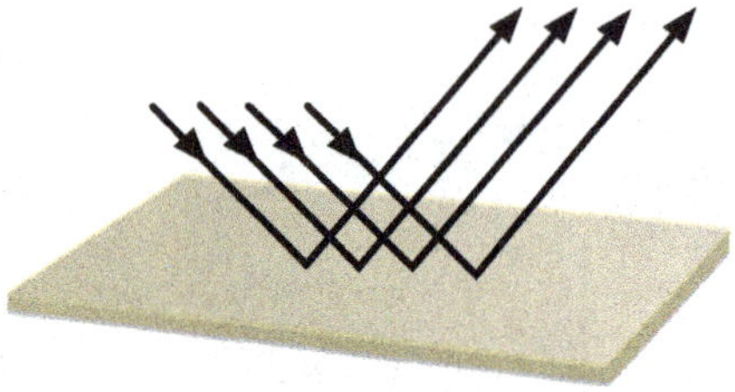

A smooth surface will reflect light at the same angle but in the opposite direction.

Rough surfaces, such as paper and snow, reflect light. But they do not make good mirrors. Mirror surfaces are shiny and smooth. A **mirror** is any surface that can reflect light to form an image, or picture, of an object.

A mirror with a flat surface is called a *plane mirror*. Most common mirrors, such as a bathroom mirror, are plane mirrors. A plane mirror shows an image that looks exactly like the real object. The only difference is that the image is reversed.

Science & The Bible

Mirrors today are usually made of glass that has a silver coating on one side. In Bible times, mirrors were sometimes made from metals, such as silver or bronze. The metals were polished until they shone.

Exodus 38:8 tells us that the brass laver in the tabernacle was made from looking glasses, or mirrors. The Israelite women willingly gave their mirrors to be used in the tabernacle.

248

Background

Mirages

An image that appears real when nothing is there is called a mirage. When two air masses have different temperatures, a mirage may occur. As light enters the different temperatures of air, it bends. Sometimes this bending makes it look like water is on the road, a ship is in the sky, or an oasis is in the desert.

Concave Mirrors

If an object is close to a concave mirror, the object's image will be upright and magnified. However, if the object is moved farther away from the concave mirror, the image appears upside down.

If the student is having difficulty differentiating between concave and convex mirrors, use a metal spoon or bowl to demonstrate the mirrors. The inside of the spoon or bowl is a concave mirror. The back or outside is a convex mirror.

Word association may also help a student having difficulty determining which mirrors are concave and which are convex. Help the student link the idea of a cave curving inward and a con*cave* mirror curving inward.

The word *ambulance* is written backward so it can be read correctly in a mirror.

You may have noticed that the front of an ambulance has the word *ambulance* written backwards. This is so a person driving a car in front of an ambulance can look in the rearview mirror and read the word clearly. This lets the person quickly know to get out of the way.

Some mirrors are curved. Curved mirrors change the way an image looks. They can make the image look larger or smaller than the object really is. You may have seen curved mirrors at a fair or an amusement park. The amount of curve can vary. Some mirrors curve a lot. Others have a curve so small that it may not even be noticeable.

A mirror that is **concave** curves inward. These mirrors reflect light toward a point in front of the mirror. If you stand close to a concave mirror, your image is *magnified*, or made larger. Many women and men use concave mirrors when they put on makeup or shave. This kind of mirror allows them to see better detail. Some telescopes use concave mirrors. These mirrors help faraway objects look bigger.

A mirror that is **convex** curves outward. These mirrors reflect light that spreads outward. Images seen in a convex mirror look smaller than the objects really are. You can use a convex mirror to help you see a large area. Many stores use convex mirrors to watch for shoplifters. An automobile's side mirrors are convex mirrors.

A **concave** mirror curves inward toward the person looking at it.

A **convex** mirror curves outward away from the person looking at it.

249

Can snow be a mirror? no Why not? It is a rough surface, so light reflects in all directions.

What are three kinds of mirrors? plane, concave, and convex

Which type of mirror has a flat surface? a plane mirror

How does the image seen in a plane mirror appear? exactly as the real object, only backward

Which kind of mirror curves inward? a concave mirror

If you stand close to a concave mirror, how does your image appear? magnified, larger

What are some uses for concave mirrors? Possible answers: as dental tools to look at people's teeth, as makeup mirrors, in telescopes

Which kind of mirror curves outward? a convex mirror

How do convex mirrors affect the way images appear? Objects appear smaller than they are.

What are some uses for convex mirrors? Possible answers: in stores to watch for shoplifters, as side mirrors on vehicles

- You may want to do the *Mirrors and the Appearance of an Image* activity at this time.
- Discuss *Science & the Bible* on page 248.

How were the mirrors used in Bible times different from modern mirrors? In Bible times, mirrors were pieces of polished metal. Today mirrors are made of glass with a silver coating.

What did the Israelite women give to be used in the tabernacle? their mirrors

How were their mirrors used? to make the brass laver

Science Misconceptions

Rough Surfaces

Although light reflects off rough surfaces in many directions, a rough surface actually reflects light just as a smooth surface does. The rough surface, though, consists of many flat surfaces that can reflect light. As light hits each of those surfaces, it bounces in the opposite direction at the same angle. Because the rough surface has so many different angles of reflecting light, the light goes in all directions.

Activity

Mirrors and the Appearance of an Image

Materials: plane (regular) mirror, concave mirror, convex mirror

Allow each student to handle each mirror. Direct him to place an object such as a pencil in front of each mirror. Ask him to try to identify the type of mirror by looking at how it changes the appearance of the image.

Allow the students to try placing objects at different distances from the mirrors. Notice how the image appears at different distances.

144
145

What is a lens? a piece of glass or other transparent object that refracts light and produces an image

- Display the prefix *tele-*. Explain that it comes from a Greek word that means "far away" or "at a distance."
- Display the word *scope*. Explain that it also comes from a Greek word and means "to look" or "an instrument for viewing."

Using the meanings of *tele-* and *scope*, what do you think *telescope* means? an instrument used to view faraway things

Can you think of other words that start with *tele-*? Possible answers: telephone, telegraph, telegram, television

What do telescopes use to collect and focus light? mirrors or lenses or both

What are two main types of telescopes? reflecting and refracting

What does a single-lens reflex camera use to collect and focus light? a mirror, a lens, and a prism

You may want to refer to Activity Manual page 171 during the discussion. Use the page to explain how a single-lens reflex camera works.

- Discuss *Science & History*.

Why were early cameras mainly used to help artists sketch pictures? They could not produce lasting images.

Who took the first actual photograph? Joseph Niépce

What did the first photograph show? the view from his window

What are some ways that photography has changed people's lives? Events can be recorded as they actually happened. Images of people can be preserved.

Light and Technology

An x-ray machine is just one tool that has been developed from the study of light. There are also many other inventions that make use of light.

What do some telescopes, lasers, and cameras have in common? They use mirrors and lenses to change the direction of light. A **lens** is a piece of glass or other transparent object that refracts light and produces an image.

Reflecting Telescope

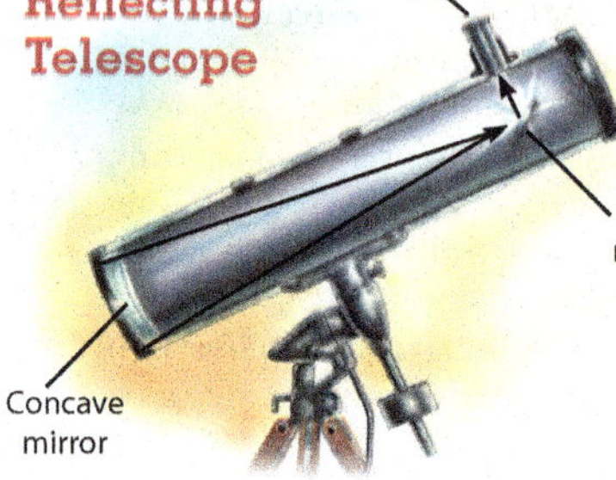

Telescopes help people see things that are far away, such as planets and stars. Some telescopes are small. Others are huge. The Subaru telescope located in Hawaii weighs 555 tons. That is more than the total weight of 56 buses! Regardless of size, however, all telescopes use mirrors or lenses or both to collect and bring the light to a point, or focus.

A *reflecting telescope* uses a concave mirror to gather and focus light. The light is reflected to the eyepiece lens, which magnifies the object. A *refracting telescope* usually uses two convex lenses. One lens gathers the light. The other magnifies the image.

Many digital cameras are single-lens reflex cameras.

SCIENCE & HISTORY

Early camera-like instruments did not produce lasting images like photographs. They were mainly used to help artists sketch their pictures. In 1826 Joseph Niépce took the first true photograph. He coated a metal plate with a light-sensitive chemical and placed it inside a camera for several hours. The photo showed the view from his window.

250

Background

Telescopes and Microscopes

Refracting telescopes use two convex lenses to magnify objects in the sky. Reflecting telescopes use a concave mirror, a plane mirror, and a convex eyepiece lens for the same purpose. Microscopes are similar to reflecting telescopes. They also use a concave mirror, a plane mirror, and a convex lens. However, they magnify objects placed on slides instead of magnifying objects in the sky.

Types of Cameras

Two common types of cameras are single-lens reflex cameras and automatic point-and-shoot cameras. The point-and-shoot cameras, such as disposable cameras and other 35mm cameras, are less expensive than single-lens reflex cameras. Point-and-shoot cameras do not use mirrors, gears, and springs. Instead, they use circuit boards and electric motors.

Laser

The term *laser* is an acronym that stands for light amplification by stimulated emission of radiation.

Ruby Laser

The color of the laser depends on the type of medium that is used. Ruby lasers have a solid ruby rod inside the laser. These lasers give off red light. Other lasers with red light may use a gas instead of a ruby. Different gases can cause lasers to give off blue light or green light.

Cameras also use mirrors and lenses. There are many kinds of cameras, though. And each works in a slightly different way. A single-lens reflex camera uses a mirror, a lens, and a prism.

Lasers use mirrors to produce a thin beam of light that is very powerful. This beam of light is only one color. Inside a laser tube is a medium for the light to travel through. This medium can be a gas, a liquid, or a solid.

At each end of the tube is a mirror. As light travels back and forth through the medium, the mirrors reflect it. One of the mirrors in the laser is semi-transparent. It reflects some light and lets some light pass through it. Some of the light "leaks" out of the tube through the mirror and produces the beam of light.

Lasers can be found almost anywhere. DVD players and CD players use lasers to read the content of a disc. Checkout scanners in many stores use lasers to read bar codes. Doctors sometimes use lasers to perform delicate surgeries or to look inside a patient's body. Some lasers cut cheese into fancy shapes. Others are powerful enough to cut through steel.

QUICK CHECK

1. What are three types of mirrors?
2. Which kind of mirror magnifies an object?
3. How is a refracting telescope different from a reflecting telescope?

251

144 145

Is a beam of laser light thin or thick? thin

What is unique about the color of laser light? It is only one color and cannot be separated into any other colors.

Through what does light travel back and forth inside a laser? a medium

Which states of matter can be used as a medium for lasers? solid, liquid, or gas

What is at each end of a laser tube? a mirror

How does light get out of a laser? Some light "leaks" out of the tube through the mirror.

What determines the color of a laser? the type of medium that is used

- Direct attention to the picture of the Ruby Laser.

Why does the Ruby laser give off red light? A ruby is used as the medium.

What are some colors of laser light? Possible answers: red, blue, green

What are some uses for lasers? Possible answers: reading CDs or DVDs, scanning bar codes, doing surgery, enabling doctors to see inside a patient's body, cutting cheese into fancy shapes, cutting steel

Why would it be unsafe to shine a laser at someone's face? It is a powerful beam of light and could cause damage to the eyes.

ANSWERS

1. plane, concave, and convex
2. a concave mirror
3. A refracting telescope uses two convex lenses. A reflecting telescope uses a concave mirror.

Activity

Make a Pinhole Camera

Materials: small box or large can (shoe box, oatmeal canister, coffee can), waxed paper or photographic paper

There are many different ways to make a pinhole camera. Some simply involve making a small hole at one end of the container and using waxed paper on the other end as a screen to show the image. Other methods include attaching film or photographic paper to the container. However, with these methods, you would need to have access to a dark room.

145

- The *Mirrors and the Appearance of an Image* and *Make a Pinhole Camera* activities found at the bottom of the pages in Lessons 144–45 may also be incorporated into the lesson.
- Suggestions for additional activities for this two-day lesson are on the Teacher's Toolkit CD.

Activity Manual

Reinforcement, page 165

146

Student Text pages 252–53
Activity Manual page 166

Objectives

- Differentiate between the angle of incidence and the angle of reflection
- Measure the angle of reflection
- Infer the relationship between the angle of reflection and the angle of incidence

Materials

- protractor (optional)

Teacher's Toolkit CD

- Instructional Aid 10.1: *Protractor*, for each group
- Rubrics: *Angles of Reflection*, for each student

Introduction

How does light reflect off a mirror? at an angle

What tool do we use to measure angles? a protractor

Protractors can be large or small. In this activity you will use a large protractor to measure some angles of light.

Preparation for Reading

- Instruct the students to read silently Student Text pages 252–53 and Activity Manual page 166 to find out what the angle of incidence and the angle of reflection are.

Teach for Understanding

Process Skills

The process skill emphasized in this lesson is defining operationally. In order to define operationally, each student should correctly use vocabulary to explain his observations or conclusions in his own words.

What new terms are used in this activity? angle of reflection, angle of incidence

What is an angle of incidence in this activity? the angle at which light hits a mirror

What does *incident* mean? something that happened, an occurrence

What is happening, or occurring, in this activity? The light is reflecting.

What is the angle of reflection? the angle at which light reflects away from a mirror

- Help each student explain his results in his own terms. The key is that he understands what is happening and is able to use the scientific terms accurately.

ACTIVITY

Angles of Reflection

Have you ever used a mirror or a glass to change the direction of light? If a ray of light hits a mirror, the light reflects away from the mirror in a different direction. The angle at which light hits a mirror is called the *angle of incidence*. The angle at which light is reflected is called the *angle of reflection*. In this activity you will be measuring the angle of reflection.

Process Skills
- Predicting
- Measuring and using numbers
- Observing
- Inferring
- Defining operationally

Purpose

Compare the angle of reflection with the angle of incidence.

Procedure

1. On the *Protractor* page, use a ruler and a blue colored pencil to trace one of the lines that marks 30°. Use red for one of the 50° lines, green for one of the 70° lines, and orange for the 90° line. These colored lines are the angles of incidence that you will use.
2. Set the *Protractor* page in front of you so that the straight edge of the protractor is at the bottom.
3. Wrap the rubber band around the center of the mirror. Have a partner stand the mirror along the straight edge of the protractor. The rubber band should touch the center point of the protractor. Hold the mirror perpendicular to the page. Be sure the mirror does not move.
4. Darken your work area.
5. Turn the flashlight on and shine the light straight down the red line to the rubber band. Look for the angle of reflection. If needed, hold the black paper around the curve of the protractor.

Materials
- *Protractor* page
- centimeter ruler
- colored pencils (red, blue, green, and orange)
- rectangular mirror with a flat surface
- rubber band
- small flashlight
- black paper or folder, optional
- Activity Manual

252

6. With the red colored pencil, circle the degree on the protractor where you see the angle of reflection. Record the number in your Activity Manual.
7. Repeat steps 5–6 for the blue, green, and orange lines. Make sure that your circles match the colors of the lines.

Conclusions
- How does the angle of incidence compare with its angle of reflection?
- What was different about the angle of reflection for 90°?

Follow-up
- Use the mirror and the light to reflect light to specific places around the room.
- Use the mirror and the light to send coded messages to a friend.

253

Procedure
- If needed, use a regular protractor to review measuring angles before beginning the activity. Guide each student as he traces the lines with the appropriate colored pencils. He may find it easier to have all the traced lines on one side of the protractor.
- Provide assistance to any student having difficulty measuring the angle of reflection.

If the student's measurements are not equal to the angle of incidence, check the position of the mirror. The student may need the optional black paper if the room cannot be thoroughly darkened. Demonstrate how to position the mirror on the *Protractor* page.

Conclusions
What did you notice about each angle of incidence and its angle of reflection? Both angles had the same measurement.

What was different about the angle of reflection for 90°? It was also 90°.

What generalization, or basic rule, could you make about the way light reflects off a smooth surface? Possible answers: The angle at which light hits a smooth surface is equal to the angle at which the light reflects from that surface. The angle of incidence is equal to the angle of reflection.

Activity Manual
Activity, page 166

Assessment
Rubrics
Select the prepared rubric, or design a rubric to include your chosen criteria.

Activity
Using God's Word as a Mirror
Materials: Instructional Aids 10.2–10.3: *Look into the Mirror*, for each student; *Look into the Mirror Answer Key*

Distribute the *Look into the Mirror* page to each student. The page discusses how God's Word can be a mirror that shows both our sinful condition and what Christ is like.

147
148

Student Text pages 254–58
Activity Manual pages 167–68

Objectives
- Identify characteristics of waves found in the electromagnetic spectrum
- Name some uses for each type of electromagnetic wave

Vocabulary
- electromagnetic spectrum

Materials
- objects that represent some of the waves of the electromagnetic spectrum (e.g., a radio, baby monitor, a cell phone, a television remote control, a light bulb, a glow-in-the-dark object, an x-ray or picture of an x-ray, a picture of the Swift observatory or of a gamma-ray telescope), laser pointer, electric space heater

Teacher's Toolkit CD
- Quiz 10-C

Two days have been allotted for this lesson. You may want to teach Lesson 147 the first day and use ideas from Lesson 148 on the second day. You may also incorporate the ideas from Lesson 148 into the text material, making the lesson two days.

Introduction
- Display the objects.
  All of these items have something in common. What is it? Answers will vary.
- Hold up the light bulb.
  This light bulb represents visible light. What kind of wave is light? electromagnetic
  All of these items use different types of electromagnetic waves. Almost all of these waves are invisible to our eyes. Yet they are always around us, and we use many of them.

Preparation for Reading
- Display the words *infrared* and *irradiation*. Pronounce the words together.
- Tell the students to mentally make questions using the headings, bold and italicized words, and captions.
- Instruct them to read silently pages 254–58 to find the answers to these questions.
  Which electromagnetic waves are used for radar?
  Which electromagnetic waves can cause something to glow?
  Which electromagnetic waves are the shortest?

Teach for Understanding
What is unique about light? It is the only electromagnetic wave that is visible to the human eye.

Where do electromagnetic waves come from? the sun and other sources

Electromagnetic Spectrum

Even though we can see light waves, there are many kinds of waves that we cannot see. Like light, these waves are electromagnetic waves, but light is the only type that is visible to the human eye. Electromagnetic waves come from the sun and other sources.

All electromagnetic waves travel at the same speed, but they have different wavelengths and frequencies. Electromagnetic waves can be arranged in order of their wavelengths. This arrangement is called the **electromagnetic spectrum**.

Light waves have a specific wavelength and frequency. Other electromagnetic waves have shorter wavelengths and higher frequencies or longer wavelengths and lower frequencies.

Electromagnetic Spectrum

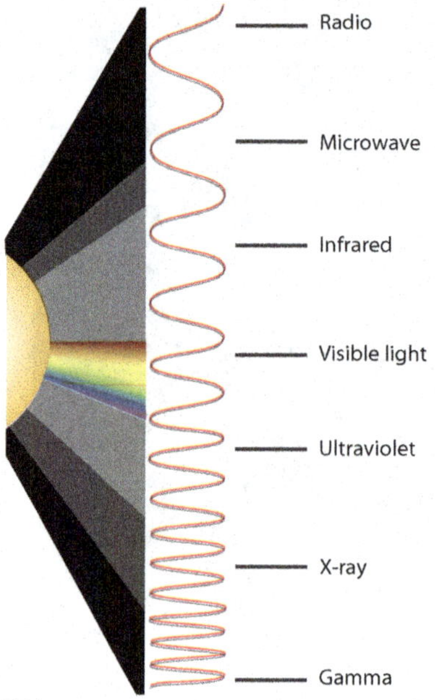

Longer Wavelengths

Radio waves are the longest kind of electromagnetic wave. Radio waves carry energy that is used for radio and television broadcasting. Sound information is added to radio waves by changing either the frequency or the amplitude of the wave.

Antennas pick up radio waves from the air and send them through wires to the radio. The radio then changes the radio waves into sound.

Radio waves are not used just for radios, though. Garage-door openers and baby monitors also use radio waves. Even wildlife-tracking collars use them. The waves help track the location of wild animals.

A remote car key uses radio waves.

254

Background

Radar and Lidar
Radar guns use microwaves to check the speed of vehicles. But lidar, also called laser radar, uses a thin beam of infrared laser light. Laser speed guns measure the time that it takes for light waves to hit a car and reflect back. Laser speed guns are very accurate, but they must be aimed at a car. Fog, rain, and snow may reduce the accuracy of a laser speed gun but do not usually affect radar.

Astronomer Sir William Herschel
Sir William Herschel (1738–1822), a musician and astronomer, was born in Germany but moved to England at age 19. With Caroline, his sister, he built telescopes to look at the night sky. He discovered the planet Uranus in 1781.

Herschel also discovered infrared light in 1800. In his experiment, he directed sunlight through a prism so that the light bent into its different colors. He measured the temperature of each color and realized that they each had a different temperature. Then he measured the temperature in the region beyond red and discovered that it had the highest temperature of all. At first he called this region *calorific rays*, but it was later called *infrared*.

Microwave towers are used by cellular phones.

Microwaves have a shorter wavelength than radio waves do. Microwave ovens use microwaves to cook and reheat food. Cellular phones also use microwaves.

Some radars use microwaves to detect the speed and location of objects. A police officer "shoots" microwaves from a radar gun. When a car passes, the waves reflect off it and return to the radar gun. The rate at which the waves return helps the police officer know how fast the car was going. Radar is also used to direct airplanes, to locate ships at sea, and to track weather systems.

In the electromagnetic spectrum, *infrared waves* are between microwaves and visible light. The heat that you feel from the sun or a fire is from infrared waves. Infrared rays transfer thermal energy better than any other kind of electromagnetic wave. We call this kind of heat transfer radiant heat.

All objects give off at least some infrared waves. Warmer things, such as people or animals, give off more infrared waves than cooler things, such as buildings. Our eyes cannot see infrared waves, but some equipment can detect them. Police officers can use this equipment to track criminals at night.

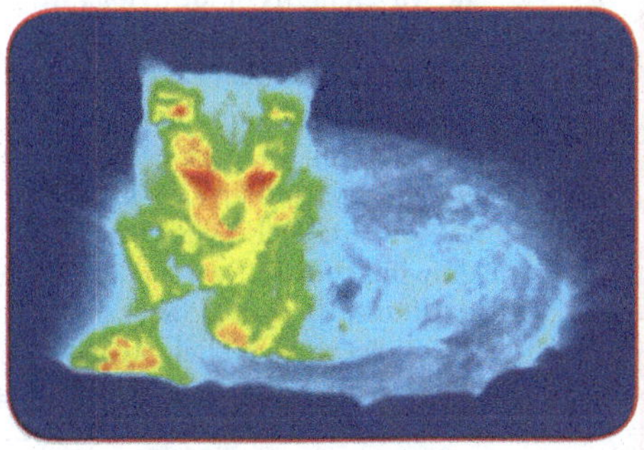
An infrared image of a cat

You also use infrared waves each time you use a remote control. You do not see the infrared waves that go between the remote control and the television, but they are there. Each time you press a button, the remote sends an infrared wave that the television picks up.

Some binoculars and cameras can change infrared waves into light that you can see. Satellites and space telescopes may use infrared cameras. Pictures from these cameras help scientists see things even when there is no visible light.

255

How are all electromagnetic waves alike? They travel at the same speed.

What are the electromagnetic waves called when they are arranged in order of wavelength? the electromagnetic spectrum

Which properties vary for each type of electromagnetic wave? wavelength and frequency

- Discuss the *Electromagnetic Spectrum* diagram on page 254.

What is a wavelength? the distance between two crests or two troughs

How do the wavelengths differ from the top of the diagram to the bottom? The wavelengths get shorter, or closer together.

What is frequency? the number of waves that pass a point in one second

How does the frequency of the waves differ from the top of the diagram to the bottom? There are more waves in the same amount of space. This shows that the frequency is increasing.

Do waves with long wavelengths have high frequencies or low frequencies? low frequencies

Which kind of electromagnetic wave is the longest? a radio wave

What are some uses for radio waves? Possible answers: radio and television broadcasting, garage-door openers, baby monitors, wildlife tracking collars, remote car keys

- Explain that sound and other information is added to radio waves by changing either the frequency or the amplitude of the wave. FM (frequency modulation) radio stations change the frequency of the wave. AM (amplitude modulation) radio stations change the amplitude.

What are some uses of microwaves? Possible answers: microwave ovens, cellular phones, radar

What are some uses of radar? to check the speed of cars, to direct airplanes, to locate ships at sea, to track weather systems

Where on the electromagnetic spectrum are infrared waves located? between microwaves and visible light

Which gives off more infrared waves, a warm object or a cool object? a warm object

How can scientists see things when there is no visible light? by using an infrared camera

- Discuss the infrared photo.

Why can you see the cat in this picture? because of the heat the cat's body produces

- You may want to do the *Find the Temperature of Infrared Rays* activity at this time.

Activity

Find the Temperature of Infrared Rays

Materials: 4 thermometers, black marker or black paint, cardboard box, white paper, tape, scissors, glass prism

Explain to the students that they are repeating William Herschel's experiment. Use the marker or paint to blacken the bulb of each thermometer. Cover the bottom of the box with white paper and tape the paper in place. Cut a notch on the side of the box for the prism. The prism should fit snugly.

Take the box and thermometers to a sunny place outside. Turn the box so that the prism faces the sun. Place the thermometers in a shady part inside the box. Wait five minutes and record the temperature of each thermometer.

Adjust the prism and box so that bands of colored light from the prism shine on the white paper. Place one thermometer in the blue light and one in the yellow light. Place the third thermometer in the blank area just beyond the red light. Leave the fourth thermometer in the shady area.

Check and record the temperature of each thermometer every minute for five minutes. Compare the temperatures.

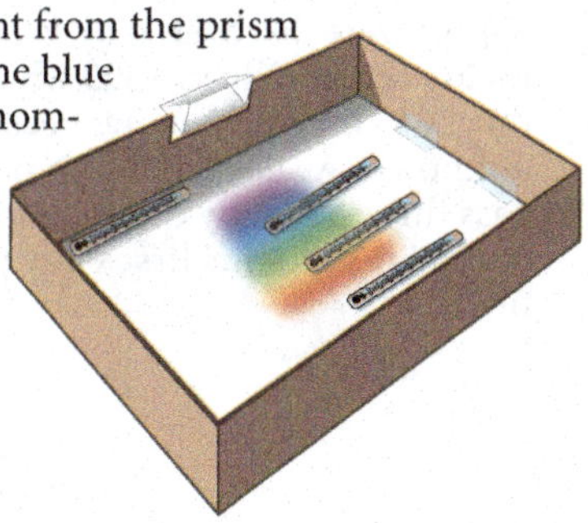

147
148

Which electromagnetic waves have wavelengths that are shorter than light? ultraviolet rays, x-rays, and gamma rays

Which type of electromagnetic wave is just a little shorter than violet light? ultraviolet

What are some uses of ultraviolet rays? Possible answers: making some items glow in the dark, detecting bacteria, killing some bacteria on food or hospital equipment

How does ultraviolet light help your skin? It helps your skin produce Vitamin D.

What may happen to a person's skin if it is exposed to too much ultraviolet light? Possible answers: wrinkles, sunburn, skin damage, skin cancer

What is one way to protect against getting too much ultraviolet light? using sunscreen with a high SPF

What does *SPF* stand for? sun protection factor

Why can people get sunburned on overcast days? Possible answer: Even though clouds filter out the visible rays of sunlight, ultraviolet rays still pass through.

- Display the prefixes *infra-* and *ultra-*. Explain that both come from Latin. *Infra-* means "below," so *infrared* refers to the wavelengths of radiation below red light. *Ultra-* means "beyond," so *ultraviolet* refers to the wavelengths of radiation beyond blue light.
- Discuss *Fantastic Facts.*

Is ultraviolet light invisible to every living organism? no

What group of animals is often able to see ultraviolet light? insects

How do honeybees use ultraviolet light? to find nectar in flowers

- Discuss the photo showing money under ultraviolet light.

What can be seen under a UV light on authentic currency? white strip

Do you think you can see this under normal light? no

Why do you think this is there? Possible answer: It helps people identify counterfeit money.

Shorter Wavelengths

Other waves in the electromagnetic spectrum have wavelengths that are shorter than that of light. Because the waves are shorter, their frequencies are higher. These waves are also invisible to humans.

The wavelengths of *ultraviolet rays* are just a little shorter than those of violet light. Ultraviolet rays cause some materials to glow. Some things, such as glow-in-the-dark balls, can absorb the rays and give off visible light. In normal light, we do not see the glowing colors of the ball, but in the darkness, the ball glows.

Ultraviolet rays can help keep people healthy. Some kinds of bacteria are harmful if found in water or food. Ultraviolet light can detect some kinds of bacteria, so it is used to test food areas. The rays can even kill some kinds of bacteria. Some hospitals use the rays to sterilize their equipment.

FANTASTIC FACTS

Although humans cannot see ultraviolet light, many insects can. In 1910 Karl von Frisch, an Austrian scientist, tested the eyesight of bees. He found that honeybees see fewer colors of visible light than humans do. But, unlike humans, the bees can see ultraviolet light. The ultraviolet light helps them find nectar. Flowers reflect ultraviolet light in certain patterns. These patterns act like a map and lead the bee to the nectar.

Small amounts of ultraviolet rays help your skin cells produce Vitamin D. But too much of these rays can cause wrinkles, sunburn, skin damage, or skin cancer. You should always be careful when outside. Most of these waves are absorbed by the ozone layer of the atmosphere. However, rays from the sun can reach the earth even on cloudy days. Using a sunscreen lotion with a high sun protection factor (SPF) can protect you from getting too much ultraviolet light.

Under UV light a white strip appears on authentic currency.

Background

Scientist Johann Ritter and Ultraviolet Light

After the discovery of infrared waves was announced, Johann Ritter (1776–1810) hypothesized that invisible light must exist beyond the blue part of the spectrum. He conducted experiments with silver chloride, a chemical that turns black when exposed to light. Ritter found that it darkened when near the violet end of the spectrum. When he tested the silver chloride beyond violet, the reaction was more intense than it had been in any of the colors tested. He proved that there is a type of invisible light beyond violet. He called the invisible light *chemical rays*, but this radiation later became known as *ultraviolet light*.

Scientist Wilhelm Conrad Röntgen and X-rays

X-rays were discovered by Wilhelm Conrad Röntgen (1845–1923). He discovered them accidentally in 1895 while he was working on another experiment. He called this new discovery *x-rays* because *X* is the mathematical symbol for something unknown. As he studied his new discovery, Röntgen realized x-rays could not only pass through many different kinds of substances but could also produce an image on film. One of his first x-ray photographs was of his wife's hand with a ring on her finger.

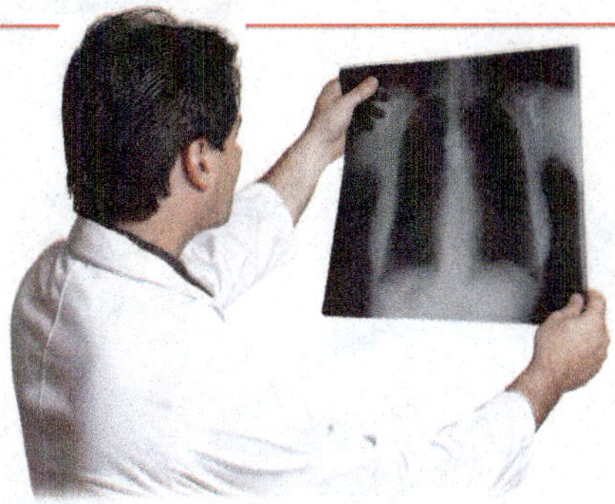

This doctor is examining an x-ray of a person's lungs.

X-rays are a type of ultraviolet light. They carry a lot of energy and can pass through a variety of materials. X-rays occur naturally. Many things in space, such as stars and comets, give off x-rays. X-rays can also be produced by machines. Marie Curie took some of the first x-ray machines to the front lines to help save lives.

Doctors use x-rays to see the bones inside someone's body. X-rays can pass through skin and muscles but are absorbed by bones. The places where the x-rays are absorbed the most show up as bright spots on the x-ray film.

Too much exposure to x-rays can cause cancer. That is why dentists and doctors use special coverings to protect the parts of the body that are not being x-rayed. Early x-ray scientists as well as those who used x-rays did not know of these dangers. In fact, Marie Curie eventually died from overexposure to x-rays.

For a long time, x-rays have been used on pieces of art. Historians can use x-rays to see whether a painting was painted on top of an earlier one. Sometimes an old canvas was reused in order to save money. At other times, the painting on top might be a forgery, a painting made to look as though someone else painted it.

The x-ray of the painting shows details of another painting under the finished work.

X-rays can also be used in industry and engineering. Some factories use x-rays to inspect packaged goods as they move along production lines. Engineers can x-ray structures made of steel or concrete. The x-rays show if there are any tiny cracks. Airports use x-rays for security and to check baggage.

257

What are some sources of x-rays? Possible answer: many things in space, such as stars and comets

What are some uses for x-rays? medicine, art, industry, engineering

How do doctors use x-rays? to look inside someone's body

- Refer the students to the picture of the x-ray and to the miniature scientist on Student Text page 237.

What could happen if a person is exposed to too many x-rays? The person could get cancer.

What do dentists and doctors use to help protect areas of the body not being x-rayed? special coverings

How are x-rays used in art? to determine if a painting has been retouched or painted on top of another painting

How do engineers sometimes use x-rays? to find out if a steel or concrete structure has any tiny cracks

How do airports use x-rays? for security and to check baggage

- Discuss the painting and its x-ray.

What part of the original painting is being shown in the x-ray picture? the faces on the right side

Look at the old man by the pillar. What is the difference between the original painting and the x-ray picture? He looks more like a soldier with a helmet in the x-ray picture.

Why might the painting look different? Answers will vary.

Activity

Effectiveness of Different Sunscreens

Materials: ultraviolet-detecting beads, sunscreens of different SPFs

The ultraviolet-detecting beads remain white when there is no ultraviolet light. They turn colors when they are exposed to sunlight or other sources of ultraviolet light. These are not thermo-sensitive beads, which change color with temperature changes.

For each sunscreen, coat several beads. Leave some beads uncoated. Expose the beads to the sun. Beads that do not change color are protected from ultraviolet light.

You may also use the beads to test the effectiveness of sunglasses or to see if fluorescent lights or other lights give off ultraviolet rays.

147
148

What are the shortest electromagnetic waves? gamma rays

Are gamma rays harmful or helpful? both

What are some ways that gamma rays are helpful? Possible answers: to treat some kinds of cancer, to kill bacteria in some foods, to sterilize bandages and other medical equipment

What has God given us to learn about how electromagnetic waves work? the command to do so and the minds to do so

What should we remember as we study God's creation? We should remember God's wisdom and power. He is the One who has created all things.

How can Christians use their knowledge about the electromagnetic spectrum? to benefit people and to glorify God for His infinite wisdom and creativity

ANSWERS

1. light
2. radio waves; gamma rays
3. Possible answers: art, engineering, medicine

148

- The *Find the Temperature of Infrared Rays* and *Effectiveness of Different Sunscreens* activities found at the bottom of the pages in Lessons 147–48 may also be incorporated into the lesson.
- Suggestions for additional activities for this two-day lesson are on the Teacher's Toolkit CD.

Activity Manual

Review, pages 167–68

These pages review Lessons 144–45 and 147–48.

Assessment

Quiz 10-C

The quiz may be given any time after completion of this lesson.

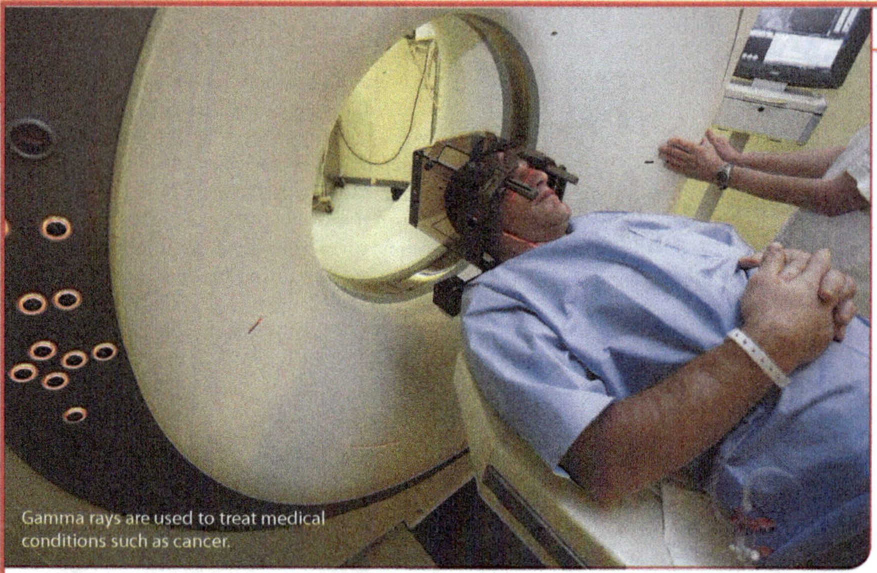
Gamma rays are used to treat medical conditions such as cancer.

The shortest waves in the electromagnetic spectrum are called *gamma rays*. The sun and other stars produce gamma rays.

Gamma rays are very powerful and can cause serious illnesses. When used in a controlled situation, though, gamma rays can be very helpful. They can treat some kinds of cancer. The rays can also be used to sterilize bandages and other medical equipment in a process called irradiation. Irradiation can help some foods stay fresh longer if gamma rays are used to kill bacteria in the foods.

God has given people the command to learn about electromagnetic energy as well as the minds to do so. Scientists and engineers have gained much knowledge about the electromagnetic spectrum. They have learned to use nearly the entire spectrum to benefit people. As we learn how people, plants, and animals use these energies of light, we can glorify God for His infinite wisdom and creativity.

QUICK CHECK

1. What part of the electromagnetic spectrum can we see?
2. Which kind of electromagnetic wave is the longest? Which is the shortest?
3. What are some ways in which x-rays are used?

258

Background

Gamma Rays Discovered

French physicist Antoine Henri Becquerel discovered gamma rays in 1896, shortly after Wilhelm Röntgen discovered x-rays. Becquerel also discovered that uranium emitted gamma rays.

Gamma Ray Uses

Gamma rays can be used to measure and control the flow of liquid in industrial situations, to measure soil density at construction sites, and to gauge the thickness of metal in steel mills. They can also be used to help law enforcement officers and museums identify the ownership of objects. The date can be added to artwork, fossils, or other objects using a special stamp called a gamma watermark. This watermark, developed in 1999 by the Lawrence Livermore National Laboratory in California, is visible only to a gamma-ray detector.

A lesson from
Answers in Genesis Name ______________________ 1:1

First Light

How would you answer the question, "How could there have been light on the earth for three days if the sun was not created until the fourth day of Creation?"

If you are a Christian and trust in the Bible, some people may think you are a little crazy for trusting in a very old book. Others will be interested in hearing why you trust the Bible. When Christians answer questions about what they believe and point people to the hope they have in Jesus, they are practicing *apologetics*. That does not mean that they are saying they are sorry for what they believe. *Apologetics* is a word that means defending the faith. Read 1 Peter 3:15–16, and you will see where the idea of defending what you believe comes from.

Returning to that question about the sun, you should look for an answer in a reliable source—the Bible.

Turn to Genesis 1 and answer the questions.

1. The first five verses in Genesis 1 talk about what happened on the first day of Creation. What is the first thing God created? *heaven and earth*
2. What did God create in verse 3? *light*
3. When light was created, what cycle began? *day and night*
4. Genesis 1:6–8 describe the second day. Was there light on the second day? *yes*
5. Genesis 1:9–13 describe the third day. Was there light on the third day? *yes*
6. In Genesis 1:14–19 what things does God create on the fourth day (verse 16)? *a greater light, a lesser light, and stars*
7. What do we call these lights? *the sun and moon*
8. Based on what God tells us in the Bible, was there light on the earth before the sun? *yes*

While the Bible does not tell us what the light was, there must have been a light source or there could not have been light and darkness and a morning and evening.

Science 5 Chapter 10 | Lesson 149 Answers in Genesis 169

Resources

The BJU Press website offers additional links from Answers in Genesis that you may find helpful with this lesson.
bjupress.com/resources/science

Background

Apologetics

While younger students may not be expected to be standing on a stage in an apologetics debate, it is important to lay a foundation of looking to the Bible for answers when they have questions or when questions come to them from others. As the entire body of Christ works together, answers in different areas of expertise can be provided. We can encourage one another in the Christian faith this way.

One common claim against the historical factuality of Genesis 1 is the light problem. What was the light source of day one if the sun was not created until day four of the Creation week? While God does not reveal what the source of the light was, He clearly indicates through Moses that there was light before the sun. Help students realize that even though God does not answer every question, His Word is sufficient to answer those things He sees as necessary. A Christian must trust Him in those things and be grateful for what He has given (Deuteronomy 29:29).

Objectives

- Contrast the naturalistic view of the sun's origin with the biblical view
- Recognize that the Bible calls Christians to defend their faith

Introduction

What questions about the Bible have your friends asked you to explain? Answers will vary.

Preparation for Reading

- Direct the students to read silently the paragraphs at the top of Activity Manual page 169 to find out what the term *apologetics* means.

Teach for Understanding

What is apologetics? defending the faith

- Ask a volunteer to read 1 Peter 3:15–16.

  What is a Christian told to do when someone asks him to give an account or reason for the hope that he has? A Christian is to give a defense or explain the hope in him.

  How is a Christian to respond or give his defense? with gentleness and respect/meekness and fear

  Let's see what the Bible says about the question, "How could there have been light on the earth for three days if the sun was not created until the fourth day of Creation?"

- Ask a volunteer to read Genesis 1:1–5. Discuss the first three questions.

  What is the first thing God created? the heavens and the earth

  This probably means that God created space and the earth, which at this time was a water-covered sphere.

  What did God create in verse 3? light

  At the creation of light, what cycle began? day and night

  In verse 5 and in later verses, you will read "evening and the morning," which means the same as night and day.

- Allow time for the students to complete the first three questions. Ask a volunteer to read Genesis 1:6–8.

  How do we know there was light on the second day? Elicit since there was a morning and an evening, there must have been light.

- Ask a volunteer to read Genesis 1:9–13.

  Was there light on the third day? yes

  How do we know? Since there was a morning and an evening, there must have been light.

- Ask a volunteer to read Genesis 1:14–19.

  What did God create on the fourth day? a greater light, a lesser light, and stars

  What do we call the greater light that rules the day? the sun

149

What do we call the lesser light that rules the night? the moon

Based on what God tells us in the Bible, was there light on the earth before the sun? yes

- Direct a student to read aloud the last sentence on page 169. Allow time for the students to complete the remainder of the questions on the page.
- Instruct the students to read silently the first paragraph on Activity Manual page 170 to discover another question that is often asked of Christians.

  How do many people believe the universe began? in an event known as the big bang
- Ask a volunteer to read aloud the first statement that the big bang theory teaches. Allow time after the discussion of each statement for the students to complete the Biblical Creation statement in the chart.

  When does the Bible say that the earth was created? on the first day

  When does the Bible say that stars were created? on the fourth day

  How long after the earth was created were the stars created? three days
- Ask a volunteer to read aloud the second statement that the big bang theory teaches.

  When does the Bible say that the sun was created? on the fourth day

  How long after the earth was created was the sun created? three days
- Ask a volunteer to read aloud the third statement that the big bang theory teaches.

  Was the earth created before or after the sun was created? before How do you know? The Bible tells us that the earth was created on the first day and the sun was created on the fourth day.
- Explain that the big bang has a different order of events. It does not line up with what the Bible says. This conflict has led many Christians to reject the text of Genesis 1 as historically accurate. It has been called poetry, fable, or myth, making the big bang theory seem more credible. This view places man in authority over God's Word rather than allowing God's Word to be the authority over our thoughts about His creation.
- Instruct the students to read the last two paragraphs on Activity Manual page 170.
- Read 1 Peter 3:15–16 aloud again.

  To what should a Christian's answers point? the hope that Christians have in Christ
- Explain that a Christian's hope is not in the accuracy of scientific models of the age of the earth or the mechanisms of explaining the Flood. A Christian's hope is in the person and work of Jesus Christ. If a Christian's answers never point to Christ, he is not giving a biblical apologetic but is defending his own thinking.

Activity Manual

Answers in Genesis, pages 169–70

Now that we know God created the sun after the earth, we also have an answer to another question. Many people believe that the universe began in an event known as the "big bang." Is the big bang a true explanation of how our universe began? We can compare it to the Bible to find out. Complete the table below by writing a statement that compares what the Bible teaches to what the big bang theory teaches.

Big Bang	Biblical Creation
Stars formed before the earth.	*The stars were created three days after the earth was.*
The sun gave light before the earth formed.	*The sun was created three days after the earth was.*
The earth formed after the sun.	*The sun was created three days after the earth was.*

9. Does the idea of a "big bang" fit with what the Bible says? *no*

When people have questions about the Bible, the best thing to do is open the Bible and read what God has said. Now, it might take you some time to learn how to answer questions, but your teachers, a parent, or a pastor can help you. That is one reason God put them in your life.

But all these answers mean nothing apart from the hope Christians have in Jesus. Read 1 Peter 3:15–16 again. Notice the first thing it says is that the Lord God should rule in your heart. It also says that your answers to questions should point to the hope that is in you. What is that hope? "Christ in you, the hope of glory" (Colossians 1:27).

The Bible is the only reliable source of truth concerning what happened during the week when God created the universe.

Light at Work

EXPLORATION

We know that we need light to see. But many kinds of technology also use light. Some types of telephone cables use light to carry communications from one telephone to another. The lasers in CD players use light to play music. Cameras use light to capture images. Reflected light is used to measure far distances, such as that from the earth to the moon.

In this exploration you will search for information about products that use light. The products may include things such as photocopiers and telescopes. Photocopiers use light to copy documents. Telescopes gather and focus light. You will need to search carefully. Sometimes it will not be obvious that a product uses light.

What to Do

1. Look through advertisements, magazines, and newspapers. Cut out at least twelve examples of different products that use light for a specific purpose. Include any product information given.
2. Make a collage or notebook of your examples. With each example, include a brief description of how the product uses light.
3. Present your collection.

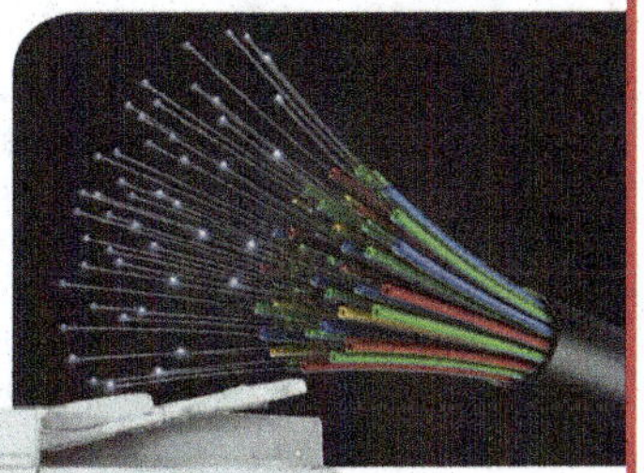
Optical fiber cables

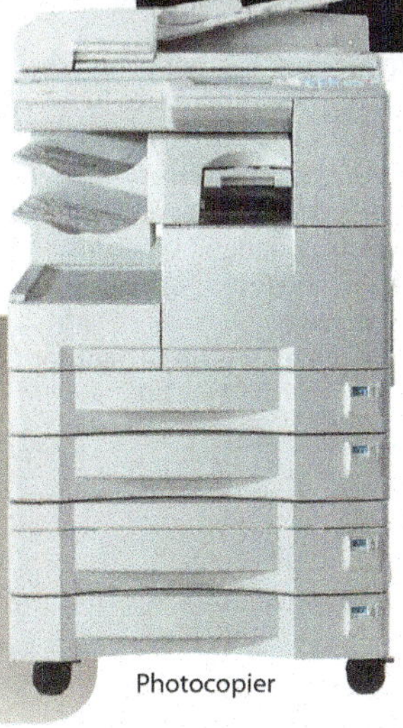
Photocopier

Telescope

Objectives

- Identify different ways that light is used in technology
- Make a collage that explains how different products use light

Materials

- ads, magazines, and newspapers

Teacher's Toolkit CD

- Rubrics: *Light at Work*, for each student

Introduction

What are some machines or technologies that you use frequently? Possible answers: telephone, hair dryer, radio, CD player, electronic game, computer

Many machines and products use light or other electromagnetic waves in order to work.

Preparation for Reading

- Instruct the students to read silently Student Text page 259 before beginning the exploration.

Teach for Understanding

What to Do

- Provide ads, magazines, and newspapers. Guide the students in finding and collecting their information.
- Products that use light, such as optical fiber cables and lasers, are often used in many different ways. For example, fiber-optic technology is used with telephone lines, cable television, Christmas trees, toys, signs, and lighting. Optical fibers are also used in many medical and industrial tools, as well as military equipment.

Activity Manual

Expansion, page 171

This page uses a diagram to explain how a single-lens reflex camera works.

Assessment

Rubrics

Select the prepared rubric, or design a rubric to include your chosen criteria.

151

Student Text page 260
Activity Manual page 172

Objectives
- Recall concepts and terms from Chapter 10
- Apply knowledge to everyday situations

Review
- Materials for the Chapter 10 Test will be taken from Student Text page 260 and Activity Manual pages 161, 163, 167–68, and 172. You may review any or all of the material during the lesson. Questions similar to the ones in Thinking It Through, Activity Manual page 172, may appear on the test.
- You may choose to review Chapter 10 by playing the review game or a game for the Game Bank on the Teacher's Toolkit CD.

Activity Manual
Review, page 172
This page requires written responses to application questions.

152

Objective
- Demonstrate knowledge of concepts taught in Chapter 10

Assessment
Chapter 10 Test

CHAPTER SUMMARY

Words to Know
light
transverse waves
wavelength
amplitude
frequency
speed
visible spectrum
refraction
reflection
primary colors
pigment
mirror
concave
convex
lens
electromagnetic spectrum

Key Ideas
Compare and contrast electromagnetic waves with mechanical waves
Describe the properties of waves
Classify light and sound waves as electromagnetic or mechanical waves
Describe the types of materials that refract light
Summarize how reflected light determines the colors we see
Identify the primary colors of light and the primary colors of pigments
Describe the relationship between primary colors of light and primary colors of pigments
Explain why it is important that a mirror's surface be smooth and shiny
Describe the differences between mirrors with concave surfaces and convex surfaces
Identify uses of mirrors and lenses
Summarize the characteristics of the waves and rays of the electromagnetic spectrum
Identify technology that uses the waves or rays of the electromagnetic spectrum
Explain why it is important for Christians to study light

260

Review Game
In Spectrum Order
Materials: set of seven cards labeled red, orange, yellow, green, blue, indigo, and violet; seven plain, dark envelopes

Place each card in an envelope.

Ask review questions. Each time a student answers correctly, he chooses an envelope, but he does not look inside. After all of the envelopes are chosen, instruct the students with the envelopes to come to the front of the room. At a signal, each opens his envelope. The students then arrange themselves from left to right in spectrum order. Once they are finished, allow the rest of the class to decide if they are in the correct order. Instruct the students to replace the cards in the envelopes. Shuffle the envelopes and repeat as time allows.

Variation: You may also choose to use cards labeled with the waves of the electromagnetic spectrum: radio waves, microwaves, infrared waves, visible light, ultraviolet rays, x-rays, and gamma rays.

Teacher Notes

153

Student Text pages 261–63
Activity Manual page 173

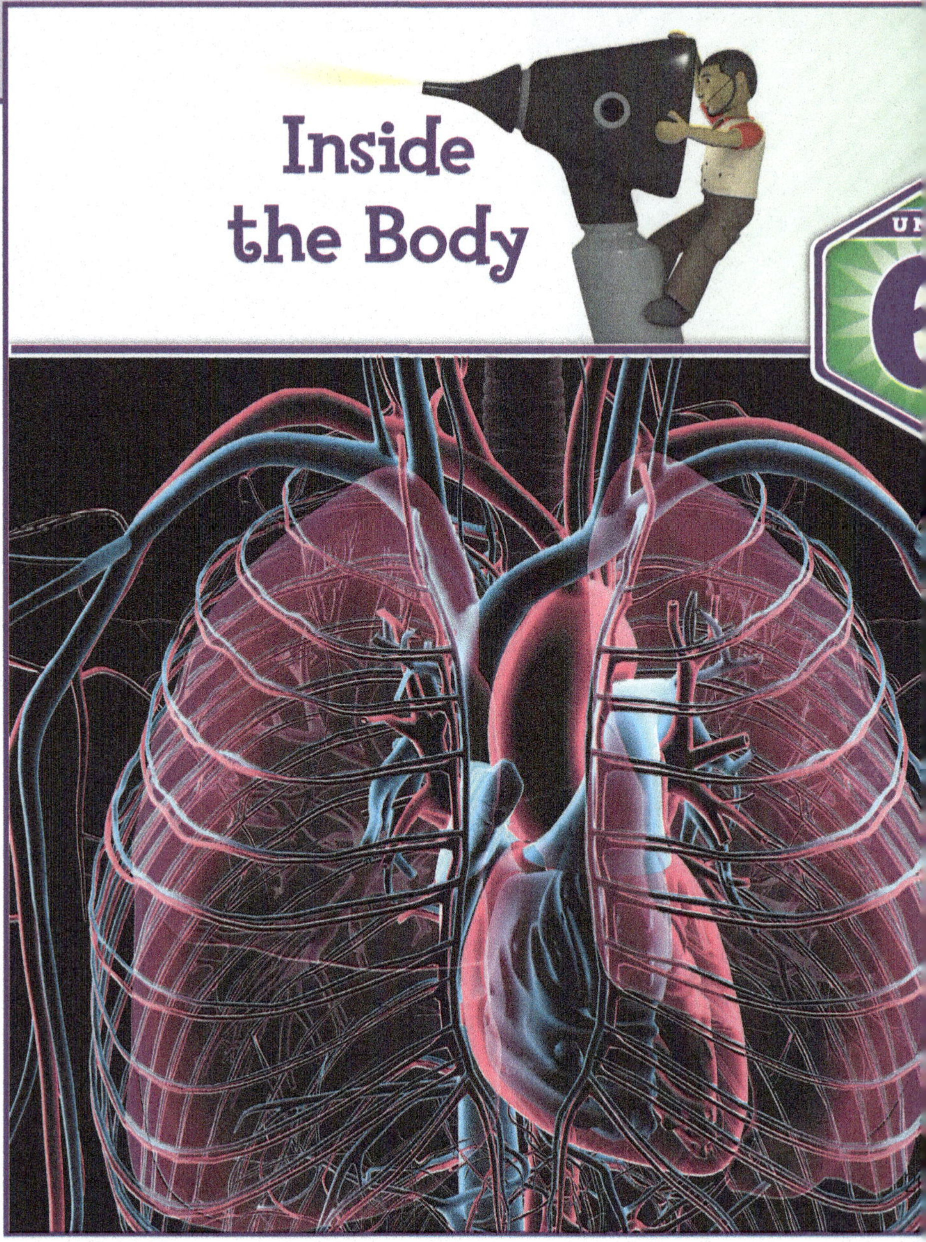

Objectives

- Contrast technology with the marvels found in the human body
- Demonstrate how people are being inspired by God's designs to develop new technology

Teacher's Toolkit CD

- Visual 11.1: *Cilia*
- Instructional Aids 11.1–11.2: *Survey Form*, for each student; *Survey Results*, for each student

Unit Introduction

- Unit 6 explores the two basic body systems that provide the essentials for human life. The Bible uses elements of these systems—breath and blood—to illustrate salvation and eternal life.
- The respiratory system discussed in Chapter 11 and the circulatory system discussed in Chapter 12 work together to provide oxygen and food to every cell in the body.

  Look at the picture on the unit page. What is it a picture of? lungs and heart

  Look closely at the x-ray picture and you can see the chambers of the heart. You can also see the bronchial tubes and the arteries and veins. You will be studying about these different parts in this unit.

  Do you know what the miniature scientist is looking through? an instrument to look inside the ear called an otoscope

- Explain that an otoscope has a light, a magnifying lens, and a funnel-shaped viewing piece with a narrow, pointed end. A pneumatic otoscope has a rubber bulb that the doctor can squeeze to send a puff of air into the ear canal. This allows him to see how the eardrum moves.

  Look through Unit 6. What kinds of topics do you think you will be studying in this unit? Possible answers: the respiratory system, the circulatory system

  Why do you think these chapters are organized into the same unit? Answers will vary. Elicit that the chapters are about the human body.

Background

Unit Photo

The photo on page 261 shows a view inside the body of the heart, lungs, bronchial tubes, arteries and veins, and circulatory system. The miniature scientist is looking into an otoscope, or auriscope, which a doctor may use to examine the ears or mouth.

Weblinks

The BJU Press website offers additional information and links you may find helpful throughout the unit.
bjupress.com/resources/science

- Ask volunteers to read the teasers on page 262. Discuss the information.

Project Idea

- The project idea presented at the beginning of each unit is designed to incorporate elements of both chapters as well as information gathered from other resources. You may choose to have the student do the project as a culminating activity at the end of the unit or as an ongoing activity while the chapters are being taught.

Unit 6: Adult Breathing and Pulse Survey

- Design a questionnaire that each student can use to survey adults that he knows. Include items such as the person's age, his occupation, and his estimated amount of physical activity per week. Also include the adult's pulse and breathing rates while sitting at rest.
- After the surveys are completed for at least five adults, each student displays his findings on graphs. The graphs can show relationships between age and breathing rates, between age and pulse rates, among occupations and breathing and pulse rates, between breathing rates and pulse rates, or between other combinations.
- The graphs may display data collected by individuals, groups, or the entire class.
- A sample survey questionnaire and graphing page are on the *Survey Form* and *Survey Results* pages. The graph shows the breathing and pulse rates for different ages. You may design additional graphs to compare other data from the survey.

This lesson continues with the Chapter 11 introduction on page 301.

Unit 6 Overview

Lesson	Teacher's Edition	Student Text	Activity Manual	Assessment	Content
Chapter 11: Respiratory System					
153	298–301	261–63	173		• Worldview thinking • Chapter preview
154	302–4	264–66	174–75	Quiz 11-A	• Involuntary and voluntary breathing • Breathing muscles
155	305	267	176	Rubric	**Activity: Breathe In, Breathe Out** • Process skill: making and using models
156–57	306–9	268–71	177–78	Quiz 11-B	• Parts of the respiratory system: nose, throat, and larynx
158–59	310–13	272–75	179–80	Quiz 11-C	• Parts of the respiratory system: trachea, bronchi, and lungs
160	314–15	276–77	181–82	Rubric	**Activity: How Much Air Is in Your Lungs?** • Process skill: collecting, recording, and using data
161	316–17		183–84		**Answers in Genesis** • Modeling the effects of crucifixion on breathing
162–63	318–21	278–81	185–87	Quiz 11-D	• Respiratory diseases • Respiratory disorders
164	322–23				• A biblical viewpoint of the dangers of smoking
165	324	282	188		**Chapter Review** • Apply knowledge to everyday situations
166	324			Chapter 11 Test	**Chapter 11 Test**

Lesson	Teacher's Edition	Student Text	Activity Manual	Assessment	Content
Chapter 12: Circulatory System					
167	325	283	189		• Worldview thinking • Chapter preview
168–69	326–29	284–87	190–92	Quiz 12-A	• Parts of the heart • How the heart works
170	330–31	288–89	193–94	Rubric	**Activity: How Fast Is the Beat?** • Process skill: measuring and using numbers
171–72	332–35	290–93	195	Quiz 12-B	• Blood vessels • William Harvey and the circulatory system
173–74	336–40	294–98	196	Quiz 12-C	• Blood contents • Blood types and blood donation
175	341	299	197–98	Rubric	**Activity: Exploring Blood Types** • Process skill: observing
176	342–43	300–301	199–200	Rubric	**Activity: Pump and Pour** • Process skill: making and using models
177	344		201–2		**Answers in Genesis** • Jesus as the Creator and Savior
178	346–49	302–5	203	Quiz 12-D	• How the body removes waste from the blood • Heart health
179	350	306	204		**Chapter Review** • Apply knowledge to everyday situations
180	350			Chapter 12 Test	**Chapter 12 Test**

Respiratory System 11

Let's take a trip inside your lungs. The tiny passageways there are lined with hairlike structures called cilia. The lining of your lungs makes a sticky mucus, which traps dust, bacteria, and other particles you breathe in. The cilia beat waves to sweep this mucus—and all it captures—out of your lungs to keep them clear and healthy.

Scientists today have learned to make artificial cilia they can control. But these cilia are not as complex as those God designed for your body. That's because no one knows how cilia are so coordinated since they have no nerves. Imagine a rowing team trying to coordinate without talking to each other! Artificial cilia have only four different parts and contain tiny magnetic particles. Scientists move them by changing magnetic fields. The actions of artificial cells are not nearly as precise or productive as the cilia in your body. As scientists and inventors look to the human body for design ideas, it draws our attention to God—the greatest designer of all.

Consider
- What is the purpose of cilia in the lungs?
- Why do you think scientists often look at the natural world to get ideas for designing things?

263

Background

Chapter Preview
Preview and prereading activities may include using a K-W-L chart, a probe, or an anticipation guide.

Chapter Photo
The photo on page 263 shows a young girl sitting against a tree, blowing a dandelion. The miniature scientist is also blowing a dandelion.

Activity

Inventions
Group the students in pairs and direct them to list things people have invented that seem to be copied from humans, animals, or plants. This activity is to encourage creative thinking by having the students come up with many varied examples.

Examples: wings of aircraft are modeled after bird wings; shapes of boats and submarines are streamlined in the shapes of fish bodies; velcro fasteners are copied from certain plant seed pods that stick to fur or clothing; the parts of a camera work together in many ways like the human eye; swim fins have webbing to move more water just like the webbed feet of ducks; windshield wipers clean a car windshield like your eyelids clean the surface of your eye.

Chapter Introduction
- Invite students to describe a robot they have seen.
- Explain that most robots are designed and engineered to perform only a single task. They may look nothing like us. They may function in the way an arm or even a single finger does. In fact, some are so tiny they could not be seen without magnification. Scientists who make robots get their inspiration and ideas from the designs that God used in nature and even inside our own bodies.

Preparation for Reading
- Instruct the students to read silently page 263 to find out what artificial cilia contain.

Teach for Understanding
- Display the *Cilia* visual. Tell the students to describe what they see.

  What is the value of having mucus in your lungs? The mucus is sticky and helps to trap unwanted material you breathe in.

  Why would having mucus but not cilia be a bad design? The foreign material would be trapped in the lungs, but then it would remain there and build up over time.

  Why would having cilia but not mucus be a bad design? The foreign material could stick to the cilia but not have a substance to be carried by.

  Does it matter which direction the cilia are moving? Yes, they must push the mucus and trapped material up out of the lungs, not deeper into them.

  Which cilia are more complex, the natural ones in your respiratory system or the ones scientists have built? The natural cilia in your lungs are much more complex in design and have many more parts.
- Guide the students to recognize that although we may never match the efficiency or greatness of God's designs, they can still inspire us with ideas for developing new technologies.

  Why do you think so many of the larger robots look like humans? If a robot is intended to perform tasks that are usually done by humans, it must have similar parts to accomplish the same purposes.

  Do you think a robot could be designed to exactly copy human movements or activities? no Why? Human actions are far too complex and involve the coordination of many more parts than a robot would have.

  Why do you think scientists often turn to the human body and God's other creations to get ideas for designing things? God made them not only beautiful, but also functional and efficient.

Activity Manual

Preview, page 173
The Looking Ahead page is intended to assess the student's prior knowledge before beginning the chapter.

Objectives

- Identify the respiratory system as the breathing system
- Differentiate between involuntary breathing and voluntary breathing
- Identify the muscles that help with breathing
- Describe the movement of the body and air when inhaling and exhaling

Materials

- balloon

Teacher's Toolkit CD

- Quiz 11-A

Vocabulary

- respiratory system
- inhale
- exhale
- involuntary breathing
- voluntary breathing
- diaphragm

Introduction

- Give a balloon to a student. Direct the student to blow up the balloon.

  What is inside the balloon? Possible answers: air, breath, gases, water vapor

  Where did the air inside the balloon come from? Possible answers: from inside the student, from the student's lungs

  In this lesson we will begin studying how air moves in and out of the body.

Preparation for Reading

- Display the words *respiratory* and *diaphragm*. Pronounce the words together.
- Tell the students to read silently pages 264–66 to find the answers to these questions.

  Do you exhale oxygen or carbon dioxide?

  What term describes breathing that you can control?

  Where is the diaphragm located in the body?

Teach for Understanding

What is the breathing system called? the respiratory system

How did each person get this breathing system? God created man and breathed life into him.

What gas is needed by every cell in the body? oxygen

What term means "to breathe in"? inhale

What term means "to breathe out"? exhale

What gas is released when a person exhales? carbon dioxide

Breathing

You need to breathe in order to stay alive. Most of the time, you do not think about breathing, though. If you did, you would not be able to do anything else. Imagine how hard it would be to talk, work, or play if you had to stop every few seconds and remind yourself to take a breath. Even when at rest, most fifth graders breathe between 15 and 25 times each minute.

In fact, you breathe in more than 14,000 L (3,698 gal) of air each day. During your life, you will breathe enough air to fill up millions of balloons.

Genesis 2:7 tells us that God created man and breathed into him the breath of life. With this breath, man became a living person. Breathing is a gift of God. The human **respiratory system**, or breathing system, is one of His amazing designs.

Why You Breathe

Every cell in your body needs oxygen in order to live and function. Without oxygen, you would not have any energy. You would not be able to think, digest food, pump blood, or move.

When you **inhale**, or breathe in, you take in air. Your blood takes the oxygen from the air to every cell in your body. The cells use the oxygen and produce carbon dioxide. Too much carbon dioxide in your body, though, can be harmful. So the carbon dioxide is sent out of your body when you **exhale**, or breathe out.

264

Background

Exchanging Gases

Inhaled air contains about 78% nitrogen, 21% oxygen, and 1% carbon dioxide and other gases. When air is exhaled, it is about 79% nitrogen, 16% oxygen, 4% carbon dioxide, and 1% other gases.

Human Respiratory Development

A baby's lungs and respiratory system start to form when the baby is only a few weeks into development. At this time, the baby is less than 2.5 cm (1 in.) long. By 28 weeks, the respiratory system can exchange gases. After 32 weeks, the lungs are able to function, but the baby cannot breathe air while in the womb. The umbilical cord, which is attached to the placenta, provides oxygen from the mother's body. At birth, the baby takes his first breath and begins using his own respiratory system.

Activity

Researching the Respiratory System

Guide the students in finding articles and pictures about current events dealing with the respiratory system. Article topics may include medical discoveries, new medications and surgical procedures, dietary guidelines, exercise benefits, and medical personnel in these fields. You may choose to expand the topics to include other health-related news items.

A person needs more oxygen when he exercises or participates in activities such as baseball.

How You Breathe

The brain controls how you breathe. Most of the time, you do not have to think about it. Your brain does it automatically. This automatic breathing is called **involuntary breathing**. You breathe involuntarily when you are unconscious, asleep, or simply not thinking about breathing.

God made us able to control our breaths at times, though. **Voluntary breathing** takes place when you think about breathing and control it. Your brain can tell your body to stop breathing for a short amount of time. Without this control, you would not be able to hold your breath when swimming. You also could not do things like play wind or brass instruments.

The brain also controls how fast and how deeply you breathe. When your brain registers that there is too much carbon dioxide, it sends a signal for you to breathe more deeply. For example, when you exercise, a signal is sent to the brain that more oxygen is needed by the cells. This causes you to breathe faster.

SCIENCE & HISTORY

History records examples of people free-diving, or diving while holding one's breath. In some cultures free-divers search for sponges or pearls. This is not always the case, though. One free-diver searched for something very interesting—an anchor!

In 1911 an Italian ship lost its anchor in a storm. So the ship's captain hired a Greek sponge diver, Yorgos Haggi Statti, to get the anchor back. Statti claimed that he could hold his breath for up to seven minutes. He tied a stone to himself and dove 77 m (252 ft) into the deep Aegean Sea to try to recover the anchor. On his third try, he ran a rope through the anchor's eye. The ship's crew was then able to pull up the anchor.

265

What part of the body controls how a person breathes? the brain

How is involuntary breathing different from voluntary breathing? Involuntary breathing occurs automatically, or without thinking. Voluntary breathing occurs when you think about breathing and control it for a time.

When the balloon was inflated at the beginning of the lesson, was involuntary breathing or voluntary breathing used? voluntary breathing

Name some other situations in which a person would use voluntary breathing. Possible answers: while swimming, while playing wind or brass musical instruments

What else does the brain control about breathing? how fast and how deeply a person breathes

Why does a person breathe faster when he exercises? because more oxygen is needed by the cells

- Discuss *Science & History*.

What kind of diving did Statti do? free-diving, or diving while holding his breath

What kind of breathing is used for free-diving? voluntary breathing

What needed to be recovered from the bottom of the Aegean Sea? a ship's anchor

What equipment did Statti use as he dove? only a rope and stone

How many dives did he make to retrieve the anchor? three

Science Misconceptions

Holding Your Breath

A person cannot die from holding his breath. If he becomes unconscious, involuntary breathing takes over.

154

- Refer to the illustration as you discuss the diaphragm and the chest muscles.
  Which muscles help a person breathe? the diaphragm and the chest muscles
  What is the diaphragm, and where in the body is it located? The diaphragm is a strong, curved muscle attached below the lungs to the lower ribs and backbone.
  What happens to the diaphragm as a person inhales? The diaphragm tightens and moves downward.
  What do the chest muscles do when the diaphragm tightens? The chest muscles lift the rib cage up and out of the way.
- Direct each student to place his hand in the area of his diaphragm. Ask him to inhale and exhale to feel the upward and inward movement of the rib cage.
  When the lungs expand inside the chest cavity, they have a larger volume. Has the air pressure inside the lungs increased or decreased? decreased
- Remind the students that air pressure is related to the density of the air. When the air particles are farther apart, as they are in the lungs when the chest expands, the air pressure decreases.
  When the diaphragm relaxes, does it push up or down? up
  When is there more space in the chest cavity, during inhaling or exhaling? during inhaling
  Why would the function of the diaphragm be important for a singer? The diaphragm helps control the amount of air in the lungs. A good singer is able to hit and hold notes by breathing correctly through use of his diaphragm.

ANSWERS

1. the respiratory system
2. oxygen
3. the diaphragm and the chest muscles

Activity Manual

Reinforcement, page 174

Review, page 175

This page reviews Lesson 154.

Assessment

Quiz 11-A

The quiz may be given any time after completion of this lesson.

Breathing Muscles

Two special sets of muscles work with your brain to help you breathe. These muscles, your diaphragm and your chest muscles, change the space and pressure inside your body to allow you to breathe. The **diaphragm** is a strong, curved muscle attached below the lungs to the lower ribs and backbone. It separates the chest from the abdomen.

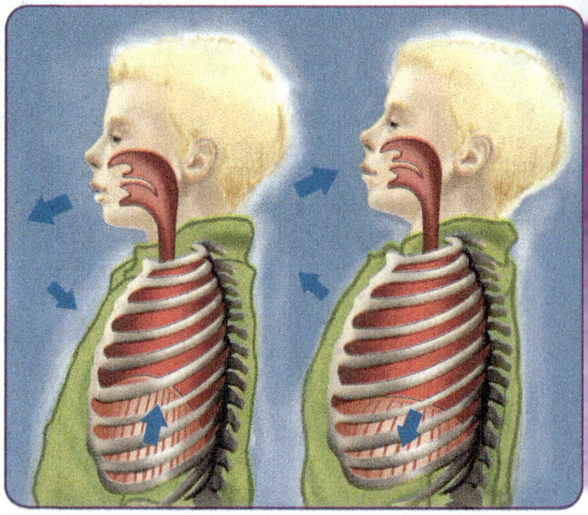

When a person inhales, the diaphragm moves downward and the rib cage moves upward. When a person exhales, the diaphragm pushes upward and the rib cage moves inward.

When you inhale, your diaphragm tightens and moves downward. At the same time, your chest muscles, which are located behind and between your ribs, lift your rib cage up and out of the way. This provides more space for your lungs. More space means the volume of your lungs expands. This, in turn, causes the air pressure inside them to decrease. The air pressure inside your lungs becomes less than the pressure outside your body. To equalize the pressure, air is sucked into your lungs.

When you exhale, the process is reversed. Your diaphragm and chest muscles relax. As your diaphragm relaxes, it pushes up. As your chest muscles relax, they push in. These actions decrease the amount of space in your chest and increase the air pressure. This forces the air in your lungs out of your body, releasing the carbon dioxide and other gases back into the air.

QUICK CHECK

1. What is your breathing system called?
2. Which gas does every cell in your body need?
3. Which muscles work together to help you breathe?

266

Background

Chest Muscles

The chest muscles are called the intercostals. These muscles are located between the ribs.

Science Misconceptions

Muscles and Lungs

The lungs are not muscles and do not have any muscles. The lungs inflate and deflate with the movement of the diaphragm and chest muscles. As the air pressure inside the lungs decreases, air is sucked in to inflate the lungs and equalize the air pressure.

You may choose to combine the activity *Breathe In, Breathe Out*, Lesson 155, with this lesson to help the students understand the breathing process.

ACTIVITY

Breathe In, Breathe Out

Take a deep breath of air. Could you feel your muscles move? In this activity you will make a model of part of your respiratory system.

Purpose

Make a model of part of the respiratory system.

Procedure

1. Slide the straw into the small balloon and secure them together with a rubber band.
2. Place the balloon and straw inside the bottle. The balloon should hang inside and most of the straw should stick out of the top.
3. Press clay around the straw to hold it in place and to seal the bottle top.
4. Cut the neck off a large balloon. Stretch the large piece of balloon over the cut end of the bottle.
5. Pull down gently on the bottom of the large balloon. Record your observations of the small balloon in your Activity Manual.

Process Skills
- Making and using models
- Inferring
- Defining operationally

Materials
drinking straw
small balloon
rubber band
20 oz clear plastic bottle, cut in half
clay
large balloon
Activity Manual

Conclusions

- What happened to the balloon inside the bottle?
- How is this model similar to your respiratory system?

Follow-up

- Use two small balloons inside the bottle to represent both lungs.

267

Objectives

- Make a model of a lung
- Use the lung model to explain how the diaphragm moves during breathing

Materials

Materials needed for this activity are listed on the Student Text page.

Teacher's Toolkit CD

- Rubrics: *Breathe In, Breathe Out*, for each student

Introduction

- Review the muscles involved in breathing and the meaning of *inhale* and *exhale* from the previous lesson.

Preparation for Reading

- Instruct the students to read silently Student Text page 267 and Activity Manual page 176 before beginning the activity.

Teach for Understanding

Procedure

- Guide the students in assembling the model.

Conclusions

What happens to the balloon inside the bottle? As the "diaphragm" moves down, the "lung" fills with air.

How is this model similar to the respiratory system? The bottle is the rib cage. The straw is the trachea. The balloon is the lung. The cut balloon is the diaphragm.

Process Skill

The process skill emphasized in this lesson is making and using models.

Why do scientists use models? Possible answers: to show things too large or too small to be seen easily; to show processes that occur inside the body

How was a model used in this activity? to demonstrate how air is sucked into the lungs

Is this an accurate model? no Why? Possible answers: You have two lungs, but this model shows only one. This model shows only the diaphragm moving, not the chest muscles.

Activity Manual

Activity, page 176

Assessment

Rubrics

Select the prepared rubric, or design a rubric to include your chosen criteria.

156–157

Student Text pages 268–71
Activity Manual pages 177–78

Objectives

- Explain how mucus and cilia help keep the respiratory system clean
- List the parts of the respiratory system from the nose to the larynx
- Describe the function of the epiglottis
- Explain how the vocal cords produce sound

Materials

- paper facemask/filter
- balloon (optional)

Teacher's Toolkit CD

- Visuals 11.1–11.2: *Cilia*; *Path of Air*
- Quiz 11-B

Vocabulary

- cilia
- pharynx
- trachea
- epiglottis
- larynx
- vocal cords

Two days have been allotted for this lesson. You may want to teach Lesson 156 the first day and use ideas from Lesson 157 on the second day. You may also incorporate the ideas from Lesson 157 into the text material, making the lesson two days.

Introduction

- Display the facemask.

  Have you ever used one of these or seen someone use one? Answers will vary.

  People that have dust or grass allergies sometimes wear a mask like this when working in the yard or a dusty place. Painters sometimes wear them. Why do you think some people wear masks? Elicit that a person may wear a mask to prevent inhaling things that may be harmful to him.

  God has given our bodies a way to help keep our respiratory systems clean.

Preparation for Reading

- Display the words *pharynx*, *trachea*, *epiglottis*, and *larynx*. Pronounce the words together.
- Tell the students to read silently pages 268–71 to find the answers to these questions.

  What helps keep dust out of the respiratory system?

  Why are men's voices usually deeper than women's voices?

Teach for Understanding

- Refer to the *Path of Air* visual as you discuss parts of the respiratory system.

  What are the two openings in the nose called? nostrils

  What is the nasal cavity? a large air space located behind the nostrils

The Path of Air

Your Nose

Most of the air you breathe comes in through your nose. Your nose has two openings that are separated by a wall of cartilage and bone. These openings are called *nostrils*. Air enters through your nostrils and passes into your nasal cavity. The *nasal cavity* is a large air space located behind your nostrils. The open spaces from the front of the nose to the beginning of the throat are called the *nasal passages*.

Your nose is good for much more than breathing. When you inhale, the air that passes through your nose is warmed, moistened, and even filtered. Special cells in your nose and

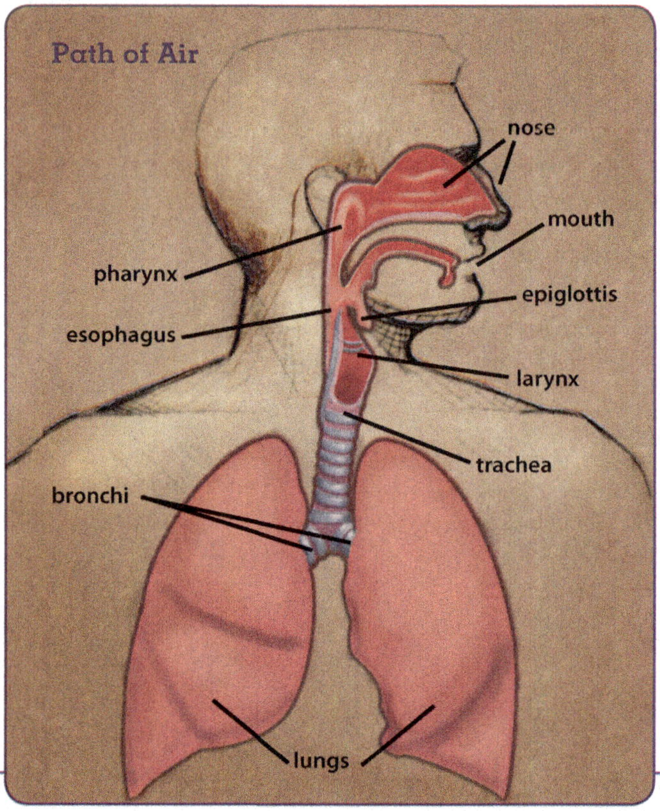

268

Background

Cilia

Cilia are microscopic hairlike projections. The constant wavelike movement of the cilia moves the mucus-trapped particles toward the throat, mouth, or nose. Cilia beat more than 1,000 times a minute, moving the mucus approximately 1 cm each minute.

Mucus

The nose and sinus cavities of a healthy adult produce about 1 L (4 cups) of mucus each day. Mucus is also produced by glands in the throat and lower respiratory system.

Electronic Noses

Scientists have developed artificial noses that can be used in a variety of ways. Some electronic noses are used to determine the quality of food or to detect food spoilage. Other electronic noses can test for diseases such as tuberculosis and lung cancer. Electronic noses are also used both on the earth and on the International Space Station to detect chemical leaks.

your lungs produce *mucus*, a sticky substance that captures bacteria, dust, and other particles in the air. Tiny hairlike structures called **cilia** line your nasal passages as well as most of the respiratory system. The cilia within your nose move in waves to transport the mucus into the body. Scientists are imitating this movement that God designed in their artifical cilia.

The cilia sweep the trapped particles toward your throat, and you swallow them. There, the acid in your stomach destroys the particles. Sometimes, though, the particles do not reach your stomach. Instead, they may irritate the lining of your nose or throat and cause you to sneeze or cough.

Air can also enter your body through your mouth. However, your mouth cannot protect your respiratory system as the cilia in your nose can. For this reason, it is best to breathe through your nose. At times, though, you may have to breathe through your mouth. When you have a cold, for instance, your nasal passage may become blocked. In this case, the only way to get air is through your mouth. You also tend to breathe through your mouth when you do strenuous exercise.

FANTASTIC FACTS

Have you ever wondered why your voice sometimes sounds different when you have a bad cold? The answer is that your sinuses are blocked. Your sinuses are the empty spaces in the bones around your nose. Like the nose, each sinus is lined with a mucus membrane. Because sinuses are empty air spaces, they help reduce the weight of the skull. They also amplify sound. When you are sick and your sinuses get blocked with extra mucus, the air cannot circulate freely through the nasal passages. This makes your voice sound different.

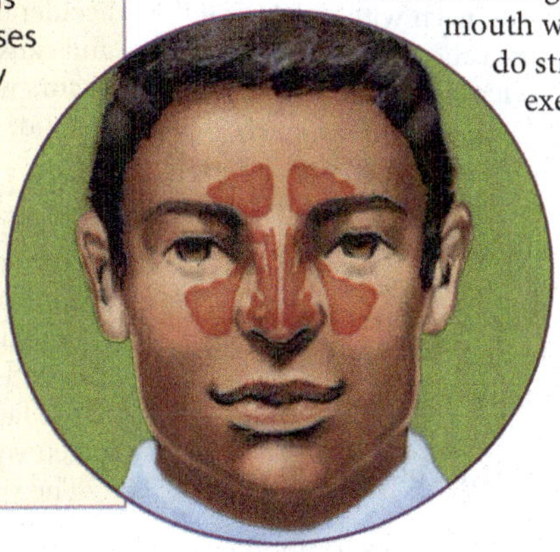

269

What separates the nasal cavity into two passages? bone and cartilage

What is cartilage? a bonelike substance that is more flexible than bone

What are these two passages in the nasal cavity called? the nasal passages

Where are the nasal passages located? from the front of the nose to the beginning of the throat

What is mucus? a sticky substance produced by special cells

- Display the *Cilia* visual. Share as much of the Background information as you would like.

What are the tiny hairs in the nose called? cilia

How do the cilia and mucus help the respiratory system? They trap dust and other particles to keep them from entering other parts of the respiratory system.

Where do the trapped dust particles go? They are swallowed or sneezed or coughed out.

Why do you think people with dust or pollen allergies often sneeze? Possible answer: Whatever they are allergic to has irritated the lining of the nose or throat. Their bodies are trying to prevent the irritant from entering the respiratory system.

Why is it better to breathe through the nose? The nose is better equipped to protect the respiratory system.

- You may want to do the *Air Particles and the Respiratory System* activity at this time.
- Discuss *Fantastic Facts*.

What are sinuses? empty spaces in the bones around the nose

What are two benefits of the sinuses? They reduce the weight of the skull and amplify sound.

Why does a person's voice sound different when he has a bad cold? Mucus blocks the sinuses and prevents air from circulating freely.

Science Misconceptions

Seeing Cilia

Sometimes nose hairs are visible at the front of the nostrils. Those hairs are not cilia. Cilia are farther back in the nose and other air passages. Cilia are also much smaller and can be seen only with strong microscopes.

Activity

Air Particles and the Respiratory System

Materials: index cards, scissors, clear packaging tape, hole punch, string, marker

Cut out the centers of several index cards. Punch two holes at the top of each card. Thread string through the holes and tie to form a loop. Write "Please do not touch" on each card. Cover one side of the open center on each index card with clear packaging tape. Hang each card in a different location inside or outside. Record the location on each card. Do not touch the sticky side of the tape.

Leave the cards undisturbed for two days. Collect the cards. For each, cover the sticky side of the tape with the plain side of a clean index card. Observe the types of particles that were collected. If needed, compare the cards with a piece of clean tape on a clean index card. Explain to the students that similar particles are collected by cilia and mucus in the respiratory system.

156
157

- Refer to the *Path of Air* visual as you discuss the upper respiratory system.

  What is another name for the throat? the pharynx

  Which tube connects the pharynx to the lungs? the trachea

  Which tube connects the pharynx to the stomach? the esophagus

  What is the function of the epiglottis? to prevent food from getting into the lungs by closing off the trachea when a person swallows

  Is the epiglottis open or closed when a person is breathing? open

  What does a person mean when he says that his food "went down the wrong pipe"? His food went down into his trachea.

  A person usually coughs when this happens. Why do you think that he coughs? Coughing is the body's way of trying to keep food from entering the lungs.

  What is another name for the larynx? the voice box

  What are stretched across the inside of the larynx? the vocal cords

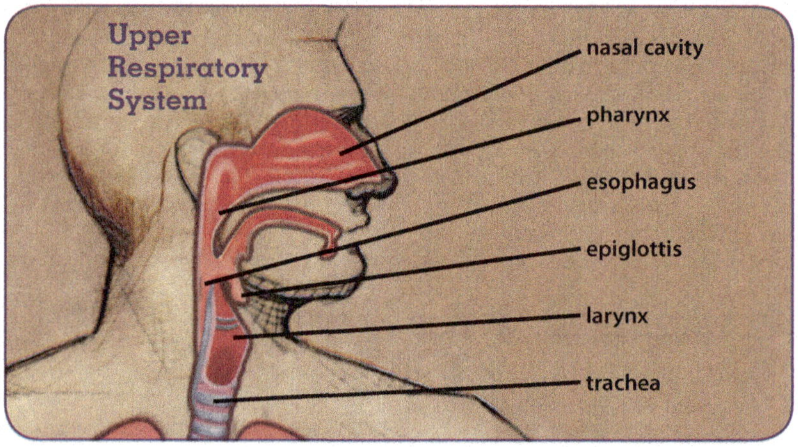

Your Throat

The air moves from your nose or mouth into your throat, or **pharynx**. At the end of your pharynx are two separate tubes, each with a different purpose. One tube is your **trachea**, or windpipe. It leads to your lungs. The other tube leads to your stomach and is called your esophagus, or food pipe.

When you swallow your food, a flap of tissue called the **epiglottis** closes automatically over your trachea. The epiglottis keeps food from going down the trachea to your lungs. When you are breathing, the epiglottis stays open. This allows the air to enter your lungs instead of going to your stomach.

Talking or laughing with food in your mouth may cause your epiglottis to work improperly. It may stay open because you are talking though it needs to close so you can swallow your food. Instead of the food going down your esophagus, the food may accidentally go down your trachea and cause you to choke. Choking occurs when an object, such as a piece of food, goes down the wrong "pipe."

Your Larynx

At the top of your trachea is your **larynx**, or voice box. It is made up of nine pieces of cartilage and many small muscles. Two small bands of elastic tissue stretch across the inside of the larynx. These soft bands are called your **vocal cords**.

The vocal cords form a V, with a small gap between them. When you are not talking, your vocal cords are relaxed. The gap stays open so that you can breathe. When you start to

270

Background

Uvula

The uvula is the rounded, fleshy mass of tissue that can be seen hanging from the soft palate at the top of the throat. It helps with the formation of sounds.

Pharynx

The pharynx of an adult is about 11.4 cm (4.5 in.) long. It has three divisions. The nasopharynx includes the air passages from the nostrils to the uvula. This part of the pharynx also connects with the ears. The oropharynx, from the uvula to the epiglottis, includes the base of the tongue and the tonsils. The part of the pharynx just above the larynx, from the epiglottis to the openings of the trachea and esophagus, is called the laryngopharynx.

Science Misconceptions

Parts of the Pharynx

The trachea and esophagus are not part of the pharynx. They begin where the pharynx ends.

156
157

Vocal Cords

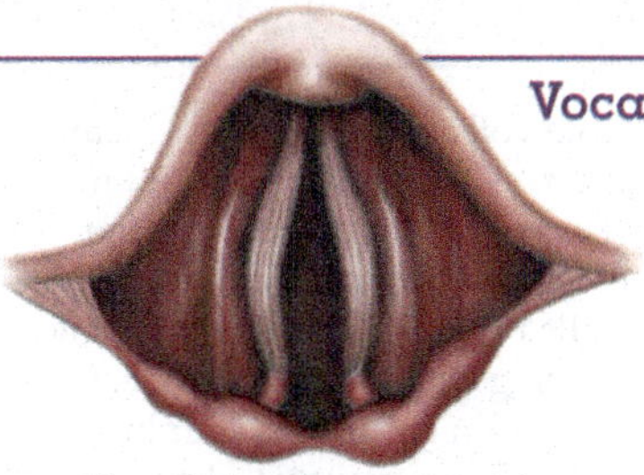
Your vocal cords are open and relaxed when you breathe.

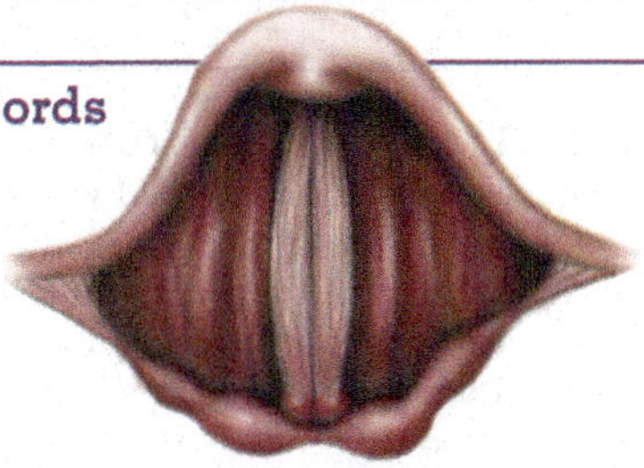
Your vocal cords tighten and narrow when you talk or sing.

talk, sing, or scream, your vocal cords tighten and narrow, leaving only a small opening.

Speech happens when air is exhaled. As you exhale, the air moves upward from your lungs and goes into your trachea. The air pushes through the closed vocal cords and makes them vibrate. This creates sound waves. Your throat, mouth, nose, sinus cavities, teeth, tongue, lips, jaw, and cheeks then work together to shape the sound into understandable speech.

The pitch of the sound depends on the thickness and length of the vocal cords. Long, thick vocal cords vibrate slowly and produce a deep sound. Short, thin vocal cords vibrate faster and produce a higher-pitched sound. Men have thicker vocal cords than women do. This is why men's voices are usually deeper than women's voices are.

TRY IT YOURSELF

Your larynx changes the pitch of your voice by changing the thickness and the length of the vocal cords. You can model this with a balloon. Inflate a balloon and let the air escape. While the air is escaping, stretch open the mouth of the balloon. This will change the length and thickness of the opening. Try this several times. Predict whether the sound will be higher or lower as the size of the opening changes.

QUICK CHECK

1. Why is it better to breathe through your nose?
2. What is the name of the flap of tissue that keeps food out of your trachea?
3. What body parts work together to produce the sound waves of your voice?

271

- Direct attention to the illustrations of the vocal cords.

  How are the vocal cords arranged? in the shape of a *V* but with a small gap between them

  When a person is silent, are his vocal cords relaxed or tightened? relaxed

  When does speech happen? when a person exhales

  How are the vocal cords involved in speech? Exhaled air pushes up through the vocal cords and makes them vibrate.

  Why would the vibrating of the vocal cords produce sound? Sound is caused by vibrations.

  Which parts of the body help shape sound from the vocal cords into recognizable speech? the throat, mouth, nose, sinus cavities, teeth, tongue, lips, jaw, and cheeks

  What affects the pitch of a sound made by vocal cords? the thickness and length of the vocal cords

  Why do men sing lower notes than women? Men's vocal cords are longer and thicker and vibrate more slowly.

  Sometimes a cold can make a person's vocal cords swell. When this happens, how does it affect the pitch of his voice? It lowers his voice. It makes his voice deeper.

- You may want to do *Try It Yourself* at this time.

ANSWERS

1. The nose is better equipped than the mouth to protect the respiratory system. The nose can filter out particles, such as dust, before they enter the respiratory system.
2. the epiglottis
3. air moves from the lungs to the vocal cords, which causes them to vibrate and form sound waves

Background

Sound and Breathing

The phonological sounds (speech) for most languages are made while exhaling. However, some languages, such as Swahili and Thai, have sounds that are made while inhaling.

157

- The *Air Particles and the Respiratory System* activity found at the bottom of the pages in Lessons 156–57 may also be incorporated into the lesson.
- Suggestions for additional activities for this two-day lesson are on the Teacher's Toolkit CD.

Activity Manual

Reinforcement, page 177

Review, page 178

This page reviews Lessons 156–57.

Assessment

Quiz 11-B

The quiz may be given any time after completion of this lesson.

158–159

Student Text pages 272–75
Activity Manual pages 179–80

Objectives

- Identify and describe the trachea, bronchi, and lungs
- Describe the function of the lungs
- Explain causes of snoring, hiccupping, coughing, and sneezing

Materials

- sheet of tissue paper

Teacher's Toolkit CD

- Visuals 11.1–11.2: *Cilia; Path of Air*
- Quiz 11-C

A graphic format of Visual 11.2 is available on the Teacher's Toolkit CD for use with interactive displays.

Vocabulary

- bronchi
- lungs
- alveoli

Two days have been allotted for this lesson. You may want to teach Lesson 158 the first day and use ideas from Lesson 159 on the second day. You may also incorporate the ideas from Lesson 159 into the text material, making the lesson two days.

Introduction

Have you ever had the hiccups?

What do you think causes hiccups? Answers will vary.

What kinds of things do you try when you want to stop hiccupping? Answers will vary.

Preparation for Reading

- Display the words *bronchi*, *bronchioles*, and *alveoli*. Pronounce the words together.
- Tell the students to mentally make questions using the headings, bold and italicized words, and captions.
- Instruct them to read silently pages 272–75 to find the answers to these questions.

  What protects the lungs?

  Why are the alveoli important?

Teach for Understanding

- Refer to the *Path of Air* visual as needed.

  What does the upper respiratory system include? the nose, throat (pharynx), and larynx

  What does the lower respiratory system include? from the trachea into the lungs, which includes the bronchial tubes, bronchioles, and alveoli

  About how long is the trachea? 10–11 cm (4 in.) in adults

Your Trachea and Bronchi

Your respiratory system is divided into two parts. The nose, throat, and larynx make up the *upper respiratory system*. Below the larynx, from the trachea into the lungs, is the *lower respiratory system*.

The trachea stretches from the larynx into the upper chest. It is about 10–11 cm (4 in.) long in most adults. The trachea branches off into two tubes called the **bronchi**. One tube goes to the left lung. The other tube enters the right lung.

Both your trachea and your bronchi have C-shaped rings of cartilage on their outsides. The rings keep them from collapsing and allow them to move as your neck and chest move. If a piece of food gets past your epiglottis, the muscles in your trachea walls react. The muscles can squeeze the rings together to try to keep the food from going into your lungs.

Just like in your nasal passages, the lining inside your trachea and bronchi has mucus and cilia. The mucus and cilia trap particles of dust and dirt. The particles are pushed back up to your throat, and you usually swallow them down your esophagus. If you have ever coughed to clear your throat, you have experienced those particles moving out of your trachea.

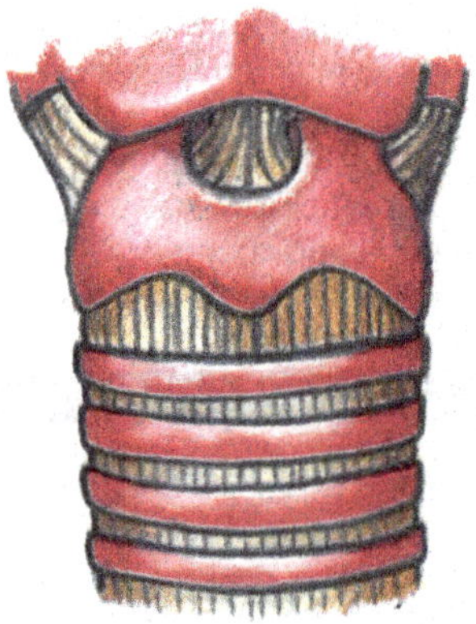

The rings of cartilage on the trachea and bronchi help keep the airways open.

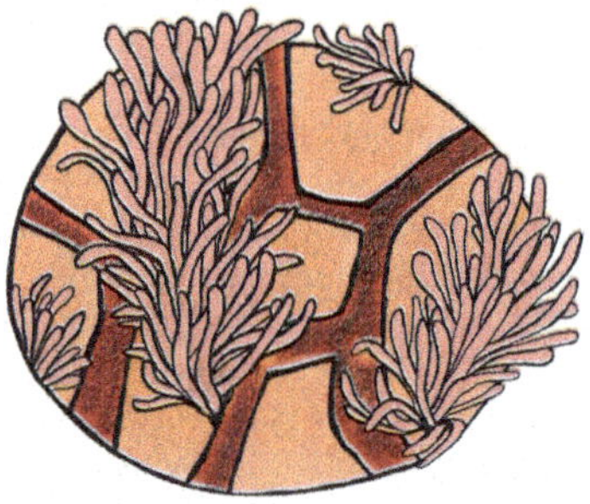

Cilia line the inside of the nasal passages, the trachea, and the bronchi.

272

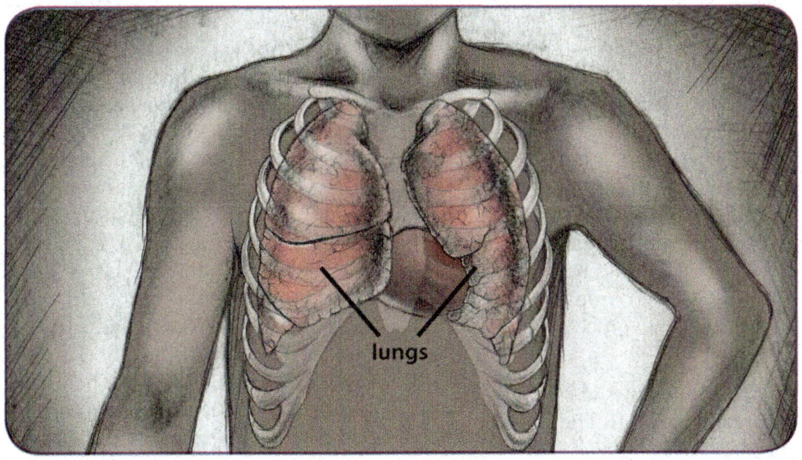

Your Lungs

Your **lungs** are two saclike organs that replace the carbon dioxide in your blood with oxygen. A flexible "cage" of bones protects your lungs. This cage is formed by your ribs, breastbone, and backbone. The top of your lungs reaches above your shoulder bones and is behind your collarbone. The base of your lungs rests on the diaphragm.

Healthy lungs are a pinkish-gray color. Each lung is about the size of a football. Your left lung is slightly smaller than your right lung. That is because your left lung has a space next to it for your heart. Your heart and main blood vessels are located between your lungs.

Lungs do not have any muscles of their own. Instead, your diaphragm and chest muscles control how your lungs move. When you breathe in, your lungs stretch and fill with air. When you breathe out, air is pushed out of your lungs, and they return to their smaller size.

The amount of air that can be taken into the lungs with one breath is called *lung capacity*. A normal breath for an average man is about half a liter of air. Lungs can hold more air if necessary, though. When a man breathes in as much air as possible, his lungs can hold about 6 L (6.4 qt) of air. A woman's lungs can hold about 4.2 L (4.5 qt) of air.

273

158
159

The trachea branches off into two tubes. What are these tubes called? the bronchi

What is at the other end of the bronchi? the lungs

- Explain that the term *bronchi* is plural. The singular form is *bronchus*.

  Why does the trachea have rings of cartilage? to keep it from collapsing and allow it to move

  What other parts of the respiratory system have cartilage rings? the bronchi

- You may want to do the *Rings of Cartilage and Trachea* demonstration at this time.

  What can be found in the trachea and bronchi that is also in the nasal passages? mucus and cilia

  What is happening when a person coughs to clear his throat? Coughing moves trapped particles from the trachea to the esophagus so they can be swallowed.

- Direct attention to the illustration on page 273.

  What are the lungs? The lungs are two saclike organs that remove carbon dioxide from the blood and replace it with oxygen.

  How are the lungs protected? They are surrounded by a flexible cage of bones.

  Where are the lungs located? from above the shoulder bones and collarbone to the diaphragm

- Using yourself as an example, show the students where the shoulder bones and the collarbone are located.

  What color are healthy lungs? pinkish-gray

  Why is the left lung slightly smaller than the right lung? The left lung has a space next to it for the heart.

  What is located between the lungs? the heart and main blood vessels

  What controls movement of the lungs? the diaphragm and chest muscles

  What is lung capacity? the amount of air that can be taken into the lungs with one breath

Demonstration

Rings of Cartilage and Trachea

Materials: a vacuum cleaner hose

Display the vacuum cleaner hose. Move the hose, showing how it bends and is flexible. Allow students to feel the hose. Discuss how the rings keep the hose from collapsing. Explain that the rings on the hose and the rings on the trachea work in similar ways.

- Discuss the illustration of the lungs.
  Are the lungs hollow like balloons? no
  Name some of the parts inside the lungs. Possible answers: bronchial tubes, bronchioles, alveoli, capillaries
  What are the smaller airways connected to the bronchi called? bronchial tubes
  What are the smallest bronchial tubes called? bronchioles
  Where does the exchange of oxygen and carbon dioxide actually take place? in the alveoli
  What are alveoli? tiny air sacs at the end of the bronchioles
- Explain that the singular form of *alveoli* is *alveolus*.
  What surrounds the alveoli? a network of capillaries
  What is the exchange of gases called? breathing, or respiration
- Display the sheet of tissue paper. Allow the students to examine its thickness.
  This piece of tissue paper is as thick as the walls of about 50 alveoli, so the wall of one alveolus is extremely thin. This thinness allows oxygen and carbon dioxide molecules to pass through the walls.

Bronchial Tubes

Inside your lungs, your bronchi branch off into many smaller *bronchial tubes*. These tubes get smaller and smaller as they spread to all the parts of your lungs. Many of these tubes are less than 1 mm wide. Some are thinner than human hairs. All of the bronchial tubes, though, have muscles and rings of cartilage to keep them open wide. The smallest tubes are called *bronchioles*.

Alveoli

The bronchioles end in tiny air sacs called **alveoli**. The alveoli look like bunches of grapes. Each lung has about 300 million alveoli. They are surrounded by a network of tiny blood vessels called *capillaries*. The walls of the alveoli and the capillaries are very thin. In some places, they are only one cell thick.

It is in the alveoli that the exchange of gases takes place. The air you inhale contains oxygen. The oxygen passes through the walls of the alveoli into the blood in the capillaries. The blood then carries it to all the parts of your body. At the same time, carbon dioxide passes from the blood in the capillaries into the alveoli. There, the carbon dioxide can be exhaled from your body. This exchange of gases is called breathing, or respiration.

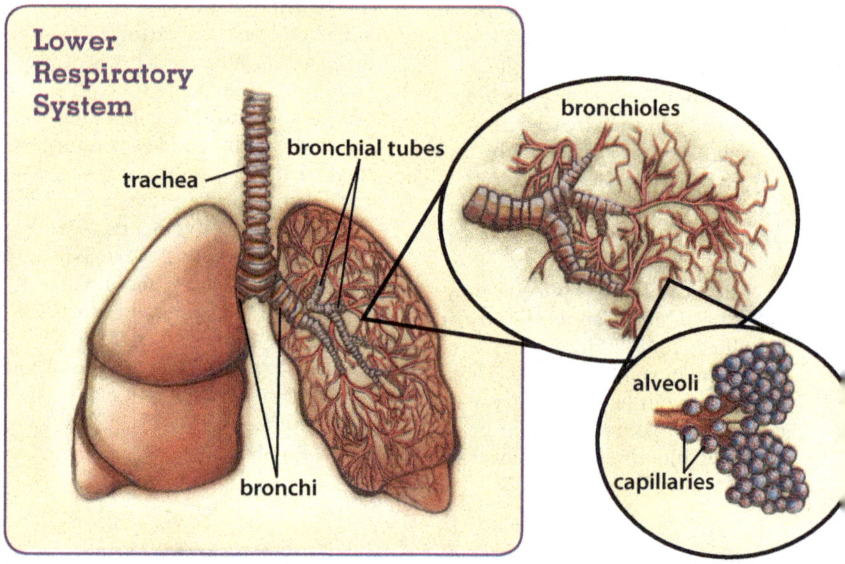

274

Background

Respiration

Respiration is classified as either external or internal. External respiration, or breathing, refers to the exchange of gases between the environment and the blood in the lungs. Internal respiration, also called cellular respiration, refers to the series of chemical reactions that release energy in each cell in the body. The respiration defined in Chapter 8, *Changes in an Ecosystem*, is cellular respiration.

Respiratory Sounds

Coughs and Sneezes

Coughs and sneezes often occur whenever something irritates the lining of the airways. The force of the cough or sneeze moves the offending particles out of the respiratory system. When you cough, your diaphragm relaxes while your other muscles contract. This action violently pushes air out through your mouth. In much the same way, a sneeze is triggered by an irritation in the nose. The air is forced out through both the nose and the mouth.

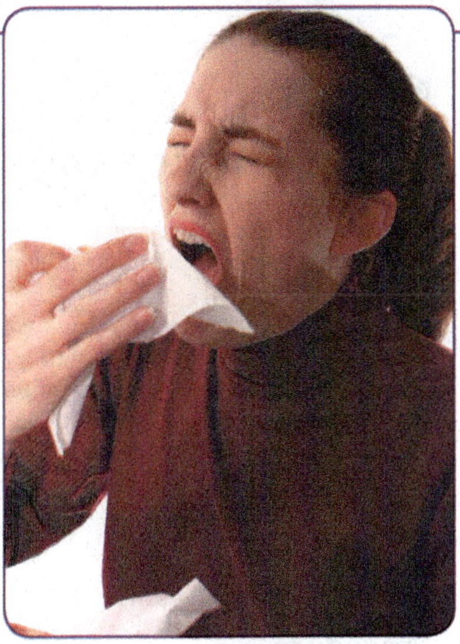

Sneezing

Hiccups

Hiccups are caused by involuntary movements of the diaphragm that disrupt normal breathing. When you have the hiccups, your diaphragm muscle contracts quickly and causes you to take quick, short breaths of air. The epiglottis closes suddenly while the diaphragm is contracting. This causes the hiccup noise.

Many times there is no obvious cause for hiccups. Some people may get hiccups when they are full or when they eat a spicy food. Strong emotions, such as fear or anger, can also cause them. Hiccups can be noisy or quiet, but they cannot be controlled.

Snoring

A person snores whenever air cannot flow freely through the back of the mouth and the nose. The snoring sound is caused when parts of the mouth and throat vibrate and hit together. There are many reasons why a person snores. The person may have an allergy or a cold. If the wall of cartilage that separates a person's nostrils is crooked, he may snore. A person may also snore if he is overweight, sleeps on his back, or has swollen tonsils.

QUICK CHECK

1. What is the name of the two air tubes that branch off the trachea?
2. What are alveoli?
3. What causes the hiccup noise?

158–159

What happens to the diaphragm and other muscles when a person coughs? His diaphragm relaxes, and his other muscles contract.

- Explain that the air expelled when a person sneezes can travel at 160 km (100 mph) and the droplets may travel 3.7 m (12 ft).

What happens to a person's respiratory system when he has hiccups? The diaphragm contracts quickly and involuntarily, causing him to take quick, short breaths of air. The epiglottis closes suddenly while the diaphragm contracts.

What causes snoring? Snoring occurs whenever air cannot flow freely through the back of the mouth and the nose.

What causes the snoring sound? Parts of the mouth and throat vibrate and hit together.

If someone is hit in the nose with a ball, he may, for a while, snore when sleeping. What may cause this snoring? Possible answers: The airway may be swollen, so the air cannot flow freely. The cartilage may be crooked or broken.

ANSWERS

1. the bronchi
2. the tiny air sacs at the end of the bronchioles where the exchange of carbon dioxide and oxygen takes place
3. the sudden closing of the epiglottis while the diaphragm is contracting

159

- The *Rings of Cartilage and Trachea* demonstration found at the bottom of the pages in Lessons 158–59 may also be incorporated into the lesson.
- Suggestions for additional activities for this two-day lesson are on the Teacher's Toolkit CD.

Activity Manual

Reinforcement, page 179

Review, page 180

This page reviews Lessons 158–59.

Assessment

Quiz 11-C

The quiz may be given any time after completion of this lesson.

160

Student Text pages 276–77
Activity Manual pages 181–82

Objectives

- Calculate the amount of air exhaled
- Identify variables that may affect the results

Materials

Materials needed for this activity are listed on the Student Text page.

Teacher's Toolkit CD

- Visual 11.3: *Vital Lung Capacity*
- Rubrics: *How Much Air Is in Your Lungs?*, for each student

Introduction

In the fable about the three little pigs, the big bad wolf was always huffing and puffing to blow their houses down. He blew down the houses of straw and sticks, but he could not blow down the house made of brick.

How do you think the big bad wolf inhaled and exhaled when he was trying to blow down the houses? Do you think he took tiny breaths or great big ones? Answers will vary.

Do you think you would have enough air in your lungs to blow down a house? no

In this activity you are going to measure how much air you can exhale in one breath.

Preparation for Reading

- Instruct the students to read Student Text pages 276–77 and Activity Manual pages 181–82 before beginning the activity.

Teach for Understanding

Process Skill

The process skill emphasized in this activity is collecting, recording, and interpreting data.

- Explain that recording data allows others to view and evaluate your results.
  What kind of data are you collecting in this activity? diameter of an inflated balloon
  How are you recording this data? in a chart
  What do you do with the data? find the average diameter
  How can we use this data or average? to interpret the vital lung capacity

ACTIVITY

How Much Air Is in Your Lungs?

Process Skills
- Hypothesizing
- Measuring and using numbers
- Collecting, recording, and interpreting data

Lung capacity is different for each person. Your age, gender, size, and level of physical fitness can all affect how much air your lungs hold.

Doctors may measure your lung volume under different conditions. Sometimes they measure your lung volume during normal, quiet breathing. Other times they may measure the maximum amount of air that you can exhale after a deep breath. This is called your *vital lung capacity.*

In this activity you will use a balloon to measure your vital lung capacity.

Problem

How much air can you exhale in one breath?

Procedure

Materials
round balloon, 12 in.
meter stick
centimeter ruler
calculator
Activity Manual

1. Complete the hypothesis in your Activity Manual.
2. Stretch a large round balloon, and then blow into it several times, letting the air out of the balloon after each time.
3. Take a deep breath. Exhale as much air as possible into the balloon. Be sure that you are exhaling only once.
4. Twist the end of the balloon, and pinch it closed with your fingers. Do not let air escape from the balloon.
5. Have your partner hold the meter stick in a vertical position. The end of the meter stick should rest on a desk or table. Place the balloon on its side against the meter stick. Rest the centimeter ruler across the top of the balloon. Do not push down on the balloon.
6. Use the ruler to measure the width of the balloon that you blew into. The place on the meter stick that the centimeter ruler points to is the diameter of the balloon. Record the diameter of the balloon that you blew into.

276

Background

Accuracy of Data

The volume of lung capacity calculated in this activity is only an estimate. The balloons are not perfectly round, and they will stretch after multiple uses. These variables will affect the results.

Peak-Flow Meters

A person who has asthma or other respiratory problems may use a peak-flow meter. This device does not measure lung capacity. Instead, it measures in liters per minute the speed at which air exits the lungs.

7. Repeat steps 3–6 two more times. Record the diameter each time.
8. Calculate and record the average diameter of the balloon.
9. Exchange places with your partner. Repeat steps 1–8 for your partner. Make sure that your partner uses a different balloon.
10. Look at the graph on your Activity Manual page. Use the average diameter to find out how many liters of air you exhaled and record the amount.

Conclusions

- Do the results support your hypothesis?
- Did the diameter of the balloon change the second and third times that you blew into the balloon? If so, why do you think it changed?

Follow-up

- Compare the lung volumes of boys and girls.
- Compare the vital lung capacity of students who play a wind or brass instrument with those who do not.

277

160

Procedure

- Remind the students to exhale only once into the balloon.
- Demonstrate how to use the meter stick and the centimeter ruler to measure the balloon's diameter. The ruler must be level and not press into the balloon.
- You may want to display the *Vital Lung Capacity* visual and demonstrate how to use the graph.

Conclusions

Did your results support your hypothesis? If not, how much difference was there between it and your actual vital lung capacity?

Did the diameter of the balloon change the second or third time that you blew into the balloon? If so, why? Possible reasons: The balloon may have stretched more. The amount inhaled may have been different.

Do you think this is a very accurate way to measure lung capacity? no Why? Possible answers: Air can escape from the balloon as you are measuring. Someone might press down on the balloon and not get an accurate measurement of the diameter. The shapes of the balloons may vary.

Activity Manual

Activity, pages 181–82

Activity Manual page 182 will be used in Lesson 161 to record the vital lung capacity.

Assessment

Rubrics

Select the prepared rubric, or design a rubric to include your chosen criteria.

Activity Manual pages 183–84

Objectives

- Describe the unique way God created man
- Relate the physical position of Jesus on the cross to His inability to breathe normally, a part of His suffering

Materials

The materials for this lesson can be found on Activity Manual page 184.

Teacher's Toolkit CD

- Visual 11.3: *Vital Lung Capacity*

Introduction

While all God's creatures are special to Him, He created mankind in a unique way. He has a special relationship with mankind.

Preparation for Reading

- Direct the students to read silently the first paragraph on Activity Manual page 183 to find out how the Bible describes the land animals.

Teach for Understanding

What is the root word for *creature*? create

How does the Bible describe the land animals? "the living creature"

- Ask volunteers to read aloud Genesis 1:24–25 and 2:7.

How did God create the animals? He spoke the animals into being.

How did God create mankind differently? God shaped mankind from the dust and breathed into him.

- Allow time for the students to complete question one on the Activity Manual page.
- Instruct the students to read silently the next two paragraphs on the page to find out who broke fellowship with God.

Did God speak to Adam? yes

How did Adam break fellowship with God? He sinned against God by disobeying His command not to eat the fruit of the tree of the knowledge of good and evil.

What are results of Adam's sin? We live in a corrupted world. All people are sinners.

- Ask volunteers to read aloud Romans 5:12, 17–19 and 1 Corinthians 15:21–22, 45–49.

What two people are referred to in these verses? Adam and Jesus

What did Adam bring to the world? sin

Are all people born into sin? yes

What does Jesus bring to the world? salvation

If Adam had not brought sin to the world, there would be no need for salvation through Jesus.

Are all people in need of redemption through Jesus? yes

A lesson from

Answers in Genesis Name ______________

1:1

The Breath of Life

You are a creature! Think about this idea for a minute. The root of the word creature is "create." So to say you are a creature is just to say that you were created by the Creator. But people are not the same as all of God's other creatures. People were made in a special way and in the image of God. Genesis 1:24 describes all the land animals with the phrase "the living creature," and verse 30 refers to these living creatures as those "wherein there is life." That phrase can also be interpreted as those having the "breath of life" in them.

Read Genesis 1:24–25 and 2:7.

1. Contrast how God created Adam (man) with how God created the animals.
   God spoke the animals into being. God shaped man from dust and breathed into him.

God created mankind in a much more personal way than He created animals. God also has a personal relationship with people. There are hints in the Bible that God spent time with Adam and Eve in the Garden of Eden, and we certainly know that God spoke with Adam. All through the Bible we read about God's desire to have fellowship with people. Sadly, Adam chose to break this fellowship. He sinned against God by disobeying His clear command not to eat the fruit of the tree of the knowledge of good and evil.

We now live in a world that has been corrupted by Adam's sin (Genesis 3). Because of Adam's disobedience, all people are sinners. The apostle Paul makes this clear in several passages.

Read Romans 5:12, 17–19; 1 Corinthians 15:21–22; 1 Corinthians 15:45–49.

2. What two people are referred to in these passages? *Adam and Jesus*
3. What did Adam bring to the world? *sin*
4. What does Jesus bring to the world? *salvation*

Adam used the breath of life God had given him to rebel against God. Jesus used His life to bring hope and restoration with God. His death on the cross was the sacrifice needed to save us from our sins.

The death that Jesus died on the cross was a horrible death. Crucifixion was meant to be cruel. One part of the torture made it hard for the victim to breathe. Let's do an activity to model a small part of what Jesus suffered on the cross.

Resources

The BJU Press website offers additional links from Answers in Genesis that you may find helpful with this lesson.
bjupress.com/resources/science

Procedure

1. Record your vital lung capacity from Activity Manual page 182.

   Vital lung capacity ________ L

2. On the playground, hang from a bar with both hands spread out a little wider than your shoulders. A partner should be ready with your balloon nearby. Take three slow, deep breaths. After inhaling a fourth time, hold your breath and drop to the ground. Immediately blow all of the air you can into the balloon. Measure the balloon as you did for the *How Much Air Is in Your Lungs* activity. Repeat two more times.

Materials
Activity Manual, page 182
round balloon, 12 in.
access to playground equipment
meter stick
centimeter ruler
calculator

3. Make a chart and record the diameter of the inflated balloon for each trial.

4. Calculate the average diameter. ________ cm

5. Use the average diameter of the balloon and the graph on page 182 to find your hanging lung capacity. Hanging lung capacity ________ L

Conclusions

1. Describe the difference between your vital lung capacity and your hanging lung capacity.

   ____

2. How would hanging affect a person's ability to breathe if they hung for several hours, like Jesus did?

   The reduced lung capacity would reduce the amount of oxygen for the blood to carry to the rest of the body. Less oxygen would affect the health and function of all parts of the body.

Believing the Bible's account of Adam's disobedience is the foundation for a person's understanding his need for salvation. Because Adam sinned, Jesus came to Earth to suffer on a cross in the place of every person. Jesus' sacrifice was needed so people could be restored to a right relationship with God. Jesus suffered many terrible things, including not being able to breathe, while He hung in the place of sinners.

- Allow time for the students to complete questions 2–4.
- Instruct the students to read the remainder of page 183.

  How did Adam use the breath of life God gave him? to rebel against God

  How did Jesus use His life? to bring hope and restoration with God

Procedure

What is the activity intended to show you? a small part of what Jesus suffered on the cross

- Direct the students to record their vital lung capacity from Activity Manual page 182.
- Guide the students as they do the activity and record the diameter of the inflated balloon in a chart.
- Allow time to calculate the average diameter.
- Display the *Vital Lung Capacity* visual or allow the students to use the chart on Activity Manual page 182 to find their hanging lung capacity.

Conclusions

- Allow time for the students to complete the questions and to read the last paragraph on the page.

  Describe the difference between your vital lung capacity and your hanging lung capacity. Answers will vary.

  How would hanging affect a person's ability to breathe if they hung for several hours, like Jesus did? The reduced lung capacity would reduce the amount of oxygen for the blood to carry to the rest of the body. Less oxygen would affect the health and function of all parts of the body.

  Christ's death on the cross was a brutal event that involved a scourging, multiple beatings, and then being hung to die.

- Explain that part of the cruelty of hanging on a cross was the inability of the victim to breathe normally because the ribs cannot expel air completely while the arms are outstretched.

  Why did Jesus come to Earth to suffer on the cross? because sin entered the world; for people to be able to restore a right relationship with God

  Jesus suffered physical pain on the cross. What else did Jesus suffer on the cross? The wrath of God for sin was poured on Jesus so that sins could be forgiven.

Activity Manual

Answers in Genesis, pages 183–84

Student Text pages 278–81
Activity Manual pages 185–87

Objectives

- Identify some diseases that make it difficult to breathe properly
- Describe what happens during an asthma attack
- Recognize that allergies are not contagious
- Name some reasons why smoking is harmful to your health

Teacher's Toolkit CD

- Quiz 11-D

Two days have been allotted for this lesson. You may want to teach Lesson 162 the first day and use ideas from Lesson 163 on the second day. You may also incorporate the ideas from Lesson 163 into the text material, making the lesson two days.

Introduction

"Wash your hands." "Cover your mouth when you sneeze or cough." "Use a tissue or handkerchief." We are often reminded to do these things. Why do you think they are important? Elicit that following this advice can prevent the spread of germs.

Preparation for Reading

- Display the words *influenza*, *pneumonia, asthma*, and *emphysema*. Pronounce the words together.
- Instruct the students to mentally make questions using the headings, bold and italicized words, and captions.
- Direct them to read silently pages 278–81 to find the answers to these questions.
  What part of the respiratory system is affected by pneumonia?
  Why do people with emphysema often feel extremely tired?

Teach for Understanding

What is one of the most common respiratory diseases? a cold

What causes colds? viruses

What parts of the respiratory system can be infected by a cold? the nose, throat, sinuses, ears, and bronchi

What other common disease is caused by viruses? the influenza, or flu

How do flu and cold viruses sometimes affect the cells in the nose and throat? The cells swell and produce more mucus than normal.

What does *contagious* mean? easy to spread to other people

Are colds and the flu contagious? yes

How do colds and the flu usually spread? through the air by coughing or sneezing

Respiratory Problems

The respiratory system involves many parts. God's design connects the different parts of our bodies to work together. When all the parts are working properly, you hardly notice any of them. However, if one part stops working well, you usually notice it very quickly. The problem may be merely an annoyance, such as a stuffy head or a runny nose. Or it could be a life-threatening problem, such as when the body is not able to get the air that it needs.

Respiratory problems vary a lot. Some affect only one part of the respiratory system. Others, however, cause trouble for many parts. Some are sicknesses that come and go, whereas others may last a person's whole life.

Diseases

Many diseases can affect the body's ability to breathe properly. The most common respiratory disease is a cold. Cold viruses usually affect the nose and throat, causing their cells to swell. The cells start producing more mucus than normal, and the person may have trouble breathing through his nose. Colds can also affect the sinuses, ears, and bronchi.

Influenza, or the flu, is also caused by a virus. Flu viruses are similar to cold viruses. They often have some of the same symptoms. Flu viruses, though, usually also cause fevers, headaches, and muscle aches. The viruses may lead to a more serious illness or infection.

Both colds and the flu are very *contagious*, or easy to spread to other

Healthy habits help keep viruses from spreading.

278

Background

Bronchitis

Bronchitis can be caused by viruses or by bacteria. Inhaling irritating fumes, dust, or smoke can also cause bronchitis.

Pneumonia

Ordinary respiratory infections can develop into pneumonia. If a person has cold or flu symptoms that last for a prolonged time, he should see a doctor to determine whether or not he has pneumonia. Walking pneumonia means that the person has a very mild case of pneumonia. Double pneumonia means that the infection is in both lungs. Pneumonia can be treated, but it may be fatal if a person's immune system is already weakened by another disease such as cancer.

Tuberculosis

Historically known as consumption, tuberculosis is a bacterial infection in the lungs. It can also infect other parts of the body. Tuberculosis is one of the most common major infectious diseases. It is spread through the air. The disease is more prevalent in developing countries than in the United States, but even in the United States there were more than 11,000 cases of tuberculosis reported in 2010. Worldwide 9,000,000 people contracted tuberculosis in 2010 and about 1,500,000 died.

people. Colds and the flu are usually spread through the air when a sick person coughs or sneezes. When another person inhales, he breathes in the virus. That particular virus may or may not make him sick. Sometimes a person may get sick by touching his mouth or nose after he has touched something that a sick person has coughed or sneezed on. Frequent hand washing can help keep viruses from spreading.

Some problems affect the throat, another part of the respiratory system. Many different things cause sore throats. Sometimes dust or smoke in the air can irritate the throat and cause it to be sore. The extra mucus produced because of a cold or flu may also irritate the lining in the throat. The throat usually has a coating that helps keep it moist. Sometimes extreme coughing will dry out the throat and cause it to hurt.

Occasionally a virus or bacteria may infect the throat or larynx. Strep throat is a type of infection that is caused by bacteria. A person with strep throat will usually have a high fever for several days. A doctor can test to see if a sore throat is strep throat. If it is, the doctor can prescribe medicine to help the body get rid of the bacteria.

Sometimes a cold or the flu can turn into bronchitis. This disease causes the bronchi to become infected and swollen. A germ, such as a virus, irritates the lining of the bronchi. This often causes the cilia to stop moving. They cannot trap or remove particles. The air passages become blocked, and more mucus is produced. The extra mucus in the bronchi makes it harder for the air to pass in and out of the lungs. This is why a person with bronchitis often sounds raspy. He may also cough a lot. Coughing is the body's way to try to remove the extra mucus.

Pneumonia is a lung infection. It is caused by viruses or bacteria and may develop from another illness. The infection causes the lungs to fill with fluid. The fluid keeps the lungs from being able to properly fill with air. Oxygen cannot reach all areas of the lungs. This makes it hard for oxygen to get from the alveoli to the bloodstream.

Streptococcus bacteria can cause strep throat.

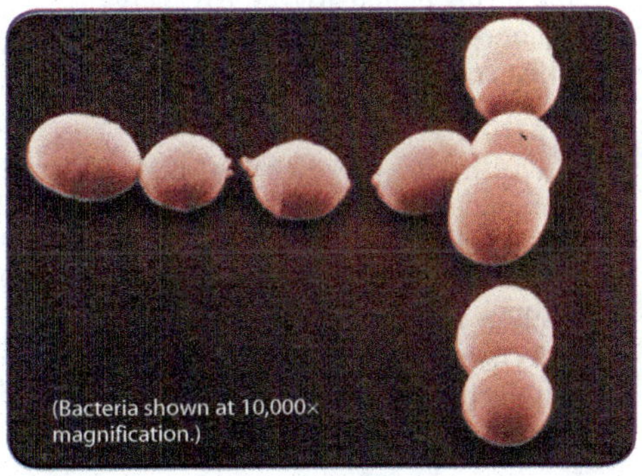

(Bacteria shown at 10,000× magnification.)

279

162
163

- Explain that most cold and flu symptoms last about 7–10 days. Doctors recommend that a person see a doctor if symptoms last longer or become worse.
What are some flu symptoms that are different from cold symptoms? fevers, headaches, muscle aches

Does a sore throat always mean that the person is sick? no

What else can cause a sore throat? Possible answers: dust, smoke, excessive coughing

What is strep throat? a throat infection caused by bacteria

What disease causes the bronchi to become irritated and swollen? bronchitis

What often stops moving when a person has bronchitis? the cilia What effect may this have? The cilia cannot trap or remove particles. More mucus is produced. This causes the air passages to become blocked.

How does the body try to remove the extra mucus produced by an infection? by coughing

What part of the respiratory system is affected by pneumonia? the lungs

When a person has pneumonia, what prevents the lungs from properly filling with air? fluid

Background

Dr. Alice Hamilton and Occupational Medicine

Dr. Alice Hamilton (1869–1970) worked with Jane Addams at the Hull House in Chicago. Through her work with immigrant families, she became aware of illnesses and deaths caused by industry. She became the first American physician to research industrial hazards. Many of the immigrant workers suffered serious health problems as a result of their jobs. Dr. Hamilton proved that inhaling and ingesting industrial toxins such as carbon monoxide, mercury, and lead caused some of these health problems and early fatalities. Her work brought about major reforms in industrial workplaces. Dr. Hamilton is known as the founder of occupational medicine in the United States.

Iron Lungs

The polio virus can affect different parts of the body. Sometimes it affects the breathing muscles. In 1928 American physicians Philip Drinker and Louis Shaw invented a machine that helped polio patients to breathe. This artificial respirator was a large cylinder that enclosed the patient's entire body except for his head. An electric motor was used to change the air pressure inside the cylinder, forcing air in and out of the patient's lungs. This machine soon began to be known as the "iron lung." While there were many problems associated with the iron lung, it did provide a way for patients to continue breathing until they could breathe again independently. The ventilators used today are much smaller and easier to use.

What are some reasons why people sneeze? Possible answers: colds, allergic reactions to dust or pollen

Does everyone have allergic reactions? no Is everyone that does have allergic reactions allergic to the same things? no

What are some symptoms of respiratory allergies? sneezing, runny nose, itchy eyes, breathing problems

Are allergies contagious? no

How do the seasons affect some allergy sufferers? Possible answer: If a person is allergic to pollen, he will probably react more in the spring and the fall.

What is asthma? a disorder that causes the small bronchial tubes to become narrow from time to time

What happens during an asthma attack? The muscles around the bronchial tubes tighten more than normal and make the bronchial tubes smaller. The airways may become swollen, causing the person to have difficulty breathing.

Is asthma contagious? no

What are two things that can trigger asthma attacks? allergies and air pollution

Does having asthma prevent people from participating in sports? usually not

What do many asthma sufferers keep with them as a precaution? an inhaler that contains medicine to help prevent an asthma attack from starting or from getting worse

- You may want to do the *Breathing Difficulty with Swollen Bronchial Tubes* demonstration at this time.

Allergies

Sneezing does not always mean that you have a cold. It may mean that you are having an allergic reaction to something in the air. The air that we inhale usually contains particles such as dust and pollen. For some people, this dust or pollen may cause breathing problems. Their bodies react to things in the air that other people never even notice.

People with respiratory allergies often sneeze. They may also have runny noses, itchy eyes, or breathing problems. Allergies often come and go with the seasons. For example, pollen is in the air the most in the spring and the fall. A person may be fine in the winter but react as soon as the spring pollen arrives. An allergy is not contagious. It cannot be spread like a cold or the flu can.

Asthma

Asthma is a disorder that causes the small bronchial tubes to become narrow from time to time. This can make it hard for a person to breathe. Asthma can vary a lot from person to person. Some people with asthma have very few problems, whereas others have severe difficulties.

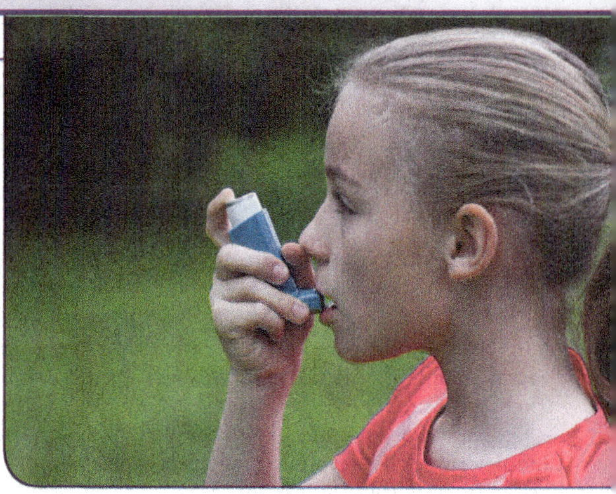

Some people need regular treatments for asthma.

Allergies and air pollution are two things that may cause someone to have an asthma attack. During an attack, the muscles around the bronchial tubes tighten more than normal. This makes the bronchial tubes smaller. The airways are often swollen and irritated. This makes it difficult for the person to breathe. The person may cough and wheeze as he breathes.

Someone with asthma may use an inhaler. The inhaler contains medicine that can be breathed in. It can help prevent an asthma attack from starting or getting worse. Asthma is not contagious. Most people with asthma are able to keep it under control. This way they can still participate in sports and other activities.

280

Background

Lung Cancer

The primary cause of lung cancer is smoking. About 85%–90% of lung cancers are caused by smoking. However, lung cancer can also be caused by other things, such as the spread of other diseases, or exposure to asbestos, radon, uranium, coal dust, or arsenic.

Emphysema

The most common cause of emphysema is smoking. However, scientists have found that some people have genetic deficiencies that put them at a higher risk for developing emphysema and other respiratory diseases. Air pollution can also increase the risk of emphysema. Because emphysema takes a long time to develop, it usually affects people over age 50.

Activity

Exploring What the Bible Says About Life

Materials: Instructional Aids 11.3–11.4: *The Breath of Life*, for each student; *The Breath of Life Answer Key*

Provide time for the students to complete the page. Discuss the truths that God gave mankind the breath of life, that He controls each breath we take, and that salvation is through Jesus.

Smoking

Smoking is harmful to your health. Cigarette smoke contains poisonous chemicals that harm the lungs and other parts of the body. Most of the smoke particles are very small. They cannot be trapped by the cilia and mucus in the airways. These poisonous chemicals then get into the lungs. As the smoke particles cool, they form a sticky tar that stays in the lungs. The tar makes the cilia clump together so that they cannot work properly. When this happens, dust, dirt, and smoke stay in the lungs. This sticky tar may cause diseases such as *lung cancer*.

Healthy lung

Smoker's lung

Another deadly disease caused by cigarette smoking is *emphysema*. Sometimes cigarette smoke causes the alveoli to stop working properly. This keeps the lungs from being able to transfer oxygen and carbon dioxide between the air and the blood. Once the alveoli are damaged, they cannot be repaired. People with emphysema have difficulty breathing. They also get tired a lot. This is because they do not get enough oxygen into their bloodstream. Even blowing out a match can be difficult for them to do.

The human body is one of God's marvelous designs. He created all the parts of our body to work together. The respiratory system works with other systems in our body to keep us alive. We should say with the psalmist in Psalm 139, "I will praise thee; for I am fearfully and wonderfully made: marvellous are thy works." Our bodies were made by God and belong to Him. Taking care of our bodies brings glory to Him.

QUICK CHECK

1. What is the most common respiratory disease?
2. If a person has pneumonia, which part of the respiratory system is infected?
3. What is a respiratory disorder that causes the bronchial tubes to narrow?
4. What are two lung diseases that are often caused by smoking?

281

162
163

What does cigarette smoke contain that is harmful? poisonous chemicals

Why are smoke particles not trapped by the cilia and mucus? They are too small.

What do smoke particles in the lungs form as they cool? a sticky tar

How does tar in the lungs affect the cilia? It makes the cilia clump together so that they cannot work properly.

What two diseases are often caused by smoking? lung cancer and emphysema

When a person has emphysema, what in his lungs are damaged? the alveoli

How is the body affected when alveoli are damaged? The lungs can no longer transfer oxygen and carbon dioxide between the air and the blood.

Can the alveoli be repaired? no

Why does a person with emphysema often feel extremely tired? He cannot get enough oxygen into his bloodstream.

- Read together as a class the portion of Psalm 139 quoted in the Student Text.

Why should we take care of our bodies? to bring God glory

ANSWERS

1. a cold
2. the lungs
3. asthma
4. lung cancer and emphysema

163

Demonstration

Breathing Difficulty with Swollen Bronchial Tubes

Materials: feather or strip of facial tissue, drinking straw, hollow coffee stirrer

Lay the feather near the edge of a table. Direct a student to inhale through the straw and then exhale through it, trying to blow the feather across the table in one breath. Observe the movement of the feather. Put the feather near the edge of the table again. Direct the student to inhale through the coffee stirrer and try to blow the feather across the table in one breath. Observe the movement of the feather.

Which feather moved the most? the one blown by the straw

Which was easier to blow through, the straw or coffee stirrer? the straw

Why do you think the straw is easier to blow through? The wider diameter allows a greater amount of air to move.

Did the narrow diameter of the coffee stirrer increase or decrease the amount of air that could be inhaled? decrease

How did decreasing the amount of air affect the amount of effort needed to blow? Decreasing the amount of air meant that the effort must increase.

Explain that the straw is like a normal airway. The air moves easily while breathing. The coffee stirrer is like an airway that is narrowed or swollen. The air movement is reduced, making breathing more difficult.

- The *Breathing Difficulty with Swollen Bronchial Tubes* demonstration found at the bottom of this page may also be incorporated into the lesson.
- Suggestions for additional activities for this two-day lesson are on the Teacher's Toolkit CD.

Activity Manual

Review, page 185

This page reviews Lessons 162–63.

Expansion, pages 186–87

These pages practice map skills as tuberculosis is discussed.

Assessment

Quiz 11-D

The quiz may be given any time after completion of this lesson.

164 Instructional Aids 11.5–11.6

Objectives

- Explain why it is hard to quit smoking
- Identify dangers of smoking
- Identify reasons people smoke
- List biblical reasons for not smoking

Teacher's Toolkit CD

- Instructional Aids: 11.5–11.7: *A Dangerous Habit,* for each student; *What Scripture Says*, for each student; *What Scripture Says Answer Key*

This lesson is designed as an interactive discussion alerting the students to the dangers of smoking. The material discussed on the pages is not all-inclusive. You may choose to incorporate additional Scripture or other resources as appropriate for your students.

Introduction

If you knew that a certain food was poisonous even though it looked attractive, would you eat it? no Why not? It would harm or kill you.

- Explain that you would not eat something that you knew was poisonous. Yet many people begin and then continue a dangerous habit, even though they know it will probably kill them. Ads make this habit appear fun and fashionable. This dangerous habit is smoking.

Preparation for Reading

- Display the words *nicotine*, *acetylene*, *formaldehyde*, and *naphthalene*. Pronounce these words together.
- Instruct the students to read silently *A Dangerous Habit* to find out what is in smoke that makes it so dangerous.

Teach for Understanding

About how much does the average smoker spend on cigarettes in one year? $1,500

Are cigarettes the only tobacco product that is harmful? no What are some others? chewing tobacco, pipe tobacco, and cigars

What does *addictive* mean? the continual need for something

What makes smoking addictive? nicotine

How does nicotine affect the body? Nicotine affects the nervous system, the brain, and the heart. It creates feelings that make the person want to smoke more.

- Direct attention to the list of the dangers of smoking. Explain that some of these dangers are health problems that can develop after a person has continued smoking for several years. But many health problems associated with smoking begin after the person has smoked just a few times.

What does the phrase *shortness of breath* mean? that the person is having trouble catching his breath, that his lungs cannot take in enough air

A Dangerous Habit

Name ______

What is smoking?

Smoking is an expensive habit that is hard to break. The average smoker spends over $1,400 a year on cigarettes.

All tobacco products—including any type of cigarette, chewing tobacco, pipe tobacco, and cigars—are harmful and damaging to the body.

Why is it hard to quit smoking?

Smoking is addictive. The nicotine in tobacco affects the nervous system, the brain, and the heart. Once a person has started smoking, the nicotine creates feelings that make the person want to smoke more. When a person tries to quit, his body craves the nicotine, and he may get irritable, have headaches, and have trouble sleeping. It is possible to quit, but it is not easy to do so.

What are some dangers of smoking?

- Causes more than 440,000 people each year to die from smoking-related illnesses
- Increases the risk of heart attacks, heart diseases, blindness, and strokes
- Increases the risk of getting many types of cancer
- Causes shortness of breath, wheezing, and coughing
- Causes a person to get tired easily when physically active
- Diminishes the ability to smell and taste
- Increases risk of gum disease and tooth loss and causes bad breath
- Makes the skin look old earlier than normal and causes wrinkles
- Causes a person to lose an average of 20–25 years of life

What is in smoke that makes it so dangerous?

There are over 4,700 chemicals in tobacco smok[e]. Hundreds of these are poisonous to humans. At lea[st] 70 of them have been identified as cancer-causing chemicals. Many of the chemicals are more concentrated in second-hand smoke than in the smoke inhaled by the smoker.

Listed below are just a few of the chemicals and poisonous gases found in tobacco and smoke.

Acetylene	usually used as fuel in welding torches
Ammonia	usually used in household cleaners, fertilizer
Benzene	usually used in motor oil
Carbon Monoxide	a poisonous gas, found in car exhaust
Formaldehyde	a fluid used to preserve dead things
Methanol	usually used as rocket fuel, antifreeze
Naphthalene	used in mothballs
Nicotine	used in insecticides
Nitrogen Oxide	a gas usually released by car exhaust or burning coal or oil

Science 5

Instructional Aid 1
For use with Lesson 1

Background

Cigars and Snuff

Cigars can be just as dangerous as cigarettes. One cigar usually has as much nicotine as several cigarettes. Snuff is finely-ground powdered tobacco that, when placed in the nose or mouth, quickly delivers high doses of nicotine. Both cigars and snuff can cause many different types of cancer as well as other health problems.

Tobacco Use Among Students

More than 600,000 middle school students and three million high-school students smoke cigarettes. Almost no one starts smoking after age 25. Nearly nine out of ten smokers started smoking by age 18. Ninety-nine percent started smoking by age 26. Young people exposed to images portraying smoking as a social norm are more likely to smoke to fit in with their peers. The tobacco companies spend more than a million dollars an hour in the United States alone to market their products. These advertisements entice many young people.

Carbon Monoxide

One of the poisonous chemicals in smoke is carbon monoxide. When carbon monoxide gets into the lungs, it can pass through the alveoli into the blood. There, the carbon monoxide attaches to the hemoglobin molecules in red blood cells and prevents the blood cells from picking up the needed oxygen.

What Scripture Says Answer Key

Name ______________________

Read the verses about a Christian's relationship to God. Match each reference with its correct sentence.

A. 1 Corinthians 3:16
B. 1 Corinthians 6:19–20
C. 1 Corinthians 10:31

__C__ 1. Everything a Christian does should be done to the glory of God.

__B__ 2. A Christian belongs to God. He should honor and glorify God with his body.

__A__ 3. The body of a Christian is the temple of God.

Some actions obey and honor God. Other actions do not. Sometimes you may find that those around you have chosen to do something that disobeys and dishonors God. You may be tempted to do the same. However, God is faithful to help you whenever you need Him. He will give you strength to do the right thing.

Read the verses about a Christian's obedience to God. Match each reference with its correct sentence.

D. Exodus 23:2
E. 1 Corinthians 6:12
F. 1 Corinthians 10:13
G. James 4:17

__E__ 4. A Christian should not be under the power or control of anyone or anything other than God.

__G__ 5. If someone knows the right thing to do but does not do it, he has sinned.

__D__ 6. A Christian should not go along with others if they are doing wrong.

__F__ 7. God will help the Christian resist and escape temptations.

God wants every Christian to be a witness for Him. The testimony that a Christian has before others is one way that he can witness. What you say and do tells others what you believe about God.

Read the verses about a Christian's testimony before others. Match each reference with its correct sentence.

H. 1 Timothy 4:12
I. Titus 2:7–8

__I__ 8. A Christian should be a good example so that no one can say anything evil about him.

__H__ 9. A Christian is to be an example to others of what he believes about God.

Choose one of the verses and sentences listed above. On your own paper, write a paragraph explaining how that truth applies to smoking.

Science 5

Instructional Aid 11.7
For use with Lesson 164

Many young people who have just begun smoking find that they have trouble breathing.

Why do you think smoking can cause a person to get tired easily when he is physically active? When a person exercises, the cells in his body need more oxygen. If he smokes, his body cannot get the needed oxygen to all the cells fast enough.

- Explain that other short-term effects of smoking include the diminishing ability to smell and taste, wheezing, coughing, bad breath, and premature aging of the skin.

What is second-hand smoke? Second-hand smoke is the smoke that a person inhales when he is around someone who is smoking.

- Point out that second-hand smoke can cause problems such as eye irritation, headaches, nausea, and dizziness. Children whose parents smoke are more likely to have problems with pneumonia, bronchitis, ear infections, coughing, and asthma. Second-hand smoke can also cause people who have never smoked to develop lung cancer and heart disease.
- Discuss the chart that lists some of the chemicals and poisonous gases found in smoke. Remind the students that this chart includes only a few of the most common chemicals found in tobacco and smoke.

Knowing how dangerous smoking can be, why do you think that some young people choose to smoke? Possible answers: They see parents or other adults doing so. They think it will make them seem older or more grown up. They think they can quit when they want to or that smoking will not hurt them. Curiosity, peer pressure, and advertising make smoking seem fashionable and exciting to some people.

- As time allows, complete *What Scripture Says* together and discuss the biblical principles. Explain that, although the Bible does not have specific verses related to smoking, Scripture does include several principles about living in a God-honoring way. Reinforce that a Christian should honor and glorify God in what he does with his body.

Background

Other Health Problems

Smoking can cause problems with a person's reproductive system and increase the risk for problems during pregnancy and birth.

Smoking also weakens a person's bones and muscles. Smokers have been found to develop osteoporosis earlier than nonsmokers do.

Advertising

Commercials for cigarettes and other tobacco products have been banned from television since 1971. Yet tobacco companies continue to advertise in more subtle ways. Movie and television actors portray characters who smoke, and brand names are often visible on billboards at televised sporting events. Additionally, advertisements that make smoking look interesting and fashionable are often included in some magazines and are on billboards in many cities.

Consult a local public library, police department, or health-care facility for more information about tobacco use. Many cancer centers and other medical associations provide information on the Internet.

You may want to invite a health-care worker or other professional to speak on this topic.

Be sensitive to any student whose parents or other family members smoke.

165

Student Text page 282
Activity Manual page 188

Objectives

- Recall concepts and terms from Chapter 11
- Apply knowledge to everyday situations

Review

- Materials for the Chapter 11 Test will be taken from Student Text page 282 and Activity Manual pages 175, 178, 180, 185, and 188. You may review any or all of the material during the lesson. Questions similar to the ones in Thinking It Through, Activity Manual page 188, may appear on the test.
- You may choose to review Chapter 11 by playing the review game or a game from the Game Bank on the Teacher's Toolkit CD.

Activity Manual

Review, page 188

This page requires written responses to application questions.

166

Objective

- Demonstrate knowledge of concepts taught in Chapter 11

Assessment

Chapter 11 Test

CHAPTER SUMMARY

Words to Know

respiratory system
inhale
exhale
diaphragm
cilia
pharynx
trachea
epiglottis
larynx
vocal cords
bronchi
lungs
alveoli

Key Ideas

Compare and contrast involuntary breathing and voluntary breathing
Identify reasons the body needs oxygen
Describe how the muscles work together to breathe
Locate the parts of the respiratory system on a diagram
Describe the function of the parts of the respiratory system and how air moves through the body
Summarize how speech happens
Identify some respiratory sounds
Describe some respiratory diseases and problems
Explain the importance of practicing good stewardship in caring for your lungs

282

Review Game

Inhaling and Exhaling

Divide the students into teams of seven. Each team represents the path that air takes in and out of the body. The first person represents the nose, the second the trachea, and so on for the bronchi, lungs, bronchial tubes, bronchioles, and alveoli. Give each team an object to pass. The object represents air molecules.

When a review question is answered correctly, the object is passed to the next person. The team receives a point for each correctly answered question. When the seventh person answers correctly, the air has been inhaled, and the team receives two bonus points. Then, as questions continue, the object is passed back up the line from person to person. When the object reaches the first person again, the air has been exhaled, and the team receives five bonus points. Continue playing until all review questions have been answered.

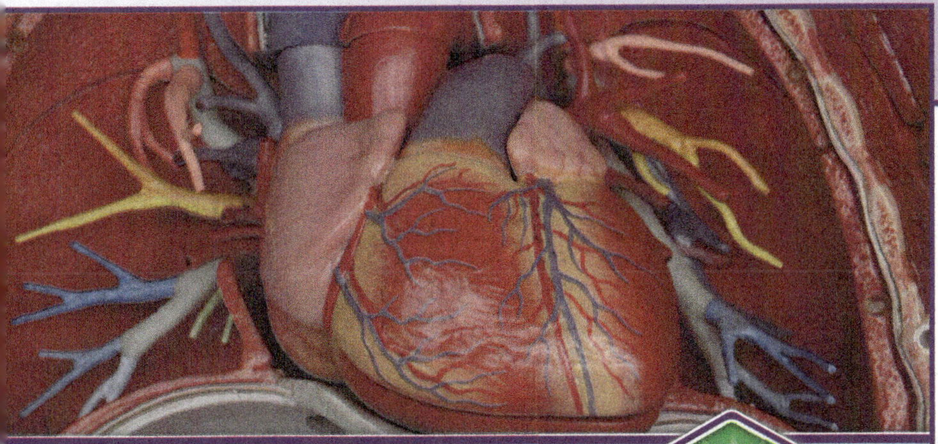

Circulatory System 12

On April 4, 1969, Dr. Denton Cooley entered an operating room in Texas. In a few hours he had placed the first artificial heart into a human body. This heart was made of rubber and a special fabric. It was run by a power unit about the size of a washing machine. But Dr. Cooley knew that the artificial heart would not last long. Three days later he removed the artificial heart and transplanted a donor heart from a person who had passed away.

Today a transplant is usually the best solution for a damaged heart. Artificial hearts are used for only weeks or months until a donor heart can be found. The longest time an artificial heart has lasted is about four years. It has taken thousands of hours and millions of dollars for researchers to create artificial hearts. With newer technology and more information, today's artificial hearts are much better than the one Dr. Cooley used. But we still have not been able to duplicate the amazing design of the heart that God created.

Consider
- How have artificial hearts been used to help people?
- How successful are artificial hearts?

283

Background

Weblinks
The BJU Press website offers additional information and links you may find helpful throughout the chapter.
bjupress.com/resources/science

Chapter Preview
Preview and prereading activities may include using a K-W-L chart, a probe, or an anticipation guide.

Chapter Photo
The photo on page 283 shows a human heart model that is open to reveal the interior. The miniature scientist is looking at an EKG screen.

Objectives
- Illustrate the superiority of God's design over mankind's technology
- Glorify God for His wisdom and power

Introduction
- Direct each student to stand on one leg without holding on to anything for as long as he can.
- Discuss how long the students were able to do this and how hard it was.

  What organ in your body pumps your blood? heart

  Does your heart ever stop working? no
- Explain that it is very hard to do anything for a long time, and yet our hearts keep right on going.

Preparation for Reading
- Instruct the students to read silently page 283 to find out about the first artificial heart.

Teach for Understanding

Why would someone receive an artificial heart? The person's heart is damaged and cannot function. The artificial heart gives the person more time to wait for a donor.

What does the word *donor* mean? a person who gives or donates something; a person who gives blood, an organ, or other tissue

How large was the power unit of the first artificial heart? the size of a washing machine

- Explain that since technology has improved, the power unit is now about the size of a briefcase or backpack.

  What do scientists use as their model when designing artificial hearts? the heart that God created

  Are scientists able to duplicate the human heart? no
- Explain that a pastor's wife in Texas who had received an artificial heart was asked to talk about her experience. She said that nothing was as good as what God could make. Recalling the three months that she had the artificial heart, she said that she put her faith in God and was glad that technology had allowed her the time needed for a donor heart to become available.

  When we see or read about the human heart, we should glorify God for his amazing design!

Activity Manual

Preview, page 189
The Looking Ahead page is intended to assess the student's prior knowledge before beginning the chapter.

168
169

Student Text pages 284–87
Activity Manual pages 190–92

Objectives
- Name the parts of the circulatory system
- Describe the path of blood through the heart
- Explain the function of the heart's pacemaker

Materials
- United States road map (optional)
- stethoscope (optional)

Teacher's Toolkit CD
- Visuals 12.1–12.3: *Circulatory System; Inside the Heart; Path of the Blood*
- Quiz 12-A

Vocabulary
- heart
- atrium
- ventricle
- valve
- pulse
- pacemaker

Two days have been allotted for this lesson. You may want to teach Lesson 168 the first day and use ideas from Lesson 169 on the second day. You may also incorporate the ideas from Lesson 169 into the text material, making the lesson two days.

Introduction
- Display the United States road map.
  Have you ever traveled on interstate highways?
  What do the interstate highways connect? many cities in the United States
  Are all cities and towns near an interstate? no
  In those cities and towns, smaller roads merge into a main road that connects to the interstate. This network of roads allows you to travel almost anywhere in the United States.
- Explain that the circulatory system is also a transportation system. It does not transfer people, though. It transfers blood.

Preparation for Reading
- Display the words *atrium* and *electrocardiograph*. Pronounce the words together.
- Instruct the students to mentally make questions using the headings, bold and italicized words, and captions.
- Direct them to read silently pages 284–87 to find the answers to these questions.
  What causes the sound of the heartbeat?
  What is a pulse?

Circulating Your Blood

The circulatory system is a transportation system. It transfers blood throughout your body. In this way, God designed your body so that the oxygen and nutrients in your blood can reach every part of your body. Without the oxygen and nutrients, not only would your cells die, but you would die! The heart is at the center of the circulatory system. Blood vessels and blood are also part of this amazing system. The circulatory system is so intricate and perfectly designed that it is almost unbelievable.

Your Heart

The **heart** is the main organ of the circulatory system. It is located in the chest, between the lungs. Both the lungs and the heart are protected by the ribs. Your heart is only about the size of your fist, but it is very strong. It needs to be strong because it never stops pumping. It keeps blood always circulating through your body.

The heart is a hollow organ that has walls made of strong muscle. A slippery sac of tissue around the heart protects it from rubbing against the lungs or the ribs.

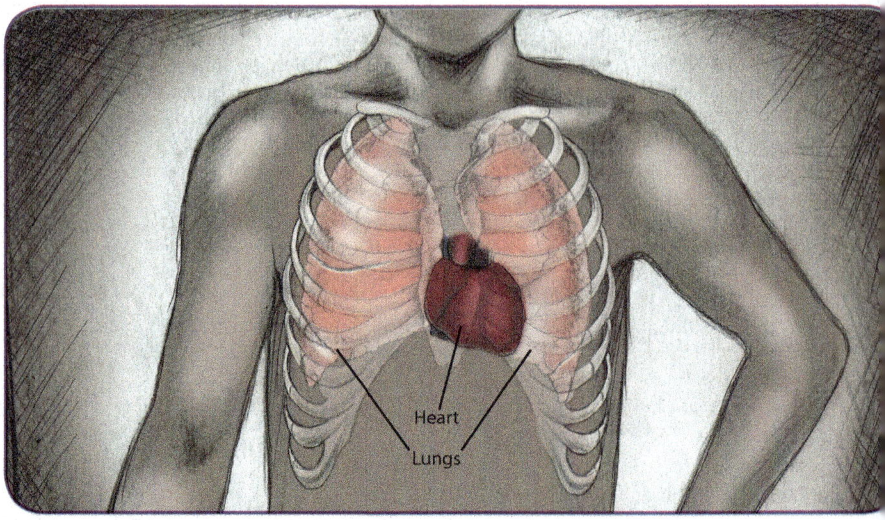

284

Background

Pericardium
The slippery sac of tissue that protects the heart is called the pericardium. It has two main layers and covers the roots of the major blood vessels. Between the layers is a small amount of lubricating fluid, called pericardial fluid. When the heart beats, these two main layers slide over each other and protect the heart from rubbing or catching on any other part of the body. Someone suffering with an "enlarged heart" has too much pericardial fluid, and the pressure from the fluid restricts the movement of the heart.

Heartstrings
When some people want to express sympathy or tenderness, they might say that something tugged at their heartstrings. The real heartstrings, or the *chordae tendineae*, anchor the valves between each atrium and ventricle.

Cardiac Muscle
The heart is made of a specific kind of muscle called cardiac muscle. This kind of muscle is not found in any other part of the body.

Inside the heart are four empty chambers, or sections. These chambers are "stacked," two on the top and two on the bottom. The upper chamber on each side is called an **atrium**. The lower chamber on each side is called a **ventricle**. The left and right sides of the heart are divided by a thick wall of muscle called the septum. The right side of the heart pumps blood to the lungs. The left side of the heart pumps blood to the rest of the body.

All four chambers of the heart work together to constantly pump blood. The upper chambers contract, or squeeze, at the same time. As they contract, they force the blood into the ventricles below them. Then both ventricles contract to push blood out into the blood vessels. While this is happening, each atrium relaxes and fills with blood.

The blood flows in only one direction. Blood enters the heart through the right atrium. When the right atrium contracts, it pushes the blood into the right ventricle. The blood goes through a small flap of tissue called a **valve**. The valve opens to let the blood through. Then the valve closes to prevent the blood from flowing backwards. The heart has valves between the atrium and ventricle on each side and where the blood vessels leave the ventricles.

The right ventricle of the heart contracts to send the blood to the lungs. In the lungs, the blood releases carbon dioxide and picks up oxygen. The blood flows from the lungs back to the heart and enters the left atrium.

Then the left atrium contracts. Its valve opens, and the blood flows down into the left ventricle. The left ventricle contracts and pumps the blood out into a large blood vessel that branches into smaller blood vessels that carry the blood throughout the rest of the body. This whole process occurs very quickly.

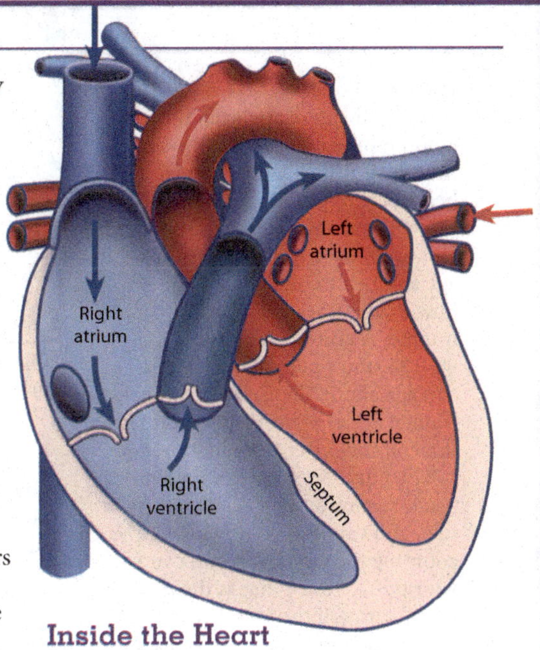

Inside the Heart

285

Background

Atria

The upper chambers of the heart are called atria. One of these chambers is an atrium. The term *atrium* comes from a Latin word meaning a "central court" or "main room."

Teach for Understanding

- Display the *Circulatory System* visual. Allow students to point out the different parts as they are discussed.
  What system did God design to deliver oxygen and nutrients to every part of the body? the circulatory system
  What are the three parts of the circulatory system? the heart, blood vessels, and blood
  What is the main organ of the circulatory system? the heart
  Where is the heart located in the body? in the chest, between the lungs
  How big is your heart? about the size of your fist
- Direct attention to the illustration of the heart on page 284.
  Are the lungs the same size? no
  Which lung is smaller? the left lung
  Why do you think the left lung is smaller? to allow room for the heart
- Display the *Inside the Heart* visual.
  How many chambers does the heart have? four
  What is the top chamber on each side of the heart called? an atrium
- Ask a student to point to each of the top chambers and name them. right atrium, left atrium
  What is each bottom chamber of the heart called? a ventricle
- Ask a student to point to each of the bottom chambers and name them. right ventricle, left ventricle
  Which part of the body receives blood from the right side of the heart? the lungs
  Which side of the heart pumps blood to the rest of the body? the left side of the heart
  What is the thick wall of muscle that divides the two sides of the heart? the septum
  Does each atrium of the heart beat separately or at the same time? at the same time
  Which chamber does the blood enter first? right atrium
  What happens when an atrium contracts? The blood goes to the ventricle below it.
  What is the purpose, or function, of the valves in the heart? The valves open to let blood flow through and then close to prevent blood from flowing backward.
- Display the *Path of the Blood* visual.
  Starting at the right atrium, describe the path of blood through the heart. Blood flows from the right atrium to the right ventricle, and then to the lungs. Next it flows back to the heart and into the left atrium. It then flows into the left ventricle and out to the rest of the body.
- Explain that it is customary to use the color red to show the blood that contains oxygen. The color blue is used to show deoxygenated or "used" blood.

When do heartbeats occur? each time the blood moves completely through the heart

Does a heart ever rest? yes When? only between beats

What is a person's heart rate? the number of times that the person's heart beats in a minute

How can you find a person's heart rate? by counting his pulse

What is a pulse? the push of blood through the blood vessels caused each time that the heart beats

Where are the easiest places to check a person's pulse? on his wrist or neck

- You may want to do the *Finding the Pulse* activity at this time. Each student will need to know how to find his pulse and calculate his heart rate before doing the *How Fast Is the Beat?* activity in Lesson 170.

What sound does the heart make? lubb-dubb

What causes the lubb-dubb sound? the valves in the heart closing

What does a doctor use to listen to a patient's heart? stethoscope

- Allow the students to listen to their heart through a stethoscope.

How does getting enough sleep provide rest for a person's heart? His heart beats more slowly while he is sleeping.

How does exercise benefit the heart? It makes the heart stronger so that it does not have to beat as often.

Why does a strong, healthy heart not have to beat as often? It is able to push a greater amount of blood into the blood vessels each time it beats.

Heartbeats

A heartbeat occurs each time the blood moves completely through the heart. With each heartbeat, the heart pushes blood into the body. The heart beats many times each minute and can rest only between beats. The age and health of a person will affect how many times his heart beats per minute. A child's heart beats about 90–120 times per minute. Adults average about 72 beats per minute. A trained athlete's heart may beat fewer than 60 times per minute.

The number of times your heart beats in a minute is called your heart rate. You can find your heart rate by counting the times that you can feel your pulse. Your **pulse** is the push of blood through the blood vessels. It occurs each time your heart beats. You can feel your pulse several places on your body. The easiest places to check your pulse are probably on your wrist or on your neck.

The heart slows down while a person sleeps.

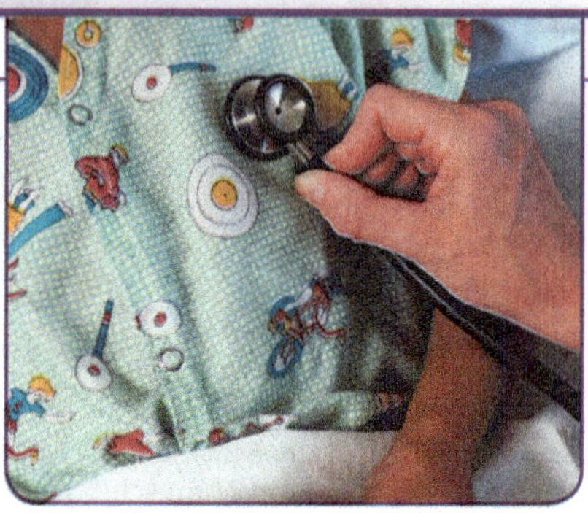

A doctor uses a stethoscope to listen to a child's heart.

Your heart makes a "lubb-dubb" sound as it beats. This is the sound that the valves in your heart make as they close. A doctor can tell much about how your heart is functioning by listening to the sounds it makes. He uses a device called a *stethoscope* to hear your heart beating. If the sound is not normal, a problem may exist. The doctor may do more tests to check it out.

Although your heart is always beating, it does not always beat at the same rate. When you are active, it beats quickly. When you are sitting still, it beats more slowly. Your heart beats even slower yet when you are sleeping. Getting enough sleep can provide more rest for your hard-working heart!

Exercise is important, too. The heart muscle needs to be exercised in order to be strong. Hearts that are healthy and strong do not have to beat as often.

286

Background

Pacemaker

The heart's natural pacemaker, also called the sinoatrial node, is located in the right atrium. Its function is influenced by many things, such as the nervous system, hormones, and mineral levels in the body.

Artificial Pacemakers

If a person's natural pacemaker is not working properly, a doctor may decide to use an artificial pacemaker. This pacemaker has a tiny computer powered by batteries that is implanted in the patient's chest and attached to the heart. The artificial pacemaker stimulates the heart to beat and keeps the heartbeats regulated.

Activity

Finding the Pulse

Demonstrate how to find the pulse on the wrist and on the neck.

Direct the student to find his own pulse. Provide assistance as needed. If a student is having difficulty, have him run in place to increase the strength of his pulse.

Direct each student to count his pulse for 15 seconds. Multiply that number by 4 to calculate the heart rate.

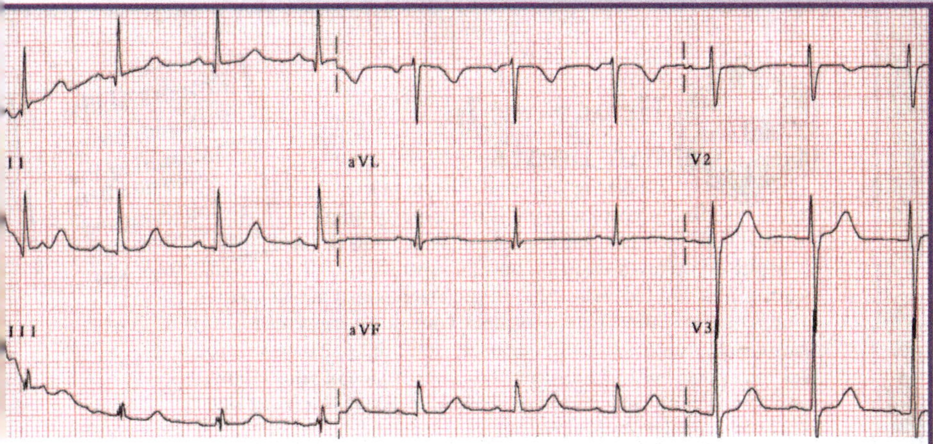

An EKG printout shows the electrical changes in a person's heart.

One part of your heart acts as a "control center." This control center, called a **pacemaker**, is a small group of cells that makes sure your heart beats at a steady pace. Your pacemaker cells send electric currents to your heart muscle. Each electric current causes your heart to contract. These cells set your heart's pace, or the number of times, that it beats.

Your body gets oxygen from the blood that your heart pumps around your body. Your pacemaker constantly receives messages from your body about how much oxygen your body needs. Your pacemaker then adjusts your heart rate to match the amount of oxygen you need. For example, when you exercise, your body uses more oxygen. So your pacemaker makes your heart beat faster. This pumps blood around your body more quickly.

The electrical signals made by your pacemaker can be read by a special machine. This machine is called an *electrocardiograph*, or EKG. This machine records the electrical changes that happen during the heartbeat cycle. Doctors can use this information to detect heart problems.

QUICK CHECK

1. What are the lower chambers of the heart called?
2. What causes the "lubb-dubb" sound of the heartbeat?
3. What is the function of the pacemaker?

287

168 169

What is the "control center" of the heart? the pacemaker

What is the heart's pacemaker? the small group of cells that sends an electric current to the heart muscle, causing it to contract

How does the pacemaker know how fast or slow the heart should beat? It receives messages from the body about how much oxygen is needed and adjusts the heart rate to match the need.

If more oxygen is needed by the body, will the heart beat faster or slower? faster

What machine can doctors use to read and record the electrical changes that happen while the heart beats? an electrocardiograph, or EKG

What is one way that doctors use information from an EKG? to detect heart problems

ANSWERS

1. ventricles
2. the valves in the heart closing
3. It sends an electric current to the heart to keep it beating, and it sets the pace of the heartbeats.

169

- The *Finding the Pulse* activity found at the bottom of page 328 may also be incorporated into the lesson.
- Suggestions for additional activities for this two-day lesson are on the Teacher's Toolkit CD.

Activity Manual

Reinforcement, page 190

Review, pages 191–92

These pages review Lessons 168–69.

Assessment

Quiz 12-A

The quiz may be given any time after completion of this lesson.

Background

Physiologist Willem Einthoven

Dutch physician and physiologist Willem Einthoven (1860–1927) developed the first electrocardiograph. Willem enjoyed sports and, as a result of a sports injury, began researching how parts of the human body worked. Eventually, he focused his research on the electrical currents of the heart. He worked for three years to develop an accurate string galvanometer that could record these currents. He called the record an electrocardiogram. Willem continued to study and research the heart and parts of the eye. In 1924 he won the Nobel Prize in Medicine for his discovery of the mechanism of the electrocardiogram.

Electrocardiograph

The first electrocardiograph, developed by Willem Einthoven, was a string galvanometer weighing about 600 lb. The string was made from a very thin thread of quartz held in a magnetic field. The thread was silvered so that it would reflect a beam of light. When an electrical current passed through the string, it would move slightly and make an image on the photographic paper below. The electrocardiograph would be attached to a patient's limbs and chest in nine places. The patient usually had to keep his hands and one foot in a strong salt solution in order to help the electrical conduction. In 1902 Einthoven published the first electrocardiogram recorded by his string galvanometer. Modern electrocardiographs are a direct result of his first electrocardiographs.

Student Text pages 288–89
Activity Manual pages 193–94

Objectives

- Calculate heart rate
- Calculate how long it takes the heart rate to return to normal after exercise
- Make a line graph using the heart-rate data

Materials

Materials needed for this activity are listed on the Student Text page.

Teacher's Toolkit CD

- Rubrics: *How Fast Is the Beat?*, for each student

Introduction

When you exercise, does your heart rate increase or decrease? increase

If you stop exercising, what happens to your heart rate? It gradually returns to normal.

In this activity you will measure how much your heart rate increases and then time how long it takes to return to normal.

Preparation for Reading

- Instruct the students to read Student Text pages 288–89 and Activity Manual pages 193–94 before beginning the activity.

Teach for Understanding

Process Skill

The process skill emphasized in this activity is measuring and using numbers.

What are some problems that may cause inaccurate measuring in this activity? Possible answers: not being able to find the pulse, not counting the pulse accurately, not watching the time

How could you avoid some of these problems? Possible answers: Once you find your pulse, keep your fingers in that place until finished with the activity. Close your eyes while counting to avoid distraction. Have the person watching the time not record while watching.

How did you have to use the measurements that you recorded? to calculate your heart rate and make a line graph

ACTIVITY

How Fast Is the Beat?

Process Skills
- Hypothesizing
- Measuring and using numbers
- Collecting and recording data

Checking your pulse is one of the first ways to find out if there is anything unusual with your circulatory system. When you check your pulse, you are measuring how fast your heart is beating. You can feel your pulse in the arteries that are closest to your skin. The easiest places to check your pulse are probably on your wrist or on the side of your neck. You might also be able to feel your pulse behind your knee or on the inside of your elbow.

In this activity you will test to see how quickly your heart rate returns to normal after exercising.

Problem

How long will it take for your heart rate to return to normal after exercising for two minutes?

Procedure

Materials
stopwatch
Activity Manual

1. Complete the hypothesis in your Activity Manual. Then sit quietly for 1 minute.
2. Find your normal heart rate. Use two fingers to find your pulse on the underside of your wrist or on the side of your neck. Count how many times you feel your pulse in 15 seconds. Record that number in the Pulse column. Calculate and record your Normal Heart Rate.
3. Stand up and run in place for 2 minutes.
4. Find your pulse. Count how many times you feel it in 15 seconds. Have your partner record that number in your chart.
5. Keep your fingers on your pulse. For the next 5 minutes, check your pulse every 30 seconds. Check for 15 seconds, then rest for 15 seconds. Have your partner record the numbers. Stop checking your pulse when it matches your normal pulse.

288

6. Switch places with your partner. Repeat steps 2–5 for your partner. Record his numbers in his Activity Manual.
7. Calculate your heart rate for each number that was recorded.
8. Make a line graph in your Activity Manual to show how your heart rate changed.

Conclusions

- Did your results support your hypothesis?
- How long did it take your heart rate to return to normal?
- How can this kind of information be useful for a person playing sports or doing other activities?

Follow-up

- Use a different physical activity for exercising.
- Change the amount of time that you spend exercising.

289

Procedure

This activity could be done in groups of three if each group has its own stopwatch. However, it may be easier to use groups of two for this activity and have the teacher monitor the time for all of the groups.

- Demonstrate how to find the pulse on the wrist or on the neck.
- Direct the students to record their hypotheses before starting the activity.
- Monitor the time as one member of each group follows steps 2–5; then allow partners to switch places.
- Provide time for the students to calculate their heart rates and to complete the Activity Manual pages.
- Before each student starts to make his line graph, direct him to plot his starting pulse on the *y*-axis (the vertical line). Then when he connects his points, he will be able to see both the increase and the decrease in his heart rate. Guide the students as needed in completing the line graph.

Conclusions

Did your results support your hypothesis?

How long did it take your heart rate to return to normal?

Healthy hearts usually have a recovery rate of about 3–5 minutes. How do you compare?

Activity Manual

Activity, pages 193–94

Assessment

Rubrics

Select the prepared rubric, or design a rubric to include your chosen criteria.

Student Text pages 290–93
Activity Manual page 195

Objectives

- Identify and describe the three types of blood vessels
- Name the largest artery and the largest veins
- Differentiate between arteries and veins
- Recognize that the exchange of gases takes place in the capillaries
- Explain why William Harvey is important as a scientist and a physician

Materials

- beanbag
- stopwatch

Teacher's Toolkit CD

- Visual 12.1: *Circulatory System*
- Quiz 12-B

Vocabulary

- arteries
- veins
- capillaries
- coronary
- pulmonary
- aorta
- superior vena cava
- inferior vena cava

Two days have been allotted for this lesson. You may want to teach Lesson 171 the first day and use ideas from Lesson 172 on the second day. You may also incorporate the ideas from Lesson 172 into the text material, making the lesson two days.

Introduction

- Direct the students to pass the beanbag from person to person around the room. Use the stopwatch to time how long it takes for the beanbag to return to the first person. Tell the students how long it took.
  The beanbag represents one drop of blood. What do you think our network of people represents? the blood vessels in the body
  How much time do you think it takes for one drop of blood to return to the heart? Answers will vary.

Preparation for Reading

- Display the words *coronary*, *aorta*, *arterioles*, *venules*, and *vena cava*. Pronounce the words together.
- Tell the students to mentally make questions using the headings, bold and italicized words, and captions.
- Instruct them to read silently pages 290–93 to find the answers to these questions.
  How long does it take for one drop of blood to travel through the circulatory system?
  Is the pulse felt in an artery or in a vein?
  What do veins have that arteries do not have?

Teach for Understanding

What are blood vessels? the "pipes" that carry blood around the body

How long does it take for one drop of blood to travel through the circulatory system? about one minute

Your Blood Vessels

Blood vessels are the "pipes" that carry blood throughout your body. There are billions of them. In fact, if all the blood vessels in your body were laid end to end, they would be long enough to wrap around the earth at least two and a half times!

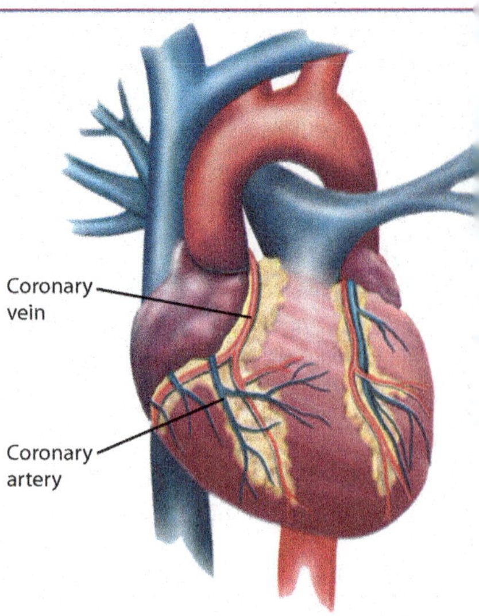

The Circulatory System

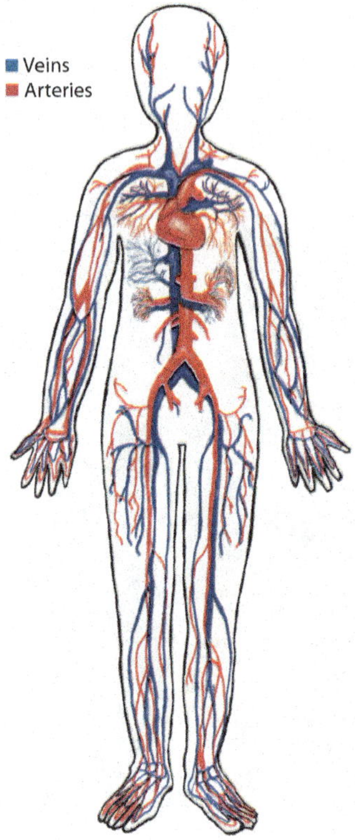

All your blood vessels are linked together. This provides the "transport system" for your blood. It is very effective. It takes only about one minute for your blood to travel through your circulatory system.

Blood vessels are classified by where they carry blood. **Arteries** carry blood from the heart to all the parts of the body. **Veins** carry the blood back to the heart. **Capillaries** are the small blood vessels that connect the arteries and veins.

All cells, even those that make up the heart, need blood. Arteries and veins are often named by where they are located in the body. **Coronary** is a word that means "relating to the heart." A network of blood vessels

290

Background

Length of Vessels

The length of the circulatory system of an adult is over 100,000 km (62,000 mi). Most of the length is the network of capillaries.

Pulmonary Blood Vessels

Arteries always take blood away from the heart. Most arteries carry oxygen-rich blood. However, the pulmonary artery carries oxygen-depleted blood from the right ventricle of the heart to the lungs.

Veins always carry blood back to the heart. Most veins carry oxygen-depleted blood. However, the pulmonary vein carries oxygen-rich blood from the lungs to the left atrium of the heart.

Science Misconceptions

Blue Blood

We often talk about blue blood, but there is really no such thing. Veins seen through the skin often appear blue because most of the light waves penetrating the skin are absorbed. The remainder are reflected back to the eye as blue.

supplies the heart with blood. The arteries and veins that provide blood for the heart are called coronary arteries and coronary veins. **Pulmonary** is a word that means "relating to the lungs." The artery that goes to the lungs is called the pulmonary artery. The pulmonary veins come from the lungs back to the heart.

Arteries

Arteries have thick, muscular walls. The artery walls can expand and relax as the heart pumps blood through them. The heart contracts and pushes blood into the arteries. This makes the artery walls expand, or stretch out. When the heart relaxes, the walls spring back into place. This causes the arteries to become smaller. The blood is pushed forward. When you feel your pulse, you are feeling the artery wall expand and then relax.

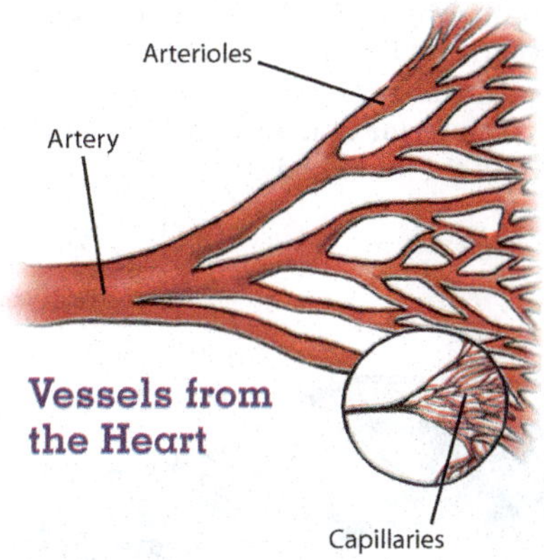

Vessels from the Heart

The **aorta** is the largest artery in the body. It is about the diameter of a garden hose. The aorta is attached to the left ventricle of the heart and carries blood away from the heart. Arteries branch off the aorta and form smaller arteries. The smallest arteries are called *arterioles*.

Capillaries

Capillaries are the smallest blood vessels. They are so small, in fact, that they can be seen only with a microscope. It takes about 10 capillaries to equal the thickness of one human hair!

Capillaries spread out from the arterioles in a web of blood vessels. The body has many of these webs. Every cell in the body is close to a capillary.

Even though they are small, capillaries have a very important job. Capillaries have thin walls. As the blood in the capillaries flows past other cells, oxygen and nutrients in the blood pass through the capillary walls into surrounding cells. At the same time, carbon dioxide and other wastes from the cells pass into the capillaries and are carried away by the blood. This process keeps the cells healthy.

291

Demonstration

How Arteries Expand and Contract

Materials: long balloon, water

Fill the balloon about one-third full of water and tie a knot to keep the balloon closed. Lay the balloon on a flat surface. Wrap your hand around one end of the balloon and gently squeeze all the water into the other end.

What happens to the water when I squeeze the balloon? The water gets pushed to the other end.

How did the balloon change as the water was pushed to the other end? The part of the balloon that now holds the water expanded.

What happened to the part of the balloon where the water used to be? That part contracted.

How does squeezing this balloon model how an artery works? When the heart pumps, it pushes the blood through the arteries. The arteries expand as the blood pushes through and contract when the heart relaxes.

Which blood vessels carry blood from the heart to all parts of the body? arteries

Which blood vessels carry blood back to the heart? veins

What are capillaries? the smallest blood vessels that connect the arteries and the veins

- Direct attention to the *Circulatory System* visual.

What color are the veins? blue

What color are the arteries? red

Why do you think red is used for the arteries? Elicit that oxygen makes blood a bright red.

In most circulatory diagrams, red indicates blood flowing from the heart and blue indicates blood flowing toward the heart.

Where are the coronary arteries and coronary veins located? in the heart

What adjective is used to describe the artery and veins that go to and from the lungs? pulmonary

What kind of walls do arteries have? thick, muscular walls

What happens to the walls of the arteries when the heart contracts and relaxes? The walls of the arteries expand when the heart contracts and pushes blood into them. When the heart relaxes, the artery walls spring back, pushing the blood along.

- You may want to do the *How Arteries Expand and Contract* demonstration at this time.

Is the pulse felt in an artery or in a vein? in an artery

What is the largest artery? the aorta Where is it located? It is attached to the left side of the heart.

What are the smallest arteries called? arterioles

In which type of blood vessel does the exchange of gases take place? capillaries

Why are capillaries visible only through a microscope? They are very small.

How many capillaries does it take to equal the thickness of one human hair? about 10 capillaries

- Discuss the *Comparing Sizes of Vessels* diagram. Point out the differences between the thicknesses of the blood vessels' walls.
- Remind the students that arteries have thick, muscular walls. Veins have thinner walls but have valves as shown in the diagram at the bottom of the page.

What are the smallest veins called? venules

Are vein walls thinner or thicker than artery walls? thinner

What helps push the blood in the veins back to the heart? muscles in the body

What do veins have that arteries do not have? valves

What is the purpose of the valves in the veins? to keep blood from flowing backward

What other part of the circulatory system has valves? the heart

What are the venae cavae? the largest veins in the body

Compare the superior vena cava and the inferior vena cava. The superior vena cava carries blood going back to the heart from the upper body. The inferior vena cava carries blood going back to the heart from the lower body.

How could you remember which vena cava carries the blood from the different parts of the body? Answers will vary. *Superior* often refers to the head of something, and the superior vena cava carries blood from the head and the upper part of the body. *Inferior* often refers to the foot of something, and the inferior vena cava carries blood from the feet and the lower body.

How large are the venae cavae? about 2.5 cm (1 in.) in diameter

Name some of the wrong ideas that people had about the circulatory system in the 1500s and 1600s. Possible answers: Air flowed through the arteries and veins. Arteries came out of the heart and veins out of the liver. Food was changed into blood by the liver. Blood was consumed by the body. The lungs moved blood through the body.

Why do you think people had these strange ideas? Possible answers: They did not understand how the human body worked. Dissecting bodies of animals or humans to learn how body systems were interconnected was not a common practice.

Comparing Sizes of Vessels

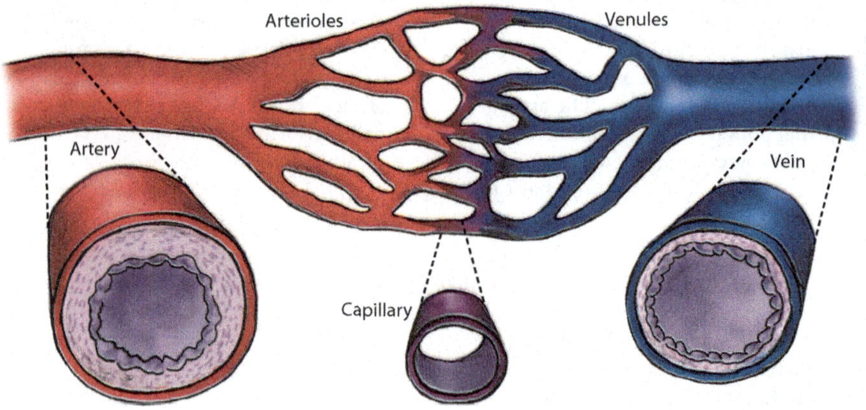

Veins

Capillaries link the smallest arteries with the smallest veins, which are called *venules*. These venules eventually become the larger blood vessels called veins. Veins return the blood to the heart.

Vein walls are thinner than artery walls are. The muscle fibers in veins are also not as strong as the muscle fibers in arteries are. Muscles in your body help push blood through your veins and back to your heart. Like the heart, veins have valves that keep the blood from flowing backward.

All the veins in the body flow into two large veins in the body. They are about 2.5 cm (1 in.) in diameter. Both veins carry blood into the right ventricle. The **superior vena cava** carries blood from the upper body. The **inferior vena cava** carries blood from the lower body.

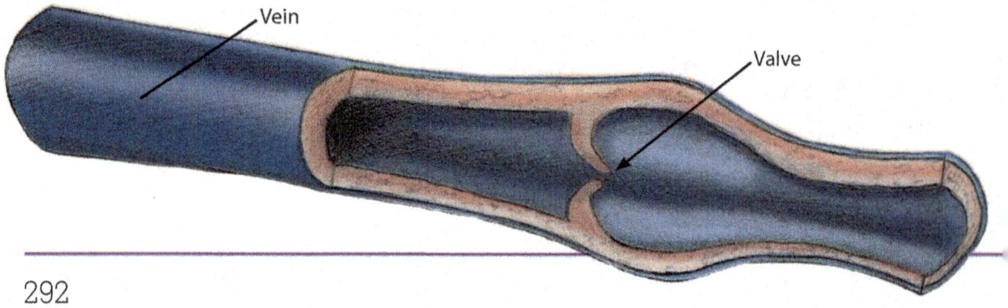

292

Background

Varicose Veins

Sometimes the valves in the veins do not work correctly. When this happens, the force of gravity may cause the blood to flow down to the legs. Sometimes the veins in the legs become larger and can be seen through the skin. This condition is called varicose veins.

171
172

William Harvey

People have not always understood how the circulatory system works. Some thought that air, not blood, flowed through arteries and veins. Later, people thought that arteries came out of the heart and veins came out of the liver. They believed that food was changed into blood by the liver and that the blood was consumed by the body. Most people in the 1500s and 1600s thought that the lungs moved blood through the body.

William Harvey, an English doctor born in 1578, was the first to understand how the circulatory system really works. By dissecting animals and studying people, Harvey knew that the theories of his day were wrong. He discovered that the arteries, veins, heart, and lungs are all part of one big circulatory system.

In 1628 Harvey published a book based on his research. He was the first to say that the heart is a pump that moves blood through the body. He explained how the blood is circulated through the body and then returns to the heart. His book had many critics at first. People said he was crazy. As man's technology improved, though, Harvey's ideas were shown to be true.

William Harvey

His ideas became the basis for all modern research on the heart and blood vessels.

QUICK CHECK

1. Which blood vessels carry blood away from the heart?
2. Which blood vessels carry blood toward the heart?
3. What keeps blood from flowing backward in the vessels and in the heart?

293

Who was the first to realize how the circulatory system actually worked? William Harvey

What did Harvey discover about the circulatory system? He realized that the arteries, veins, heart, and lungs are part of one big circulatory system.

How did Harvey describe the heart? as a pump that moves blood through the body

How did William Harvey share his theories with others? He wrote and published a book.

Did everyone agree with what Harvey had written in the book? no

What helped people change their ideas about Harvey and his theories? technology

What have Harvey's theories become? the basis for modern research on the heart and blood vessels

- Explain that science is always changing. As new technology improves people's observations, old ideas often have to be changed to fit new information. God, however, does not change. Malachi 3:6 says, "For I am the Lord, I change not." New discoveries will not change the character of God, the Creator of the universe.

ANSWERS

1. arteries
2. veins
3. valves

172

- The *How Arteries Expand and Contract* demonstration found at the bottom of page 333 may also be incorporated into the lesson.
- Suggestions for additional activities for this two-day lesson are on the Teacher's Toolkit CD.

Background

Physician William Harvey

William Harvey (1578–1657) was educated in England as well as in Italy. While he was in Italy, Harvey became interested in studying how the blood flows through the body.

When he returned to England, Harvey became a physician. His research of the circulatory system was funded by King James I of England, the same king who authorized the translation of the King James Version of the Bible. Harvey married Elizabeth Browne, the daughter of the court physician. He eventually became the personal physician of both James I and later his son Charles I, who was beheaded by Oliver Cromwell. William Harvey died in 1657.

Harvey's Book and Theory

William Harvey's book *On the Movement of the Heart and Blood in Animals* was published in 1628. Harvey had finished his research over a decade earlier but had delayed publishing his work. He knew that his theories would raise great controversy and that his medical practice would probably suffer. For many years people debated Harvey's work. But in time, people realized that his theories were correct. Harvey had hypothesized that there was a connection between arteries and veins, but it was not until about 33 years later that the Italian physician Marcello Malpighi was able to see capillaries when observing a frog's lung under a microscope.

Activity Manual

Review, page 195

This page reviews Lessons 171–72.

Assessment

Quiz 12-B

The quiz may be given any time after completion of this lesson.

173
174

Student Text pages 294–98
Activity Manual page 196

Objectives

- Identify the contents of blood
- Describe platelets, red blood cells, and white blood cells
- Name the four main blood types
- Describe a blood donation

Materials

- prepared lemon gelatin containing three different items (e.g., cherries, miniature marshmallows, and crushed pineapple)

Teacher's Toolkit CD

- Quiz 12-C

Vocabulary

- plasma
- red blood cell
- white blood cell
- platelet
- blood transfusion

Two days have been allotted for this lesson. You may want to teach Lesson 173 the first day and use ideas from Lesson 174 on the second day. You may also incorporate the ideas from Lesson 174 into the text material, making the lesson two days.

Introduction

- Display a square of the prepared gelatin.

  What is this type of salad called? a gelatin salad

  What items are in this salad? cherries, miniature marshmallows, and crushed pineapple

  What substance holds these items together? the gelatin

  Just like this salad, our blood is made up of four parts.

Preparation for Reading

- Instruct the students to mentally make questions using the headings, bold words, and captions.
- Direct them to read silently pages 294–98 to find the answers to these questions.

  What are the three jobs of blood?

  Which type of blood cell helps to fight infections?

  What does the term *universal donor* refer to?

Teach for Understanding

How much blood does an adult have in his body? about 4.7 L (5 qt)

What are the three jobs of blood? to transport substances around the body, to help keep the body's temperature at about 37 °C (98.6 °F), and to help defend the body against infections and diseases

What are the four parts of blood? plasma, red blood cells, white blood cells, and platelets

What is the liquid part of blood called? plasma

Your Blood

The circulatory system is more than a heart with a network of vessels. It carries amazing, life-giving blood. An adult usually has about 4.7 L (5 qt) of blood flowing through his body. The blood has three very important jobs. Blood transports substances around the body. As the blood circulates, it helps keep the body's temperature at about 37 °C (98.6 °F). Blood also helps to defend the body against infections and diseases.

Blood is made up of plasma, red blood cells, white blood cells, and platelets. **Plasma** is the liquid part of blood. It is a clear, yellowish liquid that is mostly water. The plasma can dissolve many substances, such as proteins, sugars, and nutrients. The blood then carries these substances to all the parts of the body. The red blood cells, white blood cells, and platelets float in the plasma. One tiny drop of blood contains at least five million red blood cells, eight thousand white blood cells, and 250 thousand platelets.

Contents of a Drop of Blood

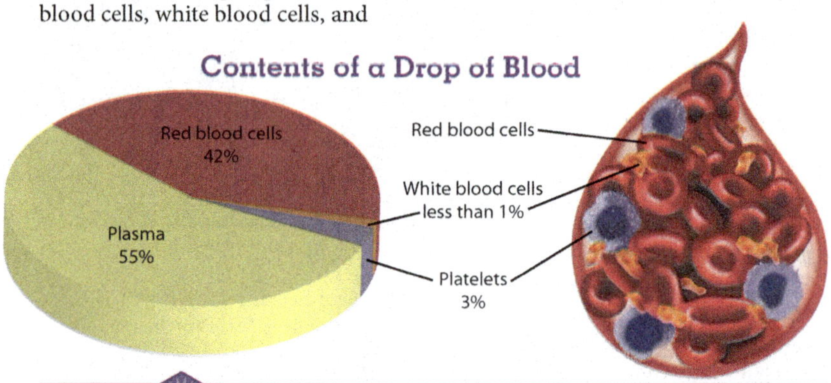

SCIENCE & THE BIBLE

Long before people understood the importance of blood, the Bible said, "the life of the flesh is in the blood" (Leviticus 17:11). In the Old Testament, atonement for sins could occur only when blood was shed. These sacrifices were a sign of a sacrifice to come. They pointed to the great sacrifice that Jesus Christ made for the sins of all people. Hebrews 9:22 says that "without shedding of blood is no remission." Only when Christ shed His precious blood could mankind's sins be forgiven.

294

Background

Making Blood

The body can make more blood to replace any blood that has been lost. Most blood cells form in the bone marrow and then move to other body organs to finish developing. A small amount of blood cells form in the spleen and in the lymph nodes.

Exchanging Gases

Although red blood cells are like buses, they do not come to a stop when they let off oxygen and pick up carbon dioxide. The blood is always moving. The oxygen and carbon dioxide are exchanged while the red blood cells are still moving.

Hemoglobin and Blood Color

The oxygen attaches to the iron in the hemoglobin molecule. This causes the oxygenated blood to be a bright red color. Carbon dioxide attaches to the globin protein in the hemoglobin molecule. The blood is not blue but is actually a dark purplish-red. Hemoglobin molecules can carry carbon dioxide at the same time as they carry oxygen. Each hemoglobin molecule is able to carry four carbon dioxide molecules and four oxygen molecules. About 70% of the carbon dioxide does not attach to hemoglobin but travels in the blood unattached.

Red Blood Cells

Red blood cells are tiny blood cells that carry oxygen to all parts of the body. Each red blood cell is concave on both sides. This shape helps them squeeze through small capillaries without bursting.

Each red blood cell is like a bus traveling on a route through the body. At the different "stops," the red blood cell "bus" lets oxygen out and picks up some carbon dioxide. When the "bus" travels through the lungs, it lets out the carbon dioxide and picks up oxygen.

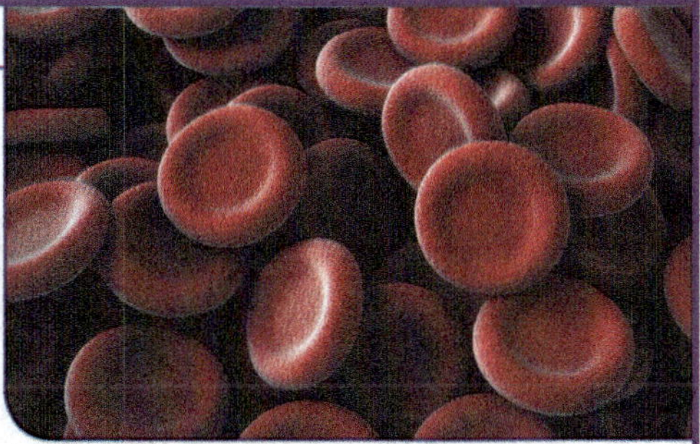

Red blood cells have a biconcave shape.

Red blood cells are made in the bone marrow in the center of bones. About 100 million of them are produced every minute. If the body does not produce enough red blood cells, a person might become anemic. People who are anemic are often tired and may feel dizzy at times. These symptoms are caused by a lack of oxygen throughout the body. With fewer red blood cells, not as much oxygen can be carried by the blood.

Each red blood cell lives about 120 days. As red blood cells get older, they lose their shape and become fragile. They shrink in size and are carried to the spleen or the liver. There they are destroyed. Most parts of the old red blood cells are recycled by the body and used to make new cells.

SCIENCE & MATH

Inside each red blood cell is a protein called *hemoglobin*. Oxygen from the lungs attaches itself to the hemoglobin molecule. In this way, oxygen is carried to each part of the body. Together, the hemoglobin and oxygen give blood its bright red color. Each red blood cell has about 250 million hemoglobin molecules. Each of these molecules can carry four molecules of oxygen. So each red blood cell carries about one billion oxygen molecules!

295

Background

Carbon Monoxide

Carbon monoxide acts differently in the body than carbon dioxide does. When a person inhales carbon monoxide, it binds to the places on the hemoglobin molecules that would usually carry oxygen. This prevents hemoglobin molecules from picking up oxygen and, in turn, can cause death.

- Discuss the *Contents of a Drop of Blood* graph on page 294.
  What percentage of the blood is made up of platelets? 3%
  Is there more plasma or red blood cells in one drop of blood? plasma
- Discuss *Science & The Bible* on page 294.
  In the Old Testament, what was the requirement for atonement, or forgiveness, for sins? Blood had to be shed.
  What were the Old Testament sacrifices pointing toward? Jesus Christ's death on the cross for our sins
  How can we have our sins forgiven? by accepting Christ's shed blood as the payment for sin
- Direct attention to the illustration of the red blood cells on page 295.
  Which part of the blood carries oxygen to all parts of the body? red blood cells
  What shape do red blood cells have? biconcave (concave on both sides)
  What does the prefix *bi-* mean? two, or double
  What does *concave* mean? curving inward
  Why is this term a good description of red blood cells? They are concave shaped on both (two) sides.
  Why is this shape important? It allows red blood cells to squeeze through capillaries without bursting.
  What gases are exchanged by red blood cells? carbon dioxide and oxygen
  In what part of the body do red blood cells pick up oxygen? lungs
  What problem can occur if a person's body does not produce enough red blood cells? anemia
  How long do red blood cells live? about 120 days
  What happens to red blood cells as they get older? They lose their shape, become fragile, shrink in size, and are carried to the spleen or the liver.
- Discuss *Science & Math.*
  What is hemoglobin? a protein found in red blood cells
  What gives blood its bright red color? the combination of hemoglobin and oxygen
  How many molecules of oxygen can one hemoglobin molecule carry? four
  How many hemoglobin molecules are in one red blood cell? about 250 million
  About how many oxygen molecules can one red blood cell carry? one billion

Which blood cells help the body fight diseases and infections? white blood cells

Are all the types of white blood cells the same? no

What are some jobs that white blood cells have? to recognize and identify germs, to make chemicals to fight the invading germs, and to surround and kill the invaders

Does the body have more red or white blood cells? more red blood cells

When might your body make more white blood cells? when you are sick

What part of the blood helps form blood clots? platelets

What are platelets? small fragments of cells

What is a blood clot? a jellylike mass of blood cells that stops the flow of blood from an injury

If the skin is injured, what do the platelets do? help plug the broken blood vessels

When this happens, what forms on the surface of the skin? a scab

What is happening underneath a scab? The body is repairing damaged tissues.

Are platelets larger or smaller than red blood cells? smaller

How long do platelets live? about nine days

Where do new platelets form? in bone marrow

- Refer to the gelatin salad used in the Introduction. Which part of blood is represented by each part of the salad? Lemon gelatin is the plasma, crushed pineapple is the platelets, cherries are the red blood cells, and marshmallows are the white blood cells.

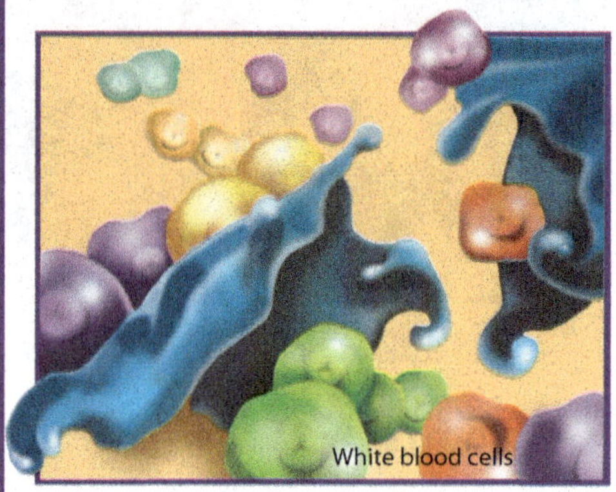

White blood cells

White Blood Cells

White blood cells help the body fight diseases and infections. The body has many types of white blood cells. Some recognize and identify germs such as bacteria or viruses. Others make chemicals that fight the invading germs. Yet another type of white blood cell surrounds and kills the invaders. White blood cells can move to wherever they are needed to fight the invading germs.

Your body has fewer white blood cells than it has red blood cells. However, your body can make more white blood cells when they are needed. For example, when you are sick, your body makes more white blood cells to fight the infection or disease. A person whose body does not produce many white blood cells gets sick easily. His body cannot protect him from invading germs.

Platelets

Platelets are small fragments of cells. These fragments help form blood clots. A blood clot is a jellylike mass of blood cells that stops the flow of blood from an injury. If the skin is injured, the platelets help plug the broken blood vessels. A scab forms on the skin. Underneath the scab, the body begins repairing the damaged tissues. The scab makes a barrier to keep out germs until the skin grows back. The scab falls off when the tissues have healed.

Platelets also make sure that blood vessels do not leak. They can stick to torn blood vessels and form a patch that slows any blood loss. A platelet is less than half the size of a red blood cell. Each usually wears out in about nine days. However, new platelets are always being formed in the bone marrow.

296

Background

Scabs and Bruises

To form a scab, platelets set off a chain of reactions that cause a protein called fibrinogen to form tiny threads called fibrin. These threads make a web that traps blood cells. The trapped blood cells harden as they dry, forming the scab. Bruises are the bleeding and the tiny clots that occur inside the body when blood vessels (capillaries) are damaged or broken.

Positive and Negative Blood Types

In addition to the proteins that identify a person's blood type, the blood may also contain another protein, often referred to as the Rhesus (Rh) factor. If a person's blood has this protein, his blood is Rh+. If he does not have this protein, it is Rh-. Like the blood types, the Rh factor is hereditary and must be matched in order for blood transfusions to be successful. A mother with Rh- blood whose baby has Rh+ blood can receive a vaccine to prevent problems during pregnancy and delivery.

Blood Types

For many years, people tried to transfer blood from a healthy person to a sick or injured person. Sometimes this transfer of blood, called a **blood transfusion**, helped the patient. Many times, however, the patient's condition got worse, sometimes ending in death. It was not until 1901 that Karl Landsteiner, an Austrian physician, discovered why blood transfusions did not always work. He found that blood is not all the same. Instead, there are different types.

Today we know that there are four main types of blood. We also know that some types cannot be successfully mixed with another type. A person who gives blood is a blood donor and a person who receives a blood transfusion is a blood receiver.

The main blood types are A, B, AB, and O. These names identify the special proteins found in the red blood cells. Some people have only the A protein or only the B protein. Other people have both proteins, and some have neither.

If a person with type A blood receives type B blood, the white blood cells in the A blood identify the B proteins as enemies. In trying to get rid of the enemy, the white blood cells cause the red blood cells to clump together. This clogs the blood vessels, which does not help the patient.

Since type O blood does not have either type of protein, it can safely mix with any other type of blood. People with type O blood are called universal donors. Those with type AB blood are called universal receivers because blood with both A and B proteins can safely receive any other type of blood.

Blood Types

Blood Type	Special Protein	Can Give Blood to	Can Receive Blood from
A	A	A and AB	A and O
B	B	B and AB	B and O
AB	A and B	AB	Anyone
O	Neither	Anyone	O

297

- Refer to the *Blood Types* chart as blood types are discussed.

What happens during a blood transfusion? An injured person receives blood from a healthy person.

Who discovered that there are different blood types? Karl Landsteiner

What are the names of the four main blood types? A, B, AB, and O

What do these letters identify about each blood type? the special proteins found in the red blood cells

Why are some blood types not able to mix with other types? Special cells in some blood may identify a protein from other blood types as an enemy, causing red blood cells to clump together and clog blood vessels.

Does every blood type have a special protein? no
Which blood type does not have either of the special proteins? type O

Why is type O blood called the universal donor? It can safely mix with any other type of blood.

Which type of blood is called the universal receiver? type AB Why? It has both the A and the B proteins, so it can safely receive any other type of blood.

Background

Artificial Blood

For many years, scientists have been experimenting with making artificial blood. Artificial blood is made from donated red blood cells and is kept in a powder form, which can be mixed into a liquid form when needed. It can be used regardless of a patient's blood type and could be useful in emergencies. It does not replace a patient's blood, but it helps the blood get enough oxygen to the different organs of the body. However, many types of artificial blood have had some unwanted side effects at times, such as collapsed capillaries and increased blood pressure. Scientists are still working on ways to make artificial blood safe and practical for use in emergency situations.

173
174

What are some reasons that a person might need a blood transfusion? Possible answers: being in an accident, having surgery

What makes it possible for people who need blood to receive a transfusion? donations of blood from other people

What are some typical requirements for donating blood? be at least seventeen years old, be healthy, and weigh at least 110 lb

What usually happens to blood after it is donated? It is tested to determine the blood type and is checked for diseases. It is often separated into its parts.

Why is donated blood often separated into parts? so that several people can use the same unit of donated blood

- Explain that people who donate their blood are called blood donors.

  Why are blood donors always needed? Most donated blood can be stored for only a short time before it is no longer useful.

- Discuss *Meet the Scientist.*

  What are two things that Dr. Charles Drew discovered about blood? He found a way to separate plasma from red blood cells, and he discovered that liquid plasma could be stored longer than whole blood.

  What did Dr. Drew do that helped soldiers in World War II? He established blood banks and found ways to preserve plasma.

  What are "bloodmobiles"? refrigerated trucks that can carry blood to where it is needed

  Why was it important that plasma be able to be shipped over long distances? Wounded soldiers needed to be able to get blood that had been donated in other places.

ANSWERS

1. plasma
2. red blood cells
3. white blood cells
4. A, B, AB, and O

174

- Suggestions for additional activities for this two-day lesson are on the Teacher's Toolkit CD.

Activity Manual

Review, page 196

This page reviews Lessons 173–74.

Assessment

Quiz 12-C

The quiz may be given any time after completion of this lesson.

Blood Donation

Blood transfusions have saved many lives. Blood is often needed for people who have been in accidents or who are having surgery. If other healthy people did not donate some of their blood, these blood transfusions would not be possible.

Usually, to donate blood, a person must be at least seventeen years old. He must be healthy and weigh at least 110 lb. When a person donates blood, his body makes more blood cells to replace the ones he has given.

After the blood is donated, it is tested to determine the blood type. It is also checked for diseases. The blood is often separated into its parts: plasma, red blood cells, white blood cells, and platelets. This way several patients can use different parts of the same unit of donated blood. The blood is stored in a blood bank until someone needs it. Most donated blood can be stored for only a short time before it is no longer useful. That is why blood donors are always needed.

QUICK CHECK

1. What is the liquid part of the blood?
2. Which blood cells carry oxygen to all parts of the body?
3. Which blood cells help fight disease?
4. What are the four main blood types?

MEET THE SCIENTIST • CHARLES DREW

Charles Drew was an African American surgeon in the early twentieth century. At that time, blood could only be stored for a few days before it could no longer be used. Drew, however, found a way to separate the plasma from the red blood cells. This way, the blood could be stored for much longer and then later mixed and used. He also discovered that plasma by itself can be used with blood of any type. This means that people who could not receive a certain type of whole blood could still receive plasma from that type of blood.

During World War II, Drew helped establish blood banks to store blood. He also found ways to preserve plasma so that it could be shipped over great distances. Many wounded soldiers were saved because of Drew's discoveries. He also created "bloodmobiles." These refrigerated trucks could safely carry donated blood to where it was needed. Bloodmobiles are still used today by organizations such as the Red Cross.

298

Background

Blood Donation

Even though many people are able to donate blood, only a few do. Less than 10% of people who could donate blood do so. A healthy person can donate blood about every fifty-six days.

Unit of Blood

One unit of blood is equal to about 450 mL. This unit of blood is often divided into its components to maximize the number of recipients as well as the length of time the blood can be stored. Red blood cells can be safely stored for forty-two days, platelets for five days, and plasma for a year if it is frozen.

Bloodmobile

The term *bloodmobile* is now usually used to refer to a mobile blood donation center.

Exploring Blood Types

ACTIVITY

During a transfusion, it is very important that a person receive blood that can successfully mix with his own. The person who is giving blood is called the donor. The one who is getting the blood is the receiver. This activity will show which blood types can safely mix with each other.

Process Skills
- Predicting
- Measuring
- Making and using models
- Observing

Problem

Which blood types can safely mix with each other?

Procedure

1. Prepare the Receivers group by labeling four cups *A* and pouring 4 oz of red water from Donor Type A into each. Follow the same procedure for Receiver Types B, AB, and O.
2. Predict which liquids will not change color when additional water from a Donors group is added to a Receivers group.
3. Pour a small amount of liquid from the Donor Type A container into the Receiver Type A cup. Observe the color. If the color does not change or only becomes diluted or concentrated, then the types are compatible. Mark the chart when the types are compatible.
4. Add liquid from Donor Type A to Receiver cups for Types B, AB, and O and mark the chart if any are compatible.
5. Repeat the procedure with Donor Types B, AB, and O until all of the combinations have been made.

Materials
- Permanent marker
- 16 clear plastic cups, 6 oz
- Donor group bottles
- White paper (optional)

Conclusions

1. Which blood type can donate to all other types?
2. Which blood type can receive from all other types?
3. Which blood type can receive blood only from itself?

Follow-up

Find out your blood type from your parents. Use the information from this activity to determine the types of blood that you can receive and can donate to.

299

Background

Activity Preparation

Prepare four large drink bottles to represent the four different blood types in the Donors group (A, B, AB, and O). Fill the four bottles with water.

- Label one bottle *A*, and add red food coloring to the water.
- Label one bottle *B*, and add blue food coloring to the water.
- Label one bottle *AB*, and add red and blue food coloring to the water, making it purple.
- Label one bottle *O*, and do not add color to the water.

Objectives

- Demonstrate which blood types can safely mix with each other

Materials

- 4 large empty drink bottles
- red and blue food coloring

To set up the Donor group bottles see the Background information.

Teacher's Toolkit CD

- Rubrics: *Exploring Blood Types*, for each student

Introduction

What are people who donate blood called? blood donors

Preparation for Reading

- Direct the students to read Student Text page 299 and Activity Manual pages 197–98 before beginning.

Teach for Understanding

Process Skill

The process skill emphasized in this activity is observing, which is using the senses to gather information.

What kinds of information can you gain by observing the combining of blood types in this activity? which blood types can safely mix together

Why are observations important to scientific research and thinking? Possible answers: to increase knowledge, to make comparisons, to see changes over time

Why is recording data from observations important in scientific research? Possible answers: to ensure accuracy, to make comparisons, for others to use the information that you observed

Procedure

- Guide the students in preparing the Receivers group as illustrated on Activity Manual page 197, using the Donor group bottles.
- Give help as needed as they complete the Predictions chart, combine the Donors and Receivers, and mark the Results chart.

Conclusions

- Discuss the results the students had and the conclusions they made.

Activity Manual

Activity, pages 197–98

Assessment

Rubrics

Select the prepared rubric, or design a rubric to include your chosen criteria.

Objectives

- Model the heart pumping blood
- Compare the model with the function of the heart

Materials

Materials needed for this activity are listed on the Student Text page. If you do not have enough buckets for all the groups, you could use deep dishpans or other large containers that can hold 5 L (1.3 gal) of water. You may also want to do this activity outside.

Teacher's Toolkit CD

- Rubrics: *Pump and Pour*, for each student

Introduction

What does the heart do? pumps blood through the body

Your heart is continuously pumping blood. It works hard with very little rest. In this activity you will see if you can work as hard as your heart.

Preparation for Reading

- Instruct the students to read Student Text pages 300–301 and Activity Manual pages 199–200 before beginning the activity.

Teach for Understanding

Process Skill

The process skill emphasized in this lesson is making and using models.

What have we used to help us examine and experience things that are too large, small, or inconvenient to handle? models

What are some models that we have made in science this year? Possible answers: plaster fossils, a model wetland, cactus sponges, a model lung

The plaster fossils, model wetland, cactus sponges, and model lung were models that you could hold. Although this activity cannot be held, it is still a type of model.

What does the activity model? the amount of blood moved (pushed) by the heart in one minute

Procedure

- If students use materials that vary from the list printed in the Activity Manual, they should put a line through each material that was not used and write the materials that were used. This reinforces accurate recording.
- Monitor the students as they work. Emphasize the importance of having the cup completely full to accurately model the heart.

ACTIVITY

Pump and Pour

Can you work as fast as your heart does? Every time your heart beats, it pushes about 60 mL of blood into your arteries. In just about one minute, your blood has been circulated throughout your entire body!

In this activity the water represents the blood in your body. The plastic cup represents the amount of blood that your heart pumps with each beat. See if you can transfer water as efficiently as your heart pumps blood.

Process Skills
- Predicting
- Measuring and using numbers
- Making and using models
- Collecting and recording data
- Defining operationally

Problem

How many cups of water can you transfer in one minute?

Materials
- 2 buckets
- 5 L water
- plastic cup, 3 oz
- stopwatch
- Activity Manual

Procedures

1. You will need at least two people for this activity—one person to time and to count and one person to transfer the water.
2. Complete the hypothesis in your Activity Manual.
3. Place the two buckets next to each other. Fill one bucket with 5 L of water.
4. One person starts timing while the other person uses the plastic cup to scoop water out of the bucket. The cup should be completely full. Pour the water into the other bucket without spilling any. The timer should count the number of cups of water that are transferred.
5. Stop the timing at one minute. Record the number of cups transferred.
6. Refill the first bucket with 5 L of water.
7. Repeat steps 4–6 for each person in your group.
8. Record the number of cups each member of the group transferred. Graph them in your Activity Manual.

300

Conclusions

- Did your results support your hypothesis?
- How does the amount of water that you transferred compare to the amount of blood that your heart pumps in one minute?
- How did your amount compare to those of others in your group?
- If you had to pump your own blood, would you be able to stay alive?

Follow-up

- Time how long it takes to transfer all 5 L of water to the other bucket.

301

176

- Assign tasks to each member of the group as needed. A group of four can have one person pouring, one person timing, one person counting, and one person recording. Smaller groups can combine tasks.
- Guide the students as they graph the results.

Conclusions

How many cups of water did you transfer in one minute? Answers will vary.

Did your results support your hypothesis?

How does transferring water between buckets model what the heart does? The heart moves about 5 L of blood through the body every minute. It pumps about 60 mL of blood with each beat. In this activity we were trying to transfer 5 L of water in one minute using a cup that was about the size of 60 mL.

How does the amount of water that you transferred compare to the amount of blood that your heart pumps in one minute? Answers will vary.

Activity Manual

Activity, pages 199–200

Assessment

Rubrics

Select the prepared rubric, or design a rubric to include your chosen criteria.

177

Activity Manual pages 201–2

Objectives

- Explain why it is important to identify the God of the Bible as the designer of our bodies
- Defend from Scripture that Jesus created the world

Introduction

What kinds of things have you made for a science fair, an art competition, or a music competition?

How did you feel about your project? Possible answers: proud of the project; wanted to show it to family and friends; happy that it has been displayed

Preparation for Reading

- Instruct the students to read the first two paragraphs on Activity Manual page 201 to find out what was overheard in the museum.

Teach for Understanding

What did the visitor say about the carved eagle? "That guy only made this to raise money for this museum."

How would you feel about your grandfather not getting credit for the real purpose of his work? Accept all reasonable answers.

Being respectful, how would you respond to the visitor? Accept all reasonable answers.

- You may allow time for the students to complete the questions as different sections are discussed or at the end of the lesson.
- Direct the students to read the last paragraph on the page to find out about God as an amazing designer.
- As the students think about the example of the grandfather getting credit for his carving, impress upon them how much more important it is that God receive the glory for His works in creation, including the amazing design of their bodies.
- Explain that William Paley used the analogy of walking through the woods and finding a pocket watch. The watch did not happen by accident. It was carefully designed. Likewise, the world must have had a designer.
  Psalm 19 and Romans 1 do not speak of just any designer but of the existence of the only true God.
- Choose volunteers to read Romans 1:18–23.
  Look at verse 20. It says that by looking at creation everyone can know God's attributes.
  What two attributes does the verse mention specifically? God's eternal power and divine nature
  Look at verses 22 and 23. In claiming to be wise, what did people exchange the glory of God for? images of man, birds, animals, and reptiles

A lesson from
Answers in Genesis Name ____________________ **1:1**

Honoring the Designer

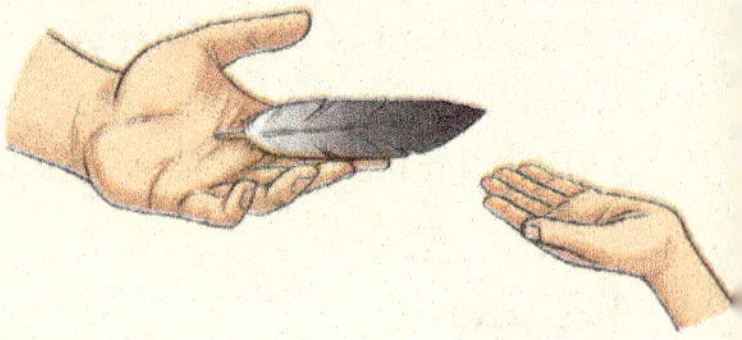

Imagine for a moment that your grandfather is an amazing wood carver. He once carved and painted a feather that looked so real that when he handed it to you, you were surprised at how heavy it was. Your grandfather was such a good craftsman that he was asked to carve a bald eagle to put on display in a museum. Your grandfather taught you how to carve and use the skill like he does, as a tool to show God's glory in His creation.

One day while you are visiting the museum, you see several people admiring his carved eagle. You are joyful, knowing that your grandfather created it to reflect God's glory in creation. But you hear one visitor say to several others, "That guy only made this to raise money for this museum."

1. How would you feel about your grandfather not getting credit for the real purpose of his work?

   ____

   ____

   ____

   ____

2. How might you respectfully respond to the visitor?

   ____

   ____

   ____

   ____

As concerned as we might be that someone we love does not get credit for his work, we should be all the more concerned when God does not get the credit He deserves. As you study the way your body works, you should be amazed by God! Could you ever make a heart of muscle? Could you ever make lungs? Could you make the two work together to pump oxygen through your body? Nobody can. Scientists have made artificial machines that copy God's design of the body, but only God can breathe life into a person and make his heart beat to sustain life.

Resources

The BJU Press website offers additional links from Answers in Genesis that you may find helpful with this lesson.
bjupress.com/resources/science

177

There are many people who believe there must be a god who created the world. But if any of these people do not acknowledge Jesus as the Savior, they will perish. However, many people do not know that the Bible teaches that Jesus is the Creator.

3. Read Hebrews 1:1–4. Who does verse 2 say created the worlds? *God's Son*
4. How does God uphold or sustain the worlds (Hebrews 1:3)?
   by the word of His power
5. What does John 1:1–5 tell you about how things were made?
   All things were made by the Word (Jesus).
6. What does Colossians 1:13–18 tell you about redemption?
   Redemption is through the blood of God's Son (Jesus).
7. What does Colossians 1:16 tell you about how things were made?
   All things were created by God's Son (Jesus).

This connection between Jesus as both the Creator and Savior is very important. The next time you hear someone talking about a creator or designer, you have the wonderful opportunity to tell him about *the* Creator and Savior, Jesus Christ. You can show him these passages, reminding him that Jesus is both the Creator God and the Savior God. Jesus will forgive him when he repents of his sins and places his faith in Christ. Remember Paul's words in Romans 1: "For I am not ashamed of the gospel of Christ, for it is the power of God unto salvation to every one that believeth."

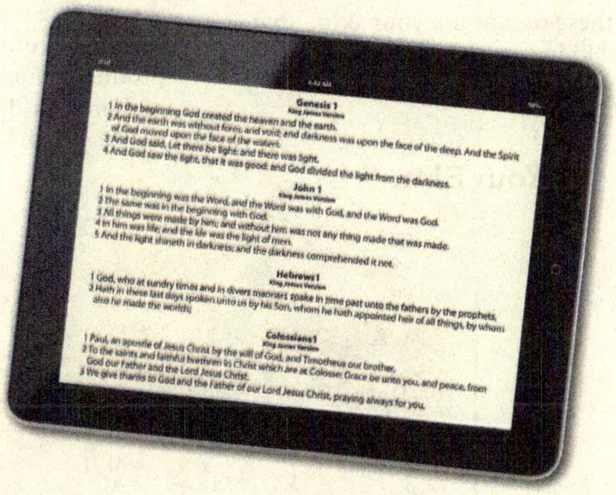

- Direct the students to read the first paragraph on Activity Manual page 202 to find out who people need to acknowledge.
  What will happen to people who do not acknowledge Jesus as their Savior? They will perish.
- Ask a volunteer to read Hebrews 1:1–4.
  Look at verse 2. Who created the worlds? God's Son
  Look at verse 3. How does God uphold or sustain the worlds? by the word of His power
- Ask volunteers to read John 1:1–5.
  What do these verses tell you about how things were made? All things were made through the Word.
  Who is the Word? Jesus
- Ask volunteers to read Colossians 1:13–18.
  What does Colossians 1:13–18 tell you about redemption? Redemption is through the blood of God's Son, Jesus.
  Look again at verse 16. What does this verse tell you about how things were made? All things were created by God's Son, Jesus.
- Explain that many people believe in Jesus as a prophet or a good teacher, but unless they acknowledge Him as Savior and Lord, they will face eternal punishment for their sins.
- Instruct the students to read the last paragraph to find out what Christians have the privilege of sharing.
  What do Christians have the privilege of sharing? Answers should include the gospel of Jesus Christ and that Jesus is Creator, Savior, and Lord.

Activity Manual

Answers in Genesis, pages 201–2

178

Student Text pages 302–5
Activity Manual page 203

Objectives

- Identify organs that help remove wastes from the body
- Recognize that the kidneys help clean the blood
- Name three ways to stay healthy
- Recognize that no inventions would be possible without God

Teacher's Toolkit CD

- Visual 12.1: *Circulatory System*
- Quiz 12-D

Vocabulary

- perspiration
- respiration
- kidney

Introduction

Have you ever heard the saying "An ounce of prevention is worth a pound of cure"? What do you think it means? Answers will vary.

In this lesson you will learn some basic ways to stay healthy.

Preparation for Reading

- Instruct the students to mentally make questions using the headings, bold words, and captions.
- Direct the students to read silently pages 302–5 to find the answers to these questions.

  Which organ in your body removes wastes and extra water from the blood?

  How might poor eating habits affect a person's arteries?

  What are exercises that are good for the heart?

Teach for Understanding

What does the circulatory system carry to every cell in the body? oxygen and nutrients

What do the cells produce as waste products? carbon dioxide and unwanted minerals

Why must the waste products be eliminated? Waste products that stay inside the body prevent the cells from working properly and poison the body.

Name some organs that work with the circulatory system to help remove waste products. Possible answers: the skin, lungs, kidneys

What is the relationship of cells, tissues, and organs? Cells form tissues. Tissues form organs.

What types of wastes are removed by the skin? water, salts, and other wastes

How does the skin release wastes? as sweat or perspiration

- Discuss the *Your Skin* diagram on page 302.

  What two layers of the skin are shown in the diagram? the epidermis and the dermis

Cleaning Your Blood

You have learned how the circulatory system works, but it does not work alone. The circulatory system carries oxygen and nutrients to every cell in your body. As your cells use the oxygen and nutrients, they produce waste products. These wastes include carbon dioxide and unwanted minerals. If the wastes stayed inside of you, your cells could not work as they should. Your body would get poisoned, and you might even die.

So God gave you organs that "clean" your body. These organs work with your circulatory system to remove waste products from your body. Some of these organs are your skin, lungs, and kidneys. These organs are always at work to help remove wastes from your body.

Your Skin

Some wastes are released from the body as **perspiration** or sweat. A mixture of water, salts, and other wastes are moved out of the blood into sweat glands. When your body temperature rises, the fluid, or perspiration, is released from your sweat glands. It travels to the surface of your skin through a tube and exits your body through the pores in your skin.

Your Lungs

Other wastes, including carbon dioxide, are removed by the lungs during **respiration**. When you inhale, you breathe in the mixture of gases that are in the air. In your lungs, the red blood cells pick up oxygen from the air. The red blood cells also release the carbon dioxide that your body produces. Your lungs then exhale the carbon dioxide and the unused gases out of your body.

Your Skin

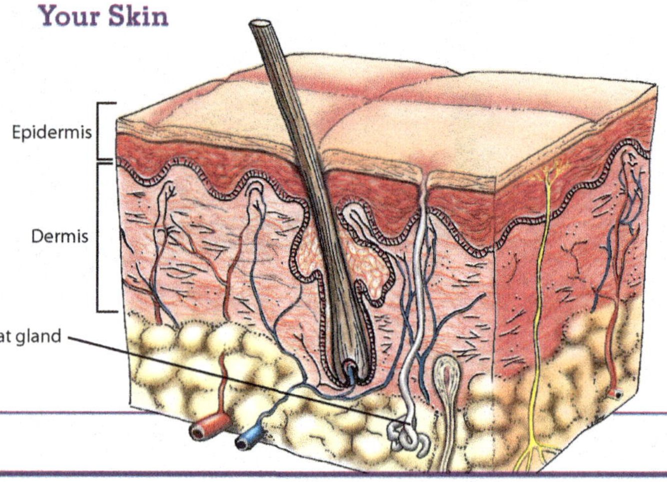

302

Background

Excretory System

The excretory system works with the circulatory system to remove waste products from the body. The excretory system includes the kidneys, the bladder, and other associated structures. The skin and lungs also help remove some wastes from the body.

Kidneys

The kidneys' main function is to filter waste products out of the bloodstream. However, the kidneys also regulate the volume of fluids and minerals in the body. The kidneys receive messages from the brain telling them to retain fluids when the body's water level is low. The kidneys also make hormones that direct the body to produce red blood cells.

Your Kidneys

Most people have two **kidneys**. These bean-shaped organs are located on either side of your spine, just under your bottom rib. Each of your kidneys is about the size of your fist. Your kidneys have many functions, but their most important job is cleaning your blood. They remove wastes and excess water from your blood.

Blood flows into the kidneys from the renal artery. Inside each kidney, there are tiny filters, called nephrons. Each kidney has more than one million nephrons. They clean the blood by removing waste material. Once the blood is cleaned, it flows out of the kidneys and back to the heart. The wastes that have been cleaned out of the blood are combined with water to make urine. These wastes are sent to the bladder and stored there until they are eliminated from the body.

Your Kidneys

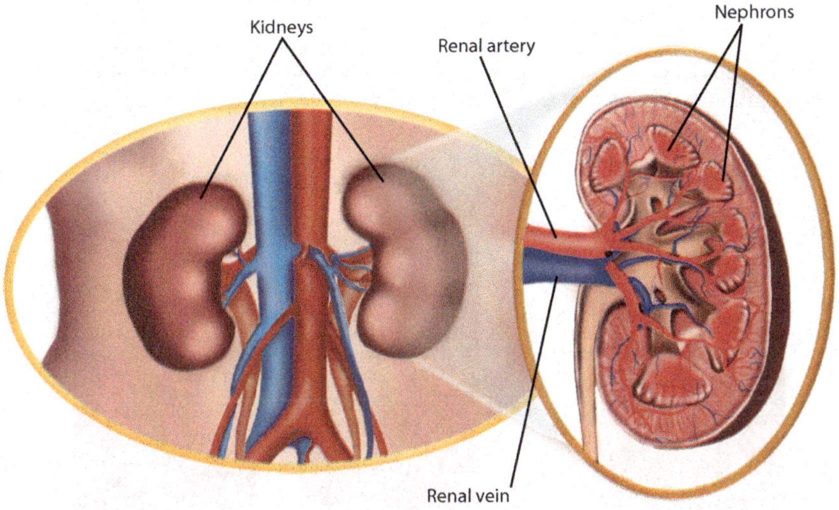

303

Where are sweat glands located? below the dermis

What types of wastes are removed by the lungs? carbon dioxide and unused gases

How do the lungs remove wastes? They are exhaled during respiration.

Which organs help filter wastes from the blood? kidneys

Where are the kidneys located? on either side of the spine, just under the bottom rib

- Display the *Circulatory System* visual. Point out where the kidneys are located in the body.
- Refer to the *Your Kidneys* diagram as the kidneys are discussed.

How many kidneys do most people have? two

How big is each kidney? about the size of your fist

What is the most important job of the kidneys? cleaning the blood

What do the kidneys remove from the blood? wastes and excess water

What is the renal artery? the blood vessel that enters the kidneys

What are nephrons? tiny filters inside the kidneys that remove the waste material from blood

How many nephrons does each kidney have? more than one million

Where does the blood go after it has been cleaned by the kidneys? back to the heart

Look at the diagram. What vessel carries blood from the kidneys? the renal vein

What is done with the wastes? They combine with water to make urine and are stored in the bladder until they are eliminated from the body.

- You may want to do the *Model How the Kidneys Filter Waste* demonstration at this time.

What is some advice that you have heard about how to stay healthy? Possible answers: to eat vegetables, to have regular checkups, to exercise, to avoid smoking

Some heart-related problems are inherited. What does *inherited* mean? passed from parents to offspring, or children

What is another cause of heart-related problems and diseases? habits and choices

Does making wise choices enable you to avoid all problems and diseases? no Why? Sometimes we are born with problems, or we are affected by the problems of others. In everything, God is in control, and He may choose for you to have a certain problem at some time in your life.

Problems and diseases often seem like bad things. Why might God choose for a person to suffer from a problem or disease? Possible answers: to be a testimony of God's grace, to comfort other people who are in similar situations, to provide opportunities for that person to witness, to help that person learn more about God

Give some biblical examples in which God made something good out of something evil. Possible answers: the life of Joseph (Genesis 37–47) and Jesus' death (Matthew 26–28)

Demonstration

Model How the Kidneys Filter Waste

Materials: sieve, large bowl, mixture of salt and rice

Hold the sieve over the bowl. Pour an equal mixture of salt and rice into the sieve. Shake the sieve back and forth. The salt should fall through the sieve, and the rice should remain in the sieve. Explain that this models how the kidneys remove the waste products from the blood but keep the nutrients that the body needs.

Activity

Inside Me

Materials: Instructional Aid 12.1: *Inside Me*, for each student; 24 × 72 in. sheet of paper, for each student

Follow the instructions on the page.

What is one of the best ways to keep the heart healthy? eating the right kinds of foods

Name some foods you think would be considered healthy. Possible answers: fruits, vegetables, dairy products

- Direct attention to the illustrations of the arteries.

What may happen to the arteries if too many unhealthy foods are eaten? The arteries can become clogged.

What kinds of foods could be considered unhealthy if they are eaten too often? Possible answers: fried foods, sweets, foods with a high oil or fat content

How do blocked blood vessels affect the body? Blocked blood vessels can prevent blood, oxygen, and nutrients from getting to important organs and other parts of the body.

Protecting Your Heart

"Stay active." "Get plenty of exercise." "Eat your vegetables." "Visit the doctor for regular checkups." You have probably heard this kind of advice many times. A basic reason to follow this advice is to help your heart stay healthy. However, the main reason is that a healthy heart and body are important to God. We should take care of everything God has given us, whether it is a biome or our bodies.

Some heart-related problems and diseases are inherited. Others are the result of habits and choices. How you live your life can affect the health of your heart. Wise choices and healthy living habits can help you avoid some heart problems and diseases.

Eating the right kind of food is one of the best ways to keep your heart healthy. Doctors usually suggest that people eat a variety of foods from each food group. Eating food that is good for you can help prevent many health problems.

Some problems can develop from poor eating habits. For example, a person who often eats high-fat foods, such as fried food and potato chips, can damage his arteries. The fatty materials build up inside the arteries. Over time, the arteries can become clogged. This can be very dangerous. The blocked blood vessels do not allow blood to get through. This can prevent the oxygen and nutrients in the blood from getting to organs and other areas of the body. Surgery may need to be done so that the blood flow is restored.

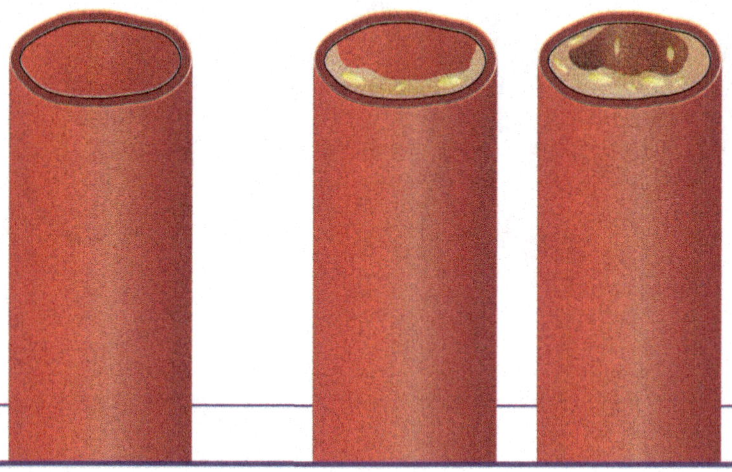

304

Background

Blood Clots

The fatty materials and other substances that can clog an artery are called plaque. The plaque reduces the blood flow through that blood vessel. Blood clots sometimes form around the plaque. If a blood clot forms, it can block the blood flow in that blood vessel. A blood clot can also travel through the blood vessels and eventually block another blood vessel. Depending on where the blood vessel is located, the blood clot may cause a heart attack or a stroke.

Exercise helps keep the heart healthy.

Another way to keep your heart healthy is to exercise. Just like other muscles, the heart muscle needs to be exercised. Walking, running, swimming, bike riding, and jumping rope are good exercises. They all will strengthen your heart.

You can help prevent many kinds of health problems simply by not smoking. Smoking leads to many heart problems. It damages the heart and blood vessels and makes them have to work harder. Smoking also reduces the amount of oxygen in the blood. A person who smokes is more likely to develop heart disease or other problems.

Remembering Our Creator

God is the perfect Creator. In His wisdom, He has given mankind an amazing transportation system. The heart pumps blood. The blood vessels carry blood to every cell in the body. Each cell receives the oxygen and nutrients that it needs. This system keeps the body alive. No manmade transportation system could ever be as efficient.

The success that scientists have had in treating heart problems has improved healthcare, made life easier, and allowed people to live longer. The more we learn about our heart and the blood that it pumps, the more amazed we are by the creation of God. Nothing this complex could happen by accident. It is the work of a loving God. Colossians 1:16–17 reminds us, "all things were created by him, and for him: And he is before all things, and by him all things consist."

QUICK CHECK

1. What are three organs that help the body "clean" itself?
2. How do the kidneys help remove wastes?
3. What are some ways to keep your heart healthy?

305

What is another way to keep the heart healthy? exercise

What are some exercises that help strengthen the heart? Possible answers: walking, running, swimming, bike riding, jumping rope

What is another way to prevent many health problems? avoid smoking

How does smoking affect the respiratory system? Possible answer: It damages the lungs and airways and can cause many diseases.

Identify several ways smoking affects the circulatory system. Smoking damages the heart and blood vessels, making them work harder. Smoking causes the blood to have less oxygen available for the cells. Smoking increases the chance of heart disease and other problems.

What is the circulatory system similar to? a transportation system

What are some benefits of inventions? improved health care, ability to do daily work more easily, awareness of things that cannot be seen

Who has made all of these inventions possible? God

What verses remind us that everything was both made by God and comes from God? Colossians 1:16–17

ANSWERS

1. the skin, lungs, and kidneys
2. They filter wastes out of the blood before the blood returns to the heart.
3. eating properly, exercising, not smoking, visiting the doctor for regular checkups

Activity Manual

Review, page 203

This page reviews Lesson 178.

Assessment

Quiz 12-D

The quiz may be given any time after completion of this lesson.

Activity

Exercise for a Healthy Heart

Materials: jump ropes, stopwatch

Find a place outside where there is enough room to jump rope. Give each student a jump rope. (You may choose to ask the students to bring their own jump ropes.) Provide time for the students to stretch and warm up. You may need to have someone demonstrate how to jump rope if some students have not used a jump rope before. Be aware of any physical problems or special health needs a student might have. If a student feels weak or wobbly while he is jumping, he probably needs to slow down his pace.

Divide the students into two equal-sized groups. Direct one group to jump rope for one minute. Ask the other group to count the number of jumps for each person jumping. If the jumpers miss a jump or get out of breath, tell them to continue where they left off. Encourage them to get as many jumps in a minute as they can. Record the number of jumps for each person. Then direct the groups to switch so that group one is counting while group two is jumping rope. Record the numbers for the second group.

A student who is physically fit should be able to get 60–70 jumps in one minute. Direct students to set reasonable individual goals for themselves and to record those goals. Over a period of several days, provide time for students to exercise and work toward reaching those goals.

179 Student Text page 306
Activity Manual page 204

Objectives
- Recall concepts and terms from Chapter 12
- Apply knowledge to everyday situations

Review
- Materials for the Chapter 12 Test will be taken from Student Text page 306 and Activity Manual pages 191–92, 195–96, and 203–4. You may review any or all of the material during the lesson. Questions similar to the ones in Thinking It Through, Activity Manual page 204, may appear on the test.
- You may choose to review Chapter 12 by playing the review game or a game from the Game Bank on the Teacher's Toolkit CD.

Activity Manual
Review, page 306
This page requires written responses to application questions.

180

Objective
- Demonstrate knowledge of concepts taught in Chapter 12

Assessment
Chapter 12 Test

CHAPTER SUMMARY

Words to Know

heart	capillary	white blood cells
atrium	coronary	platelets
ventricle	pulmonary	blood transfusion
valve	aorta	blood donation
pulse	superior vena cava	perspiration
pacemaker	interior vena cava	respiration
artery	plasma	kidneys
vein	red blood cells	

Key Ideas
Identify the parts of the heart
Summarize the movement of blood through the heart
Identify the uses of a stethoscope and an EKG
Summarize how blood circulates through the body
Identify the relationships between the different types of blood vessels
Describe the contributions of William Harvey and Karl Landsteiner
Describe the jobs of plasma, red and white blood cells, and platelets
Identify the types of blood
Describe three ways that blood is cleaned
Explain the importance of having a healthy heart
Use an example from the circulatory system to demonstrate God's intricate design

306

Review Game
Back to the Heart
Divide the class into two teams. Each student represents a drop of blood. The goal is to get all of the blood from the heart out to the rest of the body and then back to the heart.

Designate spots in the room to represent the heart and several body parts such as an arm, a foot, the liver, and the kidneys. Each time a student answers a question correctly, he moves from the heart to the first designated part of the body. He cannot move to the next spot until the next drop of blood (his team member) replaces him. The first team to have all its members return to the heart wins the game.

Variation: Set a certain number of correctly answered questions as a completed trip through the body and back to the heart. See how many trips through the body can be completed in a certain amount of time.

Appendix

Bible Action Truths

Bible Action Truths and Bible Promises are included as a resource. They provide specific goals for actions and attitudes and include Bible references for further study in identifying and applying principles of behavior.

The quality and consistency of a man's decisions reflect his character. Christian character begins with justification, but it grows throughout the lifelong process of sanctification. God's grace is sufficient for the task, and a major part of God's gracious provision is His Word. The Bible provides the very "words of life" that instruct us in salvation and Christian living. By obeying God's commands and making godly decisions based on His Word, Christians can strengthen their character.

Too often Christians live by only vague guidance—for instance, that we should "do good" to all men. While doing good is desirable, more specific guidance will lead to more consistent decisions.

Consistent decisions are made when man acts on Bible principles—or Bible Action Truths. The thirty-seven Bible Action Truths (listed under eight general principles) provide Christians with specific goals for their actions and attitudes. Study the Scriptures indicated for a fuller understanding of the principles in Bible Action Truths.

Thousands have found this format helpful in identifying and applying principles of behavior. Yet there is no "magic" in this formula. As you study the Word, you likely will find other truths that speak to you. The key for you is to study the Scriptures, look for Bible Action Truths, and be sensitive to the leading of the Holy Spirit.

1. **Salvation-Separation Principle**
   Salvation results from God's direct action. Although man is unable to work for this "gift of God," the Christian's reaction to salvation should be to separate himself from the world unto God.
   a. **Understanding Jesus Christ** (Matthew 3:17; 16:16; 1 Corinthians 15:3–4; Philippians 2:9–11) Jesus is the Son of God. He was sent to Earth to die on the cross for our sins. He was buried but rose from the dead after three days.
   b. **Repentance and faith** (Luke 13:3; Isaiah 55:7; Acts 5:30–31; Hebrews 11:6; Acts 16:31) If we believe that Jesus died for our sins, we can accept Him as our Savior. We must be sorry for our sins, turn from them, confess them to God, and believe that He will forgive us.
   c. **Separation from the world** (John 17:6, 11, 14, 18; 2 Corinthians 6:14–18; 1 John 2:15–16; James 4:4; Romans 16:17–18; 2 John 10–11) After we are saved, we should live a different life. We should try to be like Christ and not live like those who are unsaved.

2. **Sonship-Servant Principle**
   Only by an act of God the Father could sinful man become a son of God. As a son of God, however, the Christian must realize that he has been "bought with a price"; he is now Christ's servant.
   a. **Authority** (Romans 13:1–7; 1 Peter 2:13–19; 1 Timothy 6:1–5; Hebrews 13:17; Matthew 22:21; 1 Thessalonians 5:12–13) We should respect, honor, and obey those in authority over us. (attentiveness, obedience)
   b. **Servanthood** (Philippians 2:7–8; Ephesians 6:5–8) Just as Christ was a humble servant while He was on Earth, we should also be humble and obedient. (attentiveness, helpfulness, promptness, teamwork)
   c. **Faithfulness** (1 Corinthians 4:2; Matthew 25:23; Luke 9:62) We should do our work so that God and others can depend on us. (endurance, responsibility)
   d. **Goal setting** (Proverbs 13:12, 19; Philippians 3:13; Colossians 3:2; 1 Corinthians 9:24) To be faithful servants, we must set goals for our work. We should look forward to finishing a job and going on to something more. (dedication, determination, perseverance)
   e. **Work** (Ephesians 4:28; 2 Thessalonians 3:10–12) God never honors a lazy servant. He wants us to be busy and dependable workers. (cooperativeness, diligence, initiative, industriousness, thoroughness)
   f. **Enthusiasm** (Colossians 3:23; Romans 12:11) We should do all tasks with energy and with a happy, willing spirit. (cheerfulness)

3. **Uniqueness-Unity Principle**
   No one is a mere person; God has created each individual a unique being. But because God has an overall plan for His creation, each unique member must contribute to the unity of the entire body.
   a. **Self-concept** (Psalms 8:3–8; 139; 2 Corinthians 5:17; Ephesians 2:10; 4:1–3, 11–13; 2 Peter 1:10) We are special creatures in God's plan. He has given each of us special abilities to use in our lives for Him.
   b. **Mind** (Philippians 2:5; 4:8; 2 Corinthians 10:5; Proverbs 23:7; Luke 6:45; Proverbs 4:23; Romans 7:23, 25; Daniel 1:8; James 1:8) We should give our hearts and minds to God. What we do and say really begins in our minds. We should try to think of ourselves humbly as Christ did when He lived on Earth. (orderliness)
   c. **Emotional control** (Galatians 5:24; Proverbs 16:32; 25:28; 2 Timothy 1:7; Acts 20:24) With the help of God and the power of the Holy Spirit, we should have control over our feelings. We must be careful not to act out of anger. (flexibility, self-control)
   d. **Body as a temple** (1 Corinthians 3:16–17; 6:19–20) We should remember that our bodies are the dwelling place of God's Holy Spirit. We should keep ourselves pure, honest, and dedicated to God's will.
   e. **Unity of Christ and the church** (John 17:21; Ephesians 2:19–22; 5:23–32; 2 Thessalonians 3:6, 14–15) Since we are saved, we are now part of God's family and should unite ourselves with others to worship and grow as Christians. Christ is the head of His church, which includes all believers. He wants us to work together as His church in carrying out His plans, but He forbids us to work in fellowship with disobedient brethren.

4. **Holiness-Habit Principle**
Believers are declared holy as a result of Christ's finished action on the cross. Daily holiness of life, however, comes from forming godly habits. A Christian must consciously establish godly patterns of action; he must develop habits of holiness.

   a. **Sowing and reaping** (Galatians 6:7–8; Hosea 8:7; Matthew 6:1–8) We must remember that we will be rewarded according to the kind of work we have done. If we are faithful, we will be rewarded. If we are unfaithful, we will not be rewarded. We cannot fool God. (thriftiness)

   b. **Purity** (1 Thessalonians 4:1–7; 1 Peter 1:22) We should try to live lives that are free from sin. We should keep our minds, words, and deeds clean and pure.

   c. **Honesty** (2 Corinthians 8:21; Romans 12:17; Proverbs 16:8; Ephesians 4:25) We should not lie. We should be honest in every way. Even if we could gain more by being dishonest, we should still be honest. God sees all things. (fairness)

   d. **Victory** (1 Corinthians 10:13; Romans 8:37; 1 John 5:4; John 16:33; 1 Corinthians 15:57–58) If we constantly try to be pure, honest, and Christlike, with God's help we will be able to overcome temptations.

5. **Love-Life Principle**
We love God because He first loved us. God's action of manifesting His love to us through His Son demonstrates the truth that love must be exercised. Since God acted in love toward us, believers must act likewise by showing godly love to others.

   a. **Love** (1 John 3:11, 16–18; 4:7–21; Ephesians 5:2; 1 Corinthians 13; John 15:17) God's love to us was the greatest love possible. We should, in turn, show our love for others by our words and actions. (courtesy, compassion, hospitality, kindness, thankfulness to men, thoughtfulness)

   b. **Giving** (2 Corinthians 9:6–8; Proverbs 3:9–10; Luke 6:38) We should give cheerfully to God the first part of all we earn. We should also give to others unselfishly. (hospitality, generosity, sharing, unselfishness)

   c. **Evangelism and missions** (Psalm 126:5–6; Matthew 28:18–20; Romans 1:16–17; 2 Corinthians 5:11–21) We should be busy telling others about the love of God and His plan of salvation. We should share in the work of foreign missionaries by our giving and prayers.

   d. **Communication** (Ephesians 4:22–29; Colossians 4:6; James 3:2–13; Isaiah 50:4) We should have control of our tongues so that we will not say things displeasing to God. We should encourage others and be kind and helpful in what we say.

   e. **Friendliness** (Proverbs 18:24; 17:17; Psalm 119:63) We should be friendly to others, and we should be loyal to those who love and serve God. (loyalty)

6. **Communion-Consecration Principle**
Because sin separates man from God, any communion between man and God must be achieved by God's direct action of removing sin. Once communion is established, the believer's reaction should be to maintain a consciousness of this fellowship by living a consecrated life.

   a. **Bible study** (1 Peter 2:2–3; 2 Timothy 2:15; Psalm 119) To grow as Christians, we must spend time with God daily by reading His Word. (reverence for the Bible)

   b. **Prayer** (1 Chronicles 16:11; 1 Thessalonians 5:17; John 15:7, 16; 16:24; Psalm 145:18; Romans 8:26–27) We should bring all our requests to God, trusting Him to answer them in His own way.

   c. **Spirit-filled** (Ephesians 5:18–19; Galatians 5:16, 22–23; Romans 8:13–14; 1 John 1:7–9) We should let the Holy Spirit rule in our hearts and show us what to say and do. We should not say and do just what we want to, for those things are often wrong and harmful to others. (gentleness, joyfulness, patience)

   d. **Clear conscience** (1 Timothy 1:19; Acts 24:16) To be good Christians, we cannot have wrong acts or thoughts or words bothering our consciences. We must confess them to God and to those people against whom we have sinned. We cannot live lives close to God if we have guilty consciences.

   e. **Forgiveness** (Ephesians 4:30–32; Luke 17:3–4; Colossians 3:13; Matthew 18:15–17; Mark 11:25–26) We must ask forgiveness of God when we have done wrong. Just as God forgives our sins freely, we should forgive others when they do wrong things to us.

7. **Grace-Gratitude Principle**
Grace is unmerited favor. Man does not deserve God's grace. However, after God bestows His grace, believers should react with an overflow of gratitude.

   a. **Grace** (1 Corinthians 15:10; Ephesians 2:8–9) Without God's grace we would be sinners on our way to hell. He loved us when we did not deserve His love and provided for us a way to escape sin's punishment by the death of His Son on the cross.

   b. **Exaltation of Christ** (Colossians 1:12–21; Ephesians 1:17–23; Philippians 2:9–11; Galatians 6:14; Hebrews 1:2–3; John 1:1–4, 14; 5:23) We should realize and remember at all times the power, holiness, majesty, and perfection of Christ, and we should give Him the praise and glory for everything that is accomplished through us.

   c. **Praise** (Psalm 107:8; Hebrews 13:15; 1 Peter 2:9; Ephesians 1:6; 1 Chronicles 16:23–36; 29:11–13) Remembering God's great love and goodness toward us, we should continually praise His name. (thankfulness to God)

   d. **Contentment** (Philippians 4:11; 1 Timothy 6:6–8; Psalm 77:3; Proverbs 15:16; Hebrews 13:5) Money, houses, cars, and all things on Earth will last only for a little while. God has given us just what He meant for us to have. We should be happy and content with what we have, knowing that God will provide for us all that we need. We should also be happy wherever God places us.

   e. **Humility** (1 Peter 5:5–6; Philippians 2:3–4) We should not be proud and boastful but should be willing to be quiet and in the background. Our reward will come from God on Judgment Day, and men's praise to us here on Earth will not matter at all. Christ was humble when He lived on Earth, and we should be like Him.

8. **Power-Prevailing Principle**
Believers can prevail only as God gives the power. "I can do all things through Christ" (Philippians 4:13). God is the source of our power used in fighting the good fight of faith.
   a. **Faith in God's promises** (2 Peter 1:4; Philippians 4:6; Romans 4:16–21; 1 Thessalonians 5:18; Romans 8:28; 1 Peter 5:7; Hebrews 3:18; 4:11) God always remains true to His promises. Believing that He will keep all the promises in His Word, we should be determined fighters for Him.
   b. **Faith in the power of the Word of God** (Hebrews 4:12; Jeremiah 23:29; Psalm 119; 1 Peter 1:23–25) God's Word is powerful and endures forever. All other things will pass away, but God's Word shall never pass away because it is written to us from God, and God is eternal.
   c. **Fight** (Ephesians 6:11–17; 2 Timothy 4:7–8; 1 Timothy 6:12; 1 Peter 5:8–9) God does not have any use for lazy or cowardly fighters. We must work and fight against sin, using the Word of God as our weapon against the Devil. What we do for God now will determine how much He will reward us in heaven.
   d. **Courage** (1 Chronicles 28:20; Joshua 1:9; Hebrews 13:6; Ephesians 3:11–12; Acts 4:13, 31) God has promised us that He will not forsake us; therefore, we should not be afraid to speak out against sin. We should remember that we are armed with God's strength.

Bible Promises

A. **Liberty from Sin**—Born into God's spiritual kingdom, a Christian is enabled to live right and gain victory over sin through faith in Christ. (Romans 8:3–4—"For what the law could not do, in that it was weak through the flesh, God sending his own Son in the likeness of sinful flesh, and for sin, condemned sin in the flesh: that the righteousness of the law might be fulfilled in us, who walk not after the flesh, but after the Spirit.")

B. **Guiltless by the Blood**—Cleansed by the blood of Christ, the Christian is pardoned from the guilt of his sins. He does not have to brood or fret over his past because the Lord has declared him righteous. (Romans 8:33—"Who shall lay any thing to the charge of God's elect? It is God that justifieth." Isaiah 45:24—"Surely, shall one say, in the Lord have I righteousness and strength: even to him shall men come; and all that are incensed against him shall be ashamed.")

C. **Basis for Prayer**—Knowing that his righteousness comes entirely from Christ and not from himself, the Christian is free to plead the blood of Christ and to come before God in prayer at any time. (Romans 5:1–2—"Therefore being justified by faith, we have peace with God through our Lord Jesus Christ: by whom also we have access by faith into this grace wherein we stand, and rejoice in hope of the glory of God.")

D. **Identified in Christ**—The Christian has the assurance that God sees him as a son of God, perfectly united with Christ. He also knows that he has access to the strength and the grace of Christ in his daily living. (Galatians 2:20—"I am crucified with Christ: nevertheless I live; yet not I, but Christ liveth in me: and the life which I now live in the flesh I live by the faith of the Son of God, who loved me, and gave himself for me." Ephesians 1:3—"Blessed be the God and Father of our Lord Jesus Christ, who hath blessed us with all spiritual blessings in heavenly places in Christ.")

E. **Christ as Sacrifice**—Christ was a willing sacrifice for the sins of the world. His blood covers every sin of the believer and pardons the Christian for eternity. The purpose of His death and Resurrection was to redeem a people to Himself. (Isaiah 53:4–5—"Surely he hath borne our griefs, and carried our sorrows: yet we did esteem him stricken, smitten of God, and afflicted. But he was wounded for our transgressions, he was bruised for our iniquities: the chastisement of our peace was upon him; and with his stripes we are healed." John 10:27–28—"My sheep hear my voice, and I know them, and they follow me: and I give unto them eternal life; and they shall never perish, neither shall any man pluck them out of my hand.")

F. **Christ as Intercessor**—Having pardoned them through His blood, Christ performs the office of High Priest in praying for His people. (Hebrews 7:25—"Wherefore he is able also to save them to the uttermost that come unto God by him, seeing he ever liveth to make intercession for them." John 17:20—"Neither pray I for these alone, but for them also which shall believe on me through their word.")

G. **Christ as Friend**—In giving salvation to the believer, Christ enters a personal, loving relationship with the Christian that cannot be ended. This relationship is understood and enjoyed on the believer's part through fellowship with the Lord through Bible reading and prayer. (Isaiah 54:5—"For thy Maker is thine husband; the Lord of hosts is his name; and thy Redeemer the Holy One of Israel; The God of the whole earth shall he be called." Romans 8:38–39—"For I am persuaded, that neither death, nor life, nor angels, nor principalities, nor powers, nor things present, nor things to come, nor height, nor depth, nor any other creature, shall be able to separate us from the love of God, which is in Christ Jesus our Lord.")

H. **God as Father**—God has appointed Himself to be responsible for the well-being of the Christian. He both protects and nourishes the believer, and it was from Him that salvation originated. (Isaiah 54:17—"No weapon that is formed against thee shall prosper; and every tongue that shall rise against thee in judgment thou shalt condemn. This is the heritage of the servants of the Lord, and their righteousness is of me, saith the Lord." Psalm 103:13—"Like as a father pitieth his children, so the Lord pitieth them that fear him.")

I. **God as Master**—God is sovereign over all creation. He orders the lives of His people for His glory and their good. (Romans 8:28—"And we know that all things work together for good to them that love God, to them who are the called according to his purpose.")

Explaining the Gospel

One of the greatest desires of Christian teachers is to lead children to the Savior. God has called you to present the gospel to your students so that they may repent and trust Christ, thereby being acceptable to God through Christ.

Relying on the Holy Spirit, you should take advantage of the opportunities that arise during lessons for presenting the good news of Jesus Christ. Ask questions to personally apply the Ten Commandments to your students (e.g., What is sin? Have you ever (name sin)? Are you a sinner?). You may also ask questions to discern the child's sincerity or any misunderstanding he might have (e.g., What is the gospel? What does it mean to repent? Can you do anything to save yourself?). Read verses from your Bible. You may find the following outline helpful, especially when dealing individually with a child.

1. **I have sinned (Romans 3:23).**
   - Sin is disobeying God's Word (1 John 3:4). I break the Ten Commandments (Exodus 20:2–17) by loving other people or things more than I love God, worshiping other things or people, using God's name lightly, disobeying and dishonoring my parents, lying, stealing, cheating, thinking harmful and sinful thoughts, or wanting something that belongs to somebody else.
   - Therefore, I am a sinner (Psalm 51:5; 58:3; Jeremiah 17:9).
   - God is holy and must punish me for my sin (Isaiah 6:3; Romans 6:23).
   - God hates sin, and there is nothing that I can do to get rid of my sin by myself (Titus 3:5; Romans 3:20, 28). I cannot make myself become a good person.
2. **Jesus died for me (Romans 5:8).**
   - God loves me even though I am a sinner.
   - He sent His Son, Jesus Christ, to die on the cross for me. Christ is sinless and did not deserve death. Because of His love for me, Christ took my sin on Himself and was punished in my place (1 Peter 2:24a; 1 Corinthians 15:3; John 1:29).
   - God accepted Christ's death as the perfect substitution for the punishment of my sin (2 Corinthians 5:21).
   - Three days later, God raised Jesus from the dead. Jesus is alive today and offers salvation to all. This is the gospel of Jesus Christ: He died on the cross for our sins according to the Scriptures, and He rose again the third day according to the Scriptures (1 Corinthians 15:1–4; 2 Peter 3:9; 1 Timothy 2:4).
3. **I need to put my trust in Jesus (Romans 10:9–10, 13–14a).**
   - I must repent (turn away from my sin) and trust only Jesus Christ for salvation (Mark 1:15).
   - If I repent and believe in what Jesus has done, I am putting my trust in Jesus.
   - Everyone who trusts in Jesus is forgiven of sin (Acts 2:21) and will live forever with God (John 3:16). I am given His righteousness and become a new creation, with Christ living in me (2 Corinthians 5:21; Colossians 1:27).

If a child shows genuine interest and readiness, ask, "Are you ready to put your trust in Jesus and depend on only Christ for salvation?" If he says yes, then ask him to talk to God about this. Perhaps he will pray something like the following:

> God, I know that I've sinned against You and that You hate sin, but that You also love me. I believe that Jesus died to pay for my sin and that He rose from the dead, so I put my trust in Jesus to forgive me and give me a home with You forever. In Jesus' name I pray. Amen.

Show the child how to know from God's Word that he is forgiven and in God's family (1 John 5:12–13; John 3:18). Encourage him to follow Jesus by obeying Him each day. Tell him that whenever he sins, he will be forgiven as soon as he confesses those sins to God (1 John 1:9).

Materials List

Standard Science Supplies

balance
calculator
goggles
magnifying glass
meter stick
metric beakers
metric measuring cups
metric measuring spoons
protractor
spring scale
stopwatch
timer
thermometer

Volume (Liquid) Measurement Equivalents			
1 teaspoon	.33 tablespoon	5 milliliter	
1 tablespoon	.5 ounce	3 teaspoons	15 milliliters
1 cup	8 ounces	16 tablespoons	237 milliliters
1 quart	32 ounces	2 pints	946 milliliters
1 gallon	4 quarts	128 ounces	3.78 liters
1 liter	4.22 cups	1000 milliliters	
.5 liter	2.11 cups	500 milliliters	

Basic Supplies

buckets
centimeter ruler
colored pencils
construction paper
containers
crayons
dishpan
glue
newspaper
paper towels
pencils
scissors
tape
water

Mass and Weight (Dry) Measurement Equivalents		
1 ounce		30 grams
16 ounces	1 pound	454 grams
1 kilogram	2.2 pounds	1000 grams

Materials by Chapter

Materials marked with an X are optional or used for enrichment activities and demonstrations.

Materials	1	2	3	4	5	6	7	8	9	10	11	12
advertisements										X		
aerosol air freshener or hair spray					X							
aluminum foil			X	X								
apples	X											
aquarium rocks				X								
assorted items (*see* lessons)		X	X		X				X	X		
axe head			X									
bag, large, clear plastic								X				
bags, paper lunch			X									
bags, resealable plastic sandwich	X		X				X			X		
balloon, large			X		X						X	
balloon, long												X
balloon, small											X	
barometer					X							
batteries					X							
bead, pony					X							

Materials	1	2	3	4	5	6	7	8	9	10	11	12
beads, seed					X							
beads, ultraviolet-detecting										X		
beanbags				X					X			X
black paper or folder					X					X		
blanket					X							
blocks, wooden									X			
bottle, clear plastic 0.5 L											X	
bottle, clear 2 L					X			X				
bottle cap, plastic									X			
bottled water					X							
bowls	X											
box, cardboard			X							X		
bread, fresh slices			X									
brick				X								
brownies or other baked goods		X										
bubble wrap				X								
butter				X								
calorie charts				X								
can, empty				X								
candle			X									
candle holder			X									
candy with cellophane wrapper									X			
canned food with easy-open lids					X							
cardboard					X							
carpet scrap									X			
carrot pieces								X				
ceiling tile scrap									X			
cereal-and-marshmallow treats (wrapped)	X											
chalk	X											
cheesecloth			X	X		X						

Materials	1	2	3	4	5	6	7	8	9	10	11	12
clay, air dry		X										
clay, dry soil	X											
clay, modeling	X	X	X		X						X	
coffee filters	X		X			X						
coffee grinder			X									
coffee stirrer, hollow											X	
color paddles, red, green, and blue (or colored cellophane)										X		
compass, directional					X							
construction paper	X					X						X
cookies, chocolate chip	X											
correction fluid	X											
cotton batting				X								
craft foam				X								
craft materials (felt, glitter, yarn, etc.)	X											
craft sticks			X									
crayons, broken pieces				X								
cups, large foam		X			X	X						
cups, 9 oz clear plastic	X		X									X
cups, 3 oz plastic				X								X
cups, 3 oz translucent white plastic										X		
dominoes							X					
drinking straw					X						X	
dropper						X						
Epsom salt	X											
evergreen needles (on branch)						X						
feather											X	
felt		X										
flashlight					X	X				X		
flower pot, clay					X							
foil candy cups				X								

Materials	1	2	3	4	5	6	7	8	9	10	11	12
folder									X			
food coloring	X				X							X
fork, metal									X			
fossils (samples or photos)		X										
frosting container with lid					X							
funnel			X									
garden trowel							X					
gelatin, lemon containing 3 items												X
glitter	X				X							
globe						X						
grass clippings	X											
gravel	X											
hammer	X		X		X							
hole punch											X	
hot plate	X			X								
ice				X	X			X				
ice-cream maker				X								
index cards											X	
interlocking bricks, assorted colors			X									
iron filings			X									
items, hard		X										
items, small		X										
items made from minerals	X											
items to change a person's appearance			X									
jars, baby food						X						
jars, glass					X				X			
jars with lids						X						
jump ropes												X
kickball										X		
knife	X	X										

Materials	1	2	3	4	5	6	7	8	9	10	11	12
lamp, with incandescent bulb					X							
leaves, deciduous						X						
leaves, green		X				X						
light bulb								X				
loam	X											
lotion						X						
magazines		X						X		X		
magnets, assorted strengths			X									
map, United States or California								X				
map, United States road												X
map, world						X		X				
maps, weather					X							
marker, permanent		X	X		X					X	X	
markers, broad-tipped permanent										X		
match			X		X							
metal, such as a baking pan									X			
microscope slide						X						
minerals	X											
mirror, concave										X		
mirror, convex										X		
mirror, plane										X		
moss, dried						X						
nail, iron			X									
nail, thin					X							
news article about a natural disaster								X				
news magazines								X				
newspapers					X	X		X		X		
newsprint or bulletin board paper												X
nutrition labels	X											
paint, acrylic		X										

Materials	1	2	3	4	5	6	7	8	9	10	11	12
paint, watercolor					X							
paintbrush, watercolor		X			X							
pan, loaf					X							
pan, square foil		X										
paper, stiff					X							
paper, white		X			X	X		X	X	X		
paper circle					X							
paper clips		X	X			X						
paper face mask or filter											X	
paper towels	X		X									
paper tube, small						X						
peanuts, unshelled								X				
pebbles	X											
pennies	X		X									
pens, black felt-tip from several manufacturers			X									
pens, colored	X											
peppermint candy (whole and crushed)			X									
petroleum jelly						X						
picture of a cheetah										X		
picture of a CT scanner		X										
picture of a peregrine falcon										X		
pictures of a campfire				X								
pictures of a car				X								
pictures of a factory				X								
pictures of assorted plants							X					
pictures of boats and ships			X									
pictures of dragons		X										
pictures of wading birds						X						
pie plate	X					X						
pin, straight					X							

Materials	1	2	3	4	5	6	7	8	9	10	11	12
Ping-Pong ball					X							
plant pieces		X					X					
plant, potted								X				
plaster of Paris		X										
plastic insects or other animals						X						
plastic wrap				X				X				
plate, dinner	X		X									
plate, foam		X										
plate, paper			X			X						
pond water						X						
popcorn or packing peanuts							X					
potato				X								
potholder	X			X								
potting soil		X			X	X		X				
powdered drink mix, sweetened	X		X									
prism										X		
produce guide	X											
protractor					X					X		
puzzle with 10 or fewer pieces										X		
rain gauge					X							
rattle									X			
recording, music									X			
recording, sound effects or instrumental									X			
resource materials about biomes						X						
resource materials about food webs							X					
resource materials about fossils/dinosaurs		X										
resource materials about rain forests						X						
resource materials about space				X								
resource materials about stresses								X				
rice			X									X

Materials	1	2	3	4	5	6	7	8	9	10	11	12
rock and mineral field guide	X											
rocks	X											
rolling pin		X										
rope										X		
rubber bands				X	X	X		X	X	X	X	
rubbing alcohol						X						
safety pin			X									
salt, table	X											X
sand			X			X						
sandstone (sample or photo)	X											
saucepan	X		X	X								
scale, bathroom				X								
shoebox										X		
shredded lettuce								X				
skewers, wooden	X											
snack mix			X									
sponges, assorted sizes					X							
sponges, flexible thin						X						
spoon, large metal	X			X								
spoon, large wooden				X								
spoons, metal									X			
spoons, plastic	X	X	X			X		X				
spring toy, metal										X		
stakes, small							X					
stapler					X							
stethoscope									X			X
strainer			X									X
string		X	X	X		X	X	X		X	X	
sugar	X		X									
suncreens, assorted SPFs										X		

Materials	1	2	3	4	5	6	7	8	9	10	11	12
tape, clear packaging											X	
tape, duct					X							
tape, masking									X			
tape, waterproof					X							
thermometer, candy	X											
thread	X				X							
tissue paper											X	
tokens	X											
tongs				X								
toothpicks	X	X				X	X					
toy, squeaky squeeze									X			
tweezers			X									
twigs	X											
vacuum cleaner hose											X	
video, person reacting to hearing after cochlear implant									X			
waxed paper	X					X				X		
weather radio					X							
weather-safety pamphlets					X							
weather-safety posters					X							
winter jacket					X							
wood, such as a piece of board									X			
wooden dowels, ⅜ in.					X							

National Science Education Content Standards

The BJU Press science materials are developed after reviewing many standards, including the National Science Education Standards (NSES). The Science Content Standards section of the NSES groups the criteria based on grade levels K–4, 5–8, and 9–12.

The table provides an overview of the areas in which *Science 5* correlates with the NSES Science Content Standards for grades 5–8. An *X* indicates that at least one component of the standard is addressed in the chapter.

The topics listed in the table are interpreted and discussed in the *Science 5* text according to a Christian worldview.

BJU Press *Science 5* correlation with NSES Science Content Standards for grades 5–8

Chapter	1	2	3	4	5	6	7	8	9	10	11	12
Science as Inquiry												
Abilities necessary to do scientific inquiry	X	X	X	X	X	X	X	X	X	X	X	X
Understandings about scientific inquiry	X	X	X	X	X	X	X	X	X	X	X	X
Physical Science												
Properties and changes of properties in matter	X		X	X								
Motions and forces												
Transfer of energy				X	X		X	X	X	X		
Life Science												
Structure and function in living things							X	X			X	X
Reproduction and heredity		X										
Regulation and behavior		X				X	X	X			X	X
Populations and ecosystems						X	X	X				
Diversity and adaptations of organisms		X				X	X	X				
Earth and Space Science												
Structure of the earth system	X		X		X	X		X				
Earth's history	X	X										
Earth in the solar system							X	X				
Science and Technology												
Abilities of technological design	X	X	X	X	X			X	X	X	X	X
Understandings about science and technology	X	X	X	X	X		X		X	X		X
Science in Personal and Social Perspectives												
Personal health				X					X	X	X	X
Populations, resources, and environments						X	X	X				
Natural hazards	X	X			X	X	X	X				
Risks and benefits					X			X	X	X	X	X
Science and technology in society	X	X	X	X	X	X	X	X	X	X	X	X

History and Nature of Science												
Science as a human endeavor	X	X	X	X	X	X	X	X	X	X	X	X
Nature of science	X	X	X	X	X	X	X	X	X	X	X	X
History of science	X	X	X	X	X		X		X	X		X
Unifying Concepts and Processes												
Systems, order, and organization	X	X	X	X	X	X	X	X	X	X	X	X
Evidence, models, and explanation	X	X	X	X	X	X	X	X	X	X	X	X
Change, constancy, and measurement	X	X	X	X	X	X	X	X	X	X	X	X
Evolution* and equilibrium		X				X	X	X				
Form and function		X	X	X		X	X	X	X	X	X	X

* Evolution is presented as a theory to be critically examined from a Christian worldview.

Index

D

E

F

G

H

T

U

V

W

X

Teacher's Edition Index

Photograph Credits

Key: (t) top; (c) center;
(b) bottom; (l) left; (r) right;
(i) inset

Unit 1

3 © iStockphoto.com/Jodi Jacobson; **5** © Gregwbenoit | Dreamstime.com; **8t, c, 13tl, tr, 14l** (gypsum, apatite, quartz), **14r** (penny, glass), **17, 20, 21tl, 24t, b, 28, 29, 37, 39** BJU Photo Services; **8b** © iStockphoto.com/Nicola Brown; **10, 12tl, tc, tr, 13br, 14l** (diamond) www.iRocks.com photo; **11** Dorling Kindersley/Getty Images; **12b** Wally Eberhart/Visuals Unlimited/Getty Images; **13bl, 14l** (talc, topaz), **15t, c, 26tr** Copyright © Dr. Richard Busch; **14l** (calcite, fluorite, orthoclase), **21tr, bl, br** www.fabreminerals.com; **14l** (corundum) Image by Alain Darbellay, GGGems.com; **14r** Susan Perry; **14r** Aaron Dickey; **15b** mariusFM77/E+/Getty Images; **18** Art Resource, NY; **19r** debeers; **19l** Photo provided by Geolite, www.geolite.com; **21c, 23** © iStockphoto.com; **25l, 31, 41** iStockphoto /Thinkstock; **25tr** © Attila JANDI/Shutterstock.com; **25br** Joyce Landis; **26tl** Copyright © Abi Howe, American Geological Institute; **26b, 27c** PhotoDisc, Inc.; **27t** © iStockphoto.com/Christine Glade; **27b** Getty Images/Hemera/Thinkstock; **32l, r, 33 all** Brian D. Johnson; **34** © Anetlanda | Dreamstime.com; **35** © iStockphoto.com/hsvrs; **40** © John Reade/Science Source; **42** Simon Taylor; **43** © Cmlndm | Dreamstime.com; **49** "Eggs of Oviraptor (Senckenberg)" by Gerbil/Wikimedia Commons/FNU FDL 1.2, CC-BY-SA 3.0; **52** © iStockphoto.com/Hocus Focus Studio

Unit 2

55, 57, 77l, 99, 102bl iStockphoto/Thinkstock; **58** © Prillfoto | Dreamstime.com; **59l, r, 66b, 67, 71, 73tr, 74 tl, tr, 76 all, 78l, r, 80, 81, 83, 86, 91, 101** BJU Photo Services; **60** Ohaus; **62t** © iStockphoto.com/Sergey Kashkin; **62b** © iStockphoto.com/Isabelle Mory; **63t, b, 64, 69t, 102t** © 2005 HemeraTechnologies, Inc.; **66t** Dorling Kindersley/Getty Images; **69b** © iStockphoto.com /NightAndDayImages; **72, 73br, 74b** Aaron Dickey; **73tl** Copyright © Dr. Richard Busch; **73bl, 79** Sarah Ensminger; **77r** © iStockphoto.com; **85** © iStockphoto.com/intst; **89** © iStockphoto.com/Arpad Benedek; **93** Tom Uhlman/Visuals Unlimited/Getty Images; **94** © iStockphoto.com/Atilla ALTUN; **102bc** © iStockphoto.com/Bonita Hein; **102br** Army photo by Pfc. Mary Rose Xenikakis, 22nd Mobile Public Affairs Detachment; **103t, b** PhotoDisc, Inc.; **104** © iStockphoto.com/Anna Ziska; **105** © Matty Symons - Fotolia.com; **106t, b, 107** NASA

Unit 3

111, 115t, 132, 144, 146b, 151t, b, 157b iStockphoto/Thinkstock; **113** Purestock/Thinkstock; **115b, 125, 139, 155, 159, 167** BJU Photo Services; **119, 128b** Library of Congress; **123t** © iStockphoto .com/wolv; **123b** © iStockphoto.com/Laurie Knight; **126** © iStockphoto.com/edsansin; **127t** Joseph O'Brien, USDA Forest Service, Bugwood.org; **127c** Beth Hamel; **127b** C. Knight, National Center for Atmospheric Research; **128l** © iStockphoto.com; **128r** © iStockphoto.com/Todica; **129, 130t, b, 145tl, 161l, 162b, 165** PhotoDisc, Inc.; **131t** COREL Corporation; **131bl, br, 145b, 147, 148, 149l, 150** © 2005 Jupiterimages/Photos.com; **133l** © iStockphoto.com/Patrick Heagney; **133r, 136l, r** NOAA; **141** Stockbyte/Thinkstock; **145tr** Tobias Schwoerer; **149r** © iStockphoto.com/timeless; **152** © nmrtgh/Fotolia.com; **153** © iStockphoto.com/Dave L. Morgan; **157tl** © Dr. Morley Read/ Shutterstock; **157 tr** Minden Pictures/SuperStock; **160** © iStockphoto.com/magnetcreative; **161r** Dante Fenolio/Science Source; **162t** NASA; **164t, b** USFWS

Unit 4

169, 174, 175, 185b, 187t, 193, 212t iStockphoto/Thinkstock; **171** © Artur Synenko/Fotolia.com; **175i** © iStockphoto.com/Alasdair Thomson; **177, 183, 203, 209** BJU Photo Services; **181t** Hemera /Thinkstock; **181b** © mrallen/Fotolia.com; **185t** © iStockphoto.com; **186t, 189ct, b, 194r, 195l**

PhotoDisc, Inc.; **186c** Suzanne L & Joseph T. Collins/Photo Researchers/Getty Images; **186b** © Ondreicka/Dreamstime.com; **187b**, **211t** Breck Kent; **188l** © 2005 Jupiterimages/Photos.com; **188r**, **i** age fotostock/SuperStock; **189t** © Holger Karius/Dreamstime.com; **189cb** Norbert Probst/imageb/imagebroker.net/SuperStock; **194l** 2005 Hemera Technologies, Inc.; **195r** © iStockphoto.com/Steve Pepple; **198** Wally Eberhart/Visuals Unlimited/Getty Images; **200** Glenn Nagel/Bigstock.com; **204t** National Park Services/Jim Peaco; **204b** Joyce Garland; **205** Bob McMillan/FEMA News Photo; **206** Tim McCabe/USDA Natural Resources Conservation Service; **210** USDA Natural Resources Conservation Service; **211b** US Fish and Wildlife; **212c** © Denis Pepin/Dreamstime.com; **212b** floridanature.org

Unit 5

215, **218**, **227**, **233b**, **237**, **250r**, **254**, **255l** iStockphoto/Thinkstock; **217** © Mark Herreid/Fotolia .com; **223**, **229**, **247**, **253** BJU Photo Services; **224l** © iStockphoto.com/PaulWolf; **224r** Courtesy of Howard Miller; **231** Courtesy of Heather Stanley; **233t** Excerpt from "Ave Verum Corpus" by W.A. Mozart/Public Domain; **234** Jupiterimages/Photos.com; **243** Hemera/Thinkstock; **249** PhotoDisc, Inc.; **250l** © 2005 Hemera Technologies, Inc.; **255r** © Carla F. Castagno | Dreamstime.com; **256tl**, **tr** Photo Researchers/Getty Images; **256b** © Fireflyphoto | Dreamstime.com; **257l** © iStockphoto.com/Leah-Anne Thompson; **257tr** *The Visit of the Queen of Sheba to Solomon*, Jacopo Robusti, called II Tintoretto, from the Bob Jones University Collection; **257br** X-ray detail from *The Visit of the Queen of Sheba to Solomon*, Jacopo Robusti, called Il Tintoretto, from the Bob Jones University Collection; **258** ASTIER - CHRU LILLE/BSIP/SuperStock; **259l** Meade Instruments; **259c** © iStockphoto.com; **259r** © iStockphoto.com/zentilia

Unit 6

261 Hank Grebe/SuperStock; **263** Photodisc/Thinkstock; **265** Corbis Digital Stock; **267**, **277**, **289**, **301** BJU Photo Services; **275** © iStockphoto.com/Sharon Dominick; **278**, **295** iStockphoto/Thinkstock; **279** Stem Labs, Inc.; **280** © iStockphoto.com/twity1; **281** Arthur Glauberman/Science Source/Getty Images; **283** © Tinydevil | Dreamstime.com; **286t** PhotoDisc, Inc.; **286b**, **287** © iStockphoto.com; **305** Stockbyte/Thinkstock

Maps: **137**, **142–43**, **146t**, **195bl, br** BJU Press

How to Use the Teacher's Toolkit CD

Contents

The Teacher's Toolkit contains the following materials:

- Activity Manual Answer Key
- Game Bank
- Instructional Aids
- Materials List
- National Science Education Content Standards
- Quizzes and Answer Key
- Rubrics
- Science Fair Information
- Two-Day Lessons
- Visuals

Getting Started

Viewing the Teacher's Toolkit materials requires Adobe® Reader® 9.0 or higher. The most recent version of Adobe Reader may be downloaded at no charge from the Adobe website at www.adobe.com. An Internet connection is required to download Reader.

Windows

Insert the CD. The CD is designed to start automatically if your computer is set to allow it. If it does not start automatically, click "Run" if given the option. You may also choose to open the folder to view the CD's files and double-click "Startup.exe" to start the CD. Read and accept the license agreement to begin using the Teacher's Toolkit materials. Navigate within the CD using the bookmarks on the left side of the screen.

Mac

Insert the CD, click on the CD icon, and open the file "main.pdf" to begin using the Teacher's Toolkit materials.

Minimum System Requirements

Processor: Pentium IV
Operating System: Windows XP or Mac OS Leopard (version 10.5)
RAM: 256 MB
Display: 1024 × 768
Adobe Reader: version 9.0

Additional Help

Additional usage information can be found on the CD. For further assistance, call BJU Press Customer Service at 1-800-845-5731.